Combinatorial Chemistry

.7.09
.7.09

MW

200033

The Practical Approach Series

Related **Practical Approach** Series Titles

Protein-Ligand Interactions: structure and spectroscopy*

Protein-Ligand Interactions: hydrodynamic and calorimetry*

Immunoassay

Spectrophotometry and Spectrofluorimetry

Fmoc Solid Phase Peptide Synthesis

High Resolution Chromatography

Immobilized Biomolecules in Analysis

HPLC of Macromolecules 2/e

DNA Microarray Technology

* indicates a forthcoming title

Please see the **Practical Approach** series website at

http://www.oup.co.uk/pas

for full contents lists of all Practical Approach titles.

Combinatorial Chemistry

A Practical Approach

Edited by

Hicham Fenniri
1393 H. C. Brown Laboratory of Chemistry,
Purdue University, USA

OXFORD
UNIVERSITY PRESS

OXFORD

UNIVERSITY PRESS

Great Clarendon Street, Oxford OX2 6DP

Oxford University Press is a department of the University of Oxford.
It furthers the University's objective of excellence in research,
scholarship, and education by publishing worldwide in

Oxford New York

Athens Auckland Bangkok Bogotá Buenos Aires Calcutta Cape Town
Chennai Dar es Salaam Delhi Florence Hong Kong Istanbul Karachi
Kuala Lumpur Madrid Melbourne Mexico City Mumbai Nairobi Paris
São Paulo Singapore Taipei Tokyo Toronto Warsaw

with associated companies in Berlin Ibadan

Oxford is a registered trade mark of Oxford University Press in the UK
and in certain other countries

Published in the United States by Oxford University Press Inc., New York

Library of Congress Cataloguing in Publication Data
Combinatorial chemistry : a practical approach / edited by Hicham
Fenniri.
(Practical approach series ; 233)
Includes bibliographical references and index.
1. Combinatorial chemistry. I. Fenniri, Hicham. II. Series.
RS419 .C588 2000   615'.19–dc21   00–035596

1 3 5 7 9 10 8 6 4 2

ISBN 0 19 963757 1 (Hbk.)
ISBN 0 19 963754 7 (Pbk.)

Typeset in Swift by Footnote Graphics, Warminster, Wilts
Printed in Great Britain on acid-free paper
by The Bath Press, Bath, Avon

Preface

Combinatorial chemistry, the science of molecular diversity and 'rational screening' started taking shape in the mid-80's. Except for a few visionaries, no one at that time truly weighed the implications this new development would have on our way to conduct research in the chemical sciences. In fact, up to this date there seem to be more skeptics than adepts of this new technology, presumably the result of a lack of exposure by our students to its very practical aspects and the unavailability of an introductory text to its key concepts. These reasons were compelling enough to incite the compilation of this book. Many world leaders in this field, to whom I am most thankful, responded enthusiastically and accepted to contribute to this endeavor. Many more experts in this field could have taken part in this project. I anticipate, however, that what is now a single volume and a handful of chemists, will turn out to become a prelude to a series of practical volumes dealing with ongoing and yet to come exciting developments in this field. I thus wholeheartedly encourage the readers not only to apply the techniques presented here, but also to challenge their basic premises and to develop novel and even more powerful alternatives.

The specific goals of this book are to introduce the novice to:

- the key methods of generating chemical libraries using combinatorial methods and help them identify those most suitable for their own research.
- solid, liquid and solution phase organic synthesis.
- the expedient methods of library screening and evaluation.
- robotics and automation in organic synthesis.
- modern approaches to drug, catalyst, receptor, and materials development and discovery.

The general goals are to highlight how:

- the interplay between several disciplines contributed to the genesis of one of the most powerful and modern technologies.
- start-up pharmaceutical companies have had tremendous successes in only a few years using combinatorial chemistry.

Although combinatorial sciences extend well beyond chemistry, this book will focus mainly on the key combinatorial methodologies and their application in the design synthesis and evaluation of chemical libraries. This text should be seen as a starting point for whoever is willing to take some chances intellectually. After all, isn't this the true essence of combinatorial sciences?

It is safe to profess that a complete and diverse repertoire of all small organic molecules (200 < MW < 600) made possible by applying the rules of valence to carbon and its neighbors on the periodic table is in all respects inconceivable. The universe merely does not offer enough matter to explore every molecular combination. Therefore, the 'rational' design of the library is a prerequisite for its success in offering an entity with the desired properties. There are three key aspects of the library that should be addressed first:

- its ease of construction, size, and chemical diversity.

- Each library member should be equally represented and present in sufficient quantity to give rise to a measurable chemical, physical, or biological response.

- the active members of the library should be easily identified and structurally characterized.

The preparation of a chemical library involves 4 main steps, the synthesis, the evaluation, the identification of the active member(s), and the confirmation of the result(s):

- to synthesize a library effectively, the number of compounds produced should increase exponentially with the number of synthetic steps. Multiple parallel synthesis (Chapters 3, 5, 10–16) and the portion mixing method (Chapter 1), also termed the one bead-one compound approach (Chapter 2), or split synthesis (Chapters 4) are the two most widely used methods.

- the library can be built in solution without support (Chapters 9–11, 13, 15), on a soluble polymeric support (Chapter 12), or insoluble matrix (Chapters 1–7, 14, 16). In the latter case the matrix may be polymeric porous beads (Chapters 1, 2, 4, 6) or functionalized surfaces (Chapters 4, 5).

- The anchor used to attach the building blocks to the support, if any, can be acid-, base-, nucleophile-, or light-sensitive (Chapters 6)

- The building blocks and the chemistry associated with them can be oligomerizable (Chapters 1–5), non-oligomerizable (Chapters 3, 4, 6, 10), or scaffold based (Chapters 3, 4, 6, 9).

Supports. If a support is used, it must be mechanically and chemically stable to the solvents and reagents involved in the library synthesis. The support should be compatible with the milieu and the components of the subsequent screenings. Soluble polymeric supports such as dendrimers (Chapter 12) were introduced recently in combinatorial chemistry. They present the major advantages of possessing solution phase reactivity, they can be easily purified (ultrafiltration or gel filtration), and can be subjected to standard spectroscopic analysis (e.g. NMR). High yield solution phase reactions are amenable to *combinatorialization* and automation since they do not require extensive purification procedures (Chapters 9–11, 13, 15). Ugi multi-component reactions may be considered the prototype in this case (Chapter 9) since they involve one pot synthesis of amino acid derivatives from isocyanides, aldehydes, amines, and carboxylic acids. Finally, polymer bound reagents were successfully applied to *one pot* multistep solution phase combinatorial syntheses (Chapter 15).

Anchors. The chemical nature and length of the spacer between the matrix and the ligand is of the highest importance in solid phase synthesis. Like the solid support, it has to be chemically inert and compatible with the synthetic scheme. It can dramatically influence the binding and accessibility of the immobilized ligand to a soluble macromolecular receptor in subsequent biological assays. A universal linker does not exist since it is the synthetic sequence that determines which is appropriate. The linker may be considered as a semi-permanent protecting group that is introduced at the beginning of the synthesis and removed at the end without affecting the final product (Chapter 6).

Building blocks. Biopolymers such as peptides and oligonucleotides were chosen in combinatorial chemistry for their synthetic accessibility and their demonstrated pharmacological properties (Chapters 1–4). As a result of the success of solid phase peptide and DNA syntheses, most of the early reports on combinatorial chemistry involved molecules with peptidic or nucleic acid backbones. The advantages of this approach are the high average stepwise yield (hence no purification procedure), and the amenability of the scheme to automation. In addition, highly sensitive bioanalytical methods such as Edman degradation, Maxam and Gilbert microsequencing, mass spectrometry, and the PCR, are available for the detection and identification of the active molecule. It is unfortunate however, that these compounds have poor oral absorption and metabolic stability. Nevertheless, peptide libraries can provide structure activity relationships on which to base subsequent peptide mimetic library design. Carbohydrates and glycopeptides are the only classes of natural oligomers that have not seen wide application in combinatorial chemistry. The reasons behind this are their complexity and low synthetic accessibility, which make them difficult targets for automated synthesis. Nevertheless, research efforts are ongoing in several laboratories, and the first successful carbohydrate libraries have been reported (1).

The development of alternative backbones to the naturally occurring ones may be considered the second major step in library design. After the establishment of combinatorial chemistry as a viable approach to drug discovery using biopolymers (peptides, oligonucleotides), the attention turned to the question of bioavailability, chemical stability and synthetic accessibility of their libraries. A very successful approach to this problem was the design of peptide and DNA mimetics that combine chemical stability and bioavailability with the inherent synthetic accessibility of oligomeric molecules (Chapters 3, 5).

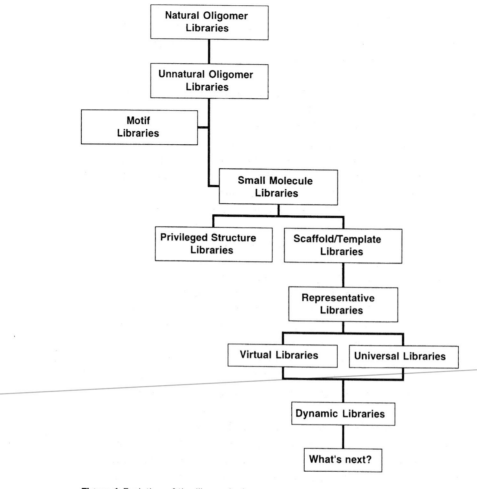

Figure 1 Evolution of the library design.

While most of the early synthetic efforts in combinatorial chemistry targeted natural oligomers (peptides, nucleic acids), the usual drug candidate is in fact a compact multifunctional small molecule with a molecular weight comprised between 200 and 600. The translation from oligomeric into non-oligomeric small molecule drug candidates is a challenging task. *Privileged structures* [95], possessing a generic scaffold found in a number of potent therapeutic agents, were used to overcome this problem (Chapters 3–5). Since the chemistry associated with this type of library differs from that developed for oligomeric molecules, research efforts are now being directed towards the exploration of several aspects of synthetic organic chemistry on solid supports (Chapter 7). Unlike biopolymer/oligomer chemistry, each time the template structure is changed, a new synthetic methodology must be developed. Hence, the synthesis must be short, high yielding and applicable to a wide range of starting building blocks. These constraints make this method best suited for the optimization of structurally similar classes of compounds at an advanced stage in the discovery process rather than in the identification of novel leads. It is noteworthy that although the first reports on solid phase synthesis of small molecules go back to the early 70's (2), this field has not seen such strong enthusiasm until the advent of 'small molecule' combinatorial chemistry.

Over the past few years, combinatorial chemistry has been drifting from the privileged structure concept to a synthetic version of it, the *scaffold/template* based libraries (Fig. 3). This approach involves the generation of smaller libraries (10^2–10^5 members) displaying a fine balance between rigidity and flexibility, high density of functional groups, exhaustive coverage of the conformational space and the universe of diversity, shapes, functional group distribution and electrostatic surfaces. This was also the precursor to a new concept in library design, the *representative* or *universal library*, which was also intended to minimize structural, conformational and electronic redundancies (Chapters 3–5). This concept in library design addressed also an important limitation imposed by solid phase chemistry on the library size. For instance, split synthesis scheme to generate a decapeptide combinatorial library using a set of 20 amino acid building blocks will require at least ~10^{14} resin beads (one bead per compound). Using 130 mm diameter beads, this number would correspond to 10.2 tons of resin, a quantity that no industrial or academic institution could possibly envision. Further developments to thwart such difficulties resulted in computer generated *virtual libraries*. *In silico* screening against a given receptor with known structure, or even a receptor for which only a structure activity relationship study is available, helps identify an optimal group of building blocks or scaffolds for the construction of the *real library*.

Chemical and structural identification of the active library members. This is clearly the bottleneck in combinatorial chemistry and its success relies entirely on the sensitivity and specificity of the assay and screening technique. The most potent candidate may not be discovered after the first screen. A lower activity ligand may be identified initially and used as a lead in a second round of selection from a smaller targeted library. The screening can be performed on tethered or soluble libraries in a physically segregated or mixed pool format. This defines the assay method to be used and ultimately determines the size of the libraries to be screened. The compounds should be present in solution or on the matrix in amounts depending on the desired affinity. High affinity receptors require small concentrations of the ligand (nM range). The large size of peptide and oligonucleotide libraries that can be readily obtained (up to 10^{15} compounds) does not allow screening of individual compounds and, for this reason, pooling strategies were developed in which sublibraries are screened for the desired activity. Small sublibraries ($>10^3$) are more successful than the larger ones ($<10^6$) for the following reasons:

- each member is present in sufficient quantities.
- artifacts resulting from non-specific interactions with low affinity ligands are minimized.

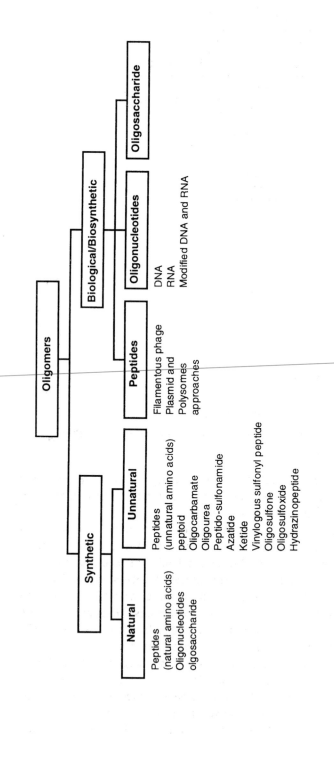

Figure 2 Oligomer based libraries.

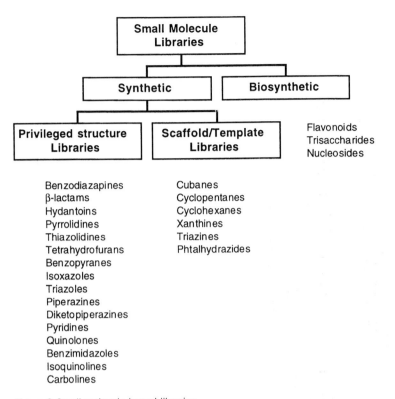

Figure 3 Small molecule based libraries.

- the quality of the library is better assessed.
- quantitative analysis is possible and qualitative analysis is more reliable.
- active components are more easily identified.

Direct screening techniques. Mass spectrometry, NMR, fluorescence and IR spectroscopies are used as direct approaches for the identification of active members or their encoding element (Chapter 8).

Non-coded screening strategies. These methods identify active members in a soluble or support-bound mixtures of compounds through iterative deconvolution schemes (Chapters 3, 7, 9). Libraries are divided into sublibraries, which are in turn broken down into smaller sublibraries. Screening for activity can then be performed from the bottom and/or the top of the tree. Although limited in size, spatially addressable libraries on porous surfaces (Chapter 5) have the advantage of presenting compounds that are physically segregated and easily identifiable through their geographic location on the surface.

Encoded screening strategies. These methods emerged because of a reorientation of the research efforts towards small molecule combinatorial libraries of non-oligomeric nature that cannot be identified with the standard procedures (e.g. Edman degradation, Sanger dideoxy sequencing, mass spectrometry, NMR) (Chapters 1, 2, 4, 8). Thus, highly sensitive indirect methods based on DNA, peptide, polyhalobenzene, polyamine (Chapters 1, 8) and radio frequency (Chapter 4) tagging allowing the identification of active members in subnanomolar quantities were developed.

Biologically generated libraries derive their proficiency from the amplifying power of phage particles and bacteria, or from PCR technology. The powerful technique of phage display is based on the construction of libraries of peptides or proteins as fusion products with proteins expresssed on the surface of the phage particle (3). This display process allows not only the selection of the peptide or protein with the desired biological activity, but also the encoding genetic material that is packaged inside the phage particle. E. Coli (4) and other bacteria (5), as well as plasmids (6) and polysomes (7) were also used as display systems.

A spectacular tour-de-force in biologically derived, yet chemically induced, combinatorial libraries is that of catalytic antibodies (8). One of the approaches for the generation of these tailored catalysts is to challenge the immune system with an antigen that resembles the transition state of a given reaction. Through the combinatorial association of variable, joining, and diversity genes, the immune system generates a tremendous number of antibodies (10^6–10^{12}) against the antigen. Through rapid screening and affinity maturation of a small subset of this library, the immune system produces highly specific antibodies that may catalyze the reaction involving the transition-state for which they were raised. Along the same Darwinian lines, a PCR based approach using synthetic oligonucleotide libraries led to the selection of RNA and DNA molecules exhibiting very high specificity and selectivity towards adenosine, ATP (9), flavin, nicotinamide cofactors (10), amino acids (11), proteins (12), aminoglycoside antibiotics (13), and other molecules (14). Finally, combinatorial biosynthesis of *unnatural* natural products is one of the latest developments in the field of biologically generated chemical libraries. The applicability of this new approach was recently demonstrated in the case of polyketides (15). The potential of this system in combinatorial chemistry was demonstrated through the genetic construction of PKS libraries that are able to synthesize an unlimited number of novel chemical entities with predetermined structures.

Conclusion and future challenges. An understanding of the fundamental rules of molecular recognition using simplified synthetic versions of biological receptors is still a tremendous challenge. Combinatorial chemistry offers an expeditious access to an inexhaustible source of information concerning the molecular basis of host-guest interactions which may lead, in the near future, to the design of synthetic receptors paralleling antibodies in terms of complex stability and specificity. The next and perhaps most challenging step to specific recognition is catalysis. Unfortunately, organic artificial catalysis did not benefit from the same outburst that led organometallic and inorganic catalysis to practical applications in industry (Chapters 14–16). The rules governing enzyme catalysis, although fairly understood, are still extremely difficult to combine productively in a single chemically accessible synthetic receptor. Combining a few features of natural enzymes is an overly simplistic approximation that has not yet led to receptors with comparable activities. Here again, the combinatorial approach might provide the needed boost. The unavailability of direct high throughput screening methods for catalytic activity has also slowed down this process. In the case of nucleic acid based catalysts, this limitation is overcome by the ability of the active members to be 'evolved' and selectively amplified using PCR technology (16).

The explosion of the field combinatorial chemistry over the past 10 years prompted several research groups in academia and industry to design robots for multiple parallel or split synthesis of large libraries of peptides, peptide mimetics and other non-oligomeric small organic molecules (Chapter 1, 4, 13). Several automated synthesizers are currently available. However, if they are not costly, they are generally limited in their capabilities. The design of an affordable yet useful robotic system for chemical synthesis remains therefore a major challenge.

Researchers in industry and academia are very optimistic about the future of combinatorial sciences, not only in medicinal chemistry but in every major area, including biotechnology, agrochemistry, materials science, molecular recognition and catalysis. We will certainly witness in the near future the combinatorialization of even more remote aspects of the chemical sciences. Several elegant solutions have been proposed for the generation of large libraries of compounds and their screening for a particular activity. We shall expect other and more original approaches to these issues to come. For instance, a very recent trend in combinatorial sciences is the concept of *dynamic libraries* that, at last, had allowed *chemistry to meet Darwin*.

References

1. (a) Kanie, O.; Barresi, F.; Ding, Y.; Labbe, J.; Otter, A.; Forsberg, L. S.; Ernst, B.; Hindsgaul, O. *Angew. Chem. Int. Ed. Engl.*, **1995**, 34, 2720. (b) Vetter, D.; Tumelty, D.; Singh, S. K.; Gallop, M. A. *Angew. Chem. Int. Ed. Engl.*, **1995**, 34, 60. (c) Schuster, M.; Wang, P.; Paulson, J. C.; Wong, C.-H. *J. Am. Chem. Soc.*, **1994**, 116, 11135. (d) Roberge, J. Y.; Beebe, X.; Dinishefsky, S. *J. Science*, **1995**, 269, 202.

2. (a) Frechet, J. M. J. *Tetrahedron*, **1981**, 37, 663. (b) Leznoff, C. C. *Acc. Chem. Res.*, **1978**, 11, 327. (c) Crowley, J. I.; Rapoport, H. *Acc. Chem. Res.*, **1976**, 9, 135.

3. (a) Scott, J. K.; Smith, G. P. *Science*, **1990**, 249, 386. (b) Delvin, J. J.; Panganiban, L. C.; Delvin, P. E. *Science*, **1990**, 249, 404. (c) McCafferty, J.; Griffiths, A. D.; Winter, G.; Chiswell, D. J. *Nature*, **1990**, 348, 552. (d) Matthews, D. J.; Wells, J. A. *Science*, **1993**, 260, 1113.

4. Wells, J. A.; Lowman, H. B. *Curr. Biol.*, **1992**, 3, 355.

5. Hasson, M.; Stahl, S.; Nguyen, T. N.; Bachi, T.; Robert, A.; Binz, H.; Sjolander, A.; Uhlen, M. *J. Bacteriol.*, **1992**, 174, 4239.

6. (a) Cull, M. G.; Miller, J. F.; Schatz, P. J. *Proc. Natl. Acad. Aci. USA*, **1992**, 89, 1865. Tuerk, C.; Gold, L. *Science*, **1990**, 249, 505. (b) Mattheakis, L. C.; Bhatt, R. R.; Dower, W. J. *Proc. Natl. Acad. Sci. USA*, **1994**, 91, 9022.

7. Schultz, P. G.; Lerner, R. A. *Science*, **1995**, 269, 1835.

8. Sassanfar, M.; Szostak, J. W. *Nature*, **1993**, 364, 550.

9. Lauhon, C. T.; Szostak, J. W. *J. Am. Chem. Soc.*, **1995**, 117, 1246.

10. Famulok, M. *J. Am. Chem. Soc.*, **1994**, 116, 1698.

11. Conrad, R. C.; Baskerville, S.; Ellignton, A. D. *Molecular Diversity*, **1995**, 1, 69.

12. (a) Lato, S. M.; Boles, A. R.; Ellington, A. D. *Chem. Biol.*, **1995**, 2, 291. Jenisson, R. D.; Gill, S. C.; Pardi, A.; Polisky, B. *Science*, **1994**, 263, 1425. (b) Morris, K. N.; Tarasow, T. M.; Julin, C. M.; Simons, S. L.; Hilvert, D.; Gold, L. *Proc. Natl. Acad. Sci. USA*, **1994**, 91, 13028. (c) Connell, G. J.; Yarus, M. *Science*, **1994**, 264, 1137. (d) Elligton, A. D.; Szostak, J. W. *Nature*, **1992**, 355, 850. (e) Ellington, A. D.; Szostak, J. W. *Nature*, **1990**, 346, 818.

13. Tsoi, C. J.; Khosla, C. *Chem. Biol.*, **1995**, 2, 355.

14. (a) Prudent, J. R.; Uno, T.; Schultz, P. G. *Science*, **1994**, 264, 1924. (b) Morris, K. N.; Tarasow, T. M.; Julin, C. M.; Simons, S. L.; Hilvert, D.; Gold, L. *Proc. Natl. Acad. Sci. USA*, **1994**, 91, 13028.

Contents

CONTENTS

Protocol list

Abbreviations

AA	amino acid
Acm	acetamidomethyl
Ac_2O	acetic anhydride
aeg	N-(2-aminoethyl)-glycine
AIBN	2,2′-azobisisobutyronitrile
Ala	alanine
All	allyl
A-MCR	Asinger multi-component reaction
APC	antigen-presenting cells
Arg	arginine
Asn	asparagine
Asn(Trt)	N-β-trityl-L-asparagine
Asp	aspartic acid
Asp(t-Bu)	L-aspartic acid β-t-butyl ester
ATP	adenosine triphosphate
$[\gamma\text{-}^{32}P]ATP$	^{32}P-labelled ATP on the γ phosphate
B_i	adenine, cytosine, or guanine (i = 2, 3, 4)
BA	benzaldehyde
$Ba(i\text{-}PrO)_2$	barium diisopropoxide
BB	bromophenol blue
BB-4CR	Bergs and Bucherer four-component reaction
$B(C_6F_6)_3$	tripentafluorophenyl borane
BCIP	5-bromo-4-chloro-3-indoyl phosphate
BCIP/NBT	5-bromo-4-chloro-3-indolyl phosphate/nitro-blue tetrazolium
Bhoc	benzhydryloxycarbonyl
$BH_3.SMe_2$	borane dimethylsulfide
Bn	benzyl
t-Boc	tertio-butyloxycarbonyl
BOP	benzotriazol-1-yl-oxy-tris(dimethylamine)-phosphonium hexafluorophosphate)
BP	benzophenone
BSA	bovine serum albumin

BTF	benzotrifluoride, trifluoromethylbenzene
t-Bu	*tertio*-butyl
n-BuLi	*n*-butyl lithium
n-Bu$_4$NI	tetrabutylammonium iodide
(*t*-Boc)$_2$O	*tertio*-butyloxycarbonyl anhydride
CAD	collisionally activated dissociation
CAN	ceric ammonium nitrate
CE-MS	capillary electrophoresis
CF	collection flask
Cha	L-cyclohexylalanine
Chg	L-cyclohexylglycine
CHex	cyclohexyl
CLEAR	crosslinked ethoxylate acrylate resins
CLND	chemiluminescent nitrogen detection
COSMOS	combinatorial synthesis on multivalent oligomeric supports
CPC	controlled pore ceramic
CPG	controlled pore glass
CP-MAS	cross-polarization magic angle spinning
CPMG	Carr–Purcell–Meiboom–Gill
CSIm$_2$	thiocarbonyldiimidazole
DABCO	1,4-diazabicyclo[2.2.2]octane
DBU	1,8-diazabicyclo [5.4.0]undec-7-ene
DCC	*N,N'*-dicyclohexylcarbodiimide
1,2-DCE	1,2-dichloroethane
DCM	dichloromethane
DCU	dicyclohexylurea
Dde	1-(4,4-dimethyl-2,6-dioxo-cyclohexylidene)3-methylbutyl
DECODE	diffusion encoded spectroscopy
DHE	dynamic hydrogen reference electrode
DIAD	diisopropylazodicarboxylate
DIC	diisopropylcarbodiimide
DIEA	*N,N*-diisopropylethylamine
DIPEA	*N,N'*-diisopropylethylenediamine
4-DMAP	4-*N,N*-dimethylaminopyridine
DMF	*N,N*-dimethylformamide
DMFC	direct methanol fuel cell
DMS	dimethyl sulfide
DMSO	dimethyl sulfoxide
DMT	dimethoxytrityl
DMT-X$_n$	5'-*O*-DMT-2'-deoxyadenosine-3'-*O*-phosphoramidite, 5'-*O*-DMT-2'-deoxycytidine-3'-*O*-phosphoramidite, or 5'-*O*-DMT-2'-deoxyguanosine-3'-*O*-phosphoramidite (n = 2, 3, 4)
DNA	deoxyribonucleic acid
DOSY	diffusion-ordered 2D NMR spectroscopy
DT	drying tube

DVB	divinylbenzene
ECD	electron capture detector
EDCI	1-(3-dimethylaminopropyl)-3-ethylcarbodiimide hydrochloride
EDT	1,2-ethanedithiol
ee	enantiomeric excess
ELISA	enzyme-linked immunosorbent assay
ELSD	evaporative light scattering detection
EOF	electro-osmotic flow
Equiv	molar equivalent
ESI-MS	electrospray ionization–MS
Et_3N	triethylamine
Et_2O	diethyl ether
EtOAc	ethyl acetate
FA	furylaldehyde
FABS	fluorescence activated bead sorting
FACS	fluorescence activated cell sorting
Fh	fluorinated hydrocarbon
FIA-MS	flow injection analysis–MS
FID	flame ionization detector
Fmoc	fluorenylmethoxycarbonyl
Nα-Fmoc-AAs	Nα-fluorenylmethoxycarbonyl amino acids
Fmoc-AA-F	Fmoc amino acid fluoride
FRP	fluorous reverse phase
FRPS	fluorous reverse phase silica
FSPE	fluorous solid phase extraction
FTICR-MS	Fourier transform ion cyclotron resonance mass spectrometry
FTIR	Fourier transform infrared
FTMS	Fourier transform mass spectrometry
FW	free weight
GLC	gas liquid chromatography
Gln	glutamine
Gln(Trt)	N-γ-trityl-L-glutamine
Glu	glutamic acid
Gly	glycine
HBTU	O-benzotriazole-N,N,N′,N′-tetramethyluronium hexafluorophosphate
HIPE	high internal phase emulsion
His	histidine
HMBA	p-hydroxymethylbenzoic acid
HMBC	heteronuclear multiple bond correlation
H-MCR	Hantzsh multi-component reaction
HMP	hydroxymethylphenoxy
^{1}H NMR	proton nuclear magnetic resonance
HOAt	1-hydroxy-7-azabenzotriazole
HOBt	N-hydroxybenzotriazole
HO-MCR	Hellmann and Opitz multi-component reaction

HONb	*N*-hydroxy-5-norbornene-2,3-dicarboximide
Hphe	L-homophenylalanine
HPLC	high performance liquid chromatography
HPLC-NMR	high performance liquid chromatography–NMR
HR-MAS	high-resolution magic angle spinning
HSQC	heteronuclear single quantum correlation
HT-NMR	high-throughput NMR
HT-MS	high-throughput MS
Hyp(*t*-Bu)	*O*-benzyl-L-hydroxyproline
IC_{50}	concentration that inhibits 50% growth or binding
i.d	internal diameter
Ile	isoleucine
IPA	isopropanol
LAC	ligand accelerated catalysis
LC-MS	liquid chromatography MS
LC-NMR	liquid chromatography NMR
LDA	lithium diisopropylamide
Leu	leucine
t-Leu	L-*t*-leucine
LPS	liquid phase synthesis
Lys	lysine
Lys(Z)	lysine amino acid with *N*-protected benzyloxycarbonyl
MALDI-TOF	matrix-assisted laser desorption and ionization–time of flight
MAS NMR	magic angle spinning NMR
MBHA	methylbenzhydrylamine
MCR	multi-component reaction
MEA	membrane electrode assemblies
mequiv	millimolar equivalents
MES	2-(*N*-morpholino)ethanesulfonic acid
Met	methionine
MIC	minimum inhibitory concentration
MOBHA	4-methoxybenzhydrilamine
m.p.	melting point
MS	mass spectrometry
MSA	methanesulfonic acid
MW	molecular weight
NA	naphthaldehyde
NaHMDS	sodium hexamethlydisilazane
NMI	*N*-methylimidazole
NMM	*N*-methylmorpholine
NMP	*N*-methylpyrrolidine
NMR	nuclear magnetic resonance
NOESY	nuclear Overhauser effect spectroscopy
OPD	*ortho*-phenylenediamine
OPfp	*O*-pentafluorophenylester

PAMAM	polyamidoamine
PA-Sc-TAD	polyallylscandium trifylamide ditriflate
PBS	phosphate-buffered saline
P-3CR	Passerini three-component reaction
Pd-C	palladium on charcoal
PEG	polyethylene glycol
PEGA	acrylamidopropyl-POE-*N*,*N*-dimethylacrylamide co-polymer
PEG/PS	polyethylene grafted polystyrene
Pfp	pentafluorophenyl
Phe	phenylalanine
Ph_3P	triphenylphsophine
PM	portioning-mixing
Pmc	2,2,5,7,8-pentamethyl chroman-6-sulfonyl
PNA	polyamide nucleic acids
POE	polyoxyethylene
Pro	proline
PS	polystyrene
PS-PEG	polystyrene–polyethylene glycol
PS-SCLs	positional scanning SCLs
PyBOP	benzotriozole-1-yl-oxy-tris-pyrrolidinophosphonium hexafluorophosphate
PyBroP	bromo-tris-pyrrolidine-phosphonium hexafluorophosphate
RBF	round-bottom flask
RDE	rotating disk electrode
REM	regenerated Michael addition
R_f	radiofrequency
RF	reaction flask
RHE	reference hydrogen electrode
RR	reservoir for reagents
RS	reservoir for solvent
r.t.	room temperature
SA	salicylaldehyde
SCAL	safety-catch amide linker
SCLs	synthetic combinatorial libraries
S-3CR	Strecker three-component reaction
SEC	size exclusion chromatography
SECSY	spin echo correlation spectroscopy
Ser	serine
Ser(*t*-Bu)	*O*-*t*-butyl-L-serine
Ser(Bn)	*O*-benzyl-L-serine
SF	separation funnel
SMPS	simultaneous multiple peptide synthesis
SPOS	solid phase organic synthesis
SPPS	solid phase peptide synthesis
SR	sample reservoir

ABBREVIATIONS

TBAF	tetrabutylammonium fluoride
TBS	Tris-buffered saline
TBTU	O-benzotriazol-1-yl-N,N,N',N'-tetramethyluronium tetrafluoroborate
Tc	critical temperature
TCA	trichloroacetic acid
TCC	T cell clone
TEA	triethylamine
TES	triethylsilane
TFA	trifluoroacetic acid
TFE	1,1,1-trifluoroethanol
TFMSA	trifluoromethanesulfonic acid
$(Tf)_2O$	trifluoromethanesulfonic acid alidanhydride
THF	tetrahydrofuran
Thr	threonine
Thr(t-Bu)	O-t-butyl-L-threonine
Thr(Bn)	O-benzyl-L-threonine
Thr(TBS)	O-t-butyldimethylsilyl-L-threonine
Thr(Trt)	O-trityl-L-threonine
$TiCl_4$	titanium tetrachloride
$Ti(i\text{-}PrO)_4$	titanium tetraisopropoxide
TIS	triisopropylsilane
TLC	thin-layer chromatography
TMOF	trimethylorthoformate
TMSCN	cyanotrimethylsilane
TMSOTf	trimethylsilyltriflate
TOCSY	total correlation spectroscopy
Tos	$para$-toluene sulfonyl protecting group
Trp	tryptophan
Trt	trityl
Tyr	tyrosine
U-4CR	Ugi four-component reaction
UF	ultrafiltration
UV	ultraviolet
Val	valine
v/v	volume/volume
WSCD	water soluble carbodiimide
w/v	weight/volume
w/w	weight/weight
XPS	X-ray photoelectron spectroscopy
XRD	X-ray powder diffraction

Chapter 1

Synthesis of combinatorial libraries using the portioning-mixing procedure

Árpad Furka, Linda K. Hamaker, and
Mark L. Peterson[†]

Advanced ChemTech, Inc., 5609 Fern Valley Road, Louisville, Kentucky 40228, USA.

†Neokimies Inc., Instit de Pharmacologie de Sherbroke, 3001, 12e Avenue Nord 25-3038, Fleurimont, Quebec J1H SN4, CANADA

1 Introduction

Combinatorial chemistry is a young, yet intensely popular branch of science. Beginning in the 1980s, several innovative synthetic methods were published (1–7) which radically changed the theory and practice of designing and preparing new substances for pharmaceutical research and other areas of application. The field of combinatorial chemistry evolved as a result of these new approaches and has revolutionized the way scientists think about drug discovery. Combinatorial processes have since spread outside of pharmaceutical research and are gradually expanding to other areas of chemistry including catalysis and materials science. One of the first and most influential combinatorial synthetic methods was the portioning-mixing (split-mix) procedure (3, 8, 9) (see also Chapters 2 and 3). In this chapter, the principles behind the portioning-mixing approach will be described along with important features to consider when applying this technology. Specific examples of how this process was put into practice and elaborated upon, including the use of encoding techniques, will also be presented.

2 The portioning-mixing (split-mix) synthesis

2.1 Basic principles

The portioning-mixing (PM) method (often termed 'split-mix' synthesis) is based on Merrifield's solid phase procedure (10). Although the method can be equally well applied for preparing peptide or other kinds of organic libraries, its principle is exemplified in *Figure 1* by the synthesis of a tripeptide library on solid support (where the larger circles represents the solid support). In each of the three coupling positions, three different amino acids (black, white, and grey circles) are used and the synthesis is executed by repetition of the following three simple operations:

(a) Dividing the solid support into equal portions.

(b) Coupling each portion individually with one of the amino acids.

(c) Mixing the portions.

In the first step, the amino acids are coupled to equal portions of the resin; the final product, after recombining and mixing the portions, is the mixture of the three amino acids bound to resin. In the second cycle, this mixture is again

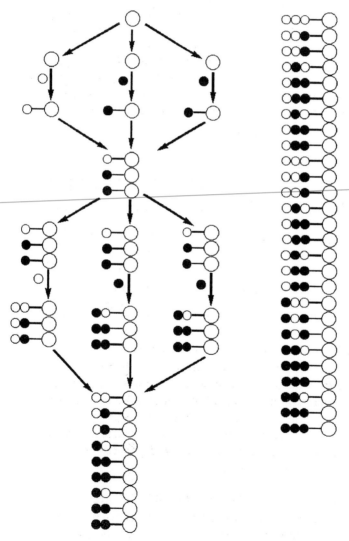

Figure 1 Scheme of the portioning-mixing synthesis.

divided into three equal portions and the three amino acids are individually coupled to these mixtures. In each coupling step, three different resin-bound dipeptides are formed, so the end-product after recombination is a mixture of nine dipeptides. The final step is executed similarly to form a mixture of 27 tripeptides. The divergent, vertical, and convergent arrows indicate portioning, coupling (with one kind of amino acid), and mixing, respectively.

2.2 Important features of the synthesis

2.2.1 Efficiency

Examining *Figure 1*, it can be seen that starting with a single substance (the resin, used as the solid support), the number of compounds is tripled after each coupling step. In the first reaction, $3 \times 1 = 3 \ (3^1)$ resin-bound amino acids are produced. Upon completion of the second coupling, $3 \times 3 = 9 \ (3^2)$ resin-bound dipeptides are formed. If peptide libraries are prepared from 20 different amino acids, the number of coupling cycles executed in the synthesis (reflecting the labour requirement) increases only linearly with the length of peptides (*Table 1*). The number of peptides formed in the procedure, however, increases exponentially. This makes the PM method a very efficient synthetic procedure for making large numbers of different compounds.

Table 1 Efficiency of the PM synthesis

Length of peptides	Number of amino acids	Number of coupling cycles	Number of peptides
n		$n \times 20$	20^n
2	20	40	400
3	20	60	8000
4	20	80	160 000
5	20	100	3 200 000
6	20	120	64 000 000

2.2.2 Formation of all possible sequence combinations

The PM synthesis has another feature that proved to be very important. Consecutive and repeated execution of the three simple operations (portioning, coupling, and mixing) ensures, with mathematical accuracy, the formation of all possible sequence combinations of amino acid building blocks used in the synthesis. This is clearly shown even in the simple example outlined in *Figure 1*. No other sequence combinations of the black, white, and grey circles can be deduced apart from those found in the figure.

2.2.3 Formation of compounds in one-to-one molar ratio

Libraries are most often prepared in order to find biologically active substances among the products. In the identification process, or screening, the goal is to find the biologically most effective component. Serious problems may arise in screening if the products are not present in equal quantities in the mixture. A low activity component, for example, if it is present in a large amount, may show a misleading more prominent effect than a highly active component present in lower quantity. It is therefore important to prepare libraries in which the constituents are present in equal molar quantities. The PM method was designed to comply with this requirement. Before each round of couplings, the resin is thoroughly mixed, then divided into homogeneous equal portions. This ensures that the previously formed intermediate products are present in equimolar quantities in each portion. For the example of peptide libraries, since couplings with the different amino acids are executed on spatially separated samples, it is possible to use appropriate chemistry to drive each coupling reaction to completion regardless of the reactivity of the different amino acids. As a result, each peptide is quantitatively transformed into an elongated new one with both the number of peptides originally present and their equimolar ratio preserved in every portion at each step. Due to the statistical nature of two of the operations (mixing and portioning), formation of compounds in a one-to-one molar ratio is expected only if the number of beads of the solid support greatly exceeds the number of compounds synthesized.

Libraries of equimolar ratio can also be obtained via a second, but less practiced, methodology; namely, chemical mixing. Instead of portioning and mixing the resin, this second technique employs a mixture of amino acids which is utilized in the coupling steps. To ensure equimolar ratio of products in the library mixture, information must be known about the coupling rates of the individual amino acids. Houghten (11) prepared a peptide library using this technique by taking advantage of published relative reaction rates (12, 13) of several protected amino acids. It was concluded that the relative coupling rates of protected amino acids were greatly independent of the resin-bound amino acid, thus allowing for the synthesis of the peptide library in near equimolarity.

2.2.4 The parallel nature of the PM synthesis and formation of individual compounds

The PM procedure has another intrinsic feature that plays an important role in screening and gives a unique character to the method: *on any individual bead of the solid support, only one kind of peptide is formed*. This may seem surprising at first glance, but becomes quite understandable upon closer examination. In *Figure 2*, the fate of a randomly selected bead is followed in a three step coupling process. Since the bead in every coupling step meets only one component, only this single component is coupled to all of its free sites. This varied component in the first, second, and third coupling step is the 'black', 'white', and 'grey' one, respectively, so the bead ends up with the black–white–grey sequence. It is also important to note that in both the Merrifield and the PM synthesis, the beads behave very much like tiny reaction vessels that do not interchange their contents with the other ones. Each of the millions of these reaction vessels preserves its content through the end of the synthesis, when they contain a single product. The structures of the substances, however, are unknown until

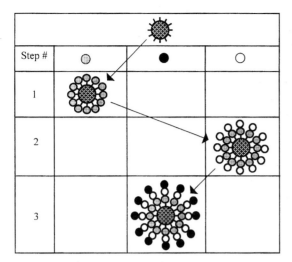

Figure 2 Formation of a single substance on a bead.

determined. The above discussion indicates that the PM synthesis can in fact be considered an exceptionally effective parallel procedure leading to individual compounds where the challenge is the identification of the individual molecules captured on a single bead.

2.2.5 Applicability of the PM method in the synthesis of 'organic' libraries

Peptides are not preferred drug candidates because of their high susceptibility to metabolic degradation. The ideal drug leads are small organic molecules due to their usually more favourable pharmacodynamic properties. The PM method is fully applicable to the synthesis of such organic libraries. Both sequential and cyclic libraries can easily be prepared as long as the solid phase reaction conditions are well developed.

3 Manual and automated synthesis

3.1 Manual synthesis

Both manual and automated methods are applicable to the realization of the PM synthesis. In *Figure 3*, a manual device is outlined that was used in one of the authors' (A. F.) laboratory (14). The device is a vacuum manifold (A) mounted on a shaker (B). It has a 55 cm long aluminium tube (b) with removable threaded end-caps (c) and rubber gaskets (e) at both ends. One of the end-caps has an outlet (f) connected to a water aspirator pump via a filter flask serving as a waste collector. The reaction vessels (a) are fitted into the holes of the tube with rubber rings (d) to hold them securely and support a vacuum seal. The reaction vessels are screw-cap glass or polypropylene tubes with a frit at the bottom. The aluminium tube can be turned on its axis so the reaction vessels can be fixed in

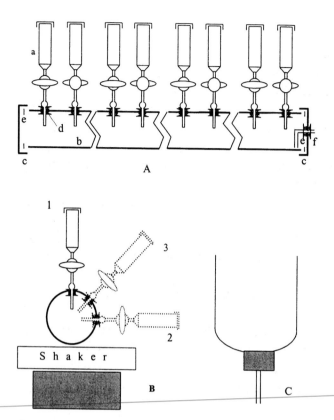

Figure 3 Manual device for PM synthesis.

any position between vertical (1) and horizontal (2). Vertical arrangement is used in all operations except shaking, when a tilted position (3) is recommended. The reaction vessel (C) is constructed of polyethylene with a frit and gas inlet tube incorporated at the bottom. Additionally, the resin suspension can be mixed with the aid of nitrogen bubbling. Portioning is carried out manually by distributing equal portions of the combined resin suspension to each reaction vessel.

Protocol 1

Manual synthesis of a resin-bound pentapeptide library

Equipment and reagents

- A 55 cm vacuum chamber (*Figure 3*) containing 20 holes with rubber rings to hold the reaction vessels
- Polypropylene syringe tubes with a polyethylene frit

- Aminomethyl resin (Advanced ChemTech)
- Boc-4-aminobutanoic acid (Advanced ChemTech)

A. Preparation of base resin

1. In a round-bottom flask, suspend aminomethyl resin (5.0 g, 200–400 mesh, sub-stitution 0.5 mmol/g) in dimethylformamide (DMF, 75 ml).

2. In a separate flask, dissolve Boc-4-aminobutanoic acid (2.5 equivalents relative to the resin) in a minimum amount of DMF.

3. Add N-hydroxybenzotriazole (HOBt, 2.5 equivalents relative to the resin) and stir until the HOBt dissolves (dichloromethane (DCM) can be added to assist in dissolution). Add this solution to the resin.

4. Add diisopropylcarbodiimide (DIC, 1 equivalent relative to the amino acid) to the resin mixture and mix on an orbital shaker for 3 h.

5. Add 2 equivalents (relative to the resin) each of acetic anhydride and pyridine to the reaction flask and mix for an additional 30 min to end-cap any residual amino groups.

6. Filter and wash[a] the resin three times with DMF, three times with DCM, and finally three times with methanol (MeOH).

7. Add trifluoroacetic acid (TFA)/DCM (1:1, v/v) to the resin and mix for 15 min. Filter and wash the resin three times with DCM and three times with MeOH.

8. Dry the resin *in vacuo* to a constant weight.

9. Couple Boc-Gly to all of the resin following steps 1–8 outlined above substituting it for Boc-4-aminobutanoic acid.

B. Portioning

1. Suspend the resin in 100 ml of DMF/DCM (2:1, v/v) and mix continuously with nitrogen bubbling.

2. Pipette equal volumes (5 ml) of the slurry into each of the 19 reaction vessels (of total volume 12 ml).

3. Dilute the remaining suspension to 100 ml with DMF/DCM (2:1, v/v) and repeat the portioning procedure.

4. Remove the solvent via a vacuum source.

C. Coupling

1. Add 5 ml DCM/TFA (2:1, v/v) to each reaction vessel and shake for 15 min to deprotect the resin.

2. Filter and wash each reaction with MeOH, DCM, MeOH, DCM, and DCM (5 ml each).[b]

3. Neutralize the resin by shaking twice with 5 ml DCM/DIEA (9:1, v/v) and wash (3 × 5 ml DCM) each sample.

4. Add the activated ester of one of the 19 common Boc-amino acids (omitting Cys) to each reaction vessel.[c]

5 Shake for 45 min, filter, and wash each reaction with MeOH, DCM, MeOH, DCM, and DCM (5 ml each).

D. Mixing

1 Combine all samples in the mixing chamber and wash with portions of DCM/DMF (2:1, v/v). This solvent can be used to assist in the transfer of resin from the reaction vessels.

2 Increase the volume to 100 ml with DCM/DMF (2:1, v/v) and mix via nitrogen bubbling for 15 min.

3 Filter and remove the solvent via a vacuum source.

E. Completion of library synthesis

1 Repeat the 'portioning', 'coupling', and 'mixing' procedures an additional three times to complete the synthesis of the pentapeptide library. This library will have Gly in all C-terminal (first coupling) positions.

2 Omit the final 'mixing' operation to provide library pools with a known residue at the final position.

3 Repeat from part A, step 9 substituting other Fmoc-amino acids for Boc-Gly to provide libraries with alternative substituents in the first position.

[a] Employ enough solvent to fully cover/swell the resin and then add an additional 20% to the slurry.

[b] The vacuum was maintained during all shaking operations to hold the reaction vessels in place.

[c] Side-chain protective groups were OBn for Asp and Glu; Bn for Ser, Thr, and Tyr; Z for Lys; Tos for Arg and His. The active esters (used in four times molar excess) were prepared in separate vessels from 0.8 mmol Boc-derivatives dissolved in 5 ml DCM/DMF (3:1, v/v)—except in the case of Boc-Gln, Boc-Asn, Boc-Arg (Tos), and Boc-His (Tos) amino acid when the solvent was 5 ml DMF—by adding with stirring, HOBt (135 mg; *c.* 1 mmol), then DIC (0.124 ml; 0.8 mmol).

Lebl and Stankova reported a manual synthesis utilizing the PM procedure which resulted in a library of 12 000 triazine-based compounds (15). The syntheses were performed manually in Wheaton glass vials on a polyethylene grafted co-polystyrene (PEG/PS) resin and involved subpooling of the last two steps which resulted in 40 pools of 300 compounds each.

3.2 Automated synthesis (see also Chapters 4 and 12)

The Advanced ChemTech Model 357 Flexible Biomolecular Synthesizer (*Figure 4*) is a computer controlled automatic instrument capable of performing the PM library synthesis (16). Its reaction block contains 36 × 9 ml reaction vessels and a larger collection vessel (600 ml) for mixing the samples. The instrument consists of two robotic arms that work in unison to dispense solvents from system fluid containers (not visible in *Figure 4*), amino acid or other monomer solutions

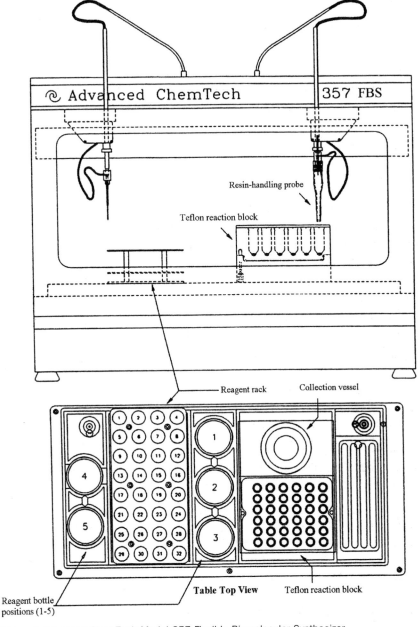

Figure 4 Advanced ChemTech Model 357 Flexible Biomolecular Synthesizer.

from the 32 containers of the reactant rack, and reagents from the five table-top bottles to the reaction vessels and the collection vessel. A resin-handling glass probe attached to the right arm provides the capability to move resin between the reaction vessels and the collection vessel. The reaction vessels and the collection vessel each contain a frit in the bottom to hold the resin during the removal of liquid. A nitrogen port at the top of each reaction vessel provides a protective atmosphere. Additionally, nitrogen can be bubbled from the bottom of the collection vessel to provide mixing. Integrated variable speed orbital mixing and operational flexibility are also notable features of this instrument.

Protocol 2

Automated construction of an organic library using split synthesis: 3-amino-5-hydroxybenzoic acid as a core structure (from ref. 17)

Equipment and reagents

- Automated synthesizer: Model 357 FBS (Advanced ChemTech, ACT 357 FBS)
- Resins: commercial suppliers of resin (Advanced ChemTech, Bachem, Novabiochem)
- Chemicals: commercial suppliers (Acros, Advanced ChemTech, Aldrich, Bachem, Fluka, Lancaster, Novabiochem, Sigma)

A. Preparation of core scaffold

1 Suspend 3-amino-5-hydroxybenzoic acid hydrochloride (0.50 g, 2.6 mmol) in 5 ml of 50% aqueous acetonitrile.

2 Add 9-fluorenylmethylsuccinimidyl carbonate (Fmoc-OSu) (0.98 g, 2.6 mmol) and diisopropylethylamine (DIEA) (0.46 ml, 2.6 mmol) then heat to 60°C for 16 h.

3 Allow the reaction mixture to cool to ambient temperature.

4 Collect the precipitated product and wash thoroughly with water.

5 Dissolve the solid in ethyl acetate.

6 Dry over anhydrous sodium sulfate.

7 Remove the desiccant by filtration and evaporate the filtrate to dryness under reduced pressure using a rotary evaporator.

8 Suspend Rink amide resin (10 g, 4.6 mmol) in 100 ml of 20% piperidine/DMF.

9 Agitate for 20 min using an orbital mixer, then filter.

10 Wash sequentially with DCM, MeOH, DCM, MeOH, DCM, then dry the resin *in vacuo.*[a]

11 Suspend the dry resin in 1-methyl-2-pyrrolidinone (NMP, 100 ml), and sequentially add the product from step 7, 1-hydroxy-7-azabenzotriazole (HOAt) (0.74 g, 6.0 mmol), 2-(1H-benzotriazole-1-yl)-1,1,3,3-tetramethyluronium hexafluorophosphate (HBTU) (2.3 g, 6.0 mmol), and DIEA (2.1 ml, 12.0 mmol).

12 Agitate the suspension for 20 h, then check the completeness of the coupling utilizing the ninhydrin test.

13 Filter the resin and wash[a] sequentially with DCM, MeOH, DCM, MeOH, DCM, then dry the resin *in vacuo.*

14 Repeat steps 8–10 to provide the deprotected core scaffold attached to the solid support.

B. Preparation of the library

1 Divide the core resin above into 36 different equivalent portions[b] utilizing the 'split' functions[c] of the ACT 357 FBS.

2 React each portion individually with either a threefold excess of carboxylic acid (with HBTU, HOAt, DIEA, NMP, 23 h) or a fourfold excess of acyl chloride (with DIEA) in DCM for 24 h.

3 Wash[a] the individual resins thoroughly with DCM, MeOH, DCM, MeOH, DCM.

4 Recombine the individual portions into the single large chamber of the reactor block utilizing the 'combine' function of the ACT 357 FBS.

5 Treat the combined resin (off instrument) with cesium carbonate (4 equivalents) in glyme for 16–19 h and then heat at 60 °C to hydrolyse any O-acylated material.

6 Filter the resin and wash[a] sequentially with DCM, MeOH, DCM, MeOH, DCM, then dry the resin in vacuo.

7 Divide the resin into 36 different equivalent portions utilizing the 'split' functions of the ACT 357 FBS.

8 Suspend the resin portions (100 mg, 0.045 mmol) in 1 ml of DMSO/NMP (1:1, v/v), an alkyl halide (0.28 mmol), and 1,8-diazabicyclo [5.4.0]undec-7-ene (DBU, 0.28 mmol).

9 Mix the reactions at room temperature for 20 h.

10 Filter the resin and wash sequentially with DCM, MeOH, DCM, MeOH, DCM, then dry the resin in vacuo.

11 Remove the products from the solid support by suspending the resin portions with 1 ml of TFA/DCM (1:1, v/v) and mixing for 1 h.

12 Filter and wash once with 1 ml of DCM.

13 Evaporate filtrate under reduced pressure to provide the cleaved library mixture.

[a] Employ enough solvent to fully cover/swell the resin and then add an additional 20% to the slurry.

[b] Resin slurry concentration should never be more concentrated than 4 ml/g of resin.

[c] To ensure accuracy, at least five passes should be made for any split process.

Dankwardt et al. (17) took advantage of automation and utilized the Advanced ChemTech Model 357 FBS to prepare a library of 2001 small organic molecules via the split-mix technique. Each pool was comprised of less than 100 compounds, thus making extensive deconvolution unnecessary. This single synthetic route provided a 69 × 29 compound library consisting of 3-amido-5-alkoxybenzamides and 3-amido-5-arylsulfonylbenzamides.

4 Preliminary considerations for library design

When experiments are planned with organic or peptide libraries prepared according to the PM procedure, several factors deserve preliminary considerations. Although these considerations will be exemplified with peptide libraries, the conclusions can be applied equally as well when working with organic libraries.

4.1 The weight of the libraries

Full peptide libraries are usually constructed from the common 19 amino acids typically omitting cysteine due to complications that arise from disulfide bond formation (18). This is most often done with the L stereoisomers although incorporation of D isomers is also fairly common. The number of peptides in such libraries increases exponentially with their length and, as a consequence, the weight of the libraries also increases exponentially. This exponential increase is reflected in the data of *Table 2*. The weight of the libraries has been calculated assigning a 1 pmol quantity for each peptide. It can be seen that if the number of varied positions is high, the weight of the libraries becomes so great as to become unmanageable with solubility problems expected in screening.

Table 2 Number and weight of peptides in libraries depending upon the number of varied positions

Number of varied positions	Number of peptides	Weights
2	361	92.00 ng
3	6859	2.58 µg
4	130 321	64.76 µg
5	2 476 099	1.53 mg
6	47 045 881	34.64 mg
7	893 871 739	765.25 mg
8	16 983 563 041	16.57 g
9	322 687 697 779	353.52 g
10	6 131 066 257 801	7.45 kg

4.2 The weight of the solid support

One of the expected difficulties of the PM procedure is the large quantity of resin needed in the preparation of libraries with a higher number of varied sequence positions (*Table 3*). In practice, these quantities are expected to be even higher than indicated since libraries are usually prepared in larger than pmol quantities. It is difficult, for example, to manipulate kilograms of resin in a laboratory and, if the number of the varied positions is high enough, it is virtually impossible to carry out the synthesis.

Table 3 Approximate weight of the resin needed to prepare libraries containing each peptide in 1 pmol quantity

Number of varied positions	Needed total capacity	Weight of the resin
2	361.00 pmol	720.0 ng
3	6.86 nmol	13.7 µg
4	130.32 nmol	260.6 µg
5	2.48 µmol	5.0 mg
6	47.05 µmol	94.1 mg
7	893.87 µmol	1.8 g
8	16.98 mmol	34.0 g
9	322.69 mmol	645.4 g
10	6.13 mol	12.3 kg

4.3 The significance of the number of resin beads

Another issue that merits consideration before beginning a synthesis is the ratio of the number of the expected peptides to the number of beads of the solid support. Since only one peptide forms on each bead, the maximum number of peptides is limited by the number of beads. Furthermore, owing to the fact that the two essential operations, mixing and portioning, are influenced by statistics, formation of all expected peptides, as well as their equal molarity, is ensured only if the number of beads greatly exceeds the number of peptides. It is a good practice to apply at least a tenfold excess of beads more than the number of expected peptides. For this reason, when very complex libraries are prepared, it is desirable to choose as small a bead size as possible; i.e. often 200–400 mesh (38–75 µm) resin. Note also that for very small bead sizes (< 20 µm), static charges and aggregation phenomena can adversely affect mechanical manipulation. Each gram of 200–400 mesh resin contains about ten million beads. The quantity of resin needed if the number of beads equals or exceeds ten times the number of peptides is illustrated in *Table 4*. This data clearly demonstrates that the number of varied positions in full libraries is limited to about six or seven for practical reasons.

5 Synthetic strategies for library construction

Most organic and peptide libraries are prepared for pharmaceutical research as previously stated, and many different methods have been developed for their screening. Concurrent with these varied protocols are the differing applications of the libraries themselves. It is easiest to consider libraries based upon the final forms in which they are utilized:

- soluble libraries
- libraries tethered to the solid support
- discrete compounds released from individual beads
- discrete compounds prepared in sorted individual containers

Table 4 Approximate weight of 200–400 mesh resin if one or ten beads are assigned to each peptide

Number of positions	Weight (one bead)	Weight (ten beads)
2	36.10 µg	361.00 µg
3	685.90 µg	6.86 mg
4	13.03 mg	130.32 mg
5	247.61 mg	2.48 g
6	4.70 g	47.05 g
7	89.39 g	893.87 g
8	1.70 kg	16.98 kg
9	32.27 kg	322.69 kg
10	613.11 kg	6.13 t

5.1 Soluble libraries (see also Chapters 3, 9–12)

When soluble libraries are prepared, the first monomer is attached to the support by a covalent bond that can be cleaved at the end of the synthesis. Since throughout the PM synthesis individual compounds are found on each bead of the solid support, a real mixture forms only upon cleavage. An interesting liquid phase variant of the PM method was also developed (18) that leads to the formation of a soluble mixture. The synthesis is carried out using a polyethylene glycol monomethyl ether (MeO-PEG) support. The approach exploits the differential solubility of the polymer in various solvents. This polymer is soluble during the course of the reaction, making it advantageous for coupling, but can be precipitated upon completion to allow for the efficient removal of excess reagents through washing.

Janda demonstrated (19) the utility of this liquid phase combinatorial synthesis (LPCS) in the construction of a 1024-membered pentapeptide library. The LPCS method was further extended to include a non-peptidyl library of sulfonamides. A definite advantage to this approach included the ease of structure determination through conventional methods (infrared spectroscopy and/or proton and carbon-13 nuclear magnetic resonance spectroscopy).

The soluble mixtures may contain thousands or millions of compounds. At the beginning of the combinatorial era, it seemed a hopeless task to identify a biologically active compound in such complex mixtures. Stepwise determination, termed deconvolution, was time-consuming, often requiring synthesis of gradually more defined pools until the active constituent was pin-pointed. Since that time, several methods have been developed which can dramatically ease the identification process. Several of the most prominent methods based on the use of partial libraries are listed below:

- iteration method (20, 21)
- positional scanning (22, 23)
- determination of monomers by omission libraries (24, 25)

The synthesis of partial libraries does not substantially differ from that of full libraries. In the peptides of the sublibraries used in the iteration method or in positional scanning, for example, one or more coupling positions are occupied by the same amino acid in all peptides. Consequently, no portioning is needed before coupling in these positions. In the synthesis of omission libraries, the same one amino acid is omitted in all positions. Information about each amino acid in the peptide can be gathered as to whether or not it is an essential biological component. For example, if alanine is an essential biological component of the peptide, then the absence of alanine in a library would generate a less active or inactive library.

5.2 Tethered libraries

5.2.1 Selective binding

Tethered libraries prepared by the PM method have an enormous advantage: they are essentially individual compounds enclosed in beads as containers. Several deconvolution methods take advantage of this fact. The first published method (26) was developed to determine bioactive sequences in peptide libraries. In this procedure, the beads are mixed with the solution of the target protein. The beads binding the target can be distinguished by colour. The coloured beads are picked out manually, then the peptide remaining on the bead after the removal of the attached protein is sequenced. When tethered libraries are prepared, it is advantageous to insert a spacer between the target products and the resin. The authors have presented the preparation of a pentapeptide sublibrary via the Boc strategy in *Protocol 1*. The spacer employed is a GABA residue and the first position in all peptides is occupied by a single amino acid: glycine. The cited example also demonstrated the applicability of the manual approach.

5.2.2 Encoding strategies

Organic libraries can easily be prepared by the PM method if the conditions of the solid phase reactions to be employed are properly optimized. The synthesized libraries are often screened as individual substances released from single beads. Since determination of the structure of the various organic compounds is not as simple as sequencing peptides, the beads must be coded in some manner. Most commonly this is done by spatial array. The resin is placed in a specific, unchanging location where the chemistry performed on it is defined and known. Alternately, the more powerful approach for the PM method is to encode the beads. In this procedure, the building blocks of the encoding tags are attached to the beads in parallel with the organic building blocks of the library. Two basic types of such encoding have been described in the literature:

- encoding with sequences (27–31)
- binary encoding (32, 33)

When encoding with sequences, the tags are either peptides or oligonucleotides. Their sequences encode both the identity of the organic reagents and the

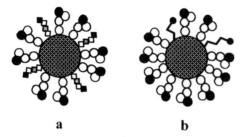

a   **b**

Figure 5 Beads encoded by sequence (a) and binary (b) coding techniques.

order of their application. Generally, a separate linker functionality is utilized for the tagging sequence to prevent interference with the target chemistry. It is possible, however, to utilize a single linker where the tag is added separately as only a small percentage of the available active sites. In *Figure 5a*, the white, grey, white, black squares encode the 'white', 'grey', and 'black' circles representing the organic reagents and their white, grey, white, black order of the reactions. Upon conclusion of the reaction series, the code can be read by cleaving the tag and determining the amino acid sequence of the peptide or the nucleotide sequence of the oligonucleotide tag utilizing commercially available instrumentation.

Nestler employed the PM procedure and the encoded library technique to recognize and discriminate various enkephalins (34). A 10^4-membered 'cholic acid' library (*Figure 6*) was generated and screened in a competitive binding assay. Two different receptors were conjugated with two different coloured dyes (red and blue) and then binding of the library was evaluated. Either a red or blue coloured bead indicated the substrates that preferentially bound to one of the receptors. Compounds that bound competitively to the receptors yielded a bead which was a blend of the two colours. This technique successfully provided several molecules with preferential binding to one receptor over the other.

In the binary encoding system, the coding units are halobenzenes carrying a hydrocarbon chain of varying length (*Figure 5b*) attached to the beads through a

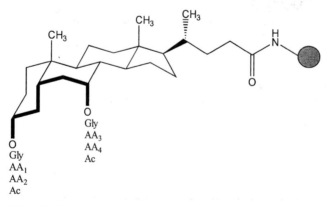

Figure 6 'Cholic acid' library.

cleavable spacer. It is characteristic of this labelling technique that the coding units do not form a sequence. It is simply their presence that codes for the organic building blocks and their position. In the original publication (31), the authors used the method for encoding peptide sequences. Utilizing 18 different coding units arranged according to a binary coding format, the authors were

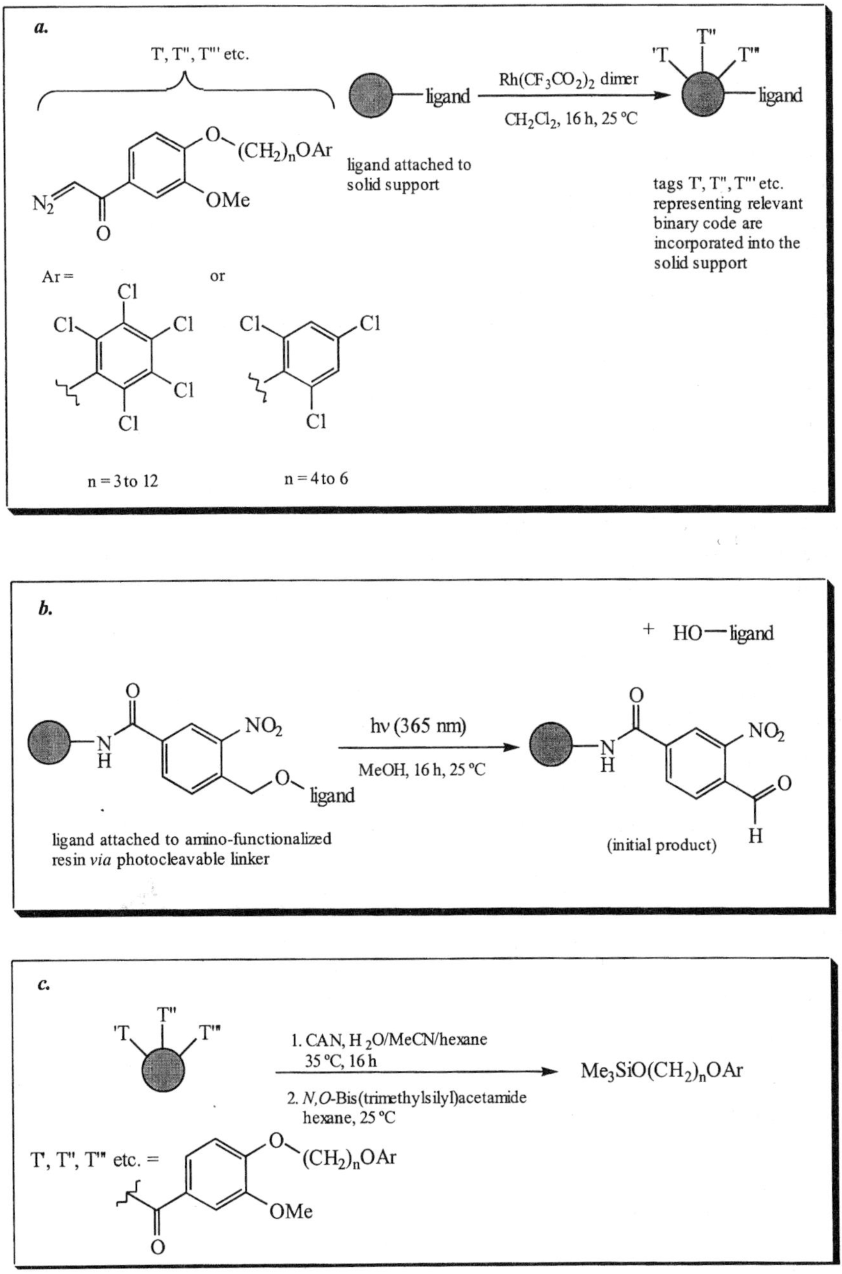

Figure 7 General tagging approach for binary encoded libraries.

Table 5 Binary encoding

Coupling No. 1	Coupling No. 2	Coupling No. 3
$A_1\,T_1$	$B_1\,T_4$	$C_1\,T_7$
$A_2\,T_2$	$B_2\,T_5$	$C_2\,T_8$
$A_3\,T_3$	$B_3\,T_6$	$C_3\,T_9$
$A_4\,T_2T_1$	$B_4\,T_5T_4$	$C_4\,T_8T_7$
$A_5\,T_3T_1$	$B_5\,T_6T_4$	$C_5\,T_9T_7$
$A_6\,T_3T_2$	$B_6\,T_6T_5$	$C_6\,T_9T_8$
$A_7\,T_3T_2T_1$	$B_7\,T_6T_5T_4$	$C_7\,T_9T_8T_7$

able to code all sequences in a 117 649-member library, formed by varying seven amino acids (D, E, I, K, L, Q, and S) in six positions. The presence of the coding units could be determined after cleavage by a single step electron capture gas chromatographic analysis. *Table 5* shows an example of how nine different tags (T_1–T_9) can be used to encode the structure of 343 organic molecules synthesized from three different types of reagents, each group containing seven members (A_1–A_7, B_1–B_7, and C_1–C_7). The codes for $A_1B_1C_3$, $A_2B_3C_4$, and $A_7B_7C_7$, for example, are, $T_9T_4T_1$, $T_8T_7T_6T_2$, and $T_9T_8T_7T_6T_5T_4T_3T_2T_1$, respectively.

Additionally, the binary encoding strategy has been very successfully applied to produce a 6727-membered library consisting of linear acyl piperidines (35, 36). The general schematic of this specific encoding example is outlined in *Figure 7*. The encoding of the resin employed a rhodium-catalysed carbene insertion (*Figure 7a*). A binary encoding scheme was utilized which ultimately resulted in products having a code consisting of the presence or absence of 13 molecular tags. Implementation of a photo-labile linker allowed for the bioassays to be performed in solution (*Figure 7b*). The decoding procedure (*Figure 7c*) was effectively carried out using ceric ammonium nitrate (CAN) and silylation to afford the molecular tags in solutions that were analysed by electron capture gas chromatography (ECGC). Upon comparison to known standard ECGC chromatograms, the identification of active ligands was determined.

Protocol 3

Portioning-mixing synthesis of a library encoded with molecular tags

Equipment and reagents

- TentaGel S NH$_2$ resin (Rapp Polymere, Advanced ChemTech)

- Gas chromatograph (GC): Hewlett Packard 5890 with electron capture detector (ECD), 25 m, 0.2 mm, crosslinked 5% phenylmethyl silicone column

- Ultraviolet lamp (365 nm): Spectroline Model XX-40

- All other reagents were obtained from standard commercial sources (Acros, Advanced ChemTech, Aldrich, Lancaster, Fluka, Sigma)

A. General procedure for the tagging of resin

1   Assign a code to the library thus forming a template for the synthesis.

2   Prepare appropriate stock solutions by dissolving a known quantity of the tagging precursor molecules[a] in DCM.

3   Add the appropriate tagging solution to portions of the resin suspended in DCM.

4   Agitate the resin for 1 h at ambient temperature.

5   Add 1.0 ml of a DCM solution of rhodium trifluoroacetate dimer (1.0 mg/ml) to each resin suspension.

6   Agitate the resin for 16 h at ambient temperature.

7   Combine the individually encoded resins into a single batch, wash with DCM (10 × 50 ml), then dry *in vacuo*.

B. Preparation of resin

1   Add a solution of compound X^b in 30 ml DCM, to 3.0 g (0.96 mmol) of TentaGel S NH_2 (0.32 mmol/g), followed by 4-dimethylaminopyridine (4-DMAP) (0.040 g, 0.30 mmol).

2   Agitate the suspension for 19 h at ambient temperature while suitably protected from atmospheric moisture.

3   Filter the resin, wash thoroughly with DCM (10 × 50 ml).

C. Preparation of photocleavable linker resin

1   Add 4-acetoxymethyl-3-nitrobenzoic acid,[a] 1.42 g (11.6 mmol), 4-DMAP, and DIC (1.81 ml, 11.6 mmol) to 3.0 g (0.96 mmol) of TentaGel S NH_2 (0.32 mmol/g) suspended in 60 ml DCM.

2   Agitate the suspension for 5 h at ambient temperature while suitably protected from atmospheric moisture.

3   Filter the resin, wash with DCM (5 × 50 ml), IPA (5 × 50 ml), DCM (5 × 50 ml), then dry *in vacuo*.

4   Wash the resin with MeOH (150 ml).

5   Suspend the resin in a 10% solution of hydrazine hydrate in MeOH (150 ml).

6   Agitate the resin for 44 h at ambient temperature.

7   Filter the resin, wash with MeOH (5 × 100 ml) and DCM (5 × 50 ml), then dry *in vacuo*.

D. Loading of photocleavable linker resin

1   Add N-Fmoc-protected substrate, (2.52 mmol) to a suspension of 3.0 g (0.84 mmol) of the photocleavable linker resin in 60 ml DCM, followed by 4-DMAP (0.031 g, 0.25 mmol), and DIC (0.40 ml, 2.52 mmol).

2   Agitate for 22 h at ambient temperature.

3   Filter the resin, then wash with DCM (5 × 50 ml), IPA (5 × 50 ml), DCM (5 × 50 ml).

Protocol 3 continued

4   Tag 1.0 g portions of each of the seven resins suspended in 20 ml DCM, according to the general procedure. Utilize 0.20 g of each tag precursor in 4 ml DCM for the stock solutions.

E. Reaction of tagged resins in second combinatorial step

1   Suspend the combined resin in 60 ml of a 50% solution of piperidine in DMF.

2   Agitate for 30 min at ambient temperature.

3   Filter the resin, wash with DMF (5 × 50 ml) and DCM (10 × 50 ml), then dry *in vacuo*.

4   Divide the resin manually into 31 equal batches of 0.21 g (approx. 0.070 mmol), then suspend each in 10 ml DCM.

5   Add 1.0 ml of a HOBt solution (1 mg/ml) in DMF to each suspension.

6   Add separately to each, a 10 ml solution (DCM, with a small amount of DMF if required for solubility) of a different Fmoc-amino acid (0.40 mmol) from a series of 31.

7   Agitate each resin for 16 h at ambient temperature.

8   Filter the resin, wash with DMF (15 ml) and DCM (10 × 15 ml).

9   Tag each of the 31 resin batches according to the general procedure. Utilize 0.160 g of each tag precursor molecule in 16 ml DCM for the stock solutions.

F. Reaction of tagged resins in third combinatorial step

1   Suspend the combined resin in 60 ml of a 50% solution of piperidine in DMF.

2   Agitate for 30 min at ambient temperature.

3   Filter the resin, wash with DMF (5 × 50 ml) and DCM (10 × 50 ml), then dry *in vacuo*.

4   Divide the resin manually into 31 equal batches of 0.15 g (approx. 0.050 mmol), then suspend each in 10 ml DCM.

5   Add to each suspension 0.40 mmol of one from a series of 31 reagents (sulfonyl chlorides, carboxylic acids, isocyanates) in 10 ml DCM or DMF, each to a different batch. Add also, 0.5 ml triethylamine and 100 mg 4-DMAP for the sulfonyl chlorides, 0.3 ml DIC and 100 mg 4-DMAP for the carboxylic acids. Nothing additional was added for the isocyanates.

6   Agitate each resin for 16 h at ambient temperature.

7   Filter the resins, wash with DMF (15 ml) and DCM (10 × 15 ml).

8   Tag each of the 31 resin batches according to the general procedure. Utilize 0.20 g or 0.60 g of each tag precursor molecule in 16 ml DCM for the stock solutions.

G. Cleavage of side-chain protection

1   Add 60 ml of a solution of TFA/thioanisole/EDT (5:5:0.5) to the combined resin.

2   Agitate at ambient temperature for 16 h.

3   Filter the resin, wash with DCM (10 × 50 ml), then dry *in vacuo*.

Protocol 3 continued

H. Removal of products from the resin for testing

1 Add resin beads from the combined, complete library, which contains a total of 6727 compounds, to each well of a standard 96-well polypropylene microplate.

2 Add 100 μl MeOH to each well.

3 Irradiate the suspension with UV light (365 nm) for 16 h.

4 Remove the MeOH *in vacuo*.

5 Add 100 μl of 50% aqueous MeOH to each well to resolubilize the cleaved product.

6 Transfer via pipette to a new microplate for biological assay.

I. Decoding procedure

1 Place a bead in a glass capillary (1.3 mm diameter) with 2 μl acetonitrile.

2 Add 2 μl of ceric ammonium nitrate and 3 μl hexane to the capillary.

3 Centrifuge the resulting two phase mixture briefly for agitation.

4 Seal the tube and heat to 35°C for 16 h.

5 Open the capillary carefully and remove the organic layer via syringe.

6 Add 1 μl of *N,O*-bis(trimethylsilyl)acetamide to each organic layer.

7 Analyse this solution via GC with ECD.

8 Compare the result with the injected standards to allow the code to be read and the identity of the attached compound to be ascertained.

[a] Nestler, H. P., Bartlett, P. A., and Still, W. C. (1994). *J. Org. Chem.*, **59**, 4723.
[b] Barany, G. and Albericio, F. (1985). *J. Am. Chem. Soc.*, **107**, 4936.

5.2.3 Radiofrequency tags (see also Chapter 4)

Although the deconvolution methods developed for screening compound mixtures prepared by the PM procedure have been very useful, the majority of pharmaceutical scientists prefer to work with discrete compounds. Furthermore, most drug entities are small molecular weight organic structures. Unfortunately, large libraries can lead to single compounds in very small quantities. The numbers in *Table 6* demonstrate that the quantity of the compound that is present on a single bead can be very low even if the size of the bead is large. If the synthesized compound is available in larger quantity, however, the screening experiments can easily be repeated or extended to other available targets. It is also desirable in some cases to obtain sufficient material on which additional purification can be performed. Consideration of this fact has led to the modification of the PM synthesis so that multi-gram quantities of product are obtained.

In 1995, two research groups independently developed methods for preparing individual compounds using small quantities of resin (about 30 mg) enclosed into permeable capsules (37, 38). In both cases, electronic transponders (also

Table 6 Approximate quantities of the synthesized compounds on a single bead of resin

Size (μm)	Capacity/bead
90	100 pmol
130	300 pmol
200	1 nmol
300	4 nmol
500	19 nmol
750	65 nmol

enclosed into capsules, see *Figure 8*) were employed for encoding the reagents and the order of their additions. Radiofrequency radiation can then be used to read the codes upon completion of the synthesis. The synthetic procedure followed the steps outlined in *Figure 1*. Instead of random portioning, however, the pooled capsules were manually sorted one by one according to the code read from their electronic tags. This approach has a significant advantage over the conventional PM synthesis. While preserving the high productivity of the method, it makes it possible to produce discrete compounds in larger quantities.

Armstrong (39, 40) extended this concept and merged the radiofrequency tagging method with the 'tea bag' approach (41) to PM synthesis. A 64 compound library was synthesized on a functionalized Wang polystyrene resin contained within a mesh bag together with a transponder. The PM synthesis followed the general scheme outlined in *Figure 1* and produced 64 small organic molecules with yields between 45–55%. The general reaction scheme for the preparation of this library is depicted in *Figure 9*.

The commercially available capsules for the radiofrequency encoding method are 0.3 ml polypropylene containers with mesh sides capable of holding (in addition to the electronic chip) 30 mg of resin. About 15–30 μmol of compound can be prepared in each capsule. This corresponds to about 7–15 mg of a low

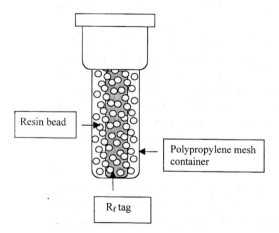

Figure 8 Microreactor with encapsulated radiofrequency tag.

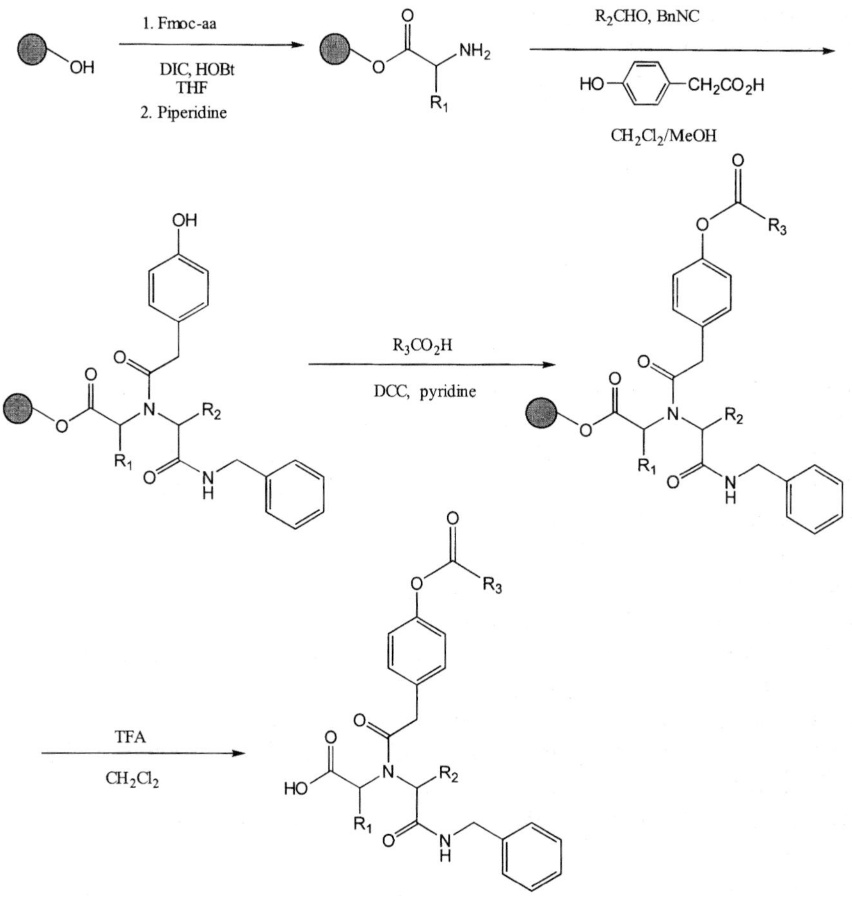

Figure 9 A small molecule library synthesized via the 'tea bag' and R_f tagging techniques.

molecular weight (500 g/mol) organic compound. The chips have a 40-bit non-volatile ID code, which is easily sufficient to encode all the compounds to be synthesized. The essential components of the electronic tags are shown in *Figure 10*. A computer controlled transceiver can read the electronic code. The antenna transmits a 125 kHz radiation frequency which is picked up by the antenna of the R_f tag and brings about the retransmission of the code. The reading process takes about 0.5 sec. The key process in application of the radiofrequency encoding method to the PM synthesis is sorting. A high speed automated procedure, which combines a feeding and reading mechanism with robotics and associated software, has been reported which can sort 10 000 tagged reactors overnight (42).

Nicolaou *et al.* prepared a library of epithilones via the application of portioning-mixing and radiofrequency encoded techniques (43) (see Chapter 4). Employing Merrifield resin-containing microreactors, a split-mix synthesis followed by cyclization/cleavage of the products led to mixtures of four 12,13-desoxy-epothilones. The resulting library, after purification and epoxidation,

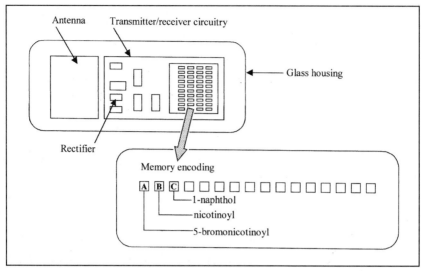

Figure 10 Representation of the radiofrequency encoding electronic chip.

was screened for induction of tubulin assembly. This reported success demonstrates the underlying potential of combining the portioning-mixing technique with appropriate encoding strategies.

Protocol 4

Synthesis of a radiofrequency encoded library (from ref. 44)

Equipment and reagents

- AccuTag™ 100 (IRORI)
- Resin encased in available containers with a radiofrequency tag (MicroKans™ IRORI)
- Knorr linker (Novabiochem)
- PyBOP® (Novabiochem)
- All other reagents were obtained from Aldrich Chemical

Method

1 Separate the containers into 18 equivalent groups (consisting of 24 reactors each) and add each group to a round-bottom flask.

2 Add 25 ml of trimethylorthoformate (TMOF) to the sets of containers.

3 Add one of the set of 18 different aldehydes (12.5 mmol) to each of the 18 reaction mixtures.

4 Add trimethylorthoformate (25 ml) and mix on an orbital shaker for 3 h.

5 Add sodium cyanoborohydride (1.26 g, 20 mmol) and continue mixing for 20 min.

6 Add a small amount (0.5 ml) of glacial acetic acid and continue mixing for an additional 3 h.

Protocol 4 continued

7 Remove solvent through aspiration.

8 Wash each set of tubes with MeOH (50 ml) once.

9 Combine all microtubes and wash two times sequentially with 200 ml DMF, DCM, and MeOH, followed by a single wash with ethyl ether (400 ml), and dry *in vacuo*.

10 Place all microtubes into a single round-bottom flask.

11 Add sufficient anhydrous DMF to allow easy agitation (400 ml).

12 To the flask add an excess of cyanoacetic acid (17 g, 200 mmol), followed by DIEA (51.6 g, 400 mmol), and DIC (32.8 g, 260 mmol).

13 Agitate the mixture overnight on an orbital shaker, then filter and wash with anhydrous DMF (400 ml).

14 Repeat steps 11–13 two additional times.

15 Wash all microtubes two times sequentially with DMF, DCM, and MeOH (400 ml each), followed by a single wash with ethyl ether (400 ml), and dry *in vacuo*.

16 Separate the containers into eight equivalent portions. The tag is read utilizing the AccuTag device to guide placement of the individual containers into the proper vessel for the next reaction.

17 To each vessel add anhydrous DMF (50 ml), MeOH (5 ml), and piperidine (850 mg, 10 mmol).

18 Add a different aldehyde (27.5 mmol) to each of the eight reaction vessels and mix for 48 h.

19 Remove the solvent by aspiration and wash with DMF (100 ml) once.

20 Combine all microtubes and wash two times sequentially with DMF, DCM, and MeOH (400 ml each), followed by a single wash with ethyl ether (400 ml), and dry *in vacuo*.

21 Separate the containers into three equivalent portions. The tag is read utilizing the AccuTag device to guide placement of the individual containers into the proper vessel for the next reaction. One portion is set aside.

22 Add triethylamine (15 g, 150 mmol), acetyl chloride (5.89 g, 75 mmol) in DCM (150 ml) to one reaction vessel.

23 To the second vessel add benzoyl chloride (10.5 g, 75 mmol) in anhydrous DMF (150 ml).

24 Agitate both mixtures overnight on an orbital shaker.

25 Remove the solvent by aspiration and wash the two sets of microtubes with 400 ml DMF, DCM, and MeOH, followed by a single wash with ethyl ether (400 ml), and dry *in vacuo*.

26 To the set-aside portions of microtubes, with free hydroxyl moieties, add 4% TFA in dioxane (3.5 ml per tube) and agitate for 1 h on an orbital shaker.

Protocol 4 continued

27 To all the microtubes reacted with acid chlorides, add 4% TFA in benzene (3.5 ml per tube) and agitate for 1 h on an orbital shaker.

28 Rinse the two sets of microtubes with dioxane or benzene, depending on which solvent was used for the cleavage, then freeze and lyophilize the resulting filtrate.

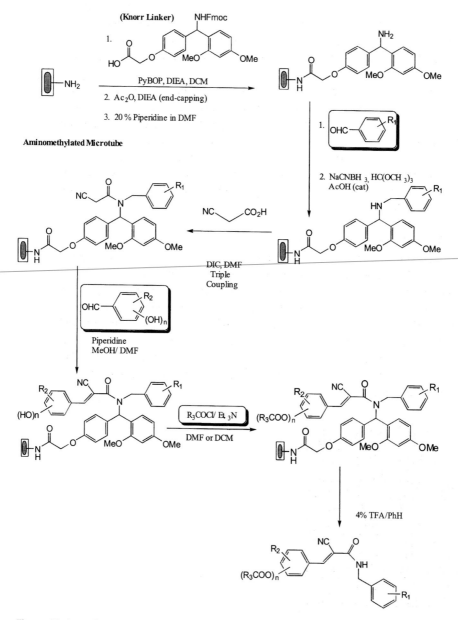

Figure 11 A small molecule library prepared with microtube reactors containing R$_f$ tags.

Extending this approach to tube-shaped reactors (44) functionalized with an aminomethylated polystyrene and containing a reusable radiofrequency tag, a small organic molecule library of 432 discrete compounds (*Figure 11*) was recently synthesized. This diversity was achieved in three separate steps. The first step implemented 18 different aldehydes in a reductive amination. These reactions were carried out in 18 separate reaction vessels each containing 24 tube reactors. The second point of diversity was obtained when different aldehydes were employed in an aldol condensation carried out in eight different reaction vessels (54 microtubes each). The last set of diversity reagents was comprised of acid halide building blocks for ester formation. Overall, a library of 432 tyrphostin derivatives was synthesized using this modification of the split-mix methodology.

Radiofrequency encoding has also been implemented in conjunction with the classical pin technology of Geysen. In this case, the radiofrequency transponder encapsulated in glass is housed in the polymeric stem that holds the functionalized crown (45). The stem, capable of multiple uses attaches to the crown, thus allowing tracking of individual crowns during a synthesis. As with the system outlined before, the radiofrequency reader is integrated with a software package that enables manual sorting of ten reaction steps and a capacity of 50 000 compounds.

5.2.4 Optical coding (see Chapter 4)

An optical coding system (46) has recently been developed involving what have been termed laser optical synthesis chips (LOSCs). The supports are 1×1 cm polystyrene grafted square plates (*Figure 12*). The medium carrying the code is a 3×3 mm ceramic plate in the centre of the synthesis support. The code is etched into the ceramic support by a CO_2 laser in the form of a two-dimensional bar code. Before each synthetic step, the pooled chips are scanned and sorted according to the principles of the PM synthesis as directed by a computer. The utility of this modified technique was demonstrated in the synthesis of a 27-membered oligonucleotide library (46).

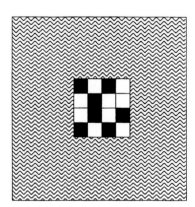

Figure 12 Representation of a laser optical synthesis chip (LOSC).

5.2.5 Visual tagging

Visual tagging is a technique recently reported by Guiles *et al.* in their efforts to simplify the microreactor technology (47). Reasoning for the simplification was based on the fact that the R_f tagging procedures require computers and readers which are not generally accessible to all laboratories. Additionally, the space inside the microreactors is limited due to the volume occupied by the trans-ponder tag. With an alternative strategy, more resin could be loaded into a reactor thereby providing more product. With these considerations, attention was turned to colour coding strategies. A set of 96 different microreactors can be visually distinguished if 12 different coloured caps are used and 8 different coloured glass beads placed among the resin beads. Depicted in *Plate 1* is an 8 × 12 matrix that would represent colour combinations for 96 individual reactions. Based upon this 96 component template, 8 reaction vessels were set up con-taining equal portions of resin for the attachment of 8 different functionalities (represented as X_{1-8} in *Plate 2*). Upon completion of these reactions, the resin in each reactor was separated into 12 equal portions (step A) and placed into capsules with a few coloured beads (the coloured beads were the same for all 12 portions but differed from each reactor). One capsule from each pool of 12 was removed and capped with the same coloured cap. A second capsule from each pool was removed and capped with a different coloured cap and so on (step B). Following sorting, the capsules were combined according to their cap colour and placed into a reactor, thus yielding 12 reactors each containing 8 capsules. The second reaction is illustrated in *Plate 2* with the diversity elements repre-sented as Y_{1-12}. All 96 capsules were combined and placed into a reactor contain-ing the third diversity element, Z. The number of 'Z' elements determined the number of reactors for this step. Upon completion, each microreactor was sep-

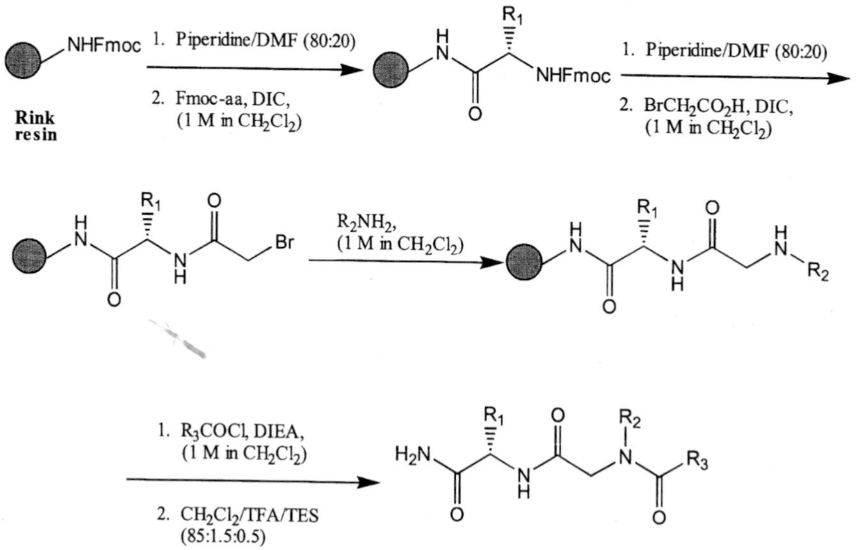

Figure 13 Synthesis of pseudopeptides via the 'visual tagging' approach.

arated by their bead and cap colour and placed into the corresponding well (refer to *Plate 1*) of a 96-well polypropylene-fritted plate. The chemical transformations reported employing this methodology gave rise to 96 different peptidomimetics (47) as outlined in *Figure 13*.

Another clever approach using visual tagging is that reported by Frank *et al.* referred to as the 'cut and combine' method of combinatorial chemistry (48). Employing a cellulose-based membrane (paper) support, a 400-membered octapeptide library was synthesized. This method of 'cut and combining' is in fact quite simple. A piece of paper is cut into several portions which are then subjected to reaction conditions for the first coupling. After washing these sections of paper and drying, they are cut into additional sections for the next set of reactions. This process is continued until the sequence is complete. For ease in tracking the separate pieces of paper, the original sheet is labelled with codes respective to each section. These labels are applied using a laser printer, providing a low cost method of 'visually' tagging the products. Details of the process used to screen the cellulose-based library generated in this way has been reported as well (48).

6 Summary

The portioning-mixing strategy is a highly effective method for producing chemical libraries of great magnitude. This field has evolved in such a way that large libraries are no longer intimidating to most scientists. That is, the necessity for tracking compounds throughout a synthesis has been experimentally addressed in several different strategies. The application of combinatorial chemistry beyond pharmaceuticals has recently become more evident. In a recent report, Burgess addresses the question of an effective combinatorial method of finding catalysts for organic syntheses (49) (see Chapter 13–15). Similarly, combinatorial approaches to new materials have been reported (50–52). To date, investigations into combinatorial library generation have successfully defined methods and tools required. This started with the synthetic procedures and has now progressed to new considerations such as library purification. As progress continues on this process, it is likely that new challenges will arise. It is equally likely that they will be met by innovative approaches that can lead to increased efficiency in discovery research.

Acknowledgements

The authors would like to extend their appreciation to Dr Monte Rhodes for his help in preparation of this manuscript.

References

1. Geysen, H. M., Meloen, R. H., and Barteling, S. J. (1984). *Proc. Natl. Acad. Sci. USA*, **81**, 3998.
2. Geysen, H. M., Rodda, S. J., and Mason, T. J. (1986). *Mol. Immunol.*, **23**, 709.

3. Furka, Á., Sebestyén, F., Asgedom, M., and Dibó, G. (1988). In *Highlights of modern biochemistry, Proceedings of the 14th International congress of biochemistry*, Vol. 5, p. 47. VSP, Utrecht.

4. Scott, J. K. and Smith, G. P. (1990). *Science*, **249**, 386.

5. Cwirla, S., Peters, E. A., Barrett, R. W., and Dower, W. J. (1990). *Proc. Natl. Acad. Sci. USA*, **87**, 6378.

6. Devlin, J. J., Panganiban, L. C., and Devlin, P. E. (1990). *Science*, **249**, 404.

7. Fodor, S. P. A., Read, J. L., Pirrung, M. C., Stryer, L., Lu, A. T., and Solas, D. (1991). *Science*, **251**, 767.

8. Furka, Á., Sebestyén, F., Asgedom, M., and Dibó, G. (1988). *Proceedings of the 10th International Symposium of medicinal chemistry, Budapest, Hungary*, p. 288.

9. Furka, Á., Sebestyén, F., Asgedom, M., and Dibó, G. (1991). *Int. J. Peptide Protein Res.*, **37**, 487.

10. Merrifield, R. B. J. (1963). *J. Am. Chem. Soc.*, **85**, 2149.

11. Ostresh, J. M., Winkle, J. H., Hamashin, V. T., and Houghten, R. A. (1994). *Biopolymers*, **34**, 1681.

12. Ragnarsson, U., Karlsson, S., and Sandberg, B. (1971). *Acta Chem. Scand.*, **25**, 1487.

13. Ragnarsson, U., Karlsson, S., and Sandberg, B. (1971). *J. Org. Chem.*, **39**, 3837.

14. Furka, Á. (1996). In *Combinatorial peptide and nonpeptide libraries* (ed. G. Jung), p. 111. VCH, Weinheim.

15. Stanková, M. and Lebl, M. (1996). *Mol. Div.*, **2**, 75.

16. Saneii, H. H., Shannon, J. D., Miceli, R. M., Fischer, H. D., and Smith, C. W. (1994). In *Peptide chemistry 1993* (ed. Y. Okada). p. 117. Protein Research Foundation, Osaka.

17. Dankwardt, S. M., Phan, T. M., and Krstenansky, J. L. (1995). *Mol. Div.*, **1**, 113.

18. Han, H., Wolfe, M. M., Brenner, S., and Janda, K. D. (1995). *Proc. Natl. Acad. Sci. USA*, **92**, 6419.

19. Gravert, D. J. and Janda, K. D. (1996). In *Molecular diversity and combinatorial chemistry: libraries and drug discovery* (ed. I. M. Chaiken and K. D. Janda), p. 118. American Chemical Society, Washington, DC.

20. Houghten, R. A., Pinilla, C., Blondelle, S. E., Appel, J. R., Dooley, C. T., and Cuervo, J. H. (1991). *Nature*, **354**, 84.

21. Erb, E., Janda, K. D., and Brenner, S. (1994). *Proc. Natl. Acad. Sci. USA*, **91**, 11422.

22. Sebestyén, F., Dibó, G., and Furka, Á. (1993). In *Peptides 1992* (ed. C. H. Schneider and A. N. Eberle), p. 63. ESCOM, Leiden.

23. Pinilla, C., Appel, J. R., and Houghten, R. A. (1993). In *Peptides 1992* (ed. C. H. Schneider and A. N. Eberle), p. 65. ESCOM, Leiden.

24. Câmpian, E., Peterson, M., Saneii, H. H., and Chou, J. (1996). In *Innovation and perspectives in solid phase synthesis and combinatorial libraries* (ed. R. Epton), p. 151. Mayflower Scientific Limited, Birmingham.

25. Câmpian, E., Peterson, M. L., Saneii, H. H., and Furka, Á. (1998). *Bioorg. Med. Chem. Lett.*, **8**, 2357.

26. Lam, K. S., Salmon, S. E., Hersh, E. M., Hruby, V. J., Kazmierski, W. M., and Knapp, R. J. (1991). *Nature*, **354**, 82; and its correction, (1992). *Nature*, **360**, 768.

27. Brenner, S. and Lerner, R. A. (1992). *Proc. Natl. Acad. Sci. USA*, **89**, 5381.

28. Needels, M. C., Jones, D. G., Tate, E. H., Heinkel, G. L., Kochersperger, L. M., Dower, W. J., *et al.* (1993). *Proc. Natl. Acad. Sci. USA*, **90**, 10700.

29. Nielsen, J., Brenner, S., and Janda, K. D. (1993). *J. Am. Chem. Soc.*, **115**, 9812.

30. Nikolaiev, V., Stierandova, A., Krchnak, V., Seligman, B., Lam, K. S., Salmon, S. E., *et al.* (1993). *Peptide Res.*, **6**, 161.

31. Kerr, J. M., Banville, S. C., and Zuckermann, R. N. J. (1993). *J. Am. Chem. Soc.*, **115**, 2529.

32. Ohlmeyer, M. H. J., Swanson, R. N., Dillard, L. W., Reader, J. C., Asouline, G., Kobayashi, R., *et al.* (1993). *Proc. Natl. Acad. Sci. USA*, **90**, 10922.

33. Borchard, A. and Still, W. C. J. (1994). *J. Am. Chem. Soc.*, **116**, 373.

34. Nestler, H. P. (1996). *Mol. Div.*, **2**, 35.

35. Baldwin, J. J., Burbaum, J. J., and Ohlmeyer, M. H. J. (1995). *J. Am. Chem. Soc.*, **117**, 5588.

36. Baldwin, J. J. (1996). *Mol. Div.*, **2**, 81.

37. Moran, E. J., Sarshar, S., Cargill, J. F., Shahbaz, M., Lio, A., Mjalli, A. M. M., *et al.* (1995). *J. Am. Chem. Soc.*, **117**, 10787.

38. Nicolaou, K. C., Xiao, X.-Y., Parandoosh, Z., Senyei, A., and Nova, M. P. (1995). *Angew. Chem. Int. Ed. Engl.*, **34**, 2289.

39. Houghten, R. A. (1985). *Proc. Natl. Acad. Sci. USA*, **82**, 5131.

40. Armstrong, R. W., Tempest, P. A., and Cargill, J. F. (1996). *Chimia*, **50**, 258.

41. Armstrong, R. W., Brown, S. D., Keating, T. A., and Tempest, P. A. (1997) In *Combinatorial chemistry: synthesis and application* (ed. S. R. Wilson and A. W. Czarnik), p. 153. John Wiley & Sons, New York.

42. Czarnik, A. and Nova, M. (1997). *Chem. Br.*, 39.

43. Nicolaou, K. C., Vourloumis, D., Li, T., Pastor, J., Winssinger, N., He, Y., *et al.* (1997). *Angew. Chem. Int. Ed. Engl.*, **36** (19), 2097.

44. Shi, S., Xiao, X., and Czarnik, A. W. (1998). *Biotechnol. Bioeng. (Comb. Chem.)*, **61**, 7.

45. Giger, R. (1997). CHIs *First European Conference on Strategies and techniques for identification of novel bioactive compounds*, 'Pin sort-and-combine method for HTP synthesis of individual compounds'. December 3–5, 1997. Barcelona, Spain. Gombert, F. O. (1998). IBCs *High throughput compound characterization and purification*, 'Semi-automated HTP preparation of libraries: single compounds using pins bearing radiofrequency tags'. September 14–15, 1998. Arlington, VA.

46. Xiao, X.-Y., Zhao, C., Potash, H., and Nova, M. P. (1997). *Angew. Chem. Int. Ed. Engl.*, **36**, 780.

47. Guiles, J. W., Lanter, C., and Rivero, R. A. (1998). *Angew. Chem. Int. Ed. Engl.*, **37**, 926.

48. Dittrich, F., Tegge, W., and Frank, R. (1998). *Bioorg. Med. Chem. Lett.*, **8**, 2351.

49. Burgess, K., Moye-Sherman, D., and Porte, A. M. (1996). In *Molecular diversity and combinatorial chemistry: libraries and drug discovery* (ed. I. M. Chaiken and K. D. Janda), p. 128. American Chemical Society, Washington, DC.

50. Xiang, X. D., Sun, X., Briceno, G., Lou, Y., Wang, K.-A., Chang, H., *et al.* (1995). *Science*, **268**, 1738.

51. Danielson, E., Golden, J. H., McFarland, E. W., Reaves, C. M., Weinberg, W. H., and Wu, X. D. (1997). *Nature*, **389**, 944.

52. Danielson, E., Devenney, M., Giaquinta, D. M., Golden, J. H., Haushalter, R. C., McFarland, E. W., *et al.* (1998). *Science*, **279**, 837.

Chapter 2

One-bead one-compound combinatorial library method

Gang Liu and Kit S. Lam

Department of Internal Medicine, University of California, UC Davis Cancer Center, 4501 X Street, Sacramento, CA 95817, USA

1 Introduction

The 'one-bead one-compound' combinatorial library method introduced by Lam *et al.* (1) involves:

(a) The generation of a large library of spatially separable synthetic compounds.

(b) The screening of this library of compounds for a specific biological, chemical, or physical property.

(c) The isolation of the positive compounds for structural determination.

The 'one-bead one-compound' combinatorial library was prepared by a 'split synthesis' method (1–3, see also Chapter 1) such that each solid phase bead displays only one chemical entity although there are approximately 10^{13} copies of the same compound on a 100 μm bead. This bead-supported library is then assayed for a specific biological activity using on-bead binding (1), or functional assays (4). Alternatively, the compound on each bead is released from the solid support via a cleavable linker for subsequent solution phase assays (5). In either solid phase or solution phase assays, the positive beads are then identified, physically isolated, and subjected to structural determination. If the compound on the bead is a peptide, each individual positive bead is inserted into a cartridge of an automatic protein sequencer for sequence analysis. Alternatively, the peptide or non-peptide compound on each bead is released chemically or photochemically and the releasate is analysed by mass spectrometry. If the structure of the compound cannot be easily determined by mass spectrometry, one can use chemical encoding (6–8) or radiofrequency tagging (9, Chapter 4) strategies.

Synthetic combinatorial chemistry was first applied to peptides (1, 2, 10). However, numerous papers on the development of methods to synthesize small molecule combinatorial libraries were reported in the literature over the past six years (11–14). A detailed account of the various synthetic methods, screening strategies, or structure determination methods are beyond the scope of this chapter. Please refer to ref. 14 for a recent comprehensive review of the 'one-bead one-compound' combinatorial library method. In this chapter we describe selected protocols for the preparation of linear and cyclic peptide libraries, and various on-bead binding as well as functional screening assays that have been employed in our laboratory.

2 Synthesis of the peptide library

2.1 Preparation of the amino acid solutions

Stock solutions of Nα-Fmoc-AAs (Nα-fluorenylmethoxycarbonyl amino acids) with HOBt (N-hydroxybenzotriazole) in DMF (N,N-dimethylformamide) or NMP (N-methylpyrrolidine) can be stored at 4°C for up to ten days. Three fold excess of Nα-Fmoc-AAs are used in each coupling cycle to ensure completion of the peptide coupling reaction. The following equations are used for the calculation of the necessary reagents.

$$W_{aa} = \frac{(MW_{aa})(3L)(g)(N)}{X} \tag{1}$$

$$W_{HOBt} = (135.1)(3L)(g)(N) \tag{2}$$

$$V_{HOBt} = (0.71)(V')(X)(N) \tag{3}$$

$$V_{DIC} = \frac{(126.1)(3L)(g)}{X(0.806)} \tag{4}$$

MW_{aa} = molecular weight of the protected amino acid;
L　　= loading of the resin in mmol/g;
g　　= mass of the resin;
N　　= total number of amino acid residues in the peptide sequence;
X　　= total number of amino acids used in each coupling step;
V'　　= volume of amino acid/HOBt solution to be added to each reaction vessel (for a reaction vial containing 100 mg TentaGel resin, we generally use 0.5 ml amino acid/HOBt solution);
0.806 = specific gravity of DIC (N,N'-diisopropylcarbodiimide);
135.1 = molecular weight of HOBt;
126.2 = molecular weight of DIC.

　　Equation 1 is used to calculate the mass of each amino acid required for the library synthesis. Equation 2 is used to calculate the amount of HOBt necessary for the whole library. Equation 3 is used to calculate the volume of HOBt solution needed. Equation 4 is used to determine the volume of DIC required for each amino acid coupling reaction.

Protocol 1

Preparation of the amino acid solutions for a heptapeptide library synthesis

Reagents

- The following $N\alpha$-Fmoc-AAs with the appropriate side-chain protecting groups are used in our library synthesis: $N\alpha$-Fmoc-Arg(Pmc)-OH, $N\alpha$-Fmoc-Asn(Trt)-OH, $N\alpha$-Fmoc-Gln(Trt)-OH, $N\alpha$-Fmoc-Asp(Ot-Bu)-OH, $N\alpha$-Fmoc-Glu(Ot-Bu)-OH, $N\alpha$-Fmoc-His(Trt)-OH, $N\alpha$-Fmoc-Lys(Boc)-OH, $N\alpha$-Fmoc-Ser(t-Bu)-OH, $N\alpha$-Fmoc-Thr(t-Bu)-OH, $N\alpha$-Fmoc-Trp(Boc)-OH, $N\alpha$-Fmoc-Tyr(t-Bu)-OH

- The other amino acids without side-chain protecting group are: $N\alpha$-Fmoc-Ala, $N\alpha$-Fmoc-Phe, $N\alpha$-Fmoc-Gly, $N\alpha$-Fmoc-Ile, $N\alpha$-Fmoc-Leu, $N\alpha$-Fmoc-Met, $N\alpha$-Fmoc-Pro, $N\alpha$-Fmoc-Val

- HOBt-H_2O

- HPLC (high performance liquid chromatography) grade DMF or NMP

- 2.0 g TentaGel S NH_2 resin (0.27 mmol/g)

Method

1. Use *Equations 1–4* to calculate the amount of reagents necessary for the synthesis of a heptapeptide library (2.0 g resin) using 19 eukaryotic[a] amino acids. $W_{aa} = 3 \times 0.27 \times 2.0 \times 7 \times MW_{aa}/19 = 0.5968 \times MW_{aa}$ mg. $W_{HOBt} = 135.1 \times 3 \times 0.27 \times 2.0 \times 7 = 1532$ mg. $V_{HOBt} = 0.7 \times 0.5 \times 19 \times 7 = 46.6$ ml.[b] $V_{DIC} = 126.1 \times 3 \times 0.27 \times 2.0/19 \times 0.806 = 13.4$ μl.

2. Pre-weigh the solid reagents into a suitable container (e.g. graduated polypropylene vials).

3. Dissolve 1532 mg HOBt in DMF until the final volume is 46.6 ml (V_{HOBt}).

4. Distribute the HOBt solution (2.45 ml each) into each of the 19 amino acid containers and add HPLC grade DMF until the final volume is $(V') \times (N)$ or 3.5 ml. Try to dissolve the amino acid by vortexing. If it does not dissolve totally, try sonication.[c]

[a] Naturally occurring amino acids.

[b] 0.5 ml of amino acid and HOBt solution (V') will be added into each reaction vessel at each coupling step.

[c] If suspension does not dissolve totally after sonication ensure that it is well mixed prior to pipetting into the reaction vessel.

2.2 Synthesis of a linear peptide library on solid phase using TentaGel resin with 19 eukaryotic amino acids (cysteine excluded)

Lam *et al.* (14) have reviewed the choice of solid supports for the 'one-bead one-compound' combinatorial chemistry methodology. TentaGel resin was generally used as the solid phase support. It consists of polystyrene matrix with a grafted polyoxyethylene (POE) linker and a functional group (amino or hydroxyl) at the end of this linker. This resin is fully compatible with both organic and aqueous conditions. The POE linker is particularly suitable for many biological assay conditions.

In principle, either the *t*-Boc/Bn (*tertio*-butyloxycarbonyl/benzyl) or Fmoc/*t*-But (fluorenylmethoxycarbonyl/*tertio*-butyl) amino acid protection strategies could be used for library synthesis on TentaGel. For the *t*-Boc/Bn chemistry, HF (hydrofluoridric acid) is needed to cleave off the side-chain protecting groups (CAUTION: HF is highly toxic and requires a special apparatus). Since it partially degrades the POE chain on the resin, we recommend the Fmoc/*t*-But strategy for the synthesis of peptide libraries on TentaGel.

Many of the commercially available activating reagents can be used for solid phase peptide library synthesis. These include DIC, BOP (benzotriazol-1-yl-oxy-tris (dimethylamine)-phosphonium hexafluorophosphate), HBTU (*O*-benzotriazole-*N*,*N*,*N*′,*N*′-tetramethyluronium hexafluorophosphate), TBTU (*O*-benzotriazol-1-yl-*N*,*N*,*N*′,*N*′-tetramethyluronium tetrafluoroborate), PyBOP (benzotriozol-1-yl-oxy-tris-pyrrolidinophosphonium hexafluorophosphate), and PyBroP (bromo-tris-pyrrolidine-phosphonium hexafluorophosphate). However, DCC (*N*,*N*′-dicyclo-hexylcarbodiimide) should be avoided because it will produce an insoluble DCU (dicyclohexylurea) side-product. An additive reagent to generate an active ester is needed to reduce racemization of the Nα-protected amino acids. HOBt and Pfp (pentafluorophenyl) esters are the most efficient. We usually use HOBt for this purpose. When BOP, PyBOP, TBTU, HBTU, or PyBroP are used as the activating reagents and HOBt as an additive, a base must be added *in situ* to neutralize the acidic side-product that may deprotect the acid-labile amino acid side-chain protecting groups such as Trt (trityl) and *t*-Bu. We recommend DIEA (*N*,*N*-diiso-popylethylamine), NMM (*N*-methylmorpholine), or TEA (triethylamine). In the following protocol we describe a procedure using DIC as an activating reagent and HOBt as an additive to synthesize a peptide library.

Protocol 2

Synthesis of a linear peptide library with 19 eukaryotic amino acids (cysteine excluded)

Reagents

- Protected amino acid stock solutions (*Protocol 1*)
- 25% piperidine/DMF (v/v)
- TFA (trifluoroacetic acid)
- Kaiser test reagents: solution A, 5% ninhydrin in ethanol (w/v); solution B, 80% crystalline phenol in ethanol (w/v); solution C, 2% 0.001 M aqueous solution of potassium cyanide in pyridine (v/v)
- DIC

- TentaGel S NH$_2$ resin (90–100 μm), loading: 0.27 mmol/g
- 1,2-ethanedithiol (EDT)
- Cleavage solution: phenol/thioanisole/H$_2$O/EDT/TFA (0.75:0.5:0.5:0.25:10, w/v/v/v/v)
- PBS (phosphate-buffered saline): 137 mM NaCl, 2.68 mM KCl, 8.0 mM Na$_2$HPO$_4$, 1.47 mM KH$_2$PO$_4$ pH 7.4

Method

1 Weigh out 2.0 g TentaGel S NH$_2$ resin (0.27 mmol/g) and swell it in HPLC grade DMF overnight.

2 Distribute the beads equally into 19 reaction vessels.[a] Allow the resin to settle for 20 min and remove the DMF carefully with a disposable polyethylene pipette down to the surface of the resin.

3 Add 0.5 ml amino acid/HOBt solution (*Protocol 1*, step 4).[b]

4 Add DIC to each reaction vessel as calculated by *Equation 4*.

5 Couple the amino acid for 1 h at room temperature with gentle shaking.

6 Transfer a small sample of the resin solution from each reaction vessel into tiny test-tubes.

7 Wash each mixture in the tubes with ethanol once and then perform the Kaiser test.[c,d]

8 After all the couplings are completed, combine the resins and transfer them to a siliconized column with a frit at the bottom.

9 Wash with DMF (5 × 10 ml).[e]

10 Remove the Nα-Fmoc protecting group by incubation twice in 10 ml of 25% piperidine/DMF for 15 min each.

11 Wash the resin again with DMF (6 × 5 ml), methanol (2 × 5 ml), and DMF (6 × 5 ml).[f]

12 Repeat the cycle of resin distribution, amino acid coupling, Kaiser test, resin mixing, resin washing, Fmoc deprotection, and resin washing again according to steps 2–11 until amino acid assembly is completed.

13 Remove the Nα-Fmoc protecting group as described in steps 10 and 11 prior to side-chain deprotection.

14 Remove the side-chain protecting groups by adding 30 ml cleavage solution to 0.5–3.0 g bead-supported library for 2.5 h at room temperature with gentle mixing.

15 Drain the deprotection solution, and wash the bead-supported library with DMF (5 × 5 ml), methanol (5 × 5 ml), DCM (5 × 5 ml), and DMF (5 × 5 ml).

16 Wash the bead-supported library with 30% H$_2$O/DMF, 60% H$_2$O/DMF, 100% H$_2$O (1 × 10 ml each), and then 50 mM PBS buffer pH 7.4 (10 × 10 ml).

17 Store the peptide-bound resin in 0.05% sodium azide/PBS at 4 °C.

[a] Polypropylene or polyethylene vials with tight-sealing cap.

[b] Use HPLC grade DMF for the coupling reaction.

[c] Transfer a minimum amount of resin beads to a small glass test-tube, wash the beads with ethanol. Add one drop each of solutions A, B, and C to the tube containing the beads. Heat the mixture to 100 °C and maintain the heat for 5 min. Blue coloured beads and solution (positive)

indicates the presence of free amino groups on the resin, i.e. incomplete coupling. Colourless resin beads indicate complete coupling. For those vessels with incomplete coupling reaction, either prolong their reaction time or remove the supernatant and perform a 'repeat coupling' by adding fresh amino acid/HOBt solution and DIC.

[d] For proline, a secondary amine, the Kaiser test will turn brown instead of blue.

[e] We suggest using technical grade DMF for washing since it is inexpensive.

[f] Use HPLC grade DMF in the last three washings.

2.3 Synthesis of disulfide cyclic peptide library

Careful choice of oxidation methods and protecting groups for cysteine are important for subsequent disulfide formation. There are three main methods to cyclize a peptide using disulfide bond formation in solution or on solid support:

(a) Air oxidation.

(b) DMSO oxidation of the free thiol (–SH) groups to form a disulfide bond (–S–S–) in polar solvents such as methanol/water or acetic acid (AcOH)/water.

(c) Iodine (I_2) oxidation of the Cys(S-Trt)/Cys(Acm) directly to form disulfide bond in methanol or AcOH.

We recommend using method (b) for free thiol oxidation and method (c) for the protected Cys(S-Trt)/Cys(S-Acm) oxidation. In the following protocol, we describe a procedure for the synthesis of disulfide peptide library. During library synthesis, Cys(S-Trt)-OH is added at the first and last residual of the library. The Trt protecting group is removed with TFA during the side-chain deprotection step. Cyclization is accomplished by disulfide formation using DMSO (method b). Since 1,2-ethanedithiol (EDT) can reduce disulfide bonds during the oxidation step, it must be replaced by triisopropylsilane (TIS) as the scavenger in the cleavage protocol.

Protocol 3

Synthesis of the disulfide cyclic peptide library on solid phase

Reagents

- Nα-Fmoc-Cys(S-Trt)-OH
- TIS
- Cleavage solution: phenol/thioanisole/H_2O/TIS/TFA (0.75:0.5:0.5:0.25:10, w/v/v/v/v)
- Ellman's reagent: 4.0 mg 5,5'-dithio-bis(2-nitrobenzoic acid) in 1.0 ml of 20 mM sodium phosphate buffer pH 8.0
- Concentrated aqueous ammonia
- Reagents for peptide synthesis (Protocol 2)

Protocol 3 continued

Method

1 Weigh out 2.0 g TentaGel S NH$_2$ resin (0.27 mmol/g) and swell it in HPLC grade DMF overnight.

2 Dissolve 3.0 Equiv. of Nα-Fmoc-Cys(S-Trt)-OH (949 mg) based on the loading of resin in 15 ml DMF.

3 Add 3.0 Equiv. of HOBt (219 mg) and 3.0 Equiv. of DIC (255 μl) to this solution. Then add the final mixture to the resin and shake overnight.

4 After the coupling is completed (Kaiser test must be negative), remove the Nα-Fmoc protecting group by incubation twice in 10 ml of 25% piperidine/DMF for 15 min each.

5 Wash the resin as in *Protocol 2*, step 11.

6 To assemble the peptide library, follow the steps outlined in *Protocol 2*, steps 2–12.

7 Couple cysteine at the desired position of the peptide library by adding 3.0 Equiv. of Nα-Fmoc-Cys(S-Trt), HOBt, and DIC as in steps 2 and 3.

8 After the peptide library assembly is completed, remove the Nα-Fmoc protecting group according to step 4.

9 Remove the side-chain protecting groups by adding 30 ml of the cleavage solution to 0.5–3.0 g of bead-supported library for 2.5 h at room temperature with gentle shaking.

10 Drain the cleavage solution, and wash the bead-supported library with 5 × 5 ml each of DMF, methanol, DCM, and DMF again.

11 Wash the bead-supported library with 10 ml each of 30% H$_2$O/DMF, 60% H$_2$O/DMF, H$_2$O, and finally test for unreacted free thiol groups on the resin using Ellman's reagent.[a]

12 Oxidize the bead-supported library in 1 litre of H$_2$O/HOAc/DMSO[b] (75:5:20) with gentle stirring in a 2 litre flask for 48 h, until the Ellman test is negative.

13 Wash the beads thoroughly with H$_2$O, followed by PBS.

14 Store the resin-bound library in 0.05% sodium azide/PBS at 4°C.

[a] In this step, Ellman test must be positive, the yellow colour indicates that 5,5'-dithio-bis(2-nitrobenzoic acid) has reacted with free thiol group on the resin to form yellow anion product.

[b] The pH is adjusted with saturated aqueous ammonium hydroxide to 6.0 before DMSO is added.

2.4 Synthesis of Dpr-Dpr cyclic peptide (oxime bond) library

Wahl and Mutter (15) first introduced two new unnatural amino acids (Nα-Fmoc-Dpr(t-Boc-Aoa)-OH and Nα-Fmoc-Dpr(Boc-Ser(Ot-Bu)-OH) as building blocks for the solution phase cyclization of peptides via oxime bond formation. The oxime bond is stable to proteolysis and represents an excellent alternative approach for constraining peptide conformation. Here we describe a method for on-resin cyclization of a peptide library using the same approach.

Protocol 4

Synthesis of Dpr-Dpr cyclic peptide (oxime bond) library

Reagents

- Nα-Fmoc-Dpr(t-Boc-Aoa)-OH and Nα-Fmoc-Dpr(Boc-Ser(Ot-Bu))-OH
- 115 mg NaIO$_4$ in 25 ml of 50% H$_2$O/DMF
- Reagents for peptide synthesis (*Protocol 2*)

Method

1 Weigh out 2.0 g TentaGel S NH$_2$ resin (0.27 mmol/g) and swell it in HPLC grade DMF overnight.

2 Dissolve 2.0 Equiv. of Nα-Fmoc-Dpr(Boc-Ser(Ot-Bu))-OH (615 mg, based on the loading of the resin) in 15 ml DMF. Add 2.0 Equiv. of HOBt (146 mg) and 2.0 Equiv. of DIC (170 μl) into this solution. Finally, add the beads and shake overnight.

3 After the Kaiser test is negative, remove the Nα-Fmoc protecting group then wash the resin as in *Protocol 2*, steps 10 and 11.

4 Assemble the peptide library according to *Protocol 2*, steps 2–12.

5 Couple 2.0 Equiv. of Nα-Fmoc-Dpr(Boc-Aoa) (539 mg) with the addition of reagents HOBt (146 mg) and DIC (170 μl).

6 Remove the side-chain protecting groups with 30 ml of cleavage solution as in *Protocol 2*.

7 Wash the resin thoroughly with DMF and then add twice 2.0 Equiv. of NaIO$_4$ (115 mg) in 25 ml of 50% H$_2$O/DMF for 10 min each to convert the serinyl to a glyoxylyl group. The beads turn red.

8 Wash the bead-supported library with 50% H$_2$O/DMF (5 × 10 ml), and DMF (3 × 10 ml).

9 Remove the last Nα-Fmoc protecting group on the N-terminus of the peptide library by incubating twice in 10 ml of 25% piperidine/DMF for 15 min each. The colour of the beads returns to light yellow.

10 Wash the beads with DMF (5 × 10 ml), 30% H$_2$O/DMF (1 × 10 ml), 60% H$_2$O/DMF (1 × 10 ml), 100% H$_2$O (1 × 10 ml), and 1 × PBS (10 × 10 ml).

11 Store the peptide-bound resin in 0.05% sodium azide/PBS at 4°C.

2.5 On-resin synthesis of a cyclic peptide library using Lys and Glu side-chains

On-resin amide bond formation between the amino and carboxyl group side-chains of Lys and Glu, respectively, is a useful approach to cyclize peptides while still attached to the solid support. Several protecting group strategies were developed for this purpose, which include:

(a) Head-to-tail cyclization using t-Boc-Asp-OFm or t-Boc-Glu-OFm (16), and Nα-Fmoc-Asp-OAll or Nα-Fmoc-Glu-OAll (17).

(b) Side-chain cyclization using Nα-Fmoc-Asp(ODmab)-OH or Nα-Fmoc-Glu(ODmab)-OH with Nα-Fmoc-Lys(Dde)-OH.

(c) Another side-chain based cyclization using t-Boc-Asp(OFm)-OH or t-Boc-Glu(OFm)-OH with t-Boc-Lys(Fmoc)-OH.

The Fm group is removed in 20% piperidine/DMF, while the allyl group is cleaved by Pd(PPh$_3$)$_4$ in CHCl$_3$:AcOH:N-methylmorpholine (37:2:1) as described in NovaBiochem Catalog Handbook (18). Both the Dmab and Dde groups are labile to 2% hydrazine in DMF. In the following protocol, we choose Nα-Fmoc-Lys(Dde)-OH and Nα-Fmoc-Glu(ODmab)-OH as building blocks for cyclization. These amino acids couple through the ε-NH$_2$ group of Lys and γ-carboxyl group of Glu.

Protocol 5

Synthesis of cyclic peptide libraries using on-resin cyclization between Lys and Glu side-chains

Reagents

- Nα-Fmoc-Lys(Dde)-OH and Nα-Fmoc-Glu(ODmab)-OH
- PyBOP
- 1-Hydroxy-7-azabenzotriazole (HOAt)

- DIEA
- 2% NH$_2$NH$_2$/DMF (v/v)
- Reagents for peptide synthesis (*Protocol 2*)

Method

1   Weigh out 2.0 g TentaGel S NH$_2$ resin (0.27 mmol/g) and swell it in HPLC grade DMF overnight.

2   Decrease the loading of the resin using the following method.

   (a) Mix 0.30 mmol of Nα-Fmoc-Gly (89 mg), 0.31 mmol of HOBt (42 mg), and 0.31 mmol of DIC (48.5 µl) in 15 ml of DMF.

   (b) Add the mixture to the TentaGel S NH$_2$ resin and shake gently overnight at room temperature.

   (c) After washing the resin with DMF and DCM, cap the exposed amino group with 15% acetic anhydride in DCM for 30 min. In this case, the final loading of the down-substituted resin will be 0.15 mmol/g.[a]

3   Cleave the Nα-Fmoc protecting group and wash the resin as in *Protocol 2*, steps 10 and 11.

4   Dissolve 2.0 Equiv. of Nα-Fmoc-Lys(Dde)-OH (320 mg based on the new loading of the resin) in 15 ml DMF. Add 2.0 Equiv. of HOBt (81 mg) and 2.0 Equiv. of DIC (94 µl) to this solution. Finally, add the beads and shake gently overnight.

5   Follow the procedure for assembly of the peptide as described in *Protocol 2*, steps 2–12, until the desired cyclization position (Glu) is reached.

Protocol 5 continued

6 Add 2.0 Equiv. of Nα-Fmoc-Glu(ODmab) (735 mg), DIC (94 µl), and HOBt (81 mg) to the beads and shake gently overnight. The Kaiser test must be negative after this coupling.

7 After washing with DMF (5 × 10 ml), incubate twice in 80 ml of 2% NH_2NH_2/DMF for 3 min each.

8 Wash the beads with DMF (6 × 10 ml).

9 Perform the cyclization step by adding 5.0 Equiv. of PyBOP and HOAt and 10 Equiv. of DIEA in HPLC grade DMF. Shake gently at room temperature overnight or until the Kaiser test is negative.

10 Remove the Nα-Fmoc protecting group (*Protocol 2*, steps 10 and 11).

11 Remove the side-chain protecting groups by incubation in 30 ml of the cleavage solution (*Protocol 2*, step 14).

12 Wash the beads with DMF (5 × 10 ml), DCM (5 × 10 ml), DMF (5 × 10 ml), 30% H_2O/DMF (1 × 10 ml), 60% H_2O/DMF (1 × 10 ml), 100% H_2O (1 × 10 ml), and finally with PBS buffer (10 × 10 ml).

13 Store the resin-bound library in 0.05% sodium azide/PBS at 4 °C.

[a] An alternate way to decrease the loading of the resin is to first couple 1.0 Equiv. of a mixture of Fmoc-Gly and Ac-Gly in a 1:1 molar ratio to the resin overnight at room temperature, then perform a coupling with 3.0 Equiv of Nα-Fmoc-Gly for 1 h. The final loading will be half of the original loading of the resin.

3 Library screening

The 'one-bead one-compound' combinatorial library method has been adapted for both on-bead solid phase as well as releasable solution phase assay (1, 4, 14). For the on-bead binding assay, the receptor molecular target (ligate) needs to be tagged with a reporter molecule such as an enzyme, a fluorescent probe, or a radionucleide. Alternatively, the ligate is biotinylated and in turn the reporter group will be linked to streptavidin which is used as a secondary reagent to detect the positive beads (19). Another approach is to use tagged antibodies that specifically recognize the ligate to label the positive beads (20). Although purified ligate is generally used for the above screening method, it is not crucial. For instance if a highly specific tagged secondary antibody is used in the assay, the ligate need not be highly purified. However, regardless of which screening method one has decided to use, it is crucial to design strategies such that the positive beads from the first screen are recycled and screened again by an additional (ortho-gonal) method to ensure that the initial positive beads are not erroneous.

Besides using soluble ligates as probes for screening, one may also use intact cells or micro-organisms. We have reported the use of intact tumour cell lines to identify integrin-specific binding peptides (21). In this method, the bead-supported library is mixed with intact cells. Positive beads rosetted by a monolayer of cells are isolated, recycled, and retested for binding to the selected beads in the presence of anti-integrin antibodies. We have also used a similar approach to identify ligands that bind specifically to yeast (*Candida albican*) cell surface (unpublished results).

In addition to on-bead binding assay, we and others have developed screening methods to identify ligands with specific function. These functional assays include identification of peptide substrate motif for protein kinases (4, 22, 23) and proteases (24, 25). In the protein kinase assay, beads with peptide substrate motif are covalently modified by ^{32}P when the bead-supported library is incubated with a specific protein kinase and [γ-^{32}P]ATP. The ^{32}P-labelled beads are then identified by autoradiography and isolated for microsequencing. As for the protease substrate identification, the library is constructed such that there is a quencher at the amino-terminus and a fluorophore at the carboxyl-terminus of the peptide library. The peptide sequence susceptible to proteolysis will be cleaved from the solid phase support thereby releasing the quenching moiety and resulting in an enhancement of the fluorescence of the resin-bound fluorophore. The highly fluorescent beads are then identified and isolated for microsequencing.

The 'one-bead one-compound' combinatorial library method has also been adapted to solution phase assay in which the ligand is attached to the bead via a cleavable linker (chemo-sensitive or photo-sensitive). Two general approaches have been used. The *in situ* approach involves the immobilization of the beads prior to release of the compounds, thus the beads from which the active compounds were released can be spatially traced back and identified (26). The second approach uses microwell plate to partition the beads (one to 500 beads/well) prior to the release of the compounds from the beads. When large numbers of beads per well are used, redistribution of beads from an active well for a second release of compounds using a multi-cleavable linker (5, 28) is required in order to identify a unique bead with the desired activity.

In the following section we illustrate the above concepts with a few selected screening methods developed in our laboratories (see also ref. 14).

Protocol 6

Enzyme-linked colorimetric assay

Equipment and reagents

- Dissecting microscope
- Automatic protein sequencer (e.g. Applied Biosystems Model ABI 477A)
- A bead-supported peptide library with a specific composition
- PBS: 8.0 mM Na_2HPO_4, 1.5 mM KH_2PO_4, 137 mM NaCl, 2.7 mM KCl pH 7.4
- TBS: 2.5 mM Tris base, 13.7 mM NaCl, 0.27 mM KCl pH 8.0
- BCIP substrate buffer: 1.65 mg 5-bromo-4-chloro-3-indoyl phosphate (BCIP) in 10 ml of 0.1 M Tris base, 0.1 M NaCl, 2.34 mM $MgCl_2$ pH 8.5–9.0
- 0.05% aqueous gelatin (w/v) (dissolve with heating in microwave oven)
- 6.0 M guanidine hydrochloride pH 1.0
- Ligate–alkaline phosphatase conjugate

Protocol 6 continued

Method

1 Transfer 1–10 ml of the bead-supported library into a disposable polypropylene column with a polyethylene frit.[a] Wash the bead-supported library with water (5 × 10 ml). Block the bead-supported library with 0.05% gelatin (w/v) in water for at least 1 h at room temperature.

2 Wash the bead-supported library with 0.1% Tween/PBS thoroughly (10 × 10 ml).

3 Incubate the bead-supported library with the ligate–alkaline phosphatase conjugate at a suitable concentration in 0.05% gelatin/0.1% Tween/PBS from 1–24 h at room temperature.

4 Wash the bead-supported library thoroughly with 0.1% Tween/PBS (10 × 10 ml) and then with TBS (10 ml).

5 Add BCIP substrate buffer.

6 Transfer and distribute the bead suspension into 10–20 polystyrene Petri dishes (10 cm diameter).

7 Develop the colour for 0.5–24 h at room temperature.

8 Stop the enzymatic reaction with a few drops of 1.0 M HCl.[b]

9 With the aid of a light box and a hand-held micropipette, transfer the turquoise beads into a small Petri dish.[c]

10 Under the dissecting microscope, pick up the deep coloured beads with a micro-pipette and carefully transfer them to a small, clean Petri dish.

11 Add the 6.0 M guanidine–HCl pH 1.0 solution to strip the protein off the bead.

12 Transfer the positive beads into a small clean Petri dish containing double distilled water.

13 Pipette each positive bead onto a glass filter and insert it into a protein sequencing cartridge for microsequencing.[d]

[a] All the incubation and washing steps are performed in this column.

[b] If there are too many positive beads, you should repeat the experiment by either decreasing the concentration of ligate–alkaline phosphatase conjugate, or shortening the incubation time.

[c] At this step, many colourless beads may be also transferred.

[d] In some experiments one may want to decolorize the beads with DMF, and recycle the bead for another round of screening with an alternate assay method. With this approach, the probability for true-positive results will be significantly higher.

Protocol 7

Unlabelled ligate detected with an enzyme-linked secondary antibody system

Equipment and reagents

- Dissecting microscope
- Automatic protein sequencer (e.g. Applied Biosystems Model ABI 477A)
- Anti-ligate antibody/alkaline phosphatase conjugate

- Binding buffer: 16 mM Na_2HPO_4, 3 mM KH_2PO_4, 274 mM NaCl, 5.4 mM KCl pH 7.2, with 0.1% Tween (v/v) and 0.05% gelatin (w/v)

Method

1 Pre-block 1–10 ml of the bead-supported library according to *Protocol* 6, step 1.

2 Wash the bead library thoroughly with 0.1% Tween/PBS (w/v).

3 Add a suitable concentration of the anti-ligate antibody/alkaline phosphatase conjugate to the bead-supported library and incubate for 3 h at room temperature.

4 Wash the bead-supported library thoroughly with PBS/Tween, and once with TBS.

5 Add the BCIP substrate buffer to the bead library and develop the colour as in *Protocol* 6, steps 5–8.

6 Remove all of the coloured beads with the aid of a micropipette and a light box.

7 Transfer the colourless beads to the washing column.

8 Treat the colourless beads with 6.0 M guanidine–HCl pH 1.0 for 20–30 min, then wash with double distilled water (5 × 10 ml).

9 Mix the beads with DMF for 1 h for decolorization.

10 Wash serially with 10 ml each of 30% H_2O/DMF, 60% H_2O/DMF, and DMF.

11 Wash with double distilled water and 0.1% Tween/PBS (5 × 10 ml each).

12 Add the unlabelled ligate at a suitable concentration to the column containing recycled bead-supported library and incubate for 5 h at room temperature.

13 Wash the bead-supported library thoroughly with 0.1% Tween/PBS.

14 Add anti-ligate antibody/alkaline phosphatase conjugate (same concentration as step 3) to the bead-supported library for 2 h at room temperature.

15 Wash the bead-supported library thoroughly with 0.1% Tween/PBS and once with TBS.

16 Add BCIP substrate buffer to the bead library, transfer the bead-supported library to Petri dishes, and allow the colour reaction to develop as outlined in *Protocol* 6, steps 5–8.[a]

17 Stop the colour reaction by adding a few drops of 1.0 M HCl into the development buffer to adjust the pH to 1–2.

18 Isolate the positive beads for microsequencing as in *Protocol* 6, step 13.

[a] The positive beads at this step are likely due to ligand–ligate interaction rather than peptide/anti-ligate antibody interaction since the latter interaction was in principle eliminated in the preceding steps.

Protocol 8

Cross-screening the library with enzyme-linked colorimetric and radiolabelled assays

Equipment and reagents

- X-ray film (e.g. Kodak X-OMAT LS)
- Glogos II autoradiogram marker (Stratagene)
- Dissecting microscope
- Automatic protein sequencer (e.g. Applied Biosystems Model ABI 477A)

- Biotinylated ligate
- [^{125}I]ligate
- 1.0% low gelling temperature agarose (w/v) in H$_2$O: melt the agarose in a microwave oven and keep at 37 °C

Method

1 Pre-mix the biotinylated ligate with streptavidin/alkaline phosphatase conjugate at a molar ratio of 4:1[a] for at least 3 h at 4 °C.

2 Treat the bead-supported library as in *Protocol 6*, steps 1 and 2.

3 Incubate the bead-supported library with the complex of biotinylated ligate and streptavidin/alkaline phosphatase conjugate in 0.05% gelatin/0.1% Tween /PBS at room temperature for 1–24 h.

4 Thoroughly wash the bead-supported library with 0.1% Tween/PBS (10 × 10 ml) and TBS (10 ml). Add BCIP as in *Protocol 6*, steps 5–8, then stop the colour development as in *Protocol 6*, step 9, and isolate the positive beads under the dissecting microscope.

5 Treat the positive beads with 6.0 M guanidine–HCl pH 1.0, and then with DMF, as in *Protocol 6*, step 11.

6 Wash the decolorized beads with 0.1% Tween/PBS, followed by double distilled water. Block the decolorized beads with 0.1% gelatin again, as in *Protocol 6*, step 1.

7 Incubate the decolorized beads with streptavidin–alkaline phosphatase alone at a concentration of 1:5000 for 2 h.

8 Wash the beads with 0.1% Tween/PBS (5 × 5 ml) and TBS (5 ml). Add BCIP for colour development as in *Protocol 6*, steps 6–8.

9 Remove and discard the coloured beads.

10 Transfer the remaining colourless beads into a 1 ml disposable polyethylene column, then wash the beads with double distilled H$_2$O (5 × 5 ml), PBS/0.1% Tween (5 × 5 ml), and TBS (2 × 5 ml).

11 Incubate the beads with a suitable concentration of [^{125}I]ligate overnight at 4 °C.

12 Wash the beads with 0.1% Tween in twice more concentrated PBS (10 × 5 ml).

13 Suspend the beads in 1.0% agarose. Carefully pour the bead suspension onto the clean glass plate and air dry it at room temperature.

14 Tape the Glogos II autoradiogram marker on the corner of the plate.

Protocol 8 continued

15 Expose the immobilized beads to X-ray film overnight at room temperature.

16 Under the microscope, carefully pick up the positive beads that correspond to the dark spots of the film by excising the embedded bead into a polypropylene container with a small amount of water.

17 Heat it in the microwave to dissolve the agar.

18 Under the microscope, carefully transfer the beads into a Petri dish containing 6.0 M guanidine–HCl pH 1.0 with the micropipette.

19 Transfer the positive beads to water and sequence the peptides as in *Protocol 6*, step 13.

[a] Each streptavidin molecule has four binding sites.

Protocol 9

Determination of peptide substrate motifs for protein kinases

Equipment and reagents

- X-ray film (e.g. Kodak X-OMAT LS)
- Dissecting microscope
- Automatic protein sequencer (e.g. Applied Biosystem Model ABI 477A)
- MES buffer: 30 mM 2-(N-morpholino)ethanesulfonic acid, 10 mM $MgCl_2$, 0.4 mg/ml bovine serum albumin (BSA) pH 6.8

- $[\gamma^{-32}P]ATP$ (25 Ci/mmol)
- Washing buffer: 0.68 M NaCl, 13 mM KCl, 40 mM Na_2HPO_4, 7 mM KH_2PO_4 pH 7.0, 0.1% Tween 20 (v/v)
- 0.1 M HCl
- 1.0% low gelling temperature agarose (w/v) in H_2O[a]

Method

1 Transfer 1 ml of bead-supported library into a disposable polypropylene column, wash the beads thoroughly with double distilled water, followed by MES buffer.

2 To the 1 ml settled bead volume add 1 ml of two times more concentrated MES buffer containing 0.2–10 μM $[\gamma^{-32}P]ATP$ (adenosine triphosphate) and a specific protein kinase. Cap the reaction column tightly and put on a rocking platform for 1–5 h at room temperature with gentle rocking.

3 Wash the resin thoroughly with washing buffer, followed by double distilled water.

4 Transfer the ^{32}P-labelled bead-supported library to a glass container with 5 ml of 0.1 M HCl, and heat it to 100°C for 15 min to hydrolyse all the residual $[\gamma^{-32}P]ATP$.

5 Transfer the ^{32}P-labelled bead-supported library back into a clean disposable polypropylene column and wash the acid treated library with the washing buffer.

Protocol 9 continued

6 Suspend 0.5 ml of the ^{32}P-labelled bead-supported library resin in 30 ml of 1.0% agarose solution (70°C). Carefully pour the suspension onto a clean glass plate (16 × 18 cm), and air dry it overnight at room temperature. Tape the Glogos II autoradiogram markers on each corner of the glass plates.

7 Repeat step 6 for the remaining 0.5 ml bead library.

8 Expose the immobilized beads to X-ray film with an intensifying screen for 20–30 h at room temperature and develop the film.

9 Align the autoradiograph with the Glogos II autoradiogram markers on the glass plates. Excise the area of the dried agar with the bead corresponding to the dark spots on the developed film.[b]

10 Collect all of the excised dried agar embedded beads in 30 ml of hot 1.0% agarose solution (70°C) for 15 min.

11 Replate the beads, expose, and develop the autoradiogram as described above.

12 Localize an individual bead that corresponds to the dark spot on the autoradiogram under the dissecting microscope, transfer the positive bead together with some attached agarose to a clean Petri dish containing a small amount of water. Heat it by microwave to dissolve the agar.

13 Under a dissecting microscope, transfer each positive bead onto a glass filter with a micropipette and insert into it the protein sequencer cartridge for microsequencing.

[a] Melt agarose with microwave oven and keep at 70°C.

[b] Generally, one dark spot corresponds to many beads.

References

1. Lam, K. S., Salmon, S. E., Hersh, E. M., Hruby, V. J., Kazmierski, W. M., and Knapp, R. J. (1991). *Nature*, **354**, 82.

2. Houghten, R. A., Pinilla, C., Blondelle, S. E., Appel, J. R., Dodey, C. T., and Cuervo, J. H. (1991). *Nature*, **354**, 84.

3. Furka, A., Sebbstyen, F., Asgedom, M., and Dibo, G. (1991). *Int. J. Pept. Protein Res.*, **37**, 487.

4. Wu, J. Z., Ma, Q. N., and Lam, K. S. (1994). *Biochemistry*, **33**, 14825.

5. Salmon, S. E., Lam, K. S., Lebl, M., Kandola, A., Khattri, P., Wade, S., et al. (1993). *Proc. Natl. Acad. Sci. USA*, **90**, 11708.

6. Nikolaiev, V., Stierandova, A., Krchnak, V., Seligmann, B., and Lam, K. S. (1993). *Peptide Res.*, **6**, 161.

7. Ohlmeyer, M. H. J., Swanson, R. N., Dillard, L. W., Reader, J. C., Asouline, G., Kobayashi, R., et al. (1993). *Proc. Natl. Acad. Sci. USA*, **90**, 10922.

8. Brenner, S. and Lerner, R. A. (1992). *Proc. Natl. Acad. Sci. USA*, **89**, 5381.

9. Nicolaou, K. C., Xiao, X. Y., Parandoosh, Z., Senyei, A., and Nova, M. P. (1995). *Angew. Chem. Int. Ed. Engl.*, **34**, 2289.

10. Geysen, H. M., Meloen, R. H., and Barteling, S. J. (1984). *Proc. Natl. Acad. Sci. USA*, **81**, 3998.

11. Thompson, L. A. and Ellman, J. A. (1996). *Chem. Rev.*, **96**, 555.

12. Fruchtel, J. S. and Jung, G. (1996). *Angew. Chem. Int. Ed. Engl.*, **35**, 17.

13. Nefzi, A., Ostresh, J. M., and Houghten, R. A. (1997). *Chem. Rev.*, **97**, 449.

14. Lam, K. S., Lebl, M., and Krchnak, V. (1997). *Chem. Rev.*, **97**, 411.

15. Wahl, F. and Mutter, M. (1996). *Tetrahedron Lett.*, **37**(38), 6861.

16. Spatola, A. F. and Romanovskis, P. (1996). *Combinatorial peptide and nonpeptide libraries: a handbook*, p 327. VCH Verlagsgesellschaft mbH, D-69451 Weinheim.

17. McMurray, J. S. (1994). *Peptide Res.*, **7**, 195.

18. Novabiochem Catalog and peptide synthesis handbook (1997/1998). p. S33.

19. Lam, K. S. and Lebl, M. (1994). *Methods: a companion to methods in enzymology*, **6**, 372.

20. Smith, M. H., Lam, K. S., Hersh, E. M., and Grimes, W. (1994). *Mol. Immunol.*, **31**, 1431.

21. Pennington, M. E., Lam, K. S., and Cress, A. E. (1996). *Mol. Div.*, **2**, 19.

22. Lam, K. S., Wu, J. Z., and Lou, Q. (1995). *Intl. J. Protein Peptide Res.*, **45**, 587.

23. Lou, Q., Leftwich, M., and Lam, K. S. (1996). *Bioorg. Med. Chem.*, **4**, 677.

24. Meldal, M., Svendsen, I., Breddam, K., and Auzanneau, F. I. (1994). *Proc. Natl. Acad. Sci. USA*, **91**, 3314.

25. Meldal, M. and Svendsen, I. (1995). *J. Chem. Soc. Perkin Trans.*, **1**, 1591.

26. Salmon, S. E., Liu-Stevens, R. H., Zhao, Y., Lebl, M., Krchnak, V., Wertman, K., *et al.* (1996). *Mol. Div.*, **2**, 46.

27. Lebl, M., Patek, M., Kocis, P., Krchnak, V., Hruby, V. J., Salmon, S. E., *et al.* (1993). *Intl. J. Protein Peptide Res.*, **41**, 201.

Chapter 3

Synthesis and screening of positional scanning synthetic combinatorial libraries

Clemencia Pinilla, Jon R. Appel, Sylvie E. Blondelle, Colette T. Dooley, Jutta Eichler, Adel Nefzi, John M. Ostresh, Roland Martin,[†] Darcy B. Wilson, and Richard A. Houghten

Torrey Pines Institute for Molecular Studies, 3550 General Atomics Court, San Diego, CA 92121, USA.

[†]National Institute of Neurological Disorders and Stroke, Neuroimmunology Branch, National Institutes of Health, Bldg. 10 Room 5B16, 10 Center Drive MSC 1400, Bethesda, MD 20892-1400, USA.

1 Introduction

Combinatorial library synthesis and screening methods, which enable the rapid identification of highly active compounds, have revolutionized basic research and drug discovery. A number of different combinatorial approaches based on the principles of solid phase synthesis have been used to generate enormous molecular diversities, including peptides, peptidomimetics, and small organic molecules. The advantage of the various combinatorial approaches compared to traditional drug synthesis and screening methods is the very large number of compounds that can be simultaneously synthesized and rapidly screened in biological assays.

In recent years, two distinct approaches for preparing combinatorial libraries have emerged. The first approach involves parallel synthesis of thousands of individual compounds. The second involves the synthesis and screening of mixture-based libraries. The number of compounds that this approach is capable of producing is virtually limitless. Innovative techniques have been developed to screen such mixtures either attached to a solid support or free in solution. This chapter covers this approach as well as the deconvolutive screening methods used to identify individual active compounds.

2 Mixture-based synthetic combinatorial libraries

Synthetic combinatorial libraries (SCLs) as described in this chapter and accompanying protocols represent systematically arranged mixtures of very

51

large numbers of synthetic compounds (1, 2). SCLs are generated using the multiple solid phase synthesis method known as the 'tea bag approach' (3), in which compartmentalized resin-bound compounds are synthesized. The characteristic feature of SCLs is the presence of individual defined building blocks at certain positions of the compound scaffold, while the remaining diversity positions are mixtures of building blocks. SCLs are cleaved from the synthetic solid support and then assayed in solution, which allows each compound within each mixture to freely interact with a given receptor. The first SCLs were made up of peptides of various lengths and amino acids (L-, D-, and unnatural). More recently, a 'libraries from libraries' approach (4) was used in the development of SCLs composed of peptidomimetics (4, 5), polyamines (5), and heterocycles (6–9). Existing peptide SCLs can be chemically transformed to yield new SCLs having entirely different physical, chemical, and biological properties relative to the original peptide SCLs used as starting materials.

Two different deconvolution methods are used to identify individual compounds from mixture-based SCLs. The first method involves an iterative deconvolution approach (3, 10, 11). Following the identification of active mixtures having defined positions, the remaining mixture positions are then defined one diversity position at a time, through a synthesis and selection process, until individual compounds are identified. The second deconvolution method is termed positional scanning (12–14), in which each diversity position is individually addressed by separate sublibraries. The building blocks of the most active mixtures at each diversity position are combined, and the resulting individual compounds are synthesized and tested to determine their activities. The advantage of the positional scanning format is that iterations are not required; in most cases a single synthesis is needed to obtain active individual compounds.

2.1 Positional scanning concept

Positional scanning SCLs (PS-SCLs) are composed of positional SCLs or sublibraries, in which each diversity position is defined with a single building block, while the remaining positions are composed of mixtures of building blocks. Each positional sublibrary represents the same collection of individual compounds. The assay screening data derived from each positional sublibrary yield information about the most important building block for every diversity position of the PS-SCL.

For instance in the case of the tripeptide combinatorial library in *Figure 1*, four different amino acids are incorporated at each of the three diversity positions resulting in a diversity of 64 (4^3) individual peptides. When the same diversity is arranged as a PS-SCL, only 12 peptide mixtures (4 amino acids × 3 positions) need to be synthesized. Each of the three positional sublibraries, namely OXX, XOX, and XXO, contains the same diversity of peptides, but differ only in the location of the position defined with a single amino acid. The O positions represent one of the four amino acids while the remaining two

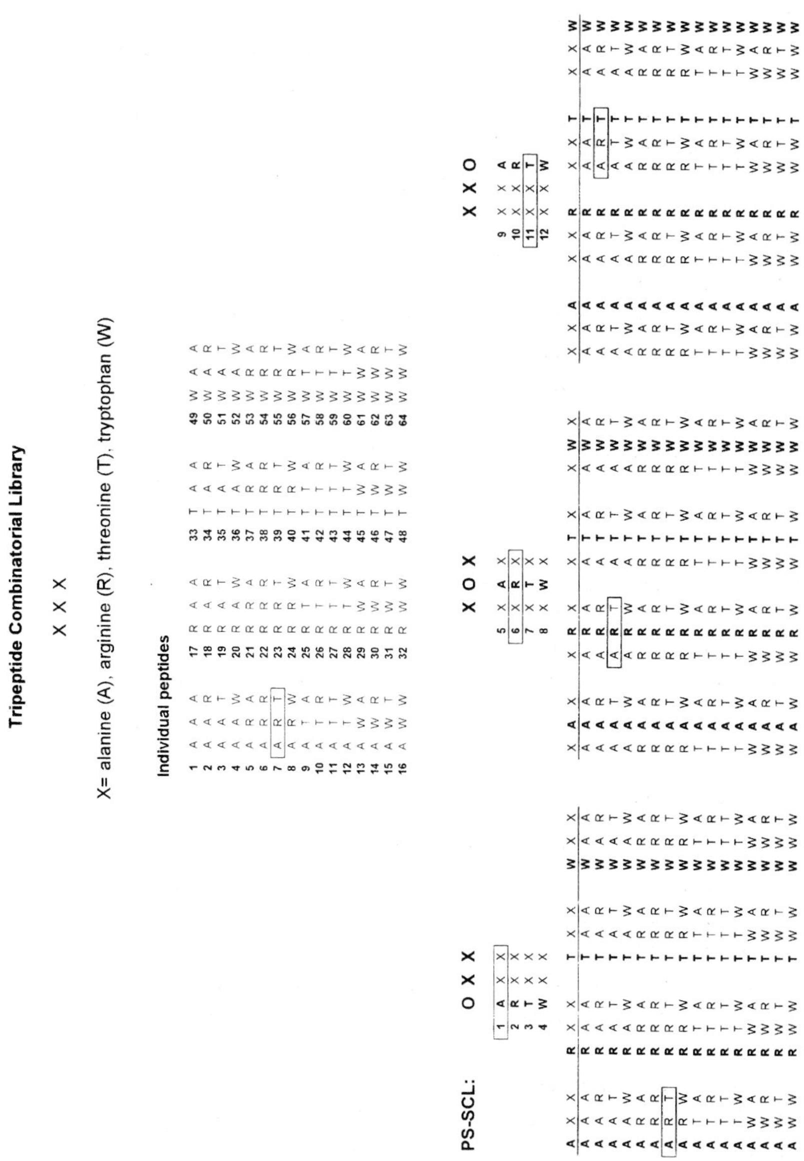
Tripeptide Combinatorial Library
X X X
X= alanine (A), arginine (R), threonine (T), tryptophan (W)
Individual peptides
PS-SCL:
O X X
X O X
X X O

Figure 1 Design and format of a tripeptide PS-SCL using four different building blocks in each of the three diversity positions. See text for detailed explanation.

positions are mixtures (X) of the same four amino acids. Shown below each mixture are the 16 peptides (4^2) that make up that mixture.

In this example, assume that ART is the only tripeptide in this library that is recognized by a receptor. Since each positional sublibrary contains the same diversity of peptides, the ART tripeptide (outlined below each sublibrary in *Figure 1*) is present in all three positional sublibraries. Thus, the only mixtures with activity are AXX, XRX, and XXT because the ART tripeptide is present only in those mixtures. The combination of these amino acids in their respective positions yields the tripeptide ART, which would then be synthesized and tested for its activity against the receptor. It should be noted that the activity observed for each of the three mixtures (AXX, XRX, and XXT) is due to the presence of the tripeptide ART within each mixture, and not due to the individual amino acids (A, R, and T) that occupy the defined positions. In more complex libraries, more than one mixture is often found to have activity at each position. Selection of the building blocks for the synthesis of individual compounds is based first on activity and then on differences in the chemical character of the building block.

Although the above example is a simple representation of the arrangement and use of a PS-SCL, the concepts described here apply to all types of mixture-based libraries having defined and mixture positions. For example, a hexa-peptide library using 20 amino acids represents a total diversity of 6.4×10^7 (20^6) individual peptides. This can be formatted into a PS-SCL of 120 mixtures (20 amino acids $\times$ 6 positions).

2.2 Synthesis of a tripeptide PS-SCL

The following two protocols describe the synthesis and cleavage of a tripeptide PS-SCL. The same protocol can be used to synthesize a PS-SCL of any length, which can be composed of either only L- or only D- amino acids, or a combination of both including unnatural amino acids. Since a predetermined ratio of amino acids are coupled to the resin-bound amino acid, it is necessary to first establish a set of ratios for the amino acids that will be used in the coupling step (15).

The following two protocols are written for a person trained and experienced in peptide chemistry, especially the method of simultaneous multiple peptide synthesis (SMPS), also known as the 'tea bag approach' (3, 16). The advantage of using SMPS is that all wash and deprotection steps are carried out in a common vessel, since the resin used for each peptide is enclosed in separate poly-propylene mesh bags (tea bags). Individual tea bags are simply placed in the appropriate amino acid solution for the coupling reaction. It is important that sufficient solvent cover the tea bags and vigorous shaking be used for each step. For the protocols described here, methylbenzhydrylamine (MBHA) polystyrene resin is used in conjunction with t-Boc chemistry. Fmoc peptide chemistry can also be used with the appropriate reagents and protected amino acids. More details on solid phase methodology for peptide synthesis have been published elsewhere (17).

Protocol 1

Synthesis of a tripeptide PS-SCL

Equipment and reagents

- Impulse sealer
- Reciprocating shaker
- Methylbenzhydrylamine (MBHA) polystyrene resin (Peninsula)
- Boc-protected amino acids (Bachem)
- Solvents and coupling reagents for peptide synthesis using Boc chemistry (Fisher, Aldrich)
- Polypropylene mesh (74 μm) for tea bags (McMaster Carr or Spectrum)

Method

1   Number polypropylene tea bags 1–60, add 100 mg MBHA resin to each bag, and seal it closed.

2   Use SMPS methodology to synthesize peptide mixtures. For the first coupling step, couple the 20 proteinogenic amino acids individually to resin-filled bags 41–60. (We use single letter code for amino acids, in alphabetical order.) To bags 1–40, couple a 19-amino acid mixture using a predetermined ratio of amino acids. Cysteine is excluded in the mixture positions to avoid the formation of disulfide aggregates. Remove the N-terminal Boc and neutralize bags before the next coupling.

3   At the second coupling step, couple 20 amino acids individually to bags 21–40. Couple the 19-amino acid mixture to bags 1–20 and 41–60. Remove the N-terminal Boc and neutralize bags before the next coupling.

4   At the third coupling step, couple 20 amino acids individually to bags 1–20. Couple the 19-amino acid mixture to bags 21–60. Remove the N-terminal Boc to have a free N-terminal amine. The N-terminal amino groups of the peptide mixtures can be N-acetylated, if desired.

Protocol 2

Cleavage and extraction of a PS-SCL

Equipment and reagents

- Multiple cleavage apparatus in chemical hood
- Vacuum pump
- Lyophilizer
- Reciprocating shaker
- 20 ml scintillation vials for extractions, 1 ml polypropylene tubes in 96-tube rack
- Hydrogen fluoride (HF) and anisole
- Acetic acid, dimethyl sulfoxide (DMSO), dimethylformamide (DMF), dichloromethane (DCM), isopropanol (IPA), methanol, dimethyl sulfide (DMS), ethylenedithiol (EDT), thiophenol, p-cresol (Aldrich, Fisher)

Method

1. Deprotect the peptide side-chains used with t-Boc chemistry. For DNP removal from histidine, add 2.5% thiophenol/DMF to bags and shake for 1 h. Wash bags three times with DMF, 12 times alternating washes of IPA and DCM. For removal of side-chain protecting groups of the amino acids, use a 'low-HF' procedure. Determine a final volume that will cover bags and add 60% DMS, 5% EDT, 10% *p*-cresol to bags. Condense HF (25%, v/v) and shake bags for 2 h at 0 °C. Wash bags eight times alternating washes of IPA and DCM, four times with DMF, three times with DCM, and once with methanol.

2. Cleave the peptide mixtures from the resin using hydrogen fluoride. Place each bag in a separate tube of the cleavage apparatus. Add 0.3 ml of anisole to each bag, making sure the anisole soaks the resin inside the bag. Condense hydrogen fluoride (5 ml) into tubes of apparatus and incubate reaction for 1 h at 0 °C. Remove hydrogen fluoride from apparatus and place tubes on vacuum pump overnight to dry the bags.

3. Extract peptide mixtures from the resin with 95% acetic acid (three times, 5 ml each) into 20 ml scintillation vials.

4. Lyophilize solutions four times in 10% acetic acid and reconstitute material in appropriate solvent for biological assays (typically a 10% organic/water solution using DMSO or DMF as solvent) and dilute with water to 10 mg/ml. Sonication is often used to dissolve peptide mixtures containing hydrophobic amino acids in the defined positions.

5. Aliquot in 1 ml polypropylene tubes in 96-tube racks for easy use in biological assays. Store one to two weeks at 4 °C while in use or indefinitely at −20 °C.

2.3 Alkylation of peptide PS-SCL

The 'libraries from libraries' concept (4, 5, 18) enables the generation of peptido-mimetic SCLs through the chemical transformation of existing peptide SCLs. A wide variety of chemical transformations permits a range of peptidomimetic libraries to be generated, thus greatly expanding the chemical diversity available. The following protocol describes the peralkylation of a tetrapeptide PS-SCL (*Figure 2*). Peralkylations have been carried out using five different alkylating reagents, namely methyl iodide, ethyl iodide, allyl bromide, benzyl bromide, and naphthylmethyl bromide. It is recommended that peptide resins having defined sequences of the same length as the peptide library always be included in the synthesis of all libraries to serve as analytical controls for the final product. Sufficient resin (> 100 mg) should be used such that multiple aliquots can be cleaved and analysed using reverse phase HPLC (RP-HPLC) and mass spectrometry (MS; an LC/MS system is preferred) as necessary to monitor the completeness of the alkylation reaction. The peralkylation reaction can be repeated as many times as necessary for the library if the control compounds show incomplete alkylation.

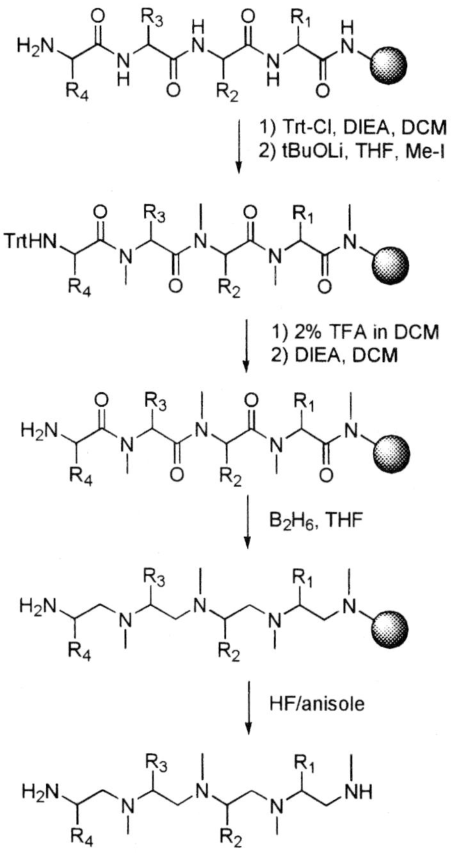

Figure 2 Reaction scheme for the chemical transformation of a tetrapeptide PS-SCL to yield a tetrapeptidomimetic and pentamine PS-SCL.

Protocol 3

Alkylation of a resin-bound peptide PS-SCL

Equipment and reagents (chemicals from Aldrich and Fisher)

- Resin-bound tetrapeptide PS-SCL
- Glove box and chemical hoods
- Diisopropylethylamine (DIEA)
- DMF, DCM, anhydrous THF, DMSO, IPA, methanol

- Triphenylmethyl chloride (trityl chloride)
- Bromophenol blue
- Lithium t-butoxide
- Alkyl bromide
- Trifluoroacetic acid (TFA)

A. Trityl protection

1   Neutralize packets containing resin-bound peptides having free N-terminal groups by washing twice with 5% DIEA in DCM (v/v). Use enough solvent to completely cover the resin packets.

Protocol 3 continued

2 Wash the resin packets once with DCM.

3 Add 0.077 M trityl chloride (5 Equiv.) in 90% DMF/10% DCM containing DIEA (29 Equiv.).

4 Shake the resin packets on a reciprocating shaker for 3 h.

5 Wash the resin packets with DMF for 1 min.

B. Bromophenol blue test

1 Wash the resin packets three times with DCM to remove excess base.

2 Cut open each resin packet and place a small aliquot of resin ($\sim$ 1 mg) into a test-tube. Reseal the resin packet.

3 Add 150 μl DCM to each resin aliquot followed by 30 μl of 0.62 mM bromophenol blue in DCM (0.42 g/litre). Vortex the resin aliquots briefly.

4 Examine the resin beads and solution. Blue beads indicate incomplete coupling. Repeat part A, steps 1–5 until the reaction is complete. Complete coupling ($>$ 99%) is indicated by yellow resin beads (normally with a slight trace of green) and yellow supernatant.

C. Peralkylation of resin-bound PS-SCL

1 Dry resin packets overnight under high vacuum.

2 All manipulations should be carried out under anhydrous conditions (nitrogen atmosphere) in a glove box. Add 0.5 M lithium *t*-butoxide (20 Equiv. per available amide) in tetrahydrofuran under anhydrous conditions to resin packets.

3 Shake for 15 min at room temperature. Remove base solution.

4 Add 1.5 M alkyl bromide (20 Equiv. per available amide) in DMSO to resin packets.

5 Shake the reaction mixture on a reciprocating shaker for 2 h at room temperature. Remove the alkylation solution.

6 Repeat steps 2–5 two times.

7 Wash the resin packets three times with DMF, then twice with IPA, three times with DCM, and once with methanol.

8 Dry the resin packets under high vacuum.

9 Check the reaction completion using control resins. Take an aliquot of resin and cleave the compound from the resin as in *Protocol 2*. Analyse the compound by RP-HPLC and MS for alkylation. Repeat steps 2–9 as necessary.

D. Removal of the trityl protecting group

1 Wash the resin packets three times with DCM.

2 Remove the trityl protecting group by two treatments with 2% TFA in DCM (once for 3 min, and then twice for 10 min).

Protocol 3 continued

3 Wash the resin packets twice with DCM, then twice with IPA, twice with DCM, and once with methanol.

4 Dry the resin packets overnight under high vacuum. The peralkylated peptides can then be cleaved from the resin and extracted as described in *Protocol 2*.

A number of amino acid side-chain functionalities undergo alteration during the chemical treatment. Modifications can vary depending on the protecting group and on the alkylating agent used. Representative examples of the modifications have been described (19). Some amino acids, such as cysteine, aspartic acid, glutamic acid, and histidine derivatives, have led to multiple products, and therefore have not been used in peralkylated libraries.

2.4 PS-SCL reduction

Peralkylated or peptide PS-SCLs can undergo a different chemical transformation to create polyamine PS-SCLs (*Figure 2*). The exhaustive reduction of the backbone carbonyl functionalities and susceptible side-chains is carried out using a borane-tetrahydrofuran procedure as described in the following protocol.

Protocol 4

Reduction of a resin-bound PS-SCL

Equipment and reagents

- Resin-bound tetrapeptide PS-SCL
- 50 ml glass conical tubes
- Heating block
- Trimethylborate, boric acid, borane, tetrahydrofuran (THF), methanol, piperidine (Aldrich)

Method

1 Add each packet containing resin-bound peptide to a 50 ml glass conical tube under nitrogen: 0.08 mequiv. resin, 100 mg of starting resin, (0.32 mequiv. carbonyl) and boric acid (312 mg, 15 ×).

2 Add trimethylborate (0.555 ml, 15 ×), followed by the slow addition of 14.4 ml borane-THF complex (1 M, 45 ×).

3 After hydrogen gas production has ceased, cap tubes and heat at 65 °C for 72 h in a heating block.

4 Decant the reaction solution, which is quenched by the slow addition to methanol. Wash the resin packet three times with 5 ml methanol, once with 5 ml THF, and twice with 5 ml piperidine.

5 Disrupt the amine-borane complex by overnight treatment (16 h) with 10 ml piperidine (400 ×) at 65 °C.

6 Decant the resulting piperidine-borane solution. Wash the resin packet twice with 5 ml DMF, twice with 5 ml DCM, and twice with 5 ml methanol. Dry under high vacuum overnight. The compound can be cleaved from the resin using *Protocol 2*. The cleavage reaction should be extended to 9 h.

2.5 From peptides to small organic and heterocyclic compounds

The general lack of oral and/or central bioavailability of peptides has led to the development of new chemistries that can be performed using established solid phase approaches to synthesize a variety of small molecule and heterocyclic libraries. To take advantage of the large diversity possible in peptide libraries (typically millions of compounds), the libraries from libraries concept (4) was developed to produce a number of different peptidomimetic and various poly-amine compounds. More recently, this approach has been used to prepare

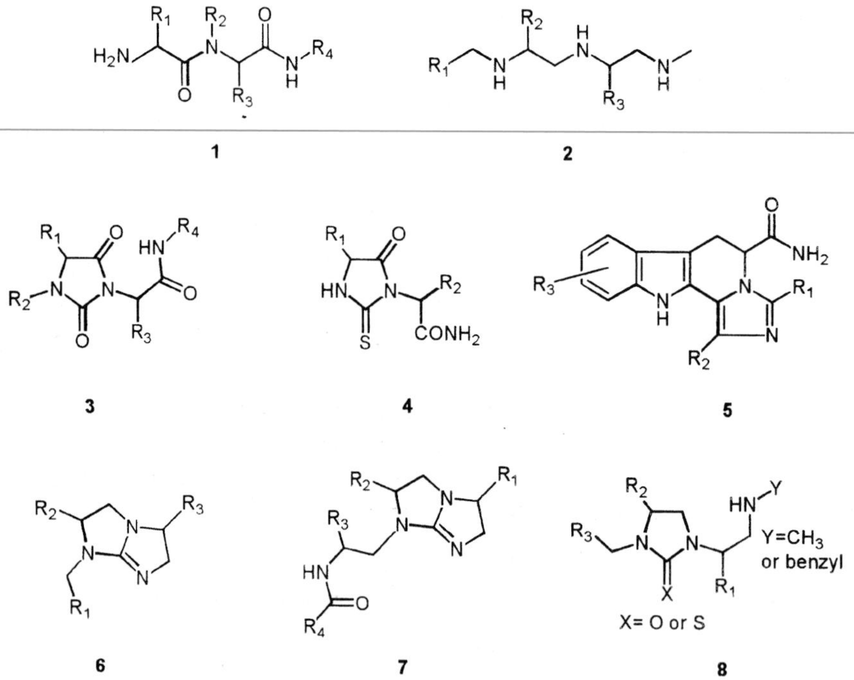

Figure 3 Examples of small molecule and heterocyclic PS-SCLs. (1) Peralkylated dipeptidomimetic, 460 mixtures/52 900 total diversity. (2) *N*-methyltriamine, 96/31 320. (3) Di-alkylated hydantoin, 375/32 400. (4) Thiohydantoin, 162/6525. (5) Indole-pyrido-imidazole, 121/25 300. (6) Bicyclic guanidine, 141/102 459. (7) Acyl bicyclic guanidine, 156/1 100 512. (8) Cyclic *N*-alkylamino urea, 4 × 157/118 400 each.

60

Figure 4 Formation of bicyclic guanidines from the cyclization of polyamines using thicarbonyldiimidazole ($CSIm_2$).

heterocyclic libraries using dipeptide SCLs as starting materials (20). Examples include hydantoin, cyclic urea, and thiourea (7), indole-pyrido-imidazole, and bicyclic guanidine SCLs (9), in which each library is composed of 10^4 to 10^6 individual compounds. *Figure 3* illustrates the various small molecule and heterocyclic scaffolds that have been prepared.

The following protocol describes the formation of a bicyclic guanidine PS-SCL. Triamines derived from acylated dipeptides can undergo further transformation to create bicyclic guanidines. *Figure 4* illustrates the cyclization of the polyamines using thiocarbonyldiimidazole.

Protocol 5

Formation of a bicyclic guanidine PS-SCL

Equipment and reagents

- Resin-bound polyamine PS-SCL derived from acylated dipeptides
- Reciprocating shaker

- Thiocarbonyldiimidazole ($CSIm_2$) (Aldrich)
- DCM, methanol (Fisher)

Method

1 For each resin-bound polyamine (0.087 mequiv. resin) derived from acylated dipeptides, add 5.2 ml of 0.37 M $CSIm_2$ in anhydrous DCM (24 ×). Shake for 15 min and decant solution.

2 Add 5.2 ml anhydrous DCM and shake for 16 h. We have found that for cyclizations involving bulkier functionalities, allowing the reaction to proceed overnight without decanting yields more complete cyclization. However, in cases where cyclization proceeds quickly, the original conditions yielded cleaner products.

3 Decant solution and repeat steps 1 and 2.

4 Wash three times (1 min each) with 5 ml DCM and twice with methanol.

5 Dry under high vacuum and cleave using *Protocol 2*.

3 Screening mixture-based libraries

One advantage to using libraries formatted as mixtures is the reduction of assay reagents and materials required for screening. For example, *Table 1* illustrates

Table 1 Comparison of the number of 96-well microtitre plates required for duplicate testing of library diversities formatted as mixtures versus individual compounds

Library	No. of mixtures	Diversity	No. of plates
Hexapeptide	120	50×10^6	4 versus 10^6
Bicyclic guanidine	141	102 459	4 versus 2135
Indole-pyrido-imidazole	121	25 300	4 versus 528
Cyclic urea	157	118 400	4 versus 2467

the number of 96-well microtitre plates that are required for testing libraries either as mixtures or as individual compounds. Four different libraries having diversities ranging from 25 000 to 50×10^6 compounds can be formatted into PS-SCLs having approximately 120–157 samples to test. Thus, each PS-SCL can be screened on several microtitre plates instead of hundreds or thousands, reducing the entire workload by several orders of magnitude. For example, the eight PS-SCLs prepared by this laboratory shown in *Figure 3* encompass a diversity of more than 1 800 000 small molecules and heterocycles, and can be screened in duplicate in a given assay on 45 microtitre plates. This can be especially useful for assays having an inherent low-throughput.

The starting concentration that is used to screen a library is assay dependent. Commonly used binding assays such as ELISA can tolerate high concentrations of a library (1–10 mg/ml). The majority of cell-based assays, however, are more sensitive to high concentrations of compounds, and screening typically starts at 0.1 mg/ml. Also, most SCLs are solubilized in an organic/water solution with either DMF or DMSO as the solvent. Since many assays are sensitive to more than 1% solvent, it is necessary to aliquot very complex libraries at the highest concentration possible (~10 mg/ml). This will ensure that individual compounds within each mixture are present at a detectable concentration without having excess solvent affecting the assay.

3.1 PS-SCL deconvolution

In order to identify the most active individual compounds from the PS-SCL, the individual compounds that correspond to the combination of the building blocks defined in the most active mixtures at each position are synthesized and tested. For practical purposes, the number of building blocks selected from each position that will be used to synthesize the individual compounds should be minimized as the number of compounds to be made rises exponentially. For example, if two amino acids were selected from each position of a hexapeptide PS-SCL, one would need to synthesize 64 peptides (2^6), and if three amino acids were chosen at each position, 729 (3^6) peptides would be required. It should be noted that different numbers of building blocks could be chosen from each position depending on the activity of those mixtures. Since PS-SCLs are composed of separate positional SCLs, each one can be considered independent of the others. Therefore, each positional SCL can be independently screened and pursued using an iterative synthesis and selection process.

Successful deconvolution of active individual compounds from mixture-based libraries is dependent on reproducible screening data and clear dose–response activities of the most active mixtures. It is most important for an assay to be optimized before screening a library. In most cases, dose–response curves can be determined for the most active mixtures, and activities based on calculated IC$_{50}$ values are used to select the building blocks that will be included in the synthesis of individual compounds. Often building blocks of similar chemical character will yield similar activities at a given position. This may indicate that a number of analogues of the same compound are responsible for the observed activity. Similar building blocks can be excluded from selection to reduce the number of final compounds needed to be synthesized. However, one can synthesize analogues of the most active individual compounds using the building blocks that were originally excluded.

In a number of examples of library screening data, the distinction between active and inactive mixtures is more difficult to determine, because either the specificities of the most active mixtures are not very clear, or the signal-to-noise ratio of the assay is less than threefold. In these cases, dose–response determinations may not be possible. Another strategy that can be useful is to compare the activity of a given mixture relative to the average mixture activity at that diversity position. As noted previously, the data analysis of complex mixtures is no different from the data obtained using individual compounds. One simply follows the activity that is significantly affecting the assay. In other words, distinguishing between active and inactive samples is independent of the complexity of these samples.

One aspect of screening PS-SCLs is that the activity of a mixture with a given building block in the defined position can be due to one or more families of compounds having that same building block at its respective position. When the combinations of the most active building blocks are synthesized and tested as individual compounds, it becomes clear if the activities of the mixtures between positions are connected. Thus, the activities of mixtures in the library are due to the activities of individual compounds, and there are several strategies that can be used for their deconvolution. Establishing the connectivity of the active mixtures at each position of diversity is the most important step in the successful identification of highly active mixtures.

3.2 Assay optimization

PS-SCLs have been used in a number of biological assays. An understanding of the various assay parameters, such as signal-to-noise ratio, variability, and sensitivity is required for the successful identification of active compounds from the library. *Table 2* illustrates the various assays in which PS-SCLs have been used and the signal ranges obtained. However, it should be noted that these parameters are inherent to the assay and are not influenced by the fact that complex mixtures are being tested instead of individual compounds. The most important parameter to control for an assay system is the variability.

Table 2 Examples of signal ranges for various assays testing PS-SCLs

Assay	Readout	Range	Ratio
Radioreceptor	Counts/min	200–1000	5
T cell proliferation	Counts/min	100–500	5
Microdilution	Optical density	0.1–0.5	5
Enzyme inhibition	Optical density	0.1–0.8	8
ELISA	Optical density	0.1–2.0	20

When screening complex mixtures, it is critical that the assay variability is known. For an assay with low variability, one has confidence that a five- to tenfold difference in observed activity between mixtures is significant. For an assay with high variability, the variation between replicates for a given mixture may obscure real differences in activities from other mixtures. The use of repeated experiments and averaged data ensures accurate deconvolution (i.e. selection of truly active mixtures), which results in the identification of individual compounds having significant activity.

4 Library screening using ELISA

An enzyme-linked immunosorbent assay (ELISA) is used to screen PS-SCLs for the inhibition of monoclonal antibodies binding to antigens to identify antigenic determinants and highly active mimics (21–23). The following protocol is separated into two parts:

(a) Direct and competitive ELISA for titring the antibody against the antigen and optimizing the conditions needed for screening the PS-SCL.

(b) The screening of the PS-SCL using the predetermined conditions.

Since the majority of linear epitopes are six residues in length, a hexapeptide PS-SCL is screened by competitive ELISA to identify specific peptides that inhibit monoclonal antibody binding to its antigen. It is important to screen the library at a high concentration (5–10 mg/ml) due to the enormous diversity of peptides within each peptide mixture. When tested at 5 mg/ml, each peptide in a mixture of the PS-SCL is present at a concentration of 3 nM. However, positional redundancy is often found in most antigen–antibody interactions. For this reason, the effective concentration of the peptide mixture recognized by the antibody would be much higher. Standardization and optimization of ELISA and screening conditions for each antigen–antibody interaction is essential for the successful identification of antigenic determinants using PS-SCLs, due to the number of peptide sequences in each peptide mixture. If most of the PS-SCL yields complete inhibition, the PS-SCL should be retested at a five- to tenfold lower concentration. In most cases, active peptide mixtures are serially diluted twofold and tested again to select the most active peptide mixtures at each position. It is also important to generate dose–response curves using non-linear regression curve-fitting software (for example, *GraphPad Prism*) and determine

IC_{50} values in order to select the most active amino acids at each position. The combinations of amino acids are then used to make the individual peptides. If the sequence of the immunogen is known, the screening results can be used to quickly locate the antigenic determinant recognized by the antibody.

Protocol 6

Screening of a PS-SCL for inhibition of mAb binding to antigen by ELISA

Equipment and reagents

- 96-well flat-bottom microtitre plates (Costar)
- Microplate spectrophotometer
- Hexapeptide PS-SCL
- Monoclonal antibody and its antigen
- Goat anti-mouse IgG–horseradish peroxidase (Calbiochem)
- o-Phenylenediamine (OPD) (Sigma)
- 3% hydrogen peroxide (Fisher)
- Bovine serum albumin (BSA) (Sigma)
- Phosphate-buffered saline (PBS)
- 0.3 M bicarbonate buffer pH 9.3
- 1% (w/v) BSA/PBS
- 4 M sulfuric acid

A. ELISA optimization

1 Coat microtitre plates with 50 μl/well antigen (1–10 μg/ml) in PBS or bicarbonate buffer (try both buffers separately to see which might give better results). Perform twofold serial dilutions of the antigen across the plate, starting with a high concentration (10 μg/ml). Incubate for either 2 h at 37°C or 18 h (overnight) at room temperature in a moistened box to avoid evaporation of reagents.

2 Shake out liquid from wells and wash plates ten times with deionized water. Remove residual water from wells by rapping plates upside down over paper towels. Avoid complete drying of wells. Repeat washing step after each incubation.

3 Block plates for non-specific binding by adding 100 μl/well of 1% BSA/PBS to microtitre plates and incubate plates for 1 h at 37°C.

4 Add 50 μl/well of antibody, performing twofold serial dilutions in 1% BSA/PBS down the plate. Incubate plates overnight at 4°C.

5 Add 50 μl/well of secondary antibody–enzyme conjugate (goat anti-mouse-peroxidase) at supplier's specified dilution in 1% BSA/PBS. Incubate plates for 1 h at 37°C.

6 Prepare developing solution for each plate by dissolving one tablet of OPD (10 mg) in 6 ml deionized water, and add 25 μl of 3% hydrogen peroxide. Add 50 μl/well of developing solution to plates and develop in the dark for 10–15 min. Terminate developing reaction with 25 μl/well of 4 M sulfuric acid.

7 Read plates on microplate spectrophotometer at 492 nm.

Protocol 6 continued

8 Choose the concentrations for antibody and antigen that give the optimum results, i.e. lowest antigen and antibody concentrations that still give high OD values (1.5–2.0).

9 Use these conditions for competitive ELISA. Coat microtitre plates with the control antigen at the predetermined concentration and incubate plates in a moist box for 2 h at 37°C.

10 Wash plates ten times with deionized water and after each subsequent incubation.

11 Block for non-specific binding as in direct ELISA.

12 Add 25 μl/well of blocking buffer to each plate. Add 25 μl/well of control antigen (10, 100, and 1000 times the amount of control antigen on the plate) to the top row and perform twofold serial dilutions down the plate. A fixed dilution of monoclonal antibody (25 μl/well) is added to each well. Do the same for antibody concentrations two to five times higher and lower than the selected concentration. Incubate plates for 18 h at 4°C.

13 Add goat anti-mouse–peroxidase conjugate, develop, and read plates as in direct ELISA.

14 Determine the concentration of antigen that inhibits 50% antibody binding (IC_{50}) of the control antigen for the three antibody concentrations. Choose the antibody concentration that gives the lowest IC_{50} while maintaining an acceptable signal-to-noise ratio (optimal is 10 to 1). Use these conditions to screen the PS-SCL.

B. Screening PS-SCL using competitive ELISA

1 Use the predetermined conditions to coat six plates, and incubate and block as described above.

2 Add the 120 peptide mixtures of the PS-SCL followed by the target monoclonal antibody using the conditions established by competitive ELISA. The first column of each microtitre plate is used for 100% antibody binding to the antigen on the plate (no inhibitor). The second column is used for the antigen inhibitor in solution serially diluted as a competitive control to ensure the assay is working in a sensitive manner. The remaining 80 wells on each plate are used to test 80 peptide mixtures at one concentration. We use 1 ml polypropylene tubes in a 96-tube rack to aliquot the peptide library, making it easy to use the same PS-SCL in a variety of different assays in the 96-well format. Add peptide mixtures (50 μl at 10 mg/ml) to the plate using a multichannel pipettor. Once the PS-SCL has been added to the plates, use a repetitive multichannel pipettor to add antibody at a fixed dilution (25 μl/well) previously determined to be optimal. Incubate plates for 18 h at 4°C.

3 Follow part A, steps 5–7 for developing and reading the plates.

4 Express inhibitory activity of peptide mixtures as optical density (OD) values

Protocol 6 continued

(inhibition = low OD) or convert peptide mixture activity to % inhibition relative to the binding of antibody to the control antigen from column 1 of each plate.

5   Retest peptide mixtures that were found to have good inhibitory activities (> 50% inhibition).

6   Calculate inhibitory concentrations at 50% of antibody binding (IC_{50}) for each peptide mixture using a curve-fitting software program (i.e. *GraphPad*). Rank order the most active peptide mixture(s) in order to select the amino acids at each position to make individual peptides. Synthesize individual peptides from combination of selected amino acids. Test peptides in above assay to determine their final activities.

5 Identification of T cell-specific ligands

Antibodies, made by B lymphocytes derived from the bone marrow, in most cases recognize discontinuous epitopes that are made up of residues that are not contiguous in the primary sequence, but are brought together by the folding of the protein. In contrast, ligands that stimulate responses by thymus-derived T lymphocytes are linear peptide fragments bound to larger cell surface molecules of the major histocompatibility complex (MHC). Generally, CD8$^+$ T cells recognize nonapeptides bound to MHC class I molecules (in humans, HLA-A, -B, and -C) whereas CD4$^+$ T cells recognize larger peptide sequences, usually 12–15-mers, bound to MHC class II molecules (HLA-DR, -DP, and -DQ).

Combinatorial libraries composed of peptides have proven useful for the identification of sequences recognized by T cell clones and for mapping T cell specificity (24, 25). In addition to defining native ligand sequences, the specificity profiles that result from screening PS-SCLs have led to the identification of cross-reactive sequences that were found from searches of sequence databases. In this respect, PS-SCLs may help to identify relevant epitopes that can lead to the design of novel vaccines for infectious diseases and cancer as well as the identification of target autoantigens involved in autoimmune diseases.

The following protocols describe how ligands can be identified for CD4$^+$ and CD8$^+$ T cells from PS-SCLs. In this protocol, the extent of T cell activation is assessed by:

(a) Proliferation assays that measure the incorporation of cell [^{3}H]thymidine into newly synthesized DNA.

(b) ^{51}Cr-release assays that measure cytotoxicity.

(c) Quantitative assessments of cytokine (IFNγ) production and release measured by standard sandwich ELISA.

Protocol 7

Screening of a PS-SCL to identify CD4$^+$ or CD8$^+$ T cell ligands

Equipment and reagents

- 96-well U-bottom microtitre plates (Costar)
- Cell harverster
- Liquid scintillation counter
- ELISA kit to measure γ-interferon production (PharMingen)

- Decapeptide PS-SCL
- T cell clone (TCC)
- Antigen-presenting cells (APC)
- Appropriate growth medium for cells
- [^{3}H]thymidine, [Na$_2$ ^{51}CrO$_4$]

Method

1 Grow TCC to large numbers. Approx. five million cells are needed to screen a decapeptide library.

2 Test proliferative (CD4$^+$) or cytotoxic (CD8$^+$; ^{51}Cr-release or cytokine production) response of the TCC to PS-SCL. Five 96-well plates can be used for a decapeptide PS-SCL, composed of 200 samples, duplicates per sample. Add 0.1 ml of TCC cell suspension (50–100 × 10^3) to each well of microtitre plates containing 5 × 10^4 APC and peptide mixtures (0.1 mg/ml final concentration).

3 Culture cells with PS-SCL for 24–72 h at 37°C (alternatively 4 h ^{51}Cr-release cytotoxic assay).

4 Determine activity of peptide mixtures against TCC.

   (a) CD4$^+$ T cell proliferation. Add 0.25–1 μCi [^{3}H]thymidine to each well during last 8 h of incubation. Harvest cells and measure incorporated radioactivity by scintillation counting.

   (b) ^{51}Cr-release assays. Transfer 150 μl from each well into counting tubes or plates and assess per cent radioactivity released to supernatant.

   (c) γ-Interferon production in CD8$^+$ T cells. Transfer the supernatants to microtitre plates that contain bound anti-interferon monoclonal antibody. Develop ELISA as specified by kit.

5 Confirm results in at least two repeat experiments. The incorporated radioactivity from CD4$^+$ cells and optical densities from CD8$^+$ cells can vary depending on the state of activation, however, the rank order of most active mixtures should be consistent.

6 Deduce optimal peptides for the TCC by combining the defined amino acids of the most active mixtures at each position of the PS-SCL.

7 Search sequence databases for potential cross-reactive ligands based on library results.

8 Synthesize individual peptides deduced from library results and database searches. Confirm library results by testing TCC responses to individual peptides.

6 Identification of antimicrobial and antifungal compounds

Since SCLs are prepared as non-support bound libraries, they can be screened in any existing biological assay. This feature allows them to be screened in a variety of cell-based assays for the identification of novel antimicrobial, antifungal, antiviral, and antitumour compounds (26–29). The following protocol describes the use of a PS-SCL in a microdilution assay to identify compounds that inhibit the growth or lyse *Staphylococcus aureus*. However, any other type of bacteria or fungi could be used in a similar manner.

The mechanism of action of antimicrobial compounds identified through a microdilution assay can be diverse. Antimicrobial peptides commonly affect cells by interacting in a non-specific manner with their membranes. Such active peptides show common conformational and/or chemico-physical characteristics. This lack of specificity, due in part to peptide flexibility, led to the identification of similar amino acids for every positional SCL (tetra- and hexapeptide) when screening PS-SCLs (30). These amino acids are not connected to each other since the individual peptides with the same amino acids in all positions showed little activity, however, the mixtures had the required chemico-physical character. In these cases, potent, individual antimicrobial peptides can be readily identified using the iterative deconvolution process. Potent individual compounds can also be identified from screening PS-SCLs composed of longer, conformationally-defined peptides (31, 32), peptidomimetics (33), and small organic compounds (34), which are less flexible than peptides.

Protocol 8

Screening of a PS-SCL for identification of antibacterial compounds

Equipment and reagents

- 96-well flat-bottom microtitre plates (Costar)
- Microplate spectrophotometer
- Tetrapeptidomimetic PS-SCL
- *S. aureus* cells (ATCC 29213) maintained in MH/19% glycerol at $-70\,°C$
- Muller-Hinton (MH) broth

Method

1 Grow *S. aureus* cells in MH broth overnight at $37\,°C$.

2 Inoculate 0.2 ml in 10 ml $2 \times$ MH broth and incubate for 2 h at $37\,°C$.

3 Dilute 60-fold in $2 \times$ MH broth to $1\text{–}5 \times 10^5$ c.f.u./ml (exponential growth state) and add 0.1 ml to each well of microtitre plate.

4 Add each mixture of PS-SCL to a well (0.1 ml at 2.5 mg/ml final concentration) and incubate at $37\,°C$ for 21 h. Typically each mixture is assayed in duplicate. Leave at least one well without mixture for 100% growth (positive control) and one well

with MH broth only for 0% growth (negative control). At least one known antibiotic is also included as a control compound.

5 Read plates on microplate spectrophotometer for optical density at 620 nm at time 0 and time 21 h.

6 Calculate percentage of bacterial growth for each peptide mixture relative to the controls.

7 Test mixtures having activities lower than 50% growth at lower concentrations of the PS-SCL, preferably using twofold dilutions, to determine the minimum inhibitory concentration (MIC) or the concentration that inhibits 50% growth (IC_{50}) for each active mixture.

8 Based on their activities (MIC or IC_{50}) and the differences in chemical character of the defined building block, select the building blocks in the defined positions of the most active mixtures. Synthesize individual tetrapeptidomimetics that result from the combination of these building blocks. Test them in the above assay to determine the activities of the individual compounds.

7 Library screening using a radioreceptor assay

SCLs have also been used in a number of different radioreceptor assays for the identification of novel agonist and antagonist compounds (35–40). Since the screening of SCLs requires a high-throughput assay, a 96-well or tube format is highly recommended. Existing radioreceptor assay protocols can be adapted to the 96-well format, however, control experiments may be required to optimize the assay conditions in the new format. There are many methods for the separation of bound from free radioligand, but due to the high sample through-put of library screening, filtration is the method of choice. Also, as stated in the ELISA protocol, it is important that there is good separation between total binding and non-specific binding (> fivefold) and little variation between replicates (1–5%). The following protocol describes the screening of a tetrapeptide PS-SCL for the identification of ligands that inhibit the binding of a radioligand to either membrane bound or soluble receptors.

Protocol 9

Screening of a PS-SCL in a radioreceptor assay

Equipment and reagents

- 1 ml polypropylene tubes in 96-tube rack or 96-well microtitre plate
- Harvester and glass fibre filters
- Counter: liquid scintillation or gamma depending on isotope
- Tetrapeptide PS-SCL
- Receptor preparation
- Radioligand for specific receptor

Method

1   Prepare receptor solution. For membrane-bound receptors, protein content of crude homogenates should be determined using standard methods (Bradford or Lowry). For soluble receptors, it is necessary to separate the bound radioactivity from the free radioactivity in solution by the addition of an adsorbing material, such as hydroxylapatite. The adsorbing material can then be harvested onto filters, washed, and counted.

2   Screening of PS-SCL at one concentration requires at least two replicates for each mixture. Add peptide mixture (50 μl of 5 mg/ml) and predetermined volumes of buffer, radioligand, and receptor preparation to each tube. Reserve one column of each rack for a standard curve, in which radioligand is incubated in the presence of a range of concentrations of a standard unlabelled ligand. Mix all reagents and incubate assay tubes under conditions previously determined in optimizing experiments.

3   Terminate the reaction by filtration through filters on a harvester. For soluble receptors, add an adsorbing material (i.e. hydroxylapatite) prior to harvesting. Wash each sample several times with assay buffer.

4   Measure bound radioactivity on counter.

5   Average replicates and express screening data as per cent bound:

$$100 - \frac{(\text{sample mean}) - (\text{non-specific binding})}{(\text{total binding}) - (\text{non-specific binding})} \times 100$$

If many mixtures are active, screen the PS-SCL again at a lower final concentration. If only a few mixtures are active, test them using serial dilutions.

6   Calculate IC_{50} values from dilution experiments using curve-fitting software. Rank order the most active mixtures at each position to select the amino acids for individual peptide synthesis.

7   Synthesize individual tetrapeptides that result from the combination of these amino acids. Test them in the above assay to determine activities.

8   Identification of enzyme inhibitors using PS-SCLs

SCLs have been screened in various enzyme inhibition assays involving proteolytic (41–43) and glycolytic enzymes (44). The following protocol describes an α-glucosidase inhibition assay, which was developed and optimized for high throughput performance in the 96-well plate format by modifying a previously described single compound assay.

Protocol 10

Screening of a PS-SCL in an α-glucosidase inhibition assay

Equipment and reagents

- Microplate spectrophotometer
- 96-well flat-bottom microtitre plates (Costar)
- PS-SCL
- 0.1 M potassium phosphate buffer pH 6.8

- α-Glucosidase from bakers yeast (4.5 U/mg, 0.03 mg/ml buffer) (Sigma)
- *p*-Nitrophenyl-α-D-glucopyranoside (1.5 mg/ml buffer) (Sigma)

Method

1 Incubate the library mixtures (50 μl, 5–10 mg/ml) with 50 μl buffer and 50 μl α-glucosidase solution in the wells of the assay plates for 30 min at room temperature.

2 Add 50 μl *p*-nitrophenyl-α-D-glucopyranoside solution to each well, incubate for 30 min at 37 °C.

3 Centrifuge plates for 3 min at 2500 r.p.m., transfer 100 μl of the supernatant to another plate, read absorbance at 405 nm.

4 Calculate [% inhibition] for each mixture using the following formula:

$$\% \text{ inhibition} = \frac{OD_{sample} - OD_{blank}}{OD_{control} - OD_{blank}} \times 100,$$

in which 'control' is a sample in which both the library sample and enzyme are replaced with buffer.

5 Determine dose–response curves and calculate IC_{50} values using serial dilutions for all mixtures with > 50% inhibition at the initial screening concentration.

6 Identify most effective building blocks at each library position, which are identical to the building blocks at the defined position in each of the positional SCLs.

7 Synthesize and test individual compounds representing all possible combinations of the most effective building blocks at the respective positions.

References

1. Houghten, R. A., Pinilla, C., Blondelle, S. E., Appel, J. R., Dooley, C. T., and Cuervo, J. H. (1991). *Nature*, **354**, 84.

2. Pinilla, C., Appel, J., Blondelle, S. E., Dooley, C. T., Dörner, B., Eichler, J., *et al.* (1995). *Biopolymers (Peptide Science)*, **37**, 221.

3. Houghten, R. A. (1985). *Proc. Natl. Acad. Sci. USA*, **82**, 5131.

4. Ostresh, J. M., Husar, G. M., Blondelle, S. E., Dörner, B., Weber, P. A., and Houghten, R. A. (1994). *Proc. Natl. Acad. Sci. USA*, **91**, 11138.

5. Dörner, B., Ostresh, J. M., Blondelle, S. E., Dooley, C. T., and Houghten, R. A. (1998). In *Advances in amino acid mimetics and peptidomimetics* (ed. A. Abell), p. 109. JAI Press, Greenwich.

6. Nefzi, A., Ostresh, J. M., and Houghten, R. A. (1997). *Chem. Rev.*, **97**, 449.

7. Nefzi, A., Ostresh, J. M., Meyer, J.-P., and Houghten, R. A. (1997). *Tetrahedron Lett.*, **38**, 931.

8. Nefzi, A., Ostresh, J. M., and Houghten, R. A. (1997). *Tetrahedron Lett.*, **38**, 4943.

9. Ostresh, J. M., Schoner, C. C., Hamashin, V. T., Meyer, J.-P., and Houghten, R. A. (1998). *J. Org. Chem.*,

10. Dooley, C. T., Chung, N. N., Schiller, P. W., and Houghten, R. A. (1993). *Proc. Natl. Acad. Sci. USA*, **90**, 10811.

11. Dooley, C. T., Chung, N. N., Wilkes, B. C., Schiller, P. W., Bidlack, J. M., Pasternak, G. W., *et al.* (1994). *Science*, **266**, 2019.

12. Pinilla, C., Appel, J. R., Blanc, P., and Houghten, R. A. (1992). *Biotechniques*, **13**, 901.

13. Dooley, C. T. and Houghten, R. A. (1993). *Life Sci.*, **52**, 1509.

14. Pinilla, C., Appel, J. R., Blondelle, S. E., Dooley, C. T., Eichler, J., Ostresh, J. M., *et al.* (1994). *Drug Dev. Res.*, **33**, 133.

15. Ostresh, J. M., Winkle, J. H., Hamashin, V. T., and Houghten, R. A. (1994). *Biopolymers*, **34**, 1681.

16. Houghten, R. A., Bray, M. K., DeGraw, S. T., and Kirby, C. J. (1986). *Int. J. Peptide Protein Res.*, **27**, 673.

17. Stewart, J. M. and Young, J. D. (1984). *Solid phase peptide synthesis*. Pierce Chemical Company, Rockford, Illinois, USA.

18. Dörner, B., Husar, G. M., Ostresh, J. M., and Houghten, R. A. (1996). *Bioorg. Med. Chem.*, **4**, 709.

19. Dörner, B., Ostresh, J. M., Husar, G. M., and Houghten, R. A. (1996). *Methods Mol. Cell. Biol.*, **6**, 35.

20. Nefzi, A., Dooley, C. T., Ostresh, J. M., and Houghten, R. A. (1998). *BioMed. Chem. Lett.*, **8**, 2273.

21. Pinilla, C., Appel, J. R., Campbell, G. D., Buencamino, J., Benkirane, N., Muller, S., *et al.* (1998). *J. Mol. Biol.*, **283**, 1013.

22. Appel, J. R., Buencamino, J., Houghten, R. A., and Pinilla, C. (1996). *Mol. Div.*, **2**, 29.

23. Pinilla, C., Appel, J. R., and Houghten, R. A. (1994). In *Current protocols in immunology* (ed. J. E. Coligan, A. M. Kruisbeek, D. H. Margulies, E. M. Shevach, and W. Strober), p. 9.8.1. John Wiley & Sons, Inc., New York.

24. Hemmer, B., Vergelli, M., Gran, B., Ling, N., Conlon, P., Pinilla, C., *et al.* (1998). *J. Immunol.*, **160**, 3631.

25. Hemmer, B., Vergelli, M., Pinilla, C., Houghten, R., and Martin, R. (1998). *Immunol. Today*, **19**, 163.

26. Blondelle, S. E. and Houghten, R. A. (1996). *Methods Mol. Cell. Biol.*, **6**, 26.

27. Blondelle, S. E., Pérez-Payá, E., and Houghten, R. A. (1996). *Antimicrob. Agents Chemother.*, **40**, 1067.

28. Lutzke, R. A. P., Eppens, N. A., Weber, P. A., Houghten, R. A., and Plasterk, R. H. (1995). *Proc. Natl. Acad. Sci. USA*, **92**, 11456.

29. Blondelle, S. E. and Houghten, R. A. (1999). In *Drug Discovery from Nature* (ed. S. Grabley and R. Thierucke) p. 311. Springer, Berlin.

30. Blondelle, S. E., Takahashi, E., Weber, P. A., and Houghten, R. A. (1994). *Antimicrob. Agents Chemother.*, **38**, 2280.

31. Pérez-Payá, E., Houghten, R. A., and Blondelle, S. E. (1996). *J. Biol. Chem.*, **271**, 4120.

32. Blondelle, S. E., Takahashi, E., Houghten, R. A., and Pérez-Payá, E. (1996). *Biochem. J.*, **313**, 141.

33. Blondelle, S. E., Pérez-Payá, E., Dooley, C. T., Pinilla, C., and Houghten, R. A. (1995). *Trends Anal. Chem.*, **14**, 83.

34. Blondelle, S. E., Nefzi, A., Ostresh, J. M., and Houghten, R. A. (1998). *Pure Appl. Chem.*, **70**, 2141.

35. Dooley, C. T. and Houghten, R. A. (1993). *Life Sci.*, **52**, 1509.

36. Dooley, C. T., Chung, N. N., Schiller, P. W., and Houghten, R. A. (1993). *Proc. Natl. Acad. Sci. USA*, **90**, 10811.

37. Dooley, C. T., Hope, S., and Houghten, R. A. (1994). *Regul. Peptide*, **54**, 87.

38. Dooley, C. T., Chung, N. N., Wilkes, B. C., Schiller, P. W., Bidlack, J. M., Pasternak, G. W., *et al.* (1994). *Science*, **266**, 2019.

39. Dooley, C. T. and Houghten, R. A. (1995). *Analgesia*, **1**, 400.

40. Dooley, C. T., Ny, P., Bidlack, J. M., and Houghten, R. A. (1998). *J. Biol. Chem.*, **273**, 18848.

41. Eichler, J. and Houghten, R. A. (1993). *Biochemistry*, **32**, 11035.

42. Eichler, J., Lucka, A. W., and Houghten, R. A. (1994). *Peptide Res.*, **7**, 300.

43. Apletalina, E., Appel, J. R., Lamango, N. S., Houghten, R. A., and Lindberg, I. (1998). *J. Biol. Chem.*, **273**, 26589.

44. Eichler, J., Lucka, A. W., Pinilla, C., and Houghten, R. A. (1996). *Mol. Div.*, **1**, 233.

Chapter 4

High-throughput combinatorial synthesis of discrete compounds in multimilligram quantities: non-chemical encoding and directed sorting

Xiao-yi Xiao* and K. C. Nicolaou†

*IRORI, 9640 Town Centre Drive, San Diego, CA 92121, USA.

†Department of Chemistry and the Skaggs Institute for Chemical Biology, The Scripps Research Institute, La Jolla, CA 92037, USA.

1 Introduction

Over the past few years the pharmaceutical community's need to generate and screen chemical libraries with large diversity has fuelled the rapid development of combinatorial chemistry (1, 2). Research areas such as high-throughput solution (3, 4) and solid phase (5–7) synthesis, the development of novel solid supports and linkers (Chapters 6 and 7) (8, 9), as well as new powerful encoding strategies (10–14) (Chapter 1) have received exceptional attention due to their central role in the development of this field. The chemical encoding strategies (15, 16), coupled with the split and pool technique (17) (Chapters 1–3) greatly increase the efficiency of combinatorial synthesis and screening. However, these strategies produce each member of the library in very small amounts (nanograms to micrograms), making subsequent multiple screenings and characterization procedures quite difficult. Although traditional parallel synthesis (18, 19) in combination with the remarkable recent progress in laboratory automation (19) produces larger quantities of material, it still lacks the efficiency of the split and pool technique particularly for the preparation of very large libraries. We have recently developed two new combinatorial synthesis strategies employing non-chemical encoding and directed sorting in a batch synthesis fashion (10, 11) that combine both the advantages of split and pool and parallel synthesis. These strategies are capable of producing large combinatorial libraries and delivering milligram quantities of each library member. This chapter discusses the development of these two non-chemical encoding techniques: radiofrequency (R_f) tagging (10) and bar coding (11) in combination with a directed sorting strategy, and their application to the synthesis of combinatorial libraries of a variety of chemical structures.

2 Non-chemical encoding and directed sorting

By giving each reaction unit (i.e. a microreactor) a non-chemically identifiable 'tag', which can be either an R_f tag (10) or a bar code (11), a group of these reaction units can be processed in a *batch* fashion along the synthetic pathway, while the identity and thus the (predetermined) destiny of each microreactor at any point of the synthesis is tracked by a synthesis manager software. Using a new concept termed *directed sorting* (11), the number of microreactors required to make all the possible combinations in the combinatorial matrix is maintained equal to the number of compounds in the library being synthesized, whereas in traditional *random* split and pool synthesis, a larger number of reaction units (i.e. resin beads in this case) have to be used to ensure a comfortable degree of confidence that all possible combinations are produced at the end of the library synthesis (20).

Figure 1 schematically illustrates how the strategy of non-chemical encoding and directed sorting actually works in a combinatorial synthesis with two diversity steps, the first with three building blocks (A, B, C) and the second with two (A, B). Precisely, six microreactors are used and each is identified with a

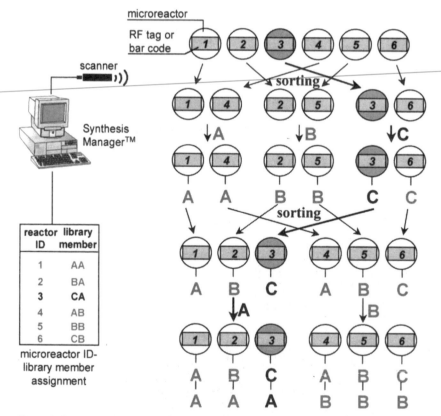

Figure 1 Directed sorting in combinatorial synthesis using microreactors and non-chemical encoding.

unique R_f tag or bar code (i.e. one through six for brevity). A software, *Synthesis Manager*, assigns each library member to one microreactor, which are then *sorted* into three reaction vessels for the first step as *directed* by the predetermined assignments (*directed sorting*). The building blocks (A, B, C) are added and the reactions are completed. The microreactors can then be pooled together and processed in one batch through common treatments like washing, drying, and any non-diversification step. A second directed sorting is then used to distribute the microreactors into two reaction vessels for the second diversity step and the addition of building blocks (A, B) is completed accordingly. At the end of the synthesis, six individual microreactors each containing one of the six library members are obtained. The microreactors can then be subjected to regular cleavage to afford six discrete compounds.

2.1 Radiofrequency tags

A radiofrequency (R_f) tag (*Figure 2*) in this application is a microelectronic device the size of a 'flea' stir bar (3×13 mm) capable of receiving and emitting radio-frequency signals from a short distance (centimetre range). It consists mainly of an antenna and a microelectronic chip encased in a glass housing. Each chip has:

(a) A unique, non-volatile 40-bit ID code permanently etched onto its memory area that allows for the encoding of up to 2^{40} unique chemical events.

(b) A transmitter/receiver circuitry.

(c) A rectifier.

(d) Some logic control circuits.

When such an R_f tag is placed in the R_f field of a transceiver's antenna, the tag's antenna picks up the R_f energy, and the rectifier converts this AC energy into

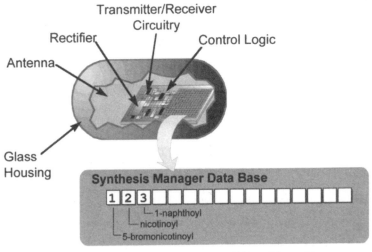

Figure 2 A radiofrequency tap. Actual size: 3 mm (OD) $\times$ 13 mm (L).

DC energy. The DC energy powers up the entire chip, which immediately reports its ID to the transceiver by means of an R_f transmission.

2.2 Radiofrequency tagged MicroKans

In order to effectively use R_f tags in combinatorial chemistry, the synthesis supports (solid phase resin beads in this case) have to be reliably associated with the R_f tags. One way to achieve this association is by using a microcapsule (MicroKan) in which an R_f tag and the resin beads are contained behind a meshed wall made of a chemically inert material such as polypropylene or fluoropolymer (*Figure 3*). The mesh size (70 μm) is such that the resin beads (larger than 75 μm in diameter) do not leak out, but solvents and reagents can freely flow through. A total of 25–30 mg of polystyrene resin can be placed in each MicroKan in addition to an R_f tag and still leave enough void space for the resin to freely swell during the chemical reactions. For a general loading level of 1 μmol/g, each MicroKan can then produce more than 25 μmol of each compound. In practice, any resin can be used provided its bead size is larger than that of the pores of the MicroKan's mesh.

2.3 Radiofrequency tagged MicroTubes

A second way to associate an R_f tag with a synthesis support is using MicroTubes (*Figure 4*), wherein an R_f tag is enclosed in an inert plastic tube (polypropylene or fluoropolymer) on which a layer of polystyrene has been radiolitically grafted (21) and chemically functionalized (22, 23). The amount of polystyrene graft in each MicroTube is about 40 mg. With a functionalization level of one styrene every ten units, each MicroTube can deliver a loading of about 30 μmol. One obvious advantage of the MicroTubes is that solvent and reagent exchange during reactions and washings is direct and effective.

2.4 2D bar codes and bar coded microreactors

Besides radiofrequency tags, bar codes can also be utilized as a means of non-chemical encoding in combinatorial synthesis (11). A miniature 2D ceramic bar

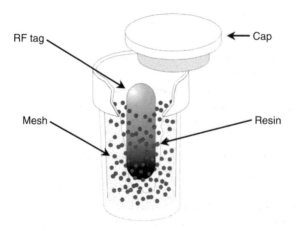

Figure 3 Radiofrequency tagged Microkan reactor. Actual size: 7 mm (OD) × 18 mm (L).

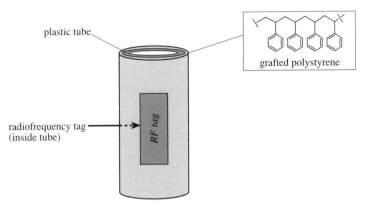

plastic tube

grafted polystyrene

radiofrequency tag
(inside tube)

RF tag

Figure 4 A radiofrequency tagged MicroTube reactor. Actual size: 6 mm (OD) × 15 mm (L).

coding methodology (3 × 3 mm) has been developed using CO_2 laser etching on a ceramic plate (*Figure 5*). The surrounding synthesis support is a chemically inert polypropylene or fluoropolymer chip grafted with polystyrene using a similar radiolitic process as with the MicroTubes (Section 2.3). The resulting bar coded microreactor has a loading of 8–10 μmol (*Figure 5*). A *Synthesis Manager* software equipped with pattern recognition capability is then used to recognize the 2D bar codes and direct the sorting of the microreactors during a combinatorial synthesis. The advantage of using bar codes is that they cost considerably less than R_f tags and could be readily miniaturized to sizes much smaller than 3 × 3 mm. Furthermore, they make the construction of much larger libraries possible and cost-effective.

3 Applications in combinatorial synthesis

3.1 R_f tagged MicroKans: synthesis of taxoid, epothilone, and muscone libraries

The application of R_f tagged MicroKans for the synthesis of small peptidic and semi-peptidic libraries (10, 12) will not be discussed in this chapter. The first application of the R_f encoding and MicroKan microreactor strategy in small molecule combinatorial synthesis was reported by our group in 1997 for the synthesis of a 400-member taxoids library (24) and a 45-member epothilones library (25). Taxol is a newly approved anticancer agent which is particularly effective against ovarian and breast cancer (26), and the recently discovered natural products epothilones possess Taxol-like mechanism of action against tumour cells through induction of tubulin assembly and microtubule stabilization (27). More importantly, it has been reported that the latter class of molecules has unique anticancer properties against *Taxol-resistant* tumour cell lines (28). The third example discussed in this section deals with the application of R_f tagged MicroKans to the synthesis of a (DL)-muscone (29, 30) library (31).

Application of combinatorial chemistry techniques to the Taxol and

2-D bar code on ceramic plate

Polystyrene grafted synthesis support

A. The Bar Coded Microreactor (10 x 10 x 2 mm)

Ceramic plate

Polymeric synthesis support

polystyrene surface graft

B. Cross-section of the Bar Coded Microreactor

Figure 5 A 2D bar coded microreactor.

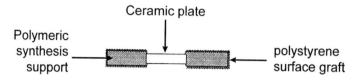

6: taxoid library

R^1 =

A B C D E F

G I J K L M N

O P Q R S T U

R^2 =

E H O Q T U

(Building blocks **A, C, D, F, G, I, J, K, L, M, N, P, R, S** for R^2 are the same as for R^1.)

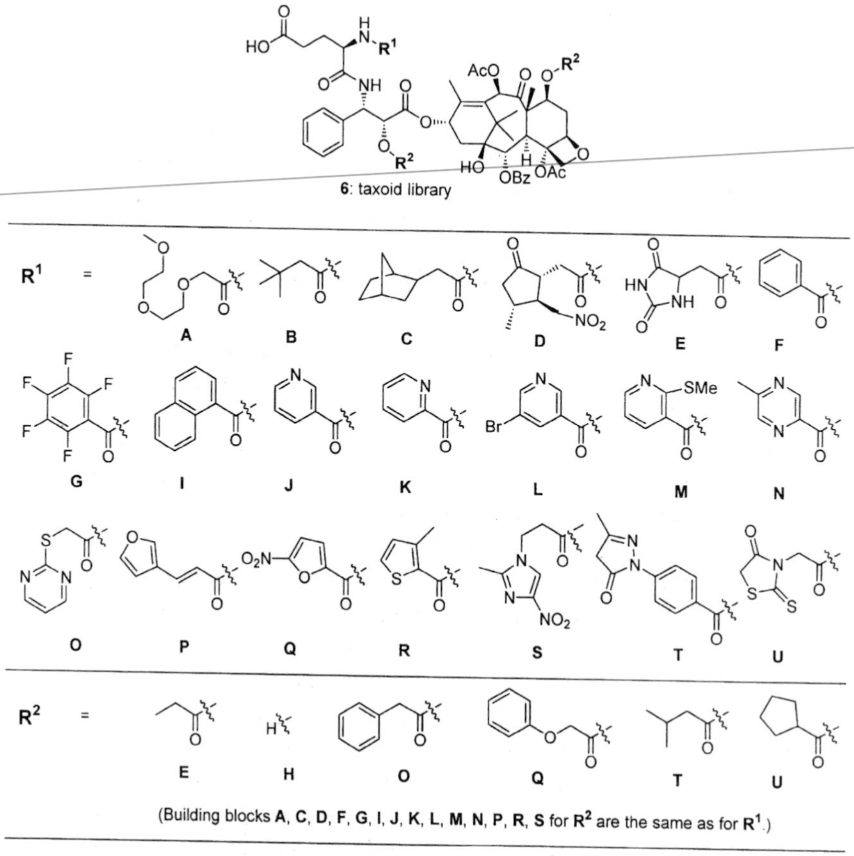

Figure 6 Building blocks used in the synthesis of the taxoid library.

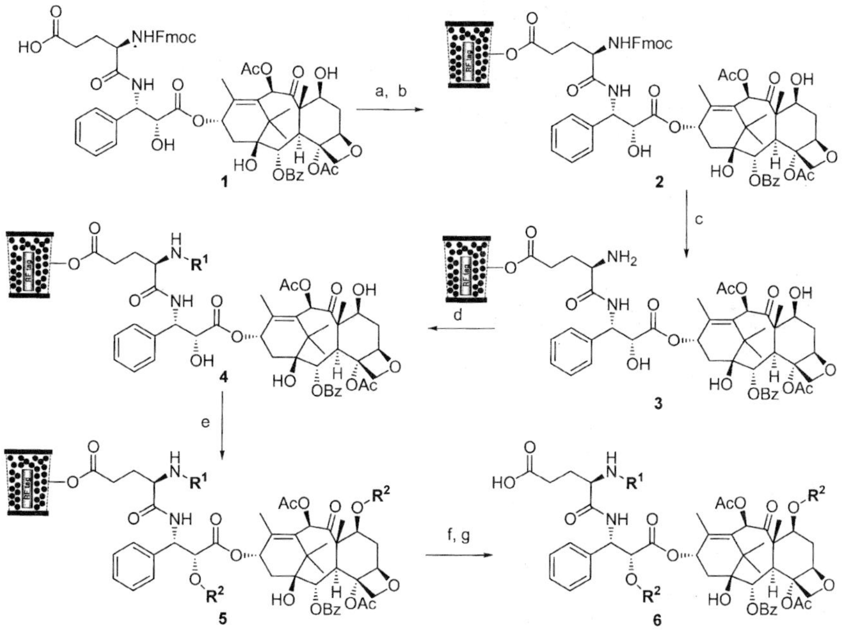

Figure 7 Synthesis of a taxoid library using R_f tagged MicroKan reactors and directed sorting. (a) Excess 2-chlorotrityl resin (8.0 g, 10 Equiv.), DIEA (20 Equiv.), DCM, r.t., 3 h; then MeOH, r.t., 0.5 h. (b) Resin is distributed into 400 microreactors. (c) 5% Piperidine in DMF, r.t., 0.5 h. (d) Microreactors are split into 20 equal groups and each group is treated with a carboxylic acid (R^1, 35 Equiv.), DIEA (70 Equiv.), PyBOP (35 Equiv.), DMF, r.t., 4 h. (e) Microreactors are re-split into 20 new groups and each group is treated with a carboxylic acid (R^2, 138 Equiv.), DIC (138 Equiv.), DMAP (120 Equiv.), DCM, r.t., 48 h. (f) Microreactors are decoded and distributed into 400 glass vials. (g) AcOH, DCM, CF_3CH_2OH, r.t., 4 h.

epothilone templates has the attractive potential of uncovering novel analogues with improved pharmaceutical properties, while the muscone synthesis may point to new directions in the cosmetics industry.

As depicted in *Figure 7*, a modified Taxol core **1** is immobilized onto 2-chloro-trityl resin and the loaded resin distributed into 400 MicroKans (loading: 2.9 μmol/MicroKan), each containing 25–30 mg of resin and an R_f tag (*Protocol 1A*). The 400 MicroKans **2** are then treated, in one single batch, with 5% piperidine in DMF to remove the Fmoc (fluorenylmethoxycarbonyl) protecting group on the glutamic acid handle. The MicroKans **3** are then thoroughly washed, dried under vacuum (*Protocol 1B*), and sorted into 20 groups as directed by *Synthesis Manager* (*Protocol 1C*). Each group of MicroKans is coupled with the corresponding carboxylic acid (R^1, *Figure 6*) to form the resin-bound amides **4** (*Protocols 1D and 1E*). After coupling, the MicroKans are pooled together, washed thoroughly, and dried under vacuum. The MicroKans are re-sorted into 20 new groups according to *Synthesis Manager* and coupled with the second set of 20 carboxylic acids (R^2, *Figure 6*) to form the diesters **5**. After pooling, washing, and drying, each MicroKan is separately subjected to mild acidic cleavage to yield

400 discrete taxoids **6** (20 × 20). The identity (or synthetic history) of each compound recorded in *Synthesis Manager* is read before the dissociation of the compounds from the R_f tags (*Protocol 1F*). Each taxoid sample is obtained in 2–4 mg quantity and with 50–100% purity as indicated by TLC (thin-layer chromatography) and HPLC (high performance liquid chromatography) analysis. The structural integrity of the library was assessed by ^{1}H NMR (proton nuclear magnetic resonance) spectroscopy on 20 randomly selected members (5% of the library), which confirmed the expected taxoid structures.

Protocol 1

Application of R_f encoded MicroKans in combinatorial synthesis

Equipment and reagents

- MicroKan (Irori)
- Radiofrenquency tags (Irori)
- MicroKan Caper (Irori)
- Automatic sorting machine (e.g. AutoSort-10 K, Irori)
- A cleavage device to release compound from microreactors (AccuCleaver, Irori) is recommended when a large number of microreactors are processed simultaneously)
- *Synthesis Manager* software and computer workstation
- Orbital shaker and stirrer
- Glassware for chemical synthesis

- 96-deep-well microplates
- Vacuum/argon line
- Resin for solid phase synthesis
- DCM (dichloromethane)
- TFA (trifluoroacetic acid)
- Hexane
- MeOH (methanol)
- DIEA (diisopropylethylamine)
- DMSO (dimethyl sulfoxide)
- DMF (dimethyformamide)
- PyBop (benzotriazol-1-yl-oxy-tris-pyrrolidinophosphonium hexafluorophosphate)

A. Loading the MicroKans
Dry loading

1 Take an empty MicroKan on a scale and weigh 25–30 mg of dry resin directly into the MicroKan.

2 Add an R_f tag.

3 Cap the reactors using a MicroKan Caper.

Wet loading[a]

1 Place 25 mg of resin in a 10 ml graduated cylinder.

2 Add 5 ml of DCM. The resin usually floats to the top.

3 Slowly add hexane while constantly mixing the resin with the solvents until an equal density point is reached (when the resin suspends). Record the amount of hexane needed.

4 Weigh the desired amount of resin for the total number of MicroKans to be filled into an Erlenmeyer flask and add a stirring bar.

5 Mix DCM and hexane according to the ratio established in steps 1–3.

6 Calculate the total amount of solvent needed to make a suspension so that 0.5 ml of the mixture contains 25 mg of resin. Add this amount of the solvent prepared in step 5 onto the resin and mix well by gentle, continuous stirring.

7 Set open MicroKans in the wells of a 96-deep-well microplate.

8 Drop an R_f tag in each MicroKan.

9 Pipette 0.5 ml of the resin suspension prepared in step 6 into each MicroKan. The solvent quickly drains to the bottom of the microplate wells.

10 Cap the MicroKans with a MicroKan Caper.

11 Dry the MicroKans under high vacuum.

B. Washing and drying microreactors (MicroKans, MicroTubes, or bar coded microreactors)

1 Place the microreactors in a glass vessel.

2 Add 1–1.5 ml/microreactor of dichloromethane (or other resin swelling solvents), and close the glass vessel.

3 Shake the glass vessel on an orbital shaker at 200 r.p.m. for 10 min.

4 Remove the solvent.

5 Add 1–1.5 ml/microreactor of methanol (or other resin shrinking solvents) and close the vessel.

6 Shake the vessel on an orbital shaker at 200 r.p.m. for 10 min.

7 Remove the solvent.

8 Repeat step 2–7 for at least four cycles.

9 Dry the microreactors under high vacuum.

C. Sorting microreactors as directed by Synthesis Manager[b]

1 Enter the synthetic route information about the number of diversity steps, and the number and identities of building blocks to be used in each diversity step into *Synthesis Manager*. The latter will automatically generate the library matrix.

2 Scan the microreactors one by one and allow the software to direct in which reaction vessel to place each microreactor. Continue until all the microreactors are sorted and placed in their designated reaction vessels.

3 The microreactors are ready to undergo the designated reaction (*Protocol 1D–F*).

D. Removing air bubbles from MicroKans

1 Place the MicroKans in a reaction vessel.

2 Add the desired solvent(s) and seal the vessel. There are usually air bubbles trapped

Protocol 1 continued

inside the MicroKans. These air bubbles slow the solvent/reagent exchange across the mesh and should be removed before starting any reaction.

3 Apply a moderate vacuum (10–20 mm Hg) briefly (2–5 sec). Immediately release the vacuum inside the reaction vessel with dry argon. The air bubbles will disappear and the reaction can proceed normally.

E. Performing an amide formation reaction in a MicroKan

1 Place a group of sorted MicroKans containing amine resin (loading: 2.9 μmol/MicroKan) in a glass reaction vessel.

2 Add anhydrous DMF (1 ml/MicroKan).

3 Add 2 Equiv. of a carboxylic acid with a final concentration of 0.1 M.

4 Add 4 Equiv. DIEA to a final concentration of 0.2 M.

5 Add 2 Equiv. PyBop to a final concentration of 0.1 M.

6 Remove the air bubbles inside the MicroKans (part D) and seal the glass vessel.

7 Shake the reaction vessel on an orbital shaker (200 r.p.m.) at room temperature for 2 h.

8 Add sequentially extra carboxylic acid, DIEA, and PyBop (half the amounts of steps 3–5).

9 Shake the reaction vessel (200 r.p.m.) at room temperature for another 2 h.

10 Add methanol (5 ml) to quench the reaction and remove the solution.

11 Pool the MicroKans together with other groups of MicroKans, then wash and dry as in part B.

F. Releasing compounds from MicroKans with TFA[c]

1 Place a microreactor in a 4 ml glass vial.

2 Add TFA (2 ml) into the vial and seal with a Teflon-lined screw cap.

3 Shake the vial on an orbital shaker (200 r.p.m.) at room temperature for the desired length of time.

4 Collect the solution and filter through a glass-wool plug into a 6 ml glass vial.

5 Scan the ID (RF code or 2D bar code) of the microreactor and label the glass vial in step 4 with this ID.

6 Add fresh TFA (1 ml) to the vial with the microreactor. Close the vial and shake briefly.

7 Collect the solution, filter, and combine with the solution of step 4.

8 Concentrate the solutions and dry under high vacuum.

[a] Moisture-sensitive resin like trityl chloride resin is best loaded by the dry loading method.

[b] An automatic sorting machine (AutoSort-10 K) is recommended when a large number of microreactors are used. It can automatically sort 1000 microreactors/hour.

[c] The use of a cleavage device (AccuCleaver) is recommended when a large number of microreactors are cleaved. AccuCleaver can process 96 microreactors simultaneously.

The synthesis of an epothilone library starts by a chain extension of Merri-field resin contained in MicroKans with 1,4-butanediol, followed by conversion of the free alcohol to the resin-bound phosphonium salt via the corresponding iodide (**8**, *Figure 9*). The MicroKans are sorted into three groups and a Wittig reaction with one of the three aldehyde fragments bearing R^1 (A, *Figure 8*) is performed to form the alkenes **9** (*Protocol 2*). In a single batch, the TBS (t-butyldimethylsilyl) ethers are deprotected and the resulting alcohols are oxidized to the corresponding aldehydes **10**. MicroKans sorting, followed by enolate addition to the aldehydes with three keto acid fragments bearing R^2 (B, *Figure 8*) yield the hydroxyacids **11**, which are coupled, after sorting, with five alcohol fragments bearing R^3 (C, *Figure 8*) to form the resin-bound dienes **12**. Macro-cyclization by olefin metathesis (32) and simultaneous cleavage of the epothi-lones **13** from the MicroKans produce 45 epothilone samples ($3 \times 3 \times 5$). From each sample four geometric/diastereomeric isomers are identified and separated. Significant structure–activity relationship information has been revealed by the biological evaluation of these epothilone analogues (33).

The synthesis of a (DL)-muscone library (**19**) starts with the phosphonate resin **14** (34) in MicroKans (*Figure 11*), which is treated with *n*-BuLi, then coupled with olefinic esters A (*Figure 11*) in THF to afford MicroKans **15**. Cross olefin meta-thesis (35) of **15** with olefinic alcohols B in the presence of $(PCy_3)_2Ru(=CHPh)Cl_2$ catalyst yields internal alkenes **16** (*Protocol 3*), which are oxidized under Dess-

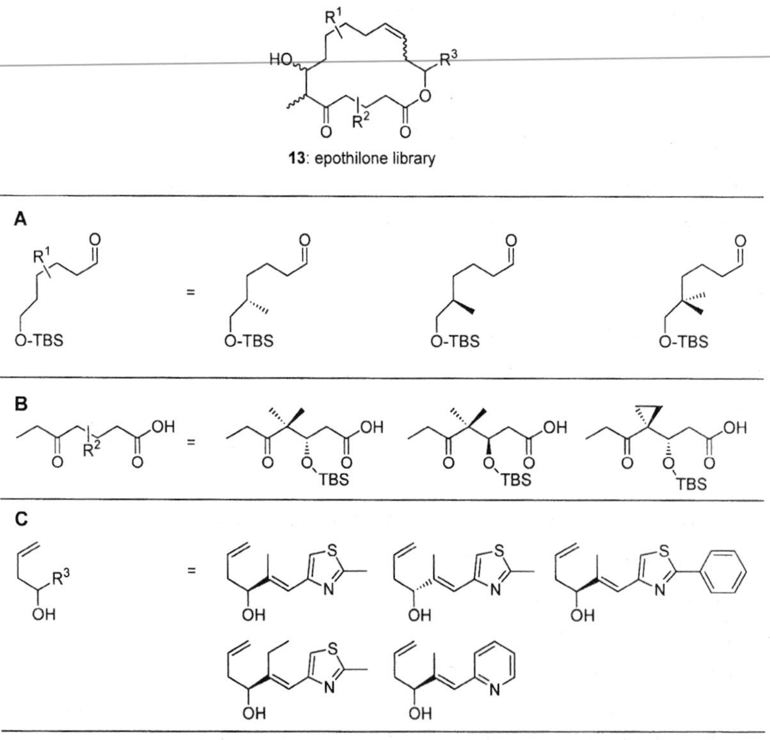

13: epothilone library

Figure 8 Building blocks used in the epothilone library.

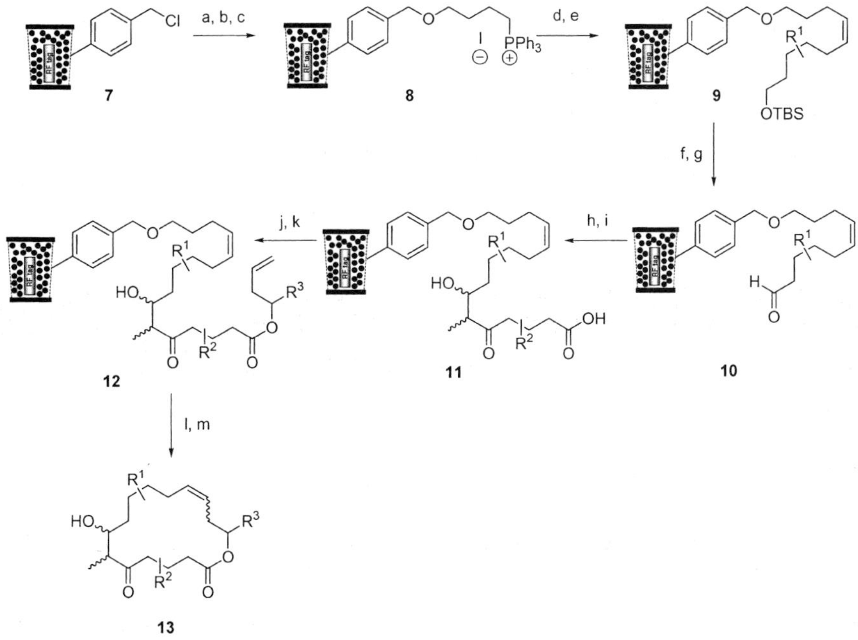

Figure 9 Synthesis of an epothilone library using R_f (rediofrequency) tagged MicroKan reactors and directed sorting. (a) 1,4-Butanediol (5 Equiv.), NaH (5 Equiv.), n-Bu$_4$NI (0.1 Equiv.), DMF, r.t., 12 h. (b) Ph$_3$P (4 Equiv.), I$_2$ (4 Equiv.), imidazole (4 Equiv.), DCM, r.t., 3 h. (c) Ph$_3$P (10 Equiv.), 90 °C, 12 h. (d) Sorting MicroKans. (e) NaHMDS (3 Equiv.), THF:DMSO (1:1), r.t., 12 h; then A (2 Equiv., see *Figure 8*), THF, 0 °C, 3 h. (f) Pooling MicroKans; then 0.2 M HCl in THF, r.t., 12 h. (g) (COCl)$_2$ (4 Equiv.), DMSO (8 Equiv.), Et$_3$N (12.5 Equiv.), −78 °C to r.t. (h) Sorting MicroKans. (i) ZnCl$_2$ (2 Equiv.) THF, enolates from B (*Figure 8*, by treating 2 Equiv. of B with 2.2 Equiv. of LDA in THF from −78 °C to −40 °C for 1 h), −78 °C to −40 °C, 2 h. (j) Sorting MicroKans. (k) C (*Figure 8*, 5 Equiv.), DCC (5 Equiv.), DMAP (5 Equiv.), r.t., 15 h. (l) Separation of individual MicroKans. (m) [RuCl$_2$(=CHPh) (PCy$_3$)$_2$] (0.2 Equiv.), DCM, r.t., 48 h.

Protocol 2

Performing reactions with MicroKans: Wittig alkene formation[a]

1 Place a group of MicroKans containing resin-bound phosphornium salt (0.3 mmol/g) in a reaction vessel.

2 Add anhydrous THF and DMSO (1:1, v/v, 1 ml/MicroKan) and remove the air bubbles according to *Protocol 1D*.

3 Add NaHMDS/THF (sodium hexamethyldisilazane in THF, 3 Equiv.) dropwise at room temperature.

4 Shake the reaction (300 r.p.m.) at room temperature for 12 h.

5 Remove the excess base via a cannula and wash the MicroKans with anhydrous THF (2 × 2 ml/MicroKan).

Protocol 2 continued

6 Add a THF solution of aldehyde fragment A (2 Equiv., *Figure 8*) dropwise at 0 °C.

7 Shake the reaction (300 r.p.m.) at 0 °C for 3 h.

8 Wash the MicroKans according to *Protocol 1B* and dry under vacuum.

[a] Step 'e' in *Figure 9*.

Protocol 3

Performing reactions with MicroKans: cross olefin metathesis[a]

1 Suspend a group of MicroKans with olefin resin (**15**, *Figure 11*) in benzene in a reaction vessel.

2 Add 5 Equiv. of terminal olefin (B, *Figure 10*), followed by the $(PCy_3)_2Ru(=CHPh)Cl_2$ catalyst (0.2 Equiv.).

3 Shake the reaction (300 r.p.m.) at room temperature for 48 h while sparging with a gentle flow of argon.

4 Remove the reaction solution, wash the MicroKans according to *Protocol 1B*, and dry under vacuum.

[a] Step 'd' in Figure 11.

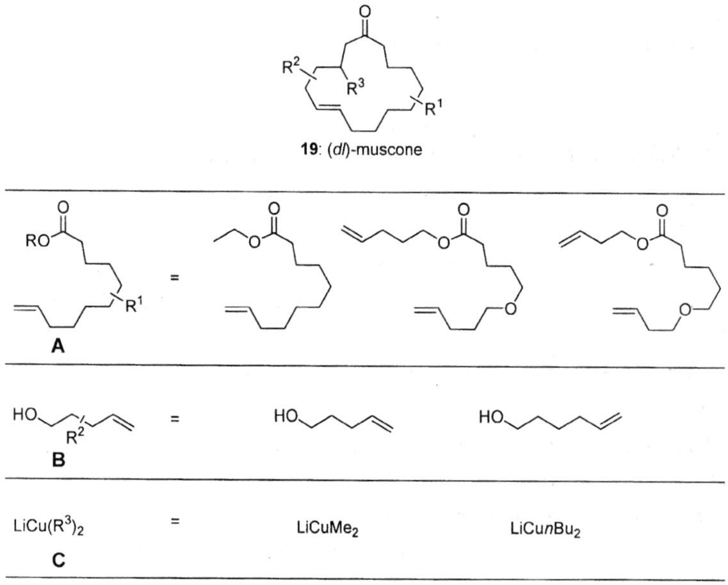

19: (*dl*)-muscone

Figure 10 Building blocks used in the (DL)-muscone library.

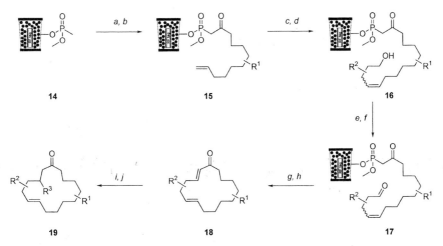

Figure 11 Synthesis of a (DL)-muscone library using R_f tagged MicroKan reactors and directed sorting. (a) Sorting MicroKans. (b) n-BuLi (1.2 Equiv.), THF, $-20\,^{\circ}$C, 10 min; then A (4 Equiv., see *Figure 10*), -20 to $25\,^{\circ}$C, 30 min. (c) Sorting MicroKans. (d) B (*Figure 10*, 5 Equiv.), $(PCy_3)_2Ru(=CHPh)Cl_2$ (0.2 Equiv.), benzene, $25\,^{\circ}$C, 48 h. (e) Pooling MicroKans. (f) Dess-Martin periodinane (1.5 Equiv.), DCM, $25\,^{\circ}$C, 12 h. (g) Sorting MicroKans. (h) K_2CO_3 (5 Equiv.), 18-crown-6 (5 Equiv.), benzene, $65\,^{\circ}$C, 12 h. (i) C (*Figure 10*, 1.2 Equiv.), Et_2O, $0\,^{\circ}$C, 1 h. (j) H_2, 5% Pd-C, MeOH, $25\,^{\circ}$C.

Martin conditions to the corresponding aldehydes **17**. Treatment of MicroKans **17** with K_2CO_3 and 18-crown-6 (36) in benzene at $65\,^{\circ}$C cycloreleases the desired macrocyclic α,β-unsaturated ketones **18** in solution with high purity and yields (35–65%), while the same macrocyclization in solution is accompanied by the formation of the undesired dimers in 10–29% yield (37). This is the result of the apparent high dilution effect induced by the solid support on the resin-bound compounds, which is beneficial to the macrocyclization step. Finally, Michael addition with cuprates C, followed by hydrogenation furnishes the 12 member ($3 \times 2 \times 2$) (DL)-muscone library **19**.

3.2 R_f tagged MicroTubes: synthesis of a tyrphostin library

Tyrphostins are a class of benzylidene malononitriles with potent inhibitory activity towards various protein tyrosine kinase (PTK) both *in vitro* and *in vivo* (38). A library strategy toward PTK inhibitors discovery has the potential to identify novel tyrphostins with a wide range of selectivities. *Figure 13* shows a synthetic route for a tyrphostin library using R_f tagged MicroTubes (39). Micro-Tubes with a Knorr linker are sorted into 18 groups and reductively alkylated with 18 aromatic aldehydes (*Figure 12*) affording thereby the MicroTube-bound secondary amines **21** (*Protocol 4*). They are then pooled, washed, dried, and acylated with cyanoacetic acid to give the cyanoamides **22**, which were condensed, after sorting into eight groups, with a set of eight aromatic hydroxyl aldehydes

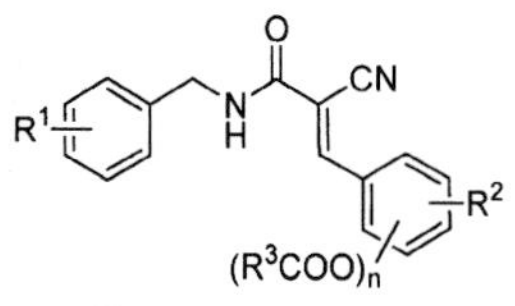

25: tyrphostin library

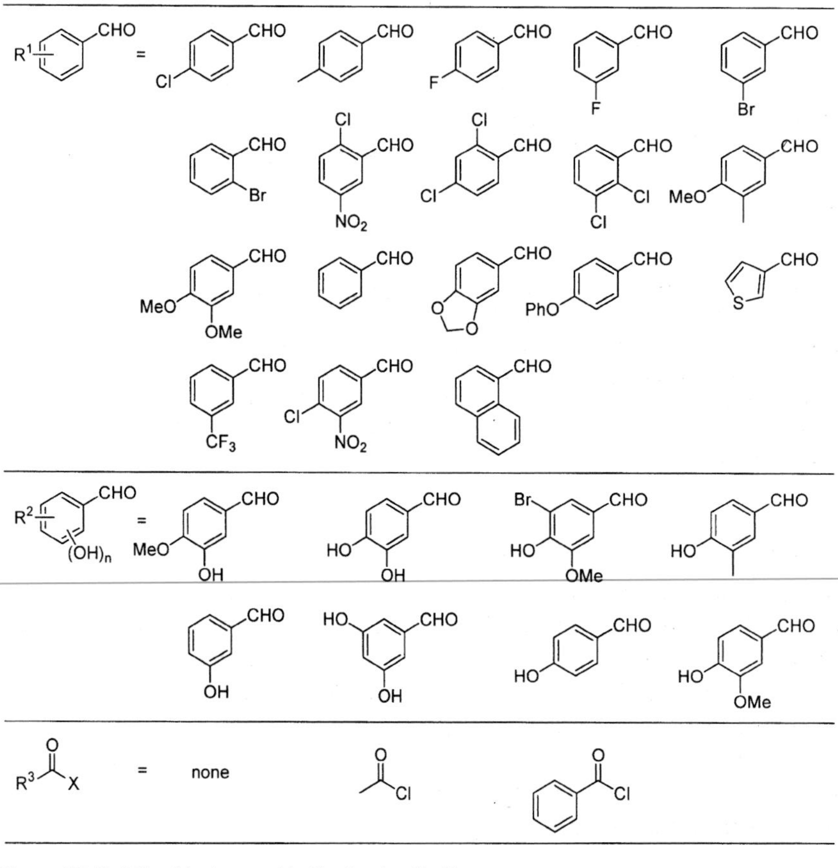

Figure 12 Building blocks used in the tyrphostin library.

to yield the MicroTube-bound phenols **23**. The MicroTubes are then sorted into three groups. One group is directly subjected to cleavage to afford the phenolic tyrphostins, while each of the remaining two groups is treated with an acyl chloride to afford the MicroTube-bound phenyl esters **24**, followed by cleavage to yield the esterified tyrphostins **25**. More than 90% of the 432 compounds synthesized (18 × 8 × 3) are obtained in 5–19 mg quantities. TLC analysis of the library shows that the majority of the members display one major spot with the expected R_f values. Twenty four randomly selected members of this library were analysed by ^1H NMR and MS (mass spectrometry), and both techniques confirmed that the desired structures were obtained.

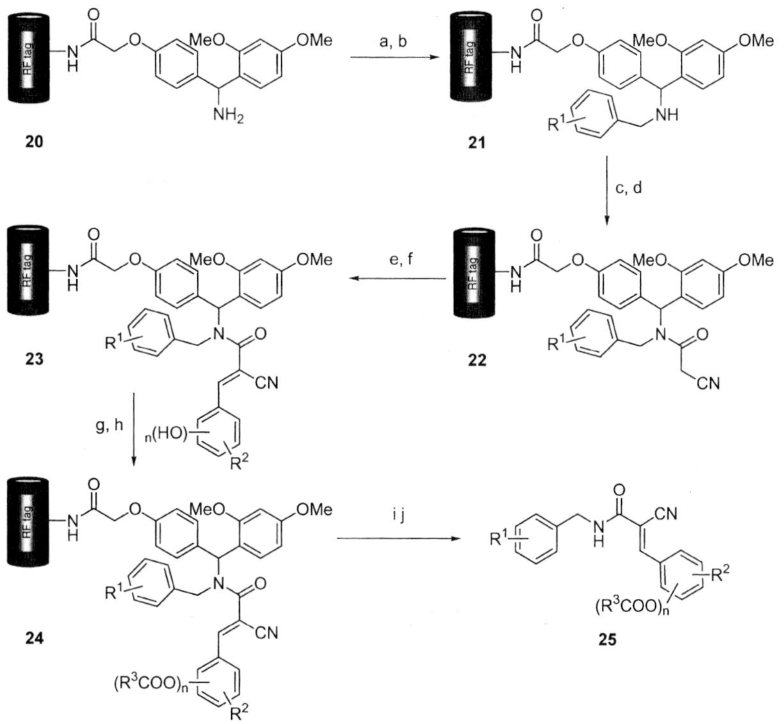

Figure 13 Synthesis of a tyrphostin library using R_f tagged MicroTube reactors and directed sorting. (a) Sorting MicroTubes. (b) Benzaldehydes bearing R^1 (10 Equiv.), CH(OMe)$_3$, r.t., 3 h; then NaCNBH$_3$ (16 Equiv.), r.t., 20 min; then AcOH (final concentration = 2%, v/v), r.t., 3 h. (c) Pooling MicroTubes. (d) Triple coupling with cyaoacetic acid (10 Equiv.), DIEA (20 Equiv.), DIC (11 Equiv.), DMF, r.t., 24 h. (e) Sorting MicroTubes. (f) Benzaldehydes bearing R^2 (10 Equiv.), piperidine (4 Equiv.), MeOH:DMF (1:10), r.t., 48 h. (g) Sorting MicroTubes. (h) R^3COCl (10 Equiv.), Et$_3$N (20 Equiv.), DCM (for R^3 = methyl) or DMF (for R^3 = phenyl), r.t., 24 h. (i) Separation of individual MicroTubes. (j) 4% TFA in benzene, r.t., 1 h.

Protocol 4

Performing reactions on MicroTubes: reductive amination[a]

1   Place a group of MicroTubes with a free amino group (average loading: 53 μmol/ MicroTube) in a reaction vessel.

2   Add dry trimethyl orthoformate (1 ml/MicroTube).

3   Add an aldehyde with a final concentration of 0.5 M (*Figure 12*).

4   Shake (200 r.p.m.) at room temperature for 3 h.

5   Add NaCNBH$_3$ to a final concentration of 0.8 M.

6   Shake (200 r.p.m.) at room temperature for 20 min.

7   Add acetic acid to a final concentration of 2% (v/v).

Protocol 4 continued

8   Shake (200 r.p.m.) at room temperature for another 3 h.

9   Remove the solution from the MicroTubes, wash and dry according to *Protocol 1B*.

a Step 'b' in *Figure 13*.

3.3 Bar coded microreactors: application to oligonucleotide synthesis

To illustrate this approach, a 27-member tetranucleotide library with the generic structure X_4-X_3-X_2-T was prepared following standard oligonucleotide synthesis procedures (11) (*Figure 14, Protocol 5*). Between 2–5 mg of oligonucleotide was

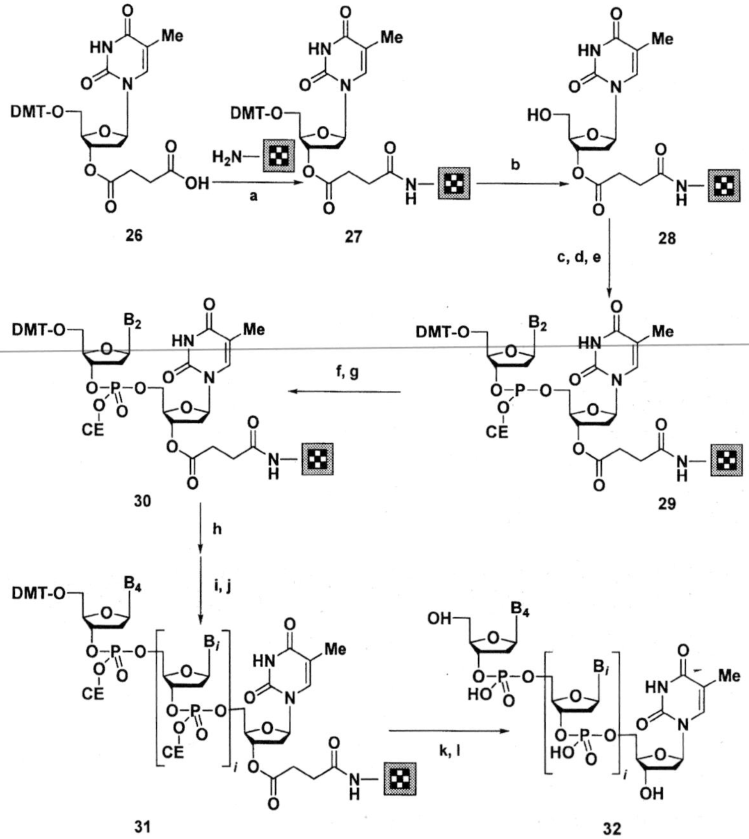

Figure 14 Synthesis of an oligonucleotide library using 2D bar coded microreactors and directed sorting. (a) 2D bar coded microreactors with free amino group, PyBOP (0.1 M), DIEA (0.2 M), DMF, r.t., 4 h. (b) 3% TCA in DCM, r.t., 4 min. (c) Sorting. (d) DMT-X_2 (0.1 M), tetrazole (0.3 M), acetonitrile:DCM (2:3), r.t., 1 h. (e) Pooling. (f) I_2 (0.1 M), THF:pyridine:H_2O (40:10:1), r.t., 20 min. (g) Ac_2O (0.5 M), 1-methylimidazole (0.5 M), 2,6-lutidine (0.5 M), THF, r.t., 20 min. (h) Repeat (b) to (g) with DMT-X_3. (i) Repeat (b) to (f) with DMT-X_4. (k) Separation of individual microreactors. (l) Concentrated ammonia:1,4-dioxane (1:1), 20 h at r.t. and 20 h at 55°C; then 3% TCA in DCM, r.t., 4 min.

produced from each bar coded microreactor. The structural and chemical integrity of the synthesized library was assessed by HPLC (67–97% purity), ^{1}H NMR, MS, and sequence analysis.

Protocol 5

Oligonucleotide synthesis on bar coded microreactors[a]

1 Place 27 2D bar coded microreactors with a free -OH group (average loading: 7 μmol/microreactor) in a reaction vessel.

2 Add dry dichloromethane (0.7 ml/microreactor).

3 Add a CE-phosphoramidite monomer to a final concentration of 0.1 M.

4 Add an acetonitrile solution of tetrazole (0.75 M) to a final concentration of 0.3 M.

5 Shake the reaction mixture (200 r.p.m.) at room temperature for 1 h.

6 Remove the solution then wash and dry the bar coded microreactors according to *Protocol 2*.

[a] Oligonucleotide coupling (step 'd' in *Figure 14*).

4 Conclusion

This chapter summarizes the application of radiofrequency (R_f) tagging (10) and bar coding (11), in combination with a directed sorting strategy, to the synthesis of combinatorial libraries of a variety of chemical structures. This strategy employs non-chemical encoding and directed sorting in a batch synthesis fashion thereby combining the advantages of both split and pool and parallel synthesis. With this methodology we have also shown that chemical libraries of complex natural product structures in which each library member is synthesized in multimilligram quantities can be readily performed.

References

1. Myers, P. (1996). *Curr. Opin. Biotech.*, **8**, 701.

2. Longman, R. (1997). *Start-up*, May.

3. Storer, R. (1996). *DDT*, **1**, 248.

4. Kung, P. P., Bharadwaj, R., Fraser, A. S., Cook, D. R., Kawasaki, A. M., and Cook, D. (1998). *J. Org. Chem.*, **63**, 1846.

5. Früchtel, J. S. and Jung, G. (1996). *Angew. Chem. Int. Ed. Engl.*, **35**, 17.

6. Balkenhohl, F., Bussche-Hünnefeld, C. von dem, Lansky, A., and Zechel, C. (1996). *Angew. Chem. Int. Ed. Engl.*, **35**, 2288.

7. Hermkens, P. H. H., Ottenheijm, H. C. J., and Rees, D. C. (1997). *Tetrahedron*, **53**, 5643.

8. Morphy, J. R., Rankovic, Z., and Rees, D. C. (1996). *Tetrahedron Lett.*, **37**, 3209.

9. Newlander, K. A., Chenera, B., Veber, D. F., Yim, N. C. F., and Moore, M. L. (1997). *J. Org. Chem.*, **62**, 6726.

10. Nicolaou, K. C., Xiao, X.-Y., Parandoosh, Z., Senyei, A., and Nova, M. P. (1995). *Angew. Chem. Int. Ed. Engl.*, **34**, 2289.

11. Xiao, X.-Y., Zhao, C., Potash, H., and Nova, M. P. (1997). *Angew. Chem. Int. Ed. Engl.*, **36**, 780.

12. Moran, E. J., Sarshar, S., Cargill, J. F., Shahbaz, M. M., Lio, A., Mjalli, A. M. M., *et al.* (1995). *J. Am. Chem. Soc.*, **117**, 10787.

13. Geysen, H. M., Wagner, C. D., Bodnar, W. M., Markworth, C. J., Parke, G. J., Schoenen, F. J., *et al.* (1996). *Chem. Biol.*, **3**, 679.

14. Czarnik, A. W. (1997). *Curr. Opin. Chem. Biol.*, **1**, 60.

15. Brenner, S. and Lerner, R. A. (1992). *Proc. Natl. Acad. Aci. USA*, **89**, 5381.

16. Ohlmeyer, M. H. J., Swanson, R. N., Dillard, L. W., Reader, J. C., Asouline, G., Kobayashi, R., *et al.* (1993). *Proc. Natl. Acad. Sci. USA*, **90**, 10922.

17. Lam, K. S., Salmon, S. E., Hersh, E. M., Hruby, V. J., Kazmierski, W. M., and Knapp, R. J. (1991). *Nature*, **354**, 82.

18. DeWitt, S. H. and Czarnik, A. W. (1996). *Acc. Chem. Res.*, **29**, 114.

19. Porco, J. A., Deegan, T., Devonport, W., Gooding, O. W., Heisler, K., Labadie, J. W., *et al.* (1996). *Mol. Div.*, **2**, 197.

20. Burgess, K. (1994). *J. Med. Chem.*, **37**, 2985.

21. Battaerd, H. A. J. and Tregear, G. W. (1967). *Graft copolymers*. Wiley, Intersicience, New York.

22. Mitchell, A. R., Kent, S. B. H., Engelhard, M., and Merrifield, R. B. (1978). *J. Org. Chem.*, **43**, 2845.

23. Farrall, A. J. and Fréchet, J. M. (1976). *J. Org. Chem.*, **41**, 3877.

24. Xiao, X.-Y., Parandoosh, Z., and Nova, M. P. (1997). *J. Org. Chem.*, **62**, 6029.

25. Nicolaou, K. C., Vourloumis, D., Li, T., Pastor, J., Winssinger, N., He, Y., *et al.* (1997). *Angew. Chem. Int. Ed. Engl.*, **36**, 2097.

26. Schiff, P. B. and Horwitz, S. B. (1980). *Proc. Natl. Acad. Sci. USA*, **77**, 1561.

27. Bollag, D. M., McQueney, P. A., Zhu, J., Hensens, O., Koupal, L., Liesch, J., *et al.* (1995). *Cancer Res.*, **55**, 2325.

28. Kowalski, R. J., Giannakakou, P., and Hamel, E. (1997). *J. Biol. Chem.*, **272**, 2534.

29. Dowd, P. and Choi, S.-C. (1992). *Tetrahedron*, **48**, 4773.

30. Takahashi, T., Machida, K., Kido, Y., Nagashima, K., Ebata, S., and Doi, T. (1997). *Chem. Lett.*, **12**, 1291.

31. Nicolaou, K. C., Pastor, J., Winssinger, N., and Murphy, F. (1998). *J. Am. Chem. Soc.*, **120**, 5132.

32. Nicolaou, K. C., He, Y., Vourloumis, D., Vallberg, H., and Yang, Z. (1996). *Angew. Chem. Int. Ed. Engl.*, **35**, 2399.

33. Nicolaou, K. C., Roschangar, F., and Vourloumis, D. (1998). *Angew. Chem. Int. Ed. Engl.*, **37**, 2014.

34. Cao, X. and Mjalli, A. M. M. (1996). *Tetrahedron Lett.*, **37**, 6073.

35. Schuster, M., Pernerstofer, J., and Blechert, S. (1996). *Angew. Chem. Int. Ed. Engl.*, **35**, 1979.

36. Aristoff, P. A. (1981). *J. Org. Chem.*, **46**, 1954.

37. Nicolaou, K. C., Seitz, S. P., and Pavia, M. R. (1979). *J. Org. Chem.*, **44**, 4011.

38. Gazit, A., Osherov, N., Gilon, C., and Levitzki, A. (1996). *J. Med. Chem.*, **39**, 4905.

39. Shi, S., Xiao, X.-Y., and Czarnik, A. W. (1998). *Biotech. Bioeng. (Combi. Chem.)*, **61**, 7.

Chapter 5

Positionally addressable parallel synthesis on continuous membranes

Holger Wenschuh, Heinrich Gausepohl,[†] Lothar Germeroth, Mathias Ulbricht,[‡] Heike Matuschewski,[§] Achim Kramer,[*] Rudolf Volkmer-Engert,[*] Niklas Heine,[*] Thomas Ast,[*] Dirk Scharn,[*] and Jens Schneider-Mergener[*]

Jerini Bio Tools GmbH, Ost-West-Kooperationszentrum, Rudower Chaussee H29, 12489 Berlin, Germany.

[†]Abimed Analysen-Technik GmbH, Raiffeisenstr. 3, 40764 Langefeld, Germany.

[‡]Institute of Chemistry, GKSS Research Centre, Kantstr. 55, 14513 Teltow, Germany.

[§]Poly-An GmbH, Rudower Chaussee 6a, 12489 Berlin, Germany.

[*]Institut für Medizinische Immunologie, Humboldt-Universität zu Berlin, Schumannstr. 20–21, 10098 Berlin, Germany.

1 Introduction

Spatially addressable high-throughput solid phase synthesis of large arrays of compounds has generated intense interest over the past few years (1). Besides parallel synthesis on resin beads (see Chapters 1–3), polymeric pins and chips (see Chapter 4), SPOT synthesis using continuous membrane supports has been shown to be an efficient solid phase synthetic alternative. The development of this approach was fuelled by the need for a facile and economical complement to the classical solid phase synthesis procedures with increased flexibility and amenability to miniaturization and automation (2).

The key feature of the SPOT method is the positionally addressed delivery of small volumes of liquids directly to the membrane support (3). The droplets dispensed form separate SPOTS and can be considered as microreactors. The volumes dispensed create a specific SPOT size determining both the scale of reaction and the absolute number of compounds that can be arranged on an area of a membrane. The compounds synthesized can be evaluated while still attached to the membrane, or in solution after release from the membrane,

using conventional high-throughput screening techniques. Semi-automated SPOT synthesis of large arrays of compound is also possible using the ABIMED ASP 222 robotic system (4).

2 SPOT synthesis of peptides on continuous cellulose surfaces

Cellulose-bound peptides assembled via the SPOT technique are well suited to detect binding to proteins (5), metals, and nucleic acids (6), or to map substrate specificity of enzymes (7). In addition, SPOT synthesis can be used for the rapid parallel microsynthesis of a large number of peptides that can be cleaved from the cellulose membrane by means of orthogonal anchor systems (see Chapter 6) (3). Two earlier accounts describe protocols for the screening of large libraries of synthetic peptides (up to 8000 compounds on a 18 × 30 cm membrane) assembled on continuous cellulose membranes (8, 9).

2.1 Preparation of the cellulose matrix

Currently, cellulose membranes can either be amino-functionalized by acylating its hydroxyl functions with activated Fmoc-β-alanine (*Protocol 1*) or derivatizing them via acid catalysed ring opening of epibromohydrine leading to amino functions attached via an ether bond to the surface (*Protocol 2*) (10). The latter method is advantageous when subsequent chemical steps involve strongly acidic, basic, or nucleophilic conditions which would hydrolyse an ester linkage. The loading of amino functions can be determined via acylation with Fmoc-Gly-OPfp and subsequent UV measurement of the absorbance at 301 nm of the dibenzofulvene/piperidine adduct obtained upon cleavage of the Fmoc group (*Protocol 3*).

Protocol 1

Preparation of amino-functionalized cellulose membranes (ester linkage)

Equipment and reagents

- Cellulose paper: Whatman 50 (Whatman)
- Stainless steel dish
- Shaking platform
- Bromophenol blue, DIC (diisopropylcarbo-diimide), DMF (*N,N*-dimethylformamide), ethanol, Fmoc-β-alanine, NMI (*N*-methylimidazole), piperidine

Method

1 Dissolve 1.25 g (4.0 mmol) Fmoc-β-alanine in 20 ml DMF, add 635 μl NMI (8.0 mmol), and 748 μl DIC (4.8 mmol).

2 Pour the acylation mixture into a stainless steel dish. Place the Whatman 50 paper

Protocol 1 continued

(18 cm × 28 cm) into this mixture. Ensure that the paper is well soaked and that the mixture is bubble-free. Close the dish and incubate the membrane for 3 h without shaking.

3   Wash the membrane with DMF (3 × 50 ml, 3 min per wash). A shaking platform is recommended.

4   Cleave the Fmoc protecting groups by shaking the membrane in 20% piperidine/DMF (2 × 50 ml, 10 min each).

5   Wash the membrane with DMF (5 × 50 ml, 3 min per wash) and ethanol (2 × 50 ml, 3 min per wash).

6   Stain the membrane with 50 ml of a 0.01% (w/v) bromophenol blue solution (in ethanol).

7   Dry the membrane.

Protocol 2

Preparation of ester-free amino-functionalized cellulose membranes

Equipment and reagents

- Cellulose paper: Whatman 50 (Whatman, Maidstone, England)
- Stainless steel dish
- Shaking platform

- Methanol, perchloric acid (60% in water), epibromohydrine, dioxane, 4,7,10-trioxa-1,13-tridecanediamine, N,N-dimethylformamide (DMF), sodium methoxide (5 M in methanol), water, dichloromethane (DCM).

Method

1.   Immerse a Whatman 50 paper (18 × 28 cm) in a stainless steel dish containing methanol (50 ml) and 60% perchloric acid (1 ml) and shake for 5 min.

2.   Remove the membrane from the tray, dry it in air then use it immediately.

3.   Carefully mix epibromohydrine (4 ml) and a solution of 60% perchloric acid (400 μl) in dioxane (36 ml) in a stainless steel dish.

4.   Place the membrane in this solution, ensure that it is well soaked and bubble-free, then cover the dish and shake for 1 h.

5.   Add methanol (100 ml) and shake for an additional 30 min.

6.   Shake the membrane in methanol 2 × 50 ml, 10 min per wash), then dry it iin air.

7.   Place the membrane in a 20% solution of 4,7,10-trioxa-1,13-tridecanediamine in DMF (60 ml), then cover the dish and shake for 1 h. Higher loadings (up to 1.4 μmol/cm^2) can be achieved using neat diamine, increased temperature (80°C), and extended incubation time (3 h).

Protocol 2 continued

8. Shake the membrane in methanol (2 × 50 ml, 10 min per wash), then dry it in air.

9. Place the membrane in a 5 M sodium methoxide solution in methanol (50 ml) and shake for 2 h.

10. Add methanol (50 ml) and shake for an additional 10 min.

11. Decant the solution carefully and shake the membrane in methanol (50 ml, 10 min wash), water (4 × 50 ml, 5 min per wash, or until pH 7), DMF (50 ml, 5 min wash), methanol (2 × 50 ml, 5 min per wash) and DCM (50 ml, 5 min wash), then dry it in air.

Protocol 3

Determining the loading of an amino-functionalized cellulose membrane

Equipment and reagents

- UV-Vis spectrophotometer
- A hole puncher
- Eppendorf tubes (2 ml)

- Amino-functionalized membrane
- DMF, Fmoc-Gly-OPfp, N-methylpyrrolidine (NMP), piperidine

Method

1 Punch out a SPOT of the prepared amino-functionalized membrane and insert it in a 2 ml Eppendorf tube.

2 Incubate the SPOT with 30 μl of a 0.3 M solution of Fmoc-glycine-OPfp in NMP for 15 min.

3 Repeat step 2.

4 Wash the SPOT with DMF (5 × 1 ml, 1–2 min per wash) and ethanol (3 × 1 ml, 3 min per wash), then dry it.

5 Cleave the Fmoc protecting groups with 20% piperidine/DMF (1 ml) and determine the membrane loading by measuring the UV absorbance at 301 nm.

6 Calculate the amount of amino functions on the SPOT using an extinction co-efficient $\varepsilon_{301} = 7800$ L × mol^{-1} × cm^{-1}. Usually the membrane carries about 300 nmol amino functions per cm^2.

2.2 Positionally addressable array synthesis on SPOT membrane

In order to generate a defined array of a specific number of SPOTS, an activated Fmoc-amino acid or an appropriate linker molecule must be coupled spotwise onto the homogeneously functionalized cellulose membrane (*Protocol 4*). The SPOT positions define the arrangement of the 'microreactors' and thus the composition of every single compound. In addition, pipetted volumes determine the SPOT size, the scale of synthesis, and the total number of compounds to be arranged on a given surface area.

Protocol 4

Definition of the SPOTS

Equipment and reagents

- Amino-functionalized cellulose membrane (see *Protocols 1* and *2*)
- Stainless steel dish
- Shaking platform

- Ac$_2$O, bromophenol blue, DIEA, DMF, ethanol, Fmoc-β-alanine-OPfp, NMP, piperidine

Method

Note: determine interspot distances by delivering pre-selected volumes on a spare membrane.

1 Spot 1 μl of a 0.3 M Fmoc-β-alanine-OPfp solution (in NMP) twice to the predefined positions on the cellulose membrane (15 min reaction time each).

2 Wash the membrane with DMF (5 × 50 ml, 3 min per wash).

3 Position the membrane carefully face down in 2% Ac$_2$O/DMF/(20 ml). Avoid shaking and air bubbles. After 2 min, shake the membrane face up in Ac$_2$O/DIEA/DMF solution (20/10/70)/(50 ml) for 30 min. The purpose of this step is to acetylate the amino functions of the membrane that had not reacted with Fmoc-β-alanine-OPfp.

4 Wash the membrane with DMF (5 × 50 ml, 3 min per wash).

5 Cleave the Fmoc protecting groups by treatment of the membrane with 20% piperidine/DMF (2 × 50 ml, 10 min each).

6 Wash the membrane with DMF (5 × 50 ml, 3 min per wash) and ethanol (2 × 50 ml, 3 min per wash). Stain the membrane with 50 ml of a 0.01% (w/v) bromophenol blue/ethanol solution until a homogeneous blue staining of the SPOTS is observed.

7 Dry the membrane.

2.3 The peptide synthesis cycle

The C- to N-terminus synthesis of cellulose-bound peptides is done by an iterative process of the following steps (*Protocol 5*):

- double coupling with activated amino acids
- washing with DMF
- cleaving of the Fmoc protecting groups
- washing with DMF and ethanol
- staining

99

- washing with ethanol

- drying

This cycle is repeated until the peptide with the desired sequence and length is obtained.

Protocol 5

The SPOT synthesis coupling cycle

Equipment and reagents

- Amino-functionalized cellulose membrane (see *Protocols 1* and *2*)
- Stainless steel dish

- Shaking platform
- Bromophenol blue, DMF, ethanol, Fmoc-amino acid-OPfp, NMP, piperidine

Method

1 Prepare 0.3 M solutions of Fmoc-amino acid-OPfp in NMP (except for Fmoc-Ser(*t*-Bu)-OPfp which is prepared as a 0.2 M solution). These solutions are stable at 20 °C for several days with the exception of the active ester of arginine that should be prepared on a daily basis. Pipette 1 μl of the appropriate amino acid solutions onto each SPOT twice (15 min reaction time each).

2 Wash the membrane with DMF (3 × 50 ml, 3 min each).

3 Cleave the Fmoc protecting groups by treatment of the membrane with 20% piperidine/DMF (2 × 50 ml, 10 min each).

4 Wash the membrane with DMF (5 × 50 ml, 3 min per wash) and ethanol (2 × 50 ml, 3 min per wash).

5 Stain the membrane with a 0.01% (w/v) bromophenol blue ethanol solution (50 ml). Do not stain after coupling the final amino acid residue.

6 Dry the membrane.

Protocol 6

Cleavage of side-chain protecting groups

Equipment and reagents

- Stainless steal dish

- DCM, DMF, ethanol, phenol, TFA, triisobutylsilane, water

Method

1 Incubate the membrane with peptide SPOTS (see *Protocol 5*) in 200 ml of a solution of TFA/triisobutylsilane/water/phenol/DCM (90:3:2:1:4) in a tightly closed stainless steel box for 30 min without shaking.

Protocol 6 continued

2 Wash with DCM (4 × 50 ml, 3 min each), DMF (3 × 50 ml, 3 min each), and ethanol (2 × 50 ml, 3 min each).

3 Dry the membrane.

4 Repeat step 1 for 150 min with a solution of TFA/TIS/H_2O/phenol/DCM (50:3:2:1:44).

5 Repeat steps 2 and 3.

2.4 Cleavage of the peptides and subsequent handling

After side-chain deprotection (*Protocol 6*) peptides can be cleaved from the cellulose membrane by dry aminolysis of the peptide–ester bond using ammonia gas. The compounds will be cleaved from the membrane as carboxamides and remain physisorbed on the cellulose membrane. The SPOTS can be punched out and the adsorbed compounds extracted with water (160 μl/SPOT) or any suitable buffer system. The amount of peptide recovered from each SPOT (0.23 cm^2) is typically around 40–50 nmol/SPOT.

Protocol 7

Cleavage of peptides from the cellulose membrane

Equipment and reagents

- Desiccator
- Puncher
- Rubber balloon

- Microtiter plate
- Ammonia, ethanol, phosphate buffer pH 7.3, water

Method

1 Wash the membrane with peptide SPOTS after cleavage of side-chain protecting groups (see *Protocol 6*) with phosphate buffer (0.1 M, pH 7.3, 3 × 1 h), water (2 × 3 min), and ethanol (2 × 3 min).

2 Dry the membrane.

3 Place the dry membrane in a desiccator and evacuate the air (vacuum pump or water aspirator).

4 Fill the desiccator with ammonia gas (use a rubber balloon).

5 Evacuate the desiccator after 12 h and subsequently flush it with air. Repeat evacuation and flushing several times to evaporate all ammonia.

6 Punch out single SPOTS with a puncher into a microtiter plate (SPOT area 0.23 cm^2).

7 Dissolve cellulose-adsorbed peptides in buffer (160 μl/SPOT).

3 Synthesis of PNA arrays using the SPOT technique

The utility of bioactive molecules synthesized in the form of arrays on a planar support is not limited to protein–ligand interactions but can extend well into other areas of molecular recognition. PNA (peptide nucleic acid) is a DNA analogue in which the sugar backbone has been replaced by an uncharged polyamide using the amino acid N-(2-aminoethyl)-glycine as building block (11). PNA oligomers hybridize strongly with complementary DNA following Watson–Crick base pairing rules. Unlike DNA–DNA interactions, however, DNA–PNA duplexes are stable at low salt concentrations. This, as well as the high thermal stability of the duplexes, make PNA a very versatile class of molecules for sequence-specific DNA recognition.

However, the little knowledge about erratic behaviour (as compared to DNA oligomers), prohibitive cost of synthesis, and low solubility of PNA oligomers currently limit its scope. Therefore, adaptation of synthesis and assay development directly on continuous cellulose membranes can greatly expedite comparative analysis of a large number of different PNA sequences. Encouraging hybridization results together with details on synthesis and characterization have been published earlier (12).

3.1 Reagents and equipment

Similar to peptide synthesis, amino-functionalized cellulose membranes (1) can be used as solid supports for PNA arrays. Due to the amino acid backbone, PNA spots can be assembled quite analogously to peptide SPOTS (13, 14). The following building blocks are commercially available (Perkin Elmer) and can be used as stock solutions (0.25 M in NMP/HOAt): Fmoc-A(Bhoc)aeg-OH, Fmoc-C(Bhoc)aeg-OH, Fmoc-G(Bhoc)aeg-OH, Fmoc-Taeg-OH.

3.2 Reagents preparation

Determine the volume required to wet the desired spot area and calculate: volume × 1.2 (to cover the border) × number of spots × 2 (for double coupling) × 1.2 (safety margin) + 20 μl (safety margin). A volume of 2 μl will cover spots of 6 mm diameter on cellulose, corresponding to about 100 nmoles (i.e. loading of 300 nmoles/cm^2). Reagent consumption per residue and cycle is then: (2 μl × 1.2 × 24 × 2 × 1.2) + 20 μl = 158 μl. At a concentration of 0.2 M this corresponds to 32 μmoles. 0.7 mmoles will therefore be sufficient for 20 cycles.

PNA monomers are dissolved in a 0.33 M solution of HOAt in NMP to a final concentration of 0.25 M. This solution is stable for several days. Prior to use, the stock solution is mixed at a ratio of 1:0.2 with a 1.25 M solution of DIC in NMP. The final concentration of activated PNA monomers is thus 0.2 M with a 10% excess of HOAt. The NMP used for preparation of the stock solutions must be free of amines and acids and of the highest purity available. For these reasons, we recommend that this solvent should be stored over molecular sieves in small bottles.

Prepare a 0.33 M solution of HOAt in NMP: 0.45 g HOAt + 9.5 ml NMP = 10 ml stock at 0.33 M. Add 2.3 ml of this solution to each bottle of monomer containing 0.7 mmoles. With approx. 500 mg of derivatives per bottle the final monomer concentration is 0.7/2.8 = 0.25 M.

Prepare 5 ml of 1.25 M DIC in NMP: 0.97 ml DIC + 4.03 ml NMP = 5 ml stock at 1.25 M.

Mix 150 μl monomer stock with 30 μl DIC stock to obtain 180 μl solution at 0.2 M. Only the required amount of solution for each cycle should be prepared. The solutions can be used after about 10 min of pre-activation and are stable for little more than 1 h.

Note: In contrast with peptide synthesis where coupling activity is generally retained for up to 24 h, a gradual loss of coupling activity after 1 h of pre-activation has been observed in the case of PNA (13).

3.3 Linker chemistry (see Chapter 6 also)

The amino-functionalized cellulose membrane (see *Protocols 1* and *2*) can be used for direct attachment of pre-activated monomers. The resulting PNA molecules can be used in hybridization assays where the PNA remains covalently attached to the support. However, since unbound PNA is required for analysis and solution phase studies, cleavable linker systems were introduced. There are four basic options for achieving the synthesis of releasable PNA:

(a) If native cellulose membranes (i.e. unmodified surfaces) are used as a support they can be derivatized with Fmoc-β-alanine via an ester bond. After the final side-chain deprotection, PNA is then released upon a 10 minute treatment with 0.1 M KOH, followed by neutralization. Unlike DNA, PNA is stable under such drastic conditions. However, the lability of the ester bond towards basic conditions during hybridization experiments leads to a gradual leakage of PNA if the pH ≥ 7.

(b) Amino-functionalized membranes can be derivatized with an acid-labile linker such as the Rink amide linker (*p*-[(R,S)-α-[1-(9H-fluoren-9-yl)-methoxy-formamido]-2,4-dimethoxybenzyl] phenoxy acetic acid). The latter can be coupled using standard peptide coupling conditions (see *Protocol 5*).

(c) A trypsin-sensitive site can be inserted between the membrane and the PNA. This requires synthesis of a short peptide sequence (e.g. Lys–Glu) that is recognized and cleaved specifically by the enzyme. The yields of enzyme catalysed cleavages are usually low and, as indicated by some preliminary experiments, the cleavage does not proceed when the PNA sequences are hybridized with their complementary strands.

(d) Amino-functionalized membranes can also be derivatized with a base-labile linker. This requires prior solution phase synthesis of a building block such as Fmoc-Gly-HMB-OH, which is a glycine amino acid coupled to the base-labile hydroxymethyl benzoic acid (HMBA) linker through a peptide bond.

3.4 **Membrane preparation**

Weigh 0.5 mmoles of each derivative used for bulk coupling into 2 ml reaction vials, and dissolve in 1.5 ml of 0.33 M HOAt stock solution. Add 0.4 ml of 1.25 M DIC solution and allow it to react for 10 min. Apply the solution to the membrane and let it react for 30 min. For normal amino acid derivatives a single coupling to linker-derivatized membranes is sufficient. Wash as described for peptide synthesis (*Protocol 5*, step 2). To prevent derivatization of less accessible sites on the membrane single couplings are recommended. Before Fmoc deprotection the entire membrane is capped with 5% Ac_2O in DMF or NMP for 5 min (*Protocol 8*).

Protocol 8

PNA assembly on cellulose membranes

Equipment and reagents

- Amino-functionalized cellulose membrane (see *Protocol 1*)
- Ac_2O, DMF, ethanol, piperidine
- Pre-activated linker, amino acids, and PNA building blocks

Method

1 Remove Fmoc protecting groups of the amino-functionalized membrane with 20% piperidine/DMF (2×10 min).

2 Wash extensively with DMF (5×50 ml) and ethanol (2×50 ml).

3 Air dry the membrane.

4 Deliver the activated monomers to their predefined positions on the membrane according to the desired PNA sequence and SPOT size, and allow for a 20 min reaction time.

5 Repeat step 4.

6 Cap unreacted amines with a 5 min treatment in 5% Ac_2O/DMF.

7 Wash extensively with DMF (5×50 ml).

3.5 **Synthesis verification**

Given the overall cost of SPOT synthesis it is highly recommended to check its progress. There are basically two complementary procedures:

(a) Control sequences included in the synthetic scheme are terminated by coupling with a monomer conjugated to rhodamine. The coloration observed will indicate that the synthesis is proceeding normally. Quantification is, however, difficult because of photobleaching during handling.

(b) The membrane is stained with bromophenol blue after Fmoc deprotection and thorough washing. Staining works only after the membrane has been thoroughly washed with amine-free DMF. Stained membranes can be photographed or scanned and kept on file for future reference.

3.6 Final work-up

The side-chain protecting groups (Bhoc) are easily cleaved off with dilute TFA. Treatment with a mixture of 25% TFA and 5% triisopropylsilane in dichloromethane for 30 minutes has been shown to work well. A similar mixture may be used to strip very tightly-bound DNA in hybridization experiments. Care must be taken, however, not to expose the cellulose membrane to these harsh conditions for prolonged periods of time. After side-chain deprotection, the membrane is washed with ethanol, dried, and stored in a dry container at low temperature.

3.7 Quality control

Quality control of the PNA synthesized can be performed either by analysis of representative SPOTS and comparison with reference sequences, or analysis of results obtained with duplicate sequences and variations thereof. Currently, mass spectrometry is the most appropriate technique for analysis of substances isolated from single SPOTS. To be able to cleave some reference sequences off the membrane, an appropriate linker must be used (see Section 3.4). Analysis by HPLC is limited because of the relatively low amount of material on the SPOT areas. Furthermore, sequence-related information cannot be obtained from HPLC analysis.

3.8 Comments

PNA is more difficult to synthesize than peptides. Couplings are incomplete and a gradual fading of the bromophenol blue coloration will be observed. Double couplings are mandatory with the methodology described here, and each cycle should be concluded with acetylation of unreacted amino groups. Due to the high sequence specificity of PNA, any truncated sequence arising from incomplete couplings is 'invisible' in hybridization experiments.

PNA tends to aggregate and is not easily solubilized. Attachment of lysine residues at each end of the sequence improves the solubility without interfering with hybridization. Since PNA degrades easily from the amino-terminus under basic conditions (15), all sequences must be acetylated or, preferably, elongated with solubilizing residues such as lysine.

Hybridization can be so strong that TFA must be used to break the hydrogen bonds, and some G-rich PNA can bind to almost any DNA sequence (12). PNA is stable to conditions that normal DNA oligonucleotides cannot withstand. This increases the number of times hybridization experiments can be performed with a PNA array. It also allows the stripping of bound DNA under drastic acidic or basic conditions.

4 Preparation of stable polymeric membranes for SPOT synthesis of organic compound libraries

Despite the advantages of cellulose membranes for SPOT synthesis and screening of peptide and PNA libraries, this material suffers from a number of limitations. The high concentration of reactive, hydrophilic hydroxyl groups of cellulose membranes is responsible for its mechanical instability. Moreover, chemical degradation of cellulose occurs under many reaction conditions that are essential for the synthesis of organic molecule libraries. Therefore, a novel polymeric membrane for solid phase syntheses was developed. The entire surface of a sheet of porous support made from a chemically, mechanically, and thermally stable polymer was evenly functionalized. Flexible polymer chains that carry reactive groups for solid phase organic synthesis and that are compatible with conventional bioassays were chemically grafted to the porous surface. A successful example for this composite concept is the photoinduced graft co-polymerization of acrylates onto polypropylene membranes. This process does not alter the membrane's porous structure nor its stability (16–18), allowing thus further chemical derivatization (*Scheme 1*).

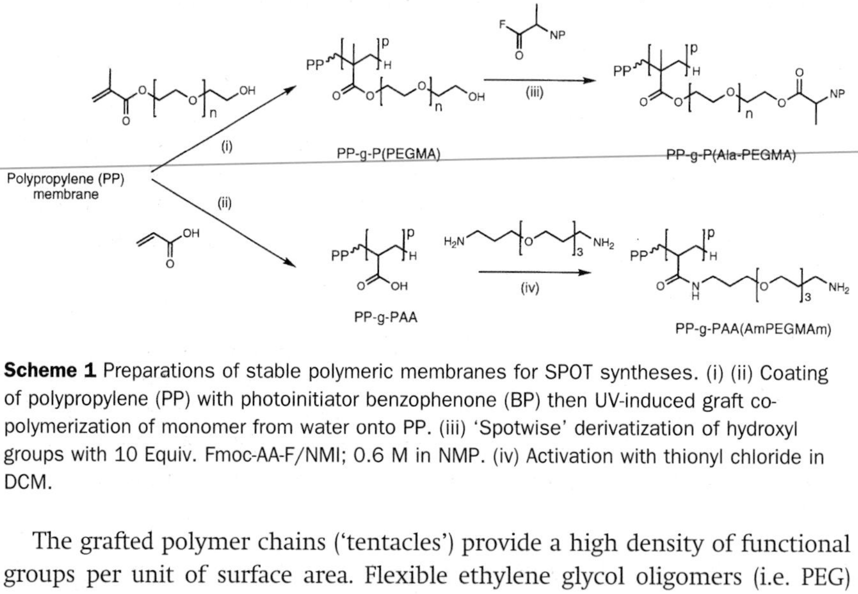

Scheme 1 Preparations of stable polymeric membranes for SPOT syntheses. (i) (ii) Coating of polypropylene (PP) with photoinitiator benzophenone (BP) then UV-induced graft co-polymerization of monomer from water onto PP. (iii) 'Spotwise' derivatization of hydroxyl groups with 10 Equiv. Fmoc-AA-F/NMI; 0.6 M in NMP. (iv) Activation with thionyl chloride in DCM.

The grafted polymer chains ('tentacles') provide a high density of functional groups per unit of surface area. Flexible ethylene glycol oligomers (i.e. PEG) were selected for this purpose and fulfil both the function of a spacer and the role of a solubilizing, biocompatible interface.

4.1 Hydroxy-functionalized PEG acrylate polypropylene membranes (PP-g-P(PEGMA))

Membranes with two different spacer lengths (n = 4 and 9, *Scheme 1*) can be synthesized in a straightforward procedure from commercially available monomers; the hydroxyl groups can then be used with established linker systems

(Chapter 6, *Scheme 1*). For further applications, the susceptibility of the ester linkage to hydrolysis must be taken into account.

4.1.1 Photoinitiated graft co-polymerization of hydroxy-functionalized PEG acrylates onto polypropylene membranes (PP-g-P(PEGMA))

The functionalization involves two essential steps (see *Protocol 9*):

(a) The support membrane is coated with the photoinitiator.

(b) Selective UV excitation of the adsorbed photoinitiator in the presence of the monomers initiates a graft co-polymerization onto the support membrane. The extent of the functionalization can be adjusted through the photo-initiator and monomer concentrations, and extent of UV irradiation. A few examples are given in *Table 1*.

Protocol 9

Preparation of PP-g-P(PEGMA) membranes

Equipment and reagents

- For laboratory scale (lab-scale): apparatus for homogeneous irradiation of 150 cm² area with parallel light (L.O.T.-Oriel), including an UV lamp HBO 350 DeepUV (Hamamatsu) and a reactor (d = 10 cm) with a glass window (cut-off: $\lambda > 310$ nm) and a Petri dish (d = 9 cm)

- For technical scale (tech-scale): UV curing system for irradiation of samples with a width of 50 cm on a transport belt, including two 500 Watts Hg lamps (Beltron GmbH) and a reaction vessel (25 × 35 cm) with a glass window (cut-off: $\lambda > 310$ nm)

- Polypropylene (PP) membranes Accurel 2E (Akzo: nominal pore diameter $d_p = 0.2$ μm, thickness $d_m = 150$ μm), and AN06 (Millipore Corp.: $d_p = 0.6$ μm, $d_m = 200$ μm)
- Filter paper (MN615, Macherey & Nagel)
- Soxhlet apparatus
- Benzophenone (Merck)
- Polyethylene glycol methacrylates PEGMA306 and PEGMA526 (the numbers correspond to average molecular weights; Polysciences Europe GmbH, Eppelheim)
- Spectral grade methanol and acetone

Method (for parameters variation see Table 1)

1. Cut circular (lab-scale, d = 8 cm) or rectangular (tech-scale, 18 × 24 cm) membrane sample and weigh.

2. Immerse and shake the membrane for 2 h at room temperature in 100 mM benzophenone in methanol.

3. Prepare monomer solution, 20 ml (lab-scale) or 150 ml (tech-scale) 50 g/l PEGMA306 or 78.5 g/l PEGMA526 in water saturated with benzophenone and degas it by flushing with nitrogen.

Protocol 9 continued

4 Put filter paper on bottom of Petri dish (lab-scale) or reaction vessel (tech-scale) and soak it with about 5 ml or 30 ml of the monomer solution, respectively. (Avoid inclusion of air bubbles!)

5 Take membrane out of the benzophenone solution and blow dry its outer surface quickly without de-wetting the pores, then put the membrane onto the filter paper, wet with monomer solution, and fix it at the edges. (Avoid inclusion of air bubbles!)

6 Immediately pour residual monomer solution onto the membrane completely covering its outer surface.

7 Close the reactor under nitrogen atmosphere (lab-scale) or reaction vessel (tech-scale) and allow 30 min equilibration time of membrane and monomer solution.

8 Proceed with UV irradiation of the membrane through the window for 30 min (lab-scale) or for 10 min (tech-scale, 10 × 1 min passage).

9 Allow the membrane to react for an additional 15 min.

10 Extract the membrane first with methanol (Soxhlet), then water, acetone, and water once again.

11 Dry the membrane at 40–50 °C in a vacuum oven.

12 Determine the degree of functionalization gravimetrically.

Table 1 Preparation conditions for selected PP-g-P(PEGMA) membranes

Membrane[a]	Solvent/ photoinitiator[b] (mM)	Monomer	Monomer (g/l)	Irradiation time (min)	loading (mg/cm²)	loading (μmol/cm²)
1	Methanol/100	PEGMA306	50	30[c]	1.12	3.5
1	Methanol/100	PEGMA526	78.5	30[c]	1.37	2.6
1	Acetone/150	PEGMA526	78.5	15[d]	1.28	2.4
2	Methanol/100	PEGMA526	78.5	10[d]	0.19	0.36
2	Methanol/100	PEGMA526	100.0	10[d]	0.34	0.65

[a] 1: 0.2 μm, Accurel 2E. 2: 0.6 μm, AN06.

[b] Photoinitiator = benzophenone.

[c] Lab-scale.

[d] Tech-scale.

4.1.2 Application of linker systems with PP-g-P(PEGMA) membranes

After derivatization of the hydroxyl group yielding an amino function (PP-g-P(Ala-PEGMA); *Scheme 1*), further functionalization of the membranes with established linker systems can be performed under SPOT synthesis conditions (see *Protocol 10*) and stepwise peptide or organic synthesis can follow.

Protocol 10

Attachment of linkers to PP-g-P(PEGMA) membranes

Equipment and reagents

- PP-g-P(PEGMA) membrane (see *Protocol 9*)
- Rink linker, photo linker (NovaBiochem, Bad Soden, Germany), imidazole linker (3)
- Fmoc-Ala-F (19)
- Bromophenol blue, DCM, DMF, ethanol, NMI, NMP, piperidine, TBTU

Method

1. Spot 2 μl of a freshly prepared 0.6 M solution of Fmoc-Ala-F in NMP containing 1 Equiv. NMI onto a PP-g-P(PEGMA) membrane, allow 30 min reaction time, and repeat the spotting.

2. Wash with DMF (2 × 5 min), methanol (2 × 5 min), DCM (2 × 5 min), and dry in air.

3. Remove Fmoc protecting groups by soaking the membrane in 20% piperidine/DMF (2 × 10 min).

4. Wash with DMF (2 × 5 min), methanol (2 × 5 min), DCM (2 × 5 min), and dry in air.

5. Stain free amino groups with bromophenol blue (0.01% in ethanol).

6. Spot solutions of activated linker molecules by applying:

   (a) 2 μL of a 0.6 M solution of Rink linker (p-[(R,S)-μ[1-(9H-fluoren-9-yl)-methoxy-formamido]2,4-dimethoxy benzyl]-phenoxy acetic acid) in NMP after 5 min pre-activation with 1 Equiv. TBTU and 2 Equiv. DIEA.

   (b) 2 μL of a 0.6 M solution of imidazole linker in NMP after 5 min pre-activation with 1 Equiv. TBTU and 1 Equiv. DIEA.

   (c) 2 μL of a 0.6 M solution of aminoethyl photo linker (4-[2-methoxy-4-(1-Fmoc-aminoethyl)-5-nitrophenoxy] butanoic acid) in NMP after 5 min pre-activation with 1 Equiv. TBTU and 2 Equiv. DIEA.

   (d) 2 μL of a 0.6 M solution of hydroxyethyl photo linker (4-[2-methoxy-4-(hydroxy-ethyl)-5-nitrophenoxy]butanoic acid) in NMP after 5 min pre-activation with 1 Equiv. TBTU and 1 Equiv. DIEA.

7. In all cases allow 30 min reaction time, then repeat spotting with freshly prepared solutions.

8. After synthesis, the compounds can be cleaved from the membranes as follows: Rink linker: 95% TFA/H$_2$O for 2 h; imidazole linker: phosphate buffer pH 7.4 for 1 h; photo-linkers: dry state cleavage via UV irradiation; 365 nm (7 mW/cm^2), for 2 h (1 h for each side).

9. For further chromatographic and mass spectrometric analysis, TFA should be evaporated before dissolving the substances in an appropriate buffer system. Adsorbed compounds (in case of photocleavage) can be directly dissolved in a buffer system and the quality of the products assessed by HPLC and mass spectrometry.

4.2 Amino-functionalized ester-free PEG methacrylamide polypropylene membranes (PP-g-P(AmPEGMAm))

Amino-functionalized PEG membranes with both covalently anchored grafted chains and a hydrolytically stable amide linkage can be synthesized in a two-step procedure (*Scheme 1*).

4.2.1 Photoinitiated graft co-polymerization of acrylic acid onto polypropylene membranes

The functionalization (*Protocol 11*) relies on the same mechanism described in Section 4.1.1, the experimental conditions of which are summarized in *Table 2*.

Protocol 11

Preparation of PP-g-PAA membranes

Equipment and reagents

- See *Protocol 9*
- Acrylic acid (> 99%, Aldrich)

Method (for parameters variation see Table 2)

1 Follow *Protocol 9*, steps 1 and 2.

2 Prepare 20 ml of a 10 g/l acrylic acid (lab-scale) or 150 ml of a 50 g/l acrylic acid (tech-scale), in water saturated with benzophenone, and degas this solution by flushing with nitrogen.

3 Follow *Protocol 9*, steps 4–7.

4 Proceed with UV irradiation for 30 min (lab-scale) or for 10 min (tech-scale, 10 × 1 min passage or, alternatively, turn the membrane after the fifth passage and continue with five more passages).

5 Follow *Protocol 9*, steps 9–12.

Table 2 Preparation conditions for selected PP-g-PAA membranes

Membrane[a]	Solvent/ photoinitiator[b] (mM)	Monomer (g/l)	Irradiation time (min)	loading (mg/cm^2)	loading (μmol/cm^2)
1	Methanol/100	10.0	30[c]	0.22	3.1
1	Acetone/150	23.3	10[d]	0.48	6.8
1	Acetone/150	50.0	10[d]	0.64	9.0
2	Methanol/100	50.0	10[d]	0.13	1.8
2	Acetone/150	50.0	5/5[e]	0.17	2.4

[a] 1: 0.2 μm, Accurel 2E. 2: 0.6 μm, AN06.
[b] Photoinitiator = benzophenone.
[c] Lab-scale.
[d] Tech-scale.
[e] Both sides.

4.2.2 Amino-functionalization of PP-g-PAA membranes via an amide linkage

Derivatization of PP-g-PAA membranes with a relatively short 'bis(amino) PEG' yields supports with hydrolytically stable and spaced reactive groups (*Protocol 12*).

Protocol 12

Synthesis of PP-g-P(AmPEGMAm) membranes

Equipment and reagents

- PP-g-PAA membrane (see *Protocol 11*)
- Stainless steel dish
- Shaking platform
- DCM, 1,13-diamino-4,7,10-trioxa-tridecane, methanol, thionyl chloride

Method

1. Treat a PP-g-PAA membrane (18 × 26 cm) with a 2 M solution of thionyl chloride in DCM 80 ml in a closed dish for 1 h at room temperature.

2. Decant the previous solution and add immediately, without de-wetting the membrane, a cold solution (4 °C) of 1,13-diamino-4,7,10-trioxa-tridecane in DCM (30%, v/v, 100 ml). Cover the dish immediately and shake the membrane carefully for 90 min. (Caution: the reaction is exothermic and generates fumes. Wear protective clothing, gloves, and goggles, and run the reaction in a fume hood.)

3. Decant the solution and wash with DCM (5 min) and methanol (5 × 5 min).

4. Dry the membrane under ambient conditions.

4.2.3 SPOT synthesis of small organic molecules on PP-g-P(AmPEGMAm) membranes

A variety of chemistries not requiring inert reaction conditions can be performed on PP-g-P(AmPEGMAm). For instance, two successful applications of the SPOT synthesis to peptoids (20) (*Protocol 13*) and 1,3,5-triazine-based compounds (*Protocol 14*) are described below.

Protocol 13

Synthesis of a tripeptoid using the SPOT synthesis method

Equipment and reagents

- PP-g-P(AmPEGMAm) membrane (see *Protocols 11* and *12*)
- Ac$_2$O, benzylamine, bromoacetic acid, DIC, DIEA, 2,6-dimethylpyridine, DMF, methanol, *n*-butylamine, NMP, piperidine, TFA

Protocol 13 continued

Method

1   Mark an array of SPOTS (1 cm² each) on a PP-g-P(AmPEGMAm) membrane using a pencil.

2   Modify the membrane with Fmoc-Rink linker as described in *Protocol 10* (2.0 μl per SPOT).

3   Acetylate unreacted amino functions (Ac$_2$O/DIEA/DMF, 1:2:7). Wash the membrane with DMF and methanol (3 × 5 min each), then dry it under ambient air.

4   Punch out a single SPOT (0.23 cm²) synthesized separately for quantitation. Determine the loading as described in *Protocol 3*, steps 5 and 6. The capacity should be in the range of 0.5–1 μmol/cm².

5   Cleave the Fmoc protecting groups by treatment of the membrane with 20% piperidine in DMF (2 × 10 min). Wash the membrane with DMF and methanol (3 × 5 min) and dry the membrane under ambient air.

6   *Acylation of the amino function.* Spot 2.0 μl of a freshly prepared solution of activated bromoacetic acid, generated by addition of 0.5 Equiv. DIC to a 0.6 M solution of bromoacetic acid in dry NMP followed by addition of 0.5 Equiv. 2,6-dimethyl-pyridine after 5 min pre-activation time. Repeat the procedure once after 15 min. Wash the membrane with DMF and methanol (3 × 5 min) and dry the membrane under ambient conditions.

7   *Nucleophilic substitution of the bromine.* Spot a 50% solution of *n*-butylamine in NMP (2.0 μl, 3 × 15 min). Wash the membrane with DMF and methanol (3 × 5 min) and dry the membrane at ambient temperature.

8   Repeat the procedure of acylation and nucleophilic substitution (steps 6 and 7) with benzylamine and piperidine as nucleophiles.

9   Cleave the peptoid (*N*-benzyl-*N*-[butyl-(carbamoylmethyl carbamoyl-methyl)-2-[piperidine-1-yl]acetamide) with TFA as described in Section 4.1.2 (*Protocol 10*). Isolate and analyse the product (*Figure 1* please see page 114).

Protocol 14

Synthesis of a 1,3,5-triazine derivative via SPOT synthesis method

Equipment and reagents

- PP-g-P(AmPEGMAm) membrane (see *Protocols 11* and *12*)
- Microwave oven (600 w)
- Teflon microtitre plate for cleavage procedure

- 2-aminoethyl-morpholine Cyanuric chloride, DABCO (1,4-diazabicyclo[2,2,2]octane), DCM, Fmoc-Phe-OPfp, methanol, NMP, 3-picolylamine piperidine, TFA

Protocol 14 continued

Method

1 Follow *Protocol 13*, steps 1–5.

2 Spot 2.0 μl of a 0.3 M solution of Fmoc-Phe-OPfp in DMF. Repeat the procedure once after 15 min. Wash the membrane with DMF and methanol (3 × 5 min) and dry the membrane under ambient conditions.

3 Follow *Protocol 13*, step 5.

4 Bath the membrane in a 3 M solution of cyanuric chloride in DCM containing 5 mol% DABCO (4°C, 20 min). Remove the excess of cyanuric chloride and precipitated salts by washing with DMF, methanol, and DCM.

5 Replace the second chlorine of the triazine by reacting the spots with 1.5 μl of a 4 M solution of the 3-picolylamine in NMP at room temperature. Wash the membrane after 30 min with DMF, methanol, and DCM.

6 Spot 2.0 μl of a 80% solution of 2-aminoethylmorpholine on the air dried membrane and heat the wet membrane in a microwave oven (3 min/600 W) to yield the fully substituted polymer-bound 1,3,5-triazine.

7 For cleavage of the 1,3,5-triazine derivative (2-phenylalanine-4-N-(2-aminoethylmorpholino)-6-(3-picolyl)-s-triazine) from the solid support the SPOT is punched out into a Teflon microtiter plate and treated with 80–100 μl of 90% TFA/DCM for 45 min. Removal of the cleavage solution by a gentle nitrogen stream yields the 1,3,5-triazine derivatives.

8 Dissolve and analyse the product by LC-MS (*Figure 2* please see page 115).

4.3 Comments on the functionalization procedures and applications of the novel polymeric membranes

(a) Selection of the support membrane material involves finding a compromise between (i) *high mechanical stability*, best achieved by a robust woven structure and a (ii) *large specific surface area* yielding high binding capacity, best achieved by a sponge-like phase inversion morphology. The first criteria imposes a limit on the specific surface area and pore size, and the second criteria is limited by the inherent mechanical instability of highly porous materials.

(b) The efficiency of the photoinitiator coating along with its stability in presence of the monomer solutions (best achieved with water, saturated with photoinitiator) and the selective UV excitation conditions (UV absorption only by the photoinitiator) are essential for effective surface functionalization.

(c) For a homogeneous modification wetting of the hydrophobic PP is critical; this is achieved by solvent exchange during the photografting procedure (*Protocols 9* and *11*).

(d) If feasible, UV irradiation should be performed under anaerobic conditions in order to minimize crosslinking of the grafted polymer chains via per-

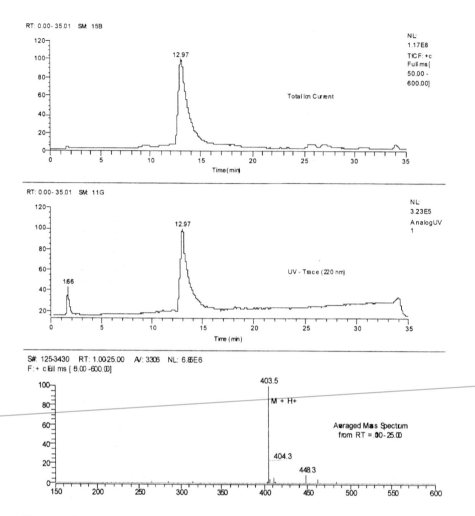

Figure 1 LC-MS profile of *N*-benzyl-*N*-[butyl-(carbamoylmethyl carbamoyl-methyl)-2-[piperidine-1-yl]acetamide assembled via the SPOT synthesis method on a polymeric membrane.

oxides (this crosslinking will reduce the amount of reactive groups accessible for solid phase synthesis). For the technical scale functionalization this is not feasible, but alternatively an oxygen scavenger can be added to the monomer solution.

(e) Besides gravimetry, the functionalization and its efficiency for solid phase synthesis can be monitored qualitatively by FT-IR spectroscopy, and both qualitatively and quantitatively by staining via ion exchange or reactive binding of dyes.

(f) When larger areas of continuous polymeric membranes are used, a stainless steel support frame is advantageous in maintaining the spots alignment during the synthesis.

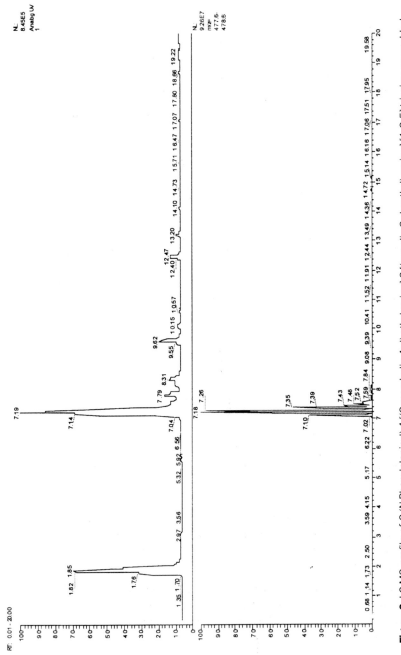

Figure 2 LC-MS profile of 2-(N-Phenylalaninyl)-4-[(2-morpholin-4-yl)-ethylamino]-6-[(pyrudin-3-ylmethyl)-amino]-[1,3,5]-triazine assembled via the SPOT synthesis method on a polymeric membrane.

(g) According to the results obtained so far on the synthesis of PNA-, peptoid-, and 1,3,5-triazine-based libraries on polymeric membranes, the increased chemical and mechanical stability together with the defined number of functional groups of the novel carrier clearly increase the scope of the SPOT technology in solid phase organic synthesis.

References

1. Pirrung, M. C. (1997). *Chem. Rev.*, **97**, 4738.
2. Frank, R. (1992). *Tetrahedron*, **48**, 9217.
3. Frank, R., Hoffmann, S., Kieß, M., Lahmann, H., Tegge, W., Behn, C., *et al.* (1996). In *Combinatorial peptide and nonpeptide libraries* (ed. G. Jung), p. 363. VCH, Weinheim.
4. Abimed Analysen-Technik GmbH, Raiffeisenstr. 3, D-40736 Langenfeld, Germany. User manual of the ASP 222 Automated Spot Robot.
5. Reineke, U., Sabat, R., Volk, D., and Schneider-Mergener, J. (1998). *Protein Sci.*, **7**, 951.
6. Kramer, A., Volkmer-Engert, R., Malin, R., Reineke, U., and Schneider-Mergener, J. (1993). *Peptide Res.*, **6**, 314.
7. Tegge, W. and Frank, R. (1998). *Methods Mol. Biol.*, **87**, 99.
8. Frank, R. and Overwin, H. (1996). *Methods Mol. Biol.*, **66**, 149.
9. Kramer, A. and Schneider Mergener, J. (1998). *Methods Mol. Biol.*, **87**, 25.
10. Ast, T., Heine, N., Germeroth, L., Schneider-Mergener, J., and Wenshuh, H. (1999). *Tetrahedron Lett.*, **40**, 4317.
11. Nielsen, P. E., Egholm, M., Berg, R. H., and Buchardt, O. (1991). *Science*, **254**, 1497.
12. Weiler, J., Gausepohl, H., Hauser, N., Jensen, O. N., and Hoheisel, J. D. (1997). *Nucleic Acids Res.*, **25**, 2792.
13. Gausepohl, H., Weiler, J., Schwarz, S., Fitzpattrick, R., and Hoheisel, J. (1998). In *Peptides 1996* (ed. R. Ramage and R. Epton), p. 411. Mayflower Scientific Ltd., Kingswinford.
14. Gausepohl, H. and Behn, C. (1998). In *Peptides 1996* (ed. R. Ramage and R. Epton), p. 409. Mayflower Scientific Ltd., Kingswinford.
15. PerSeptive Biosystems, Framingham, USA (now Perkin Elmer, Applied Biosystems Division, Foster City, CA). Manual for PNA synthesis on the Expedite oligonucleotide synthesizer.
16. Ulbricht, M., Oechel, A., Lehmann, C., Tomaschewski, G., and Hicke, H. G. (1995). *J. Appl. Polym. Sci.*, **55**, 1707.
17. Ulbricht, M., Matuschewski, H., Oechel, A., and Hicke, H. G. (1996). *J. Membr. Sci.*, **115**, 31.
18. Ulbricht, M. (1996). *React. Funct. Polym.*, **31**, 165.
19. Carpino, L. A., Sadat-Aalaee, D., Chao, H. G., and DeSelms, R. H. (1990). *J. Am. Chem. Soc.*, **112**, 9651.
20. Zuckermann, R. N., Kerr, J. N., Kent, S. B. H., and Moos, W. H. (1992). *J. Am. Chem. Soc.*, **114**, 10646.

Chapter 6

Resins and anchors for solid phase organic synthesis

Martin Winter and Ralf Warrass[†]

University of Tuebingen, Institute of Organic Chemistry, Auf der Morgenstelle 18, D-72076 Tuebingen, Germany.
[†]Institut de Biologie et Institut Pasteur de Lille, Département V - URA CNRS 1309, 1 rue du Prof. Calmette - BP 447, 59021 Lille, France.

1 Introduction

1.1 Why do organic synthesis on a solid support?

Isolation of the pure intermediates and products in a multi-step synthetic scheme is often the most time-consuming and uneconomical part in the overall process. This effort increases proportionally with the number of synthetic steps. The extent of this problem is greater when several different compounds are to be made in parallel as is the case in combinatorial chemistry. Differences in the physical and chemical properties of members of a chemical library, such as solubility, pK_a values, polarity of the building blocks and intermediate products, aggravate to a great extent the standardization of a procedure or may even make it completely impracticable.

Since its inception in 1961 by Merrifield (1), solid phase peptide synthesis (SPPS) has not only allowed great progress in immunology (2), but extended beyond into combinatorial chemistry to include the synthesis of small organic molecules with medicinal properties. The enormous advantage of solid phase synthesis is the result of the simplified purification of intermediate products in between repetitive synthetic steps and their amenability to automation. Excess reagent may be removed by simple filtration and washing of the solid support, while the substrate remains covalently bound to the support. After the synthesis is completed the target compounds can be released from the support by a cleavage reagent and are thus usually isolated from the resulting cleavage filtrate as pure compounds.

1.2 The orthogonality principle

From the synthetic chemical point of view the solid support is comparable with the use of a semi-permanent protecting group. In this context the word *semi-*

117

permanent indicates that this protective group, present at the beginning of the solid phase synthesis, will not be removed before the last step of the synthesis. In order for a solid support to fulfil the orthogonality principle, none of the chemical steps of the solid phase synthetic scheme may chemically modify or otherwise affect the support nor prematurely release the compound from it. This principle will thus dictate the chemical nature of the polymeric support as well as the type of bond that links it to the synthesized molecule. Various linker and anchor strategies are described in Section 3.

1.3 The practice of solid phase synthesis

1.3.1 Manual synthesis on solid support

The growing importance of single compound arrays led to the development of multiple parallel synthesis techniques allowing the handling of a large number of parallel reactions. The ideal reactors for such applications must be:

- affordable
- robust
- easy to handle in filtration, reagent addition, mixing, and heating/cooling steps

Simple filter columns derived from standard injection syringes meet these requirements (*Figure 1*).

1.3.2 Automated synthesis

Due to the increased number of parallel syntheses in the production of single-compound arrays, pipetting and filtration steps become excessively time-consuming. Moreover, work-up procedures involving evaporation, extraction, weighing, and characterization require even more effort, and have to be scheduled and interconnected by transport and liquid handling systems.

In recent years, robotic systems derived from peptide synthesizer were developed that allow the preparation of single-compound arrays as well as parallel product purification. They are comprised of a central robotic arm and peripheral devices for pipetting, filtration, heating/cooling the reaction chambers under inert atmosphere, evaporation, extraction, and weighing steps.

2 Solid supports

2.1 Physical properties of solid supports

2.1.1 Swelling

The expansion of the resin beads' volume by solvation of its polymeric network is called 'swelling'. The tendency of the polymer to dissolve in an appropriate solvent depends mainly on the polarity of the solvent and the degree of polymer crosslinking (3). Furthermore, when the polymer is subjected to an external mechanical pressure, such as through packing in a continuous-flow column, the swelling is prevented and as a result the solvation process is attenuated.

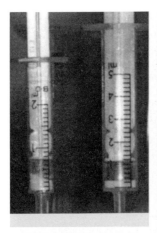

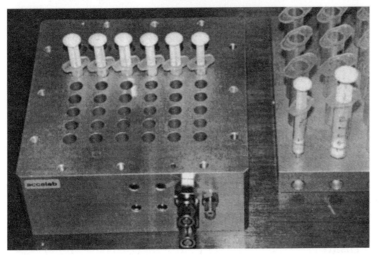

Figure 1 Filter columns (top left) and reactor blocks (top right and bottom) for solid phase synthesis with heating/cooling, mixing, and under inert atmosphere capabilities.

2.1.2 Crosslinking

Depending on the degree of crosslinking the support could be either a microporous gel-like polymer (low crosslinking) or a rigid macroporous solid (high crosslinking, > 5%). The number and accessibility of anchor positions (reacting sites) are considerably higher with the former type. Gel-like polymer undergo better solvation allowing nearly free mobility of the polymer chains, which in turn provide a solution-like environment for the resin-bound substrates (3) (see *Figure 2*).

2.1.3 Diffusion

Provided that a sufficiently fast exchange of low molecular weight reagents prevails between the polymer particle interior and the surrounding reagent solution through sufficient swelling of the support, the rate of solid phase reaction is

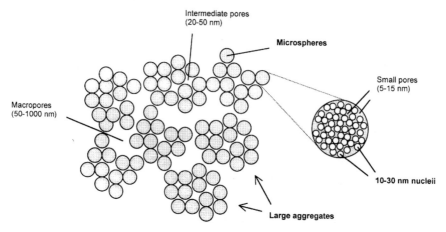

Figure 2 Structural domains of macroporous polystyrene (3).

Table 1 Reaction conditions determining diffusion in solid phase synthesis

High diffusion rates (fast reactions)	Low diffusion rates (slow reactions)
Highly swollen polymer	Unswollen polymer
Similarity between reagent and polymer interior polarity	Difference between reagent and polymer interior polarity
Low molecular weight reagents	High molecular weight reagents

generally not diffusion-controlled. The dependency on particle dimensions, reagent polarity and size, solvents, and support material is summarized in *Table 1*.

2.1.4 Spacer effects

A measure of the similarity of a solid phase system to a solution phase system may be quantified through the measurement of the NMR (nuclear magnetic resonance) relaxation time of support-bound substrates. Rapp and co-workers have measured ^{13}C NMR relaxation rates of a tripeptide bound to different polymers, and established the role of the spacer between the support's polymeric core and the bound substrate. For instance, it was found that the effect of a POE (polyoxyethylene) confers a physical mobility to the resin-bound tripeptide similar to that of a solution phase tripeptide methyl ester (5) (see *Figure 3*).

2.1.5 Polarity effects

When suspended in an appropriate solvent the polymer should both swell and allow the penetration of solvated reagents and reactants into the polymer network. If these chemicals are not compatible with the polymer they will be found at lower concentrations in the interior of the resin beads (3). The choice of the appropriate solvent or combination of solvents is therefore of utmost importance.

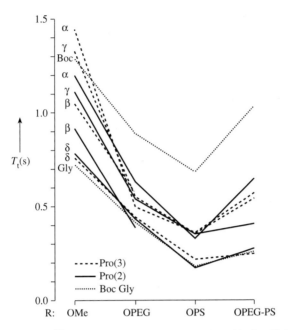

Figure 3 ^{13}C NMR relaxation times of the tripeptide Boc-Gly-Pro-Pro methyl ester (OMe), POE ester, and polystyrene-bound POE ester (TentaGel). The spacer effect of the POE moiety in TentaGel leads to similar relaxation times as those of the POE ester in solution (5). (Reprinted with permission, WILEY-VCH.)

2.2 Families of solid supports and their specific properties

2.2.1 Polystyrene

The polystyrene supports used for solid phase organic synthesis (SPOS) are normally crosslinked by addition of 1% divinylbenzene (DVB) to the polymerization mixture. This degree of crosslinking offers the best compromise between mechanical stability and swelling performance. The term 'degree of crosslinking' is often used synonymously with the portion of DVB in the monomer mixture. When the degree of crosslinking is above 5% an increasing amount of vinyl moieties remain unreacted and thus the portion of DVB in the monomers mixture does no longer reflect the degree of crosslinking. Highly crosslinked polystyrenes have a higher mechanical and thermal stability, but their swelling performance and loading capacity are clearly reduced.

Polystyrene is presently the most used support material in SPOS. These supports generally distinguish themselves from others by their good swelling performance (up to 8 ml/g of resin depending of the solvent) and high loading, but are limited in terms of mechanical and thermal stability. Although in many cases the functionalization with a linker has a significant influence on the physical properties of PS-based resins, the properties of the polymer usually dominate. PS-based resins are not suited for continuous-flow synthesis (4) or reaction temperatures above 100 °C. Solvents such as dichloromethane, dimethyl-

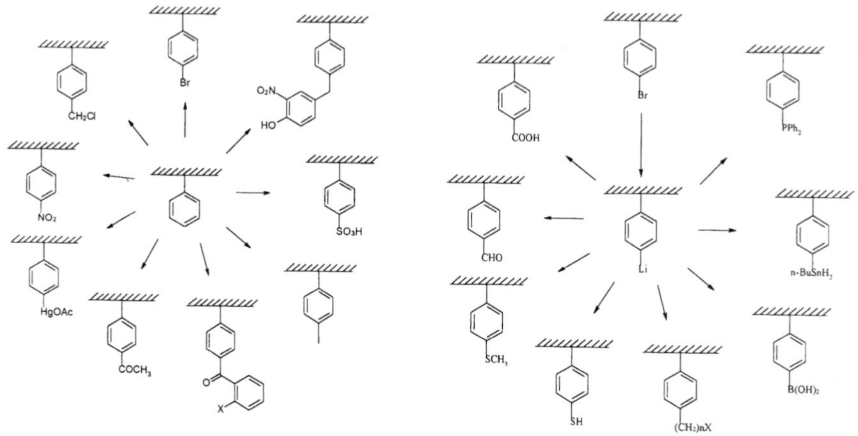

Figure 4 Synthetic pathways to functionalized polystyrene.

formamide, swell PS resins while solvents like methanol, ether, or water 'shrink' it and induce a loss of its gel-like properties (5); a property that can be useful during filtration and washing steps.

The chemical properties of polystyrene are dominated by the phenyl residues of the polymeric backbone. Over 99% of the phenyl residues are unaffected by the crosslinking process and remain chemically accessible. As a result, a large number of aromatic substitution reactions allowing the chemical modification of the polymer with anchors and linker residues (3) can be performed. The frequently used functionalization reactions are shown in *Figure 4*.

Merrifield's resin is a 2% divinylbenzene (DVB)/polystyrene (PS) co-polymer, which upon chloromethylation yields reactive benzyl chloride functional groups (3). The polymeric support is available as small, mostly beaded particles of 20–150 μm in diameter. In SPPS The first *N*-protected amino acid could be coupled via its carboxyl group. The resulting ester bond is to a great extent stable against subsequent deprotection and coupling steps involved in peptide chemistry but can be cleaved at the end of the synthesis by fluorhidric acid (HF) or trifluoromethanesulfonic acid (TFMSA). Although the strongly acidic cleavage conditions are not completely harmless to the resulting peptides, the ortho-gonality principle is somewhat preserved.

The demand for higher yields and product purity fuelled the development of new polymeric supports with improved polarity, swelling, diffusion, and mobility; these properties are crucial for the immobilized reacting molecules and incoming reagents to react in a solution phase-like environment.

2.2.2 Polystyrene–polyoxyethylene (POE) co-polymers

TentaGel

The polystyrene–polyoxaethylene (PS-POE) graft co-polymer known as 'TentaGel' consists of a PS scaffold functionalized with hydroxyethyl residues on which a

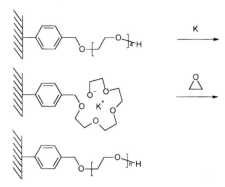

Figure 5 Synthesis of TentaGel resins.

subsequent anionic polymerization of ethyl oxirane leads to the formation of polyoxyethylene tentacles. The physico-chemical and mechanical properties of the parent polymer are thereby altered by the up to 70% in weight of the POE chain (5, 7) (*Figure 5*).

Compared to PS the substantially better swelling properties of TentaGel in organic and polar solvents, including water, make this support a powerful alternative to PS in SPOS. The exceptional swelling properties of this support result from the amphiphilic properties of the POE. These tentacles exist in various helical 'secondary structures' that can present both polar and apolar surfaces depending on the solvent. It is interesting, however, that diethyl ether is a unique exception in which PS-POE is not solvated. The mechanical stability of TentaGel resins resulting from the strong solvation of the POE chains is re-markably increased compared to standard polystyrene resins, allowing thereby its use in continuous-flow procedures. TentaGel's intermediate loading capacity (0.1–0.4 mmol/g) between the high load PS resins and lower load of other continuous flow supports (see below) makes it particularly suitable for the large scale solid phase synthesis of oligonucleotides. Like PS-based resins, TentaGel is also available with various linkers allowing the coupling with a wide variety of building blocks.

The hygroscopic character of the POE tentacles (propensity to absorb and retain moisture), and their chemical instability under strongly basic conditions may however compromise certain SPOS reactions on this support.

2.2.3 Polyamide supports

i. Pepsyn

Pepsyn is a polyamide support which has been developed especially for peptide synthesis (6). It is a co-polymer of N,N-dimethylacrylic amide, *bis*(acrylamido-ethane), and N-acryloylsarcosine methyl ester. The latter monomer serves to introduce anchor points for SPPS and SPOS (*Figure 6*).

The development of Pepsyn was caused by the apprehension that substrate and polymer should have similar polarity in order to obtain the best synthetic

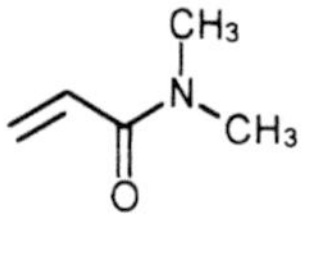

N,N-dimethylacrylamide (monomer)

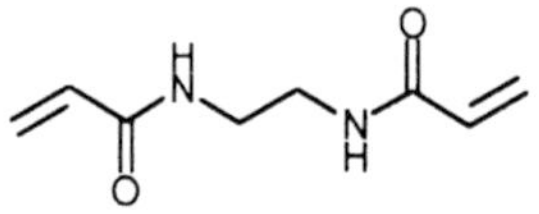

Bis(acrylamido)ethane (crosslinker)

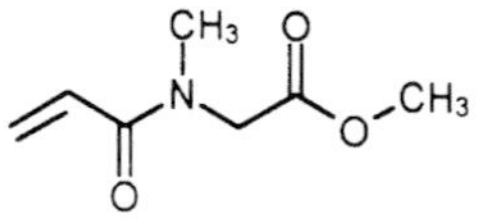

N-Acryloyl sarcosine methyl ester (functionalisation)

Figure 6 Monomers for Pepsyn preparation.

results. The polyamide material fulfils very well this requirement. Nevertheless, as a result of their lower mechanical stability, Pepsyn supports were generally unsuccessful in routine SPPS/SPOS. This limitation was overcome by polymerizing N,N-dimethylacrylamide bis(acrylamidoethane) and N-acryloylsarcosine methyl ester into macroporous inorganic (Kieselguhr) or organic (polystyrene) particles (12).

ii. Pepsyn K (Kieselguhr-supported polyamide)

The first continuous-flow SPPS support resins consisted of Kieselguhr-supported polyamide. The macroporous inorganic Kieselguhr particles provide the necessary mechanical stability for continuous-flow applications, while the built-in polyamide serves as a support and determines the solvation of the bound peptide (12).

The size of the pores of the Kieselguhr particles impose a limit on the loading of the support. Thus, when Kieselguhr particles are heavily loaded, they become inaccessible to reagents and lead to problems of solvent flow in continuous-flow columns. The optimal loading capacities are relatively low (< 0.1 mmol/g). For this reason Kieselguhr-based supports are increasingly rivalled by their organic counterpart (e.g. polyHIPE, TentaGel).

iii. PolyHIPE

PolyHIPE (HIPE = high internal phase emulsion) is a stable, porous network of PS-DVB with very low density (90% of the total volume is void space) in which the precursors of polyamide were co-polymerized. The advantages of polyHIPE over Kieselguhr are its higher loading capacity (up to 5 mmol/g) (8), its compatibility with a large number of solvents, and the better anchorage of the resin (polyamide) to the support matrix (PS/DVB).

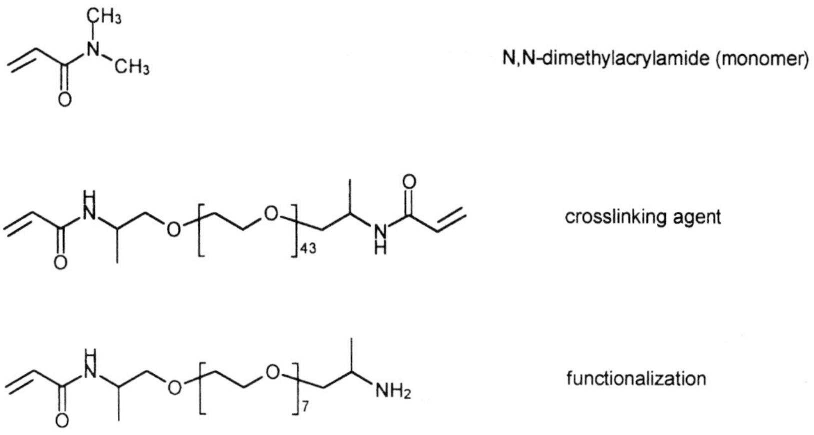

Figure 7 Monomers for the synthesis of PEGA resins.

2.2.4 Poly(acrylic amide–ethylene glycol) co-polymers

i. PEGA

Among the continuous-flow supports one of the most recent and successful developments is the acrylamidopropyl-POE-*N,N*-dimethylacrylamide co-polymer PEGA (9) (*Figure 7*). Due to its amphiphilic properties, PEGA resin is outstandingly suitable for the SPOS of polar compounds.

PEGAs high degree of crosslinking and its optical transparency in the aromatic absorption region are particularly useful properties in continuous-flow SPOS and *in situ* monitoring. Furthermore, the high degree of crosslinking does not lead to reduced swelling properties. Equal swelling volumes are observed in DCM, alcohols, and water (~ 6 ml/g) and up to 8 ml/g in DMF. The latter property meets particularly the requirements for the synthesis of large molecules (e.g. long peptides). The support material shows a weak adhesion to glass and metal but not to plastic or Teflon surfaces (9).

ii. CLEAR

Crosslinked ethoxylate acrylate resin (CLEAR) (10), consist of highly crosslinked polyoxyethylene and various acrylate units (*Figure 8*).

Due to their high crosslinking and high solvation energy, these supports are suitable for continuous-flow procedures and have good swelling properties in polar as well as non-polar solvents.

iii. POEPS and POEPOP

Further development of the PEGA crosslinking principle led to POEPS and POEPOP resins (11). Polyoxyethylene chains are used as crosslinking agents by *bis*-terminal functionalization with vinylbenzylic chloride (in the case of POEPS) or epichlorohydrin (in the case of POEPOP) prior to polymerization (*Figure 9*).

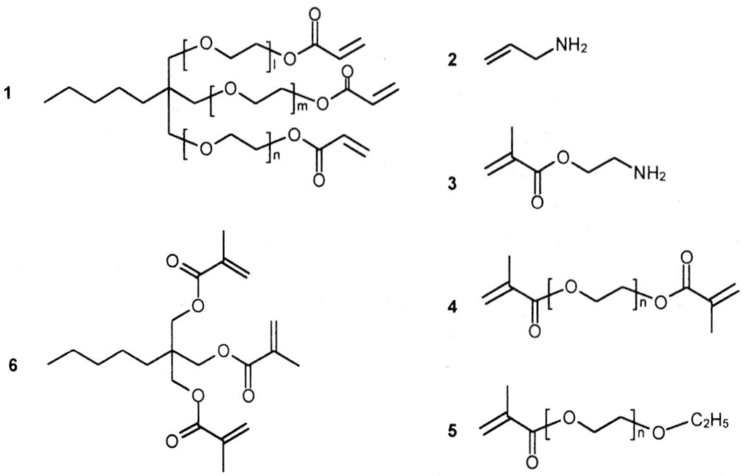

Figure 8 CLEAR monomers. 1, 4, and 6 are crosslinking units and 2, 3, and 5 are functionalization units.

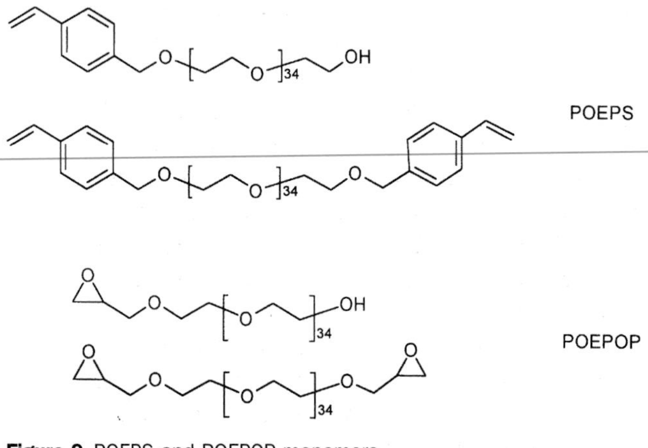

Figure 9 POEPS and POEPOP monomers.

These resins are distinguished by their chemical and mechanical stability and, owing to the POE units, a broad solvent compatibility. The high solvation energy of the POE units results in enormous swelling volumes (up to 50 times the dry volume when long-chain POE are employed), and a very high mechanical stability of the polymer beads (when short-chain POE are employed). Crosslinking by large bifunctional polymeric agents enables macromolecules the size of enzymes to access the interior of the bead. For this reason, POEPS and POEPOP resins are increasingly used in support-bound enzyme assays (11). Although these polymers are commercially available, their preparation is relatively straightforward.

126

2.2.5 Inorganic supports

Controlled pore glass/controlled pore ceramics

Although organic polymers are the most widely used supports, they suffer from limited thermal and mechanical stability, whereas inorganic materials do not show such limitations. Controlled pore glass (CPG) and controlled pore ceramic (CPC) supports, originally developed as chromatographic supports, are typical examples of such materials. The pore size of the support particles, which determine the accessibility of the inner domain of the bead, can be exquisitely controlled during the production process. These supports are particularly attractive in continuous-flow synthesis and were for that reason heavily used in oligonucleotide synthesis since the 1980s. Their loading capacities (>0.2 mmol/g) exceeds that of Pepsyn K supports.

3 Linkers and anchors in SPOS

The higher yields achieved through multiple couplings and addition of excess reagents, the easy purification achieved by simple filtration of the resin, and its amenability to automation are the key advantages of SPOS. This technique requires, however, the temporary attachment of a portion of the compound to be synthesized to the solid support. Once the synthesis is completed, the compound can be submitted to a bioassay while still attached to the matrix or in most cases released from the support for solution phase assay. Thus, the chemistry associated with the linker is central in the design of a SPOS scheme.

A linker is usually a bifunctional molecule that serves to connect the first building block of the synthesis to the matrix in a stable but reversible manner. The role of the linker is also to minimize the effects of the resin on the progress of the synthesis or the subsequent biological assays to be performed on the compound. A linker should be chemically stable throughout the synthesis, chemically compatible with both the matrix and the compound synthesized, and selectively cleavable to allow the release of the compound from the resin at any time of the synthesis. Furthermore, the linker should be chosen so as to minimize its structural and chemical effects on the sought after properties of the synthesized compound.

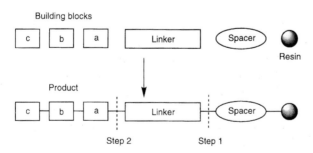

Figure 10 Components of a solid phase organic synthetic scheme (resin, spacer, linker, anchor, building blocks, immobilized product).

3.1 Choosing the right linker

The resin/linker may be considered as a protecting group of one of the functional groups of the final product. Over the past thirty years, a large variety of acid- and base-sensitive linkers have been developed. But, since the outburst of combinatorial chemistry, this area has seen the development of more sophisticated and elegant anchoring systems. Thus, besides the classical acid/base and nucleophilic mechanisms of cleavage, the newly developed anchors can be cleaved photolytically, enzymatically, by hydrogenolysis, or with transition metals (12–15).

Silicon-based anchors were introduced as 'traceless' points of attachment because they do not leave any chemical trace on the synthesized compound. 'Safety-catch' linkers are a class of linkers that require chemical modification, which renders the linker chemically labile to a second cleavage reagent. This approach makes product release more specific and the linker more stable throughout the synthetic scheme. Below is a selection of anchors that shows the variety available to a solid phase organic chemist.

3.1.1 Acid-labile anchors

Besides the linkers based on 2,3-dihydro-4H-pyran (16–18), well-known acid-labile protecting group for alcohols, most of the acid-labile linkers can be classified according to their ability to stabilize the benzylic carbocations generated during the cleavage reaction. Generally, better stabilization results in easier cleavage. As a result, a broad spectrum of acidic conditions can be used depending on the linker chosen and the chemical stability and synthetic requirements of the target molecule. Anchors requiring strongly acidic cleavage conditions give the most stable attachments of the synthesized products to the polymer and hence premature leakage of the product is minimized. When the compound bears various acid-labile side-chain protecting groups, linkers that are labile under weakly acidic conditions become advantageous.

Table 2 reveals the acid lability of various linkers and the structural basis that led to it. From left to right (benzyl-, benzhydryl-, and trityl-), the electron delocalization increases along with the acid lability. From top to bottom the influence of electron donating substituents is unveiled. With the help of this table one can fine-tune the conditions under which the linker can be cleaved. The most frequently used support in SPPS is the Wang resin which is endowed with a HMP (hydroxymethylphenoxy-) linker that was designed to synthesize carboxylate-terminated compounds. Further developments of this resin resulted in the generation of alcohol- (19), amine- (20), phenol- (21), and carboxamide-terminated (22) compounds. For amide- or sulfonamide-terminated (23) compounds the Rink amide resin is the most popular choice.

3.1.2 Nucleophile-labile anchors

Nucleophile-labile anchors were introduced to overcome the sensitivity of certain target molecules to acidic or basic conditions, or the acid/base conditions

Table 2 Acid-labile linkers and their cleavage conditions

Benzyl anchors	Benzhydryl anchors	Trityl anchors

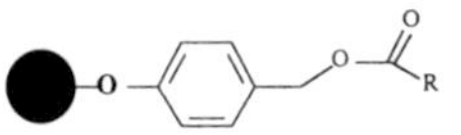

Merrifield resin
(HF) (24) |

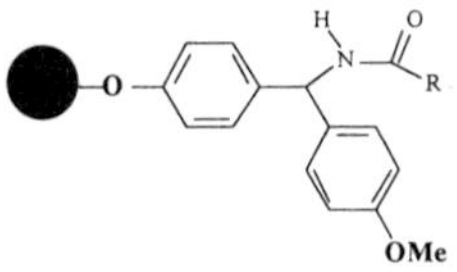

BHA resin
(TFMSA) (25) |

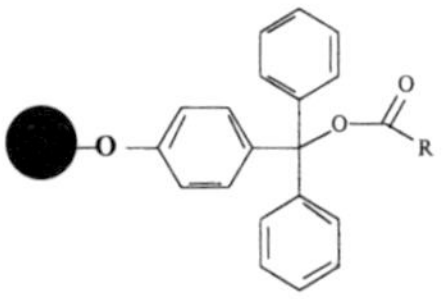

Trityl resin
(2% TFA) (26) |
|

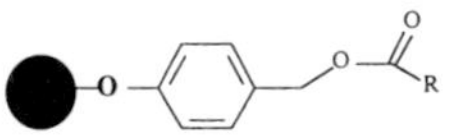

Wang resin
(95% TFA) (27) |

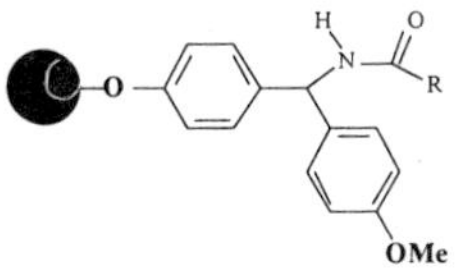

MOBHA resin
(HF) (28) |

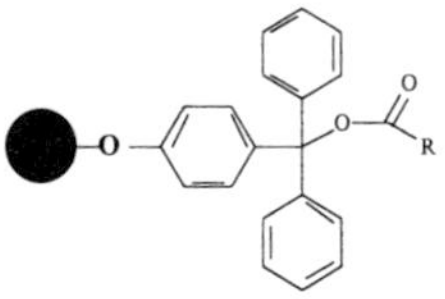

4-Hydroxytrityl resin
(1% TFA) (29) |
|

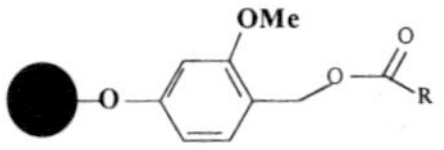

SASRIN
(1% TFA) (30, 31) |

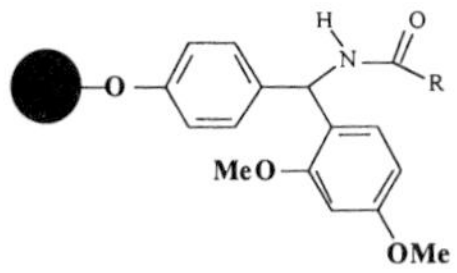

Rink-amide resin
(TFA) (32) |

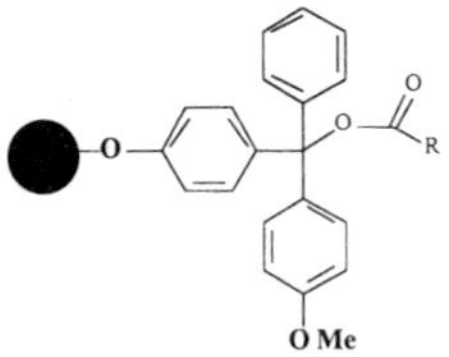

4-Methoxytrityl resin
(1% TFA) (26) |
|

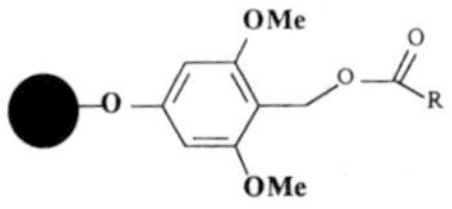

HAL resin
(0.1% TFA) (33) |

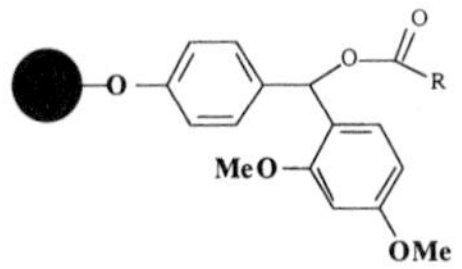

Rink-acid resin
(0.2% TFA) (32) | |

imposed on the linker by the target's synthetic scheme. It is a process where a nucleophilic displacement is used to release the compound from the resin, a process that results, generally, in a chemically modified compound (see *Table 3*).

The regenerated Michael addition (REM) resin for instance is compatible with acid/base conditions and allows the anchoring of primary or secondary amines through Michael addition. The cleavage is effected by Hoffmann elimination which results in the release of the compound as a tertiary amine. Both HMBA and bromoacetyl resins are stable toward strong acids. By treating these resins with NaOH, MeOH/Et$_3$N, NH$_3$ aq., or hydrazinhydrate the product is liberated as carboxylic acid, carboxyester, carboxamide, or as carboxyhydrazide respectively.

1-(9-Hydroxmethylfluorenyl) resin is a typical example of the application of a known protecting group as an anchorage system (36). Primary, secondary, as

129

Table 3 Nucleophilic-labile linkers and their cleavage reagents

REM resin
(DIEA) (34, 35)

1-(9-Hydroxmethyl)-
fluorenyl resin
(15% piperidine) (36)

N-methoxypropionic acid
amide resin
(LiAlH₄) (37)

HMBA resin
(NaOH, MeOH, NH₃,
hydrazine) (38)

Bromacetyl resin
(NaOH, hydrazine,
amine) (39)

Kaiser oxime resin
(amine, hydrazine) (40)

well as tertiary amines can abstract the 9H-atom of the fluorenyl ring, thus re-leasing the product from the resin as an amino-terminated compound (*Table 3*).

N-methoxypropionic acid amide resin can be used to immobilize carboxylic acids, which can than be released as their corresponding aldehyde using Weinreb's method (37). This resin anchor shows a remarkable resistance to strong acids (TFA, HF) and amines (piperidine).

Kaiser's oxime resin (*Table 3*) yields hydrazones, amides, and carboxylic acids upon cleavage with hydrazine, amines, or hydroxypiperidine/Zn/AcOH, respectively.

3.1.3 Others

When photo-labile anchors are used (see *Table 4*) the product is released by irradiation at λ = 320–365 nm, thus providing an elegant alternative strategy. The photocleavage can also be performed in aqueous solutions, which renders this approach particularly attractive for biological assays. Carboxylic acid-, carboxamide-, amidine- (41), and hydroxy-terminated compounds can be liberated as final products. The 2-methoxy-5-nitrophenoxy resins are stable to acids and bases while the bromoethyl-3-nitrophenyl resin is stable only to acids and can be cleaved by hydrolysis. The methyl group in the proximity of the cleaving site is necessary since a ketone instead of an aldehyde is formed after photolysis. The formation of an aldehyde would lead to absorption of the necessary radiation for cleavage, thus prolonging the time required to accomplish the photocleavage.

The safety-catch anchor liberates the synthesized product following a sequence of two discrete steps. The first reaction can be regarded as a switch that turns the secured product into an activated and cleavable one. The second

Table 4 Photolabile anchors

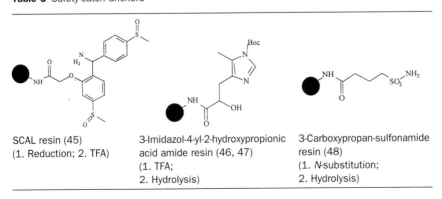

1-Bromoethyl-3-nitrophenyl resin (42, 43)	4-(2-Aminoethyl)-2-methoxy-5-nitrophenoxy resin (44)	4-(2-Hydroxyethyl)-2-methoxy-5-nitrophenoxy resin (41, 44)

step liberates the product. Many different and elegant two-step release strategies have been applied in SPOS and three of them are shown in *Table 5*. The acid/base-resistant safety-catch amide linker (SCAL) is first activated by the reduction of the electron withdrawing sulfoxy groups to their corresponding sulfides. This activation step causes the benzhydrylamine bond to become acid-labile. 3-Imidazol-4-yl-2-hydorxypropionamid resin takes advantage of the powerful intramolecular catalysis provided by the imidazol group. Upon treating the resin with TFA, the Boc-protecting group of the imidazol is removed and an intramolecularly catalysed hydrolysis of the ester bond linkage by the imidazole leads to the liberation of the product under neutral, aqueous conditions. The direct cleavage into a bioassay medium is equally feasible. In the third example, carboxylic acids are immobilized on 3-carboxypropan-sulfonamide resin to form stable amide bonds. At the end of the synthesis, the sulfonamide is activated by reaction with bromoacetonitrile forming a tertiary amide. This structure can then be cleaved by aminolysis or hydrolysis to form carboxylic acids or carboxamides, respectively.

Generally, when a compound is released from a solid support it carries with it the chemical information that served to keep it bonded to the resin. Prompted by the necessity to minimize the effect of such functional groups on the subsequent physico-chemical and pharmacological properties of the synthesized compound, so-called 'traceless' anchors were developed. These are silicon-based linkers (*Table 6*) that do not result in the formation of a new

Table 5 Safety-catch anchors[a]

SCAL resin (45) (1. Reduction; 2. TFA)	3-Imidazol-4-yl-2-hydroxypropionic acid amide resin (46, 47) (1. TFA; 2. Hydrolysis)	3-Carboxypropan-sulfonamide resin (48) (1. *N*-substitution; 2. Hydrolysis)

[a] The reagents for activation (1.) and cleavage (2.) are given in parentheses.

131

Table 6 Traceless anchors and their cleaving conditions

4-(4-(Hydroxymethyl)-phenoxymethyl-
dimethylsilyl-bromo benzene resin (49)
(TFA or CsF)

4-Methoxymethyl-oxyphenyl-diisopropyl
-hydroxysilyl resin (50)
(TBAF)

functional group on the synthesized compound upon cleavage. In this case the cleavage is often achieved by fluoride salts or by ipso-substitution with C–H, C–I, or C–Br at the Si–C bond.

Although hydrogenolysis is a very selective and mild cleavage method it is not widely applied in solid phase synthesis (*Table 7*). Upon hydrogenolysis with dihydrogen or dicyclohexa-1,4-diene in the presence of Pd(OAc)$_2$, the *p*-nitrobenzyhydrylamine yields carboxamide-terminated compounds. Carboxylic acids, immobilized on Merrifield resin, can also be cleaved by hydrogenolysis or, alternatively, by ammonium formate/Pd(OAc)$_2$.

Enzyme promoted release of the product from the solid support offers an interesting alternative to the chemically promoted ones (see *Table 7*). The specificity of enzymes ensures that no other part of the synthesized molecule is altered during the cleavage process.

Allylfunctionalized anchors (see *Table 7*) are stable to strong acids (TFA), and bases (pyridine, NaOCH$_3$). The mechanism of cleavage involves transition metals that form an intermediate π-complex with the double bond. Upon adding an H$^+$-donor, the unsaturated bond is opened and the product released from the support. More recently, syntheses have been reported wherein a second alkene has been generated during or at the end of the solid phase synthesis. Following

Table 7 Anchors cleaved by hydrogenolysis, Pd catalysed, and enzymes[a]

p-Nitrobenzhydrylamine resin (51)
(H$_2$ or dicyclohexa-1,4-diene (Pd(OAc)$_2$)

Phenylalanine ester resin (52)
(α-chymotrypsin)

Hycron resin (53, 54)
(Pd(0)/*N*-methylaniline)

4-Hydroxy-2-en-oxy-acidic acid
amide resin (55)
(PdCl$_2$ + *n*Bu$_3$SnH + H$^+$ donor)

[a] Cleaving reagents are given in parentheses.

132

the addition of a ruthenium complex, a ring closing olefin metathesis leads to the release of the product from the solid support (56).

3.2 Introduction of the first building block

Two different strategies are available:

- The consecutive attachment of the linker and the first building block to the resin.
- Combination of the linker and the first building block in solution prior to their attachment to the resin.

The latter is chosen when the attachment of the first building block proves inefficient. *Protocols 1–4* present procedures describing the attachment of the first building block to the most commonly used resin–linker combinations.

Protocol 1

Attachment of carboxylic acids to PS/DVB-HMBA resin

Equipment and reagents

- Round-bottom flask (RBF)
- HMBA resin
- Dichloromethane (DCM)
- Dimethylformamide (DMF)
- Diisopropylcarbodiimide (DIC)
- 4-Dimethylaminopyridine (DMAP)

- Diisopropylethylamine (DIEA)
- 2-(1H-benzotriazole-1-yl)-1,1,3,4,-tetramethyluronium tetrafluoroborate (TBTU)
- N-hydroxybenzotriazole (HOBt)

Method

1 Place 5 g of HMBA resin in a 100 ml RBF.

2 Swell resin in 30 ml of DCM and filter off the solvent (three times).

3 In an RBF dissolve 20 Equiv.[a] of a carboxylic acid in DCM/DMF (10:1) to a final concentration of 0.5–1 mM.

4 Add 10 Equiv. of DIC[a] to the acid and stir the solution for 30 min at 0 °C under N_2 atmosphere. If necessary, add small amounts of DMF to maintain the fluidity of the solution.

5 Filter off the precipitate, rinse with DCM, then evaporate the DCM under reduced pressure.

6 Add half of the remaining DMF solution to the swollen resin.

7 Add DMAP (0.5 Equiv.) to the resin and stir for 1 h at room temperature.

8 Wash the resin with DMF (2 × 40 ml) and DCM (2 × 40 ml).

9 Repeat steps 6–8.

10 Wash the resin with DCM (2 × 40 ml), diethyl ether (2 × 40 ml), and dry under reduced pressure at room temperature.

[a] With respect to the loading of the resin.

Protocol 2

Attachment of nucleophiles to 2-chlortrityl chloride PS/1% DVB resin

Equipment and reagents

• See *Protocol 1*

Method

1. Place 5 g of 2-chlortrityl chloride PS/1% DVB resin in a 100 ml RBF.[a]
2. Swell resin in 30 ml of DCM and filter off the solvent (three times).
3. Add to the filtered resin 0.8 Equiv. of the nucleophile, 1.5 Equiv. of DIEA, and 30 ml of DCM.[b]
4. Shake thoroughly for 2 h.
5. Add directly to the reaction mixture 10 Equiv. of MeOH and 10 Equiv. of DIEA[c] and shake for another 30 min.
6. Wash the resin with DCM (2 × 40 ml), MeOH (2 × 40 ml), and diethyl ether (2 × 40 ml), and dry the resin under reduced pressure at room temperature.[d]

[a] Resins with chloride loading of 0.8–1.6 mmol/g. Always use dry conditions in order to avoid hydrolysis of the reaction sites.

[b] Small amounts of DMF can be added to help dissolve the reagents.

[c] Remaining reaction sites are capped with methanol. It should be remembered that methanol is regenerated when the product is cleaved off the resin.

[d] Resin should be stored in firmly closed containers under N_2 atmosphere. For amino acids as nucleophiles it is advisable to cleave the Fmoc protecting groups. The basic amine protects the resin from acid catalysed cleavage.

Protocol 3

Attachment of carboxylic acids to PS/DVB-Wang resin

Equipment and reagents

• See *Protocol 1*

Method

1 Place 5 g of PS/DVB-Wang resin in a 100 ml reactor.

Protocol 3 continued

2 Swell resin in 30 ml of DMF and filter off the solvent (three times).[a]

3 Add to the filtered resin 0.8 Equiv. each of carboxylic acid, TBTU, HOBt, and 1.2 Equiv. of DIEA.

4 Shake thoroughly while adding DMF until a viscous slurry is obtained.

5 Shake the reactor for 3 h then add an additional 1.2 Equiv. of DIEA and shake for 2 h.

6 Wash the resin with DMF (3 × 20 ml) and DCM (2 × 20 ml).

7 Add 10 Equiv. of pyridine, 10 Equiv. of acetic anhydride and DCM until a viscous slurry is obtained, and shake for 30 min.[b]

8 Wash the resin with DCM (3 × 20 ml), MeOH (2 × 20 ml), diethyl ether (2 × 20 ml).

9 Cover the resin with a layer of *tert*-BuOH and shake until a fine suspension is obtained, then lyophilize.

[a] Use freshly opened DMF bottle to avoid the presence of amines which may catalyse the cleavage of the Fmoc protecting group and lead to unwanted side-reactions.

[b] This is a capping step that prevents the inaccessible sites from generating side-products during subsequent steps.

3.3 Determination of the loading of the resin

The loading of the resin can be determined by measuring its weight change, or after cleaving and quantification of the resin-bound compound. The method described in *Protocol 4* allows the determination of the resin loading through the titration of chromophoric group quantitatively released from the resin. This reporter molecule is usually a protecting group that is cleaved in between solid phase chemical steps.

Protocol 4

Quantification of coupling sites on solid support by Fmoc (fluorenylmethoxycarbonyl) cleavage (57)

1 Place 10–15 mg of dry resin in a test-tube.

2 Add three drops of piperidine, seven drops of DCM, and stir for 30 min.

3 Filter off the solution into a 25 ml round-bottom flask and wash the resin twice with 200 µl of DCM.

4 Evaporate the combined solutions under reduced pressure.

5 Add 1–2 ml of *tert*-BuOH/H$_2$O, sonicate the suspension, then lyophilize.[a]

6 Dissolve the solid obtained in 10 ml of DCM; transfer by means of a volumetric pipette 1 ml of the solution into a 10 ml measuring flask, and add 9 ml of DCM.

Protocol 4 continued

7   Fill solution into a UV cuvette and measure the UV absorbance at 301 nm, 290 nm, and 267 nm against DCM.

8   The loading of the resin (x) is calculated.[b]

$x\ (mmol/g) = Abs \cdot 100\,000 \cdot \varepsilon\lambda^{-1} \cdot g^{-1}.$

The mean value of the three measurements gives the loading (x) of the resin.

[a] Prolonged storage of the dry solid, even under reduced pressure, should be avoided.

[b] Abs = absorbance. Extinction coefficient $\varepsilon\ (\lambda)$ = extinction coefficients: $\varepsilon\ (301)$ = 7800; $\varepsilon\ (290)$ = 5800; $\varepsilon\ (267)$ = 17 500. g = amount of resin used.

3.4 Cleavage

Cleavage from the resin can be performed in three different ways (see *Figure 10*).

(a) The physical/chemical conditions of the environment (pH, temperature, light irradiation) can be changed to break the sensitive attachment.

(b) A cleavage cocktail containing a reagent which will react with the sensitive functional group is used resulting in the liberation of the product from the resin.

(c) Inclusion of a cleavage mechanism at the last step of the solid phase synthetic scheme in which the synthesized compound would react with itself and induce its own release. Truncated compounds lacking the functional group necessary for self-release remain attached to the resin.

Acknowledgements

We would like to thank S. Brooks and U. Stroeher for careful reading of the manuscript.

References

1. Merrifield, R. B. (1963). *J. Am. Chem. Soc.*, **85**, 2149.
2. Jung, G. and Beck-Sickinger, A. G. (1992). *Angew. Chem.*, **104**, 375. *Angew. Chem. Int. Ed. Engl.*, **31**, 367.
3. (a) Guyot, A. (1988). In *Syntheses and separations using functional polymers* (ed. D. C. Sherrington and P. Hodge). John Wiley & Sons, New York.
   (b) Hodge, P. (1988). In *Syntheses and separations using functional polymers* (ed. D. C. Sherrington and P. Hodge). John Wiley & Sons, New York.
4. Gausepohl, H. (1990). PhD Thesis, University of Tübingen.
5. Bayer, E. (1991). *Angew. Chem.*, **103**, 117.
6. Atherton, E., Sheppard, R. C., and Rosevar, A. J. (1981). *J. Chem. Soc. Chem. Commun.*, 1151.
7. (a) Bayer, E. and Rapp, W. (1988). DE-A 3714258. (b) Bayer, E. and Rapp, W. (1986). *Chem. Peptides Protein*, **3**, 3.
8. Small, P. W. and Sherrington, D. C. (1989). *J. Chem. Soc. Chem. Commun.*, 1589.

9. Meldal, M. (1992). *Tetrahedron Lett.*, **33**, 3077.

10. Kempe, M. and Barany, G. (1996). *J. Am. Chem. Soc.*, **118**, 7083.

11. Renil, M. and Meldal, M. (1996). *Tetrahedron Lett.*, **37**, 6185.

12. Arshady, R., Atherton, E., Clive, D. L. J., and Sheppard, R. C. (1981). *J. Chem. Soc. Perkin Trans. 1*, 529.

13. Backes, B. J. and Ellman, J. A. (1998). *Curr. Opin. Chem. Biol.*, **1**, 86.

14. Früchtel, J. S. and Jung, G. (1996). *Angew. Chem. Int. Ed. Engl.*, **35**, 17.

15. Balkenhohl, F., von dem Bussche-Hünnefeld, C., Lansky, A., and Zechel, C. (1996). *Angew. Chem. Int. Ed. Engl.*, **35**, 2289.

16. Winter, M. (1996). In *Combinatorial peptide and nonpeptide libraries* (ed. G. Jung), p. 465. VCH–Wiley, Weinheim.

17. Thompson, L. A. and Ellman, J. A. (1994). *Tetrahedron Lett.*, **35**, 9333.

18. Liu, G. and Ellman, J. A. (1995). *J. Org. Chem.*, **60**, 7712.

19. Kick, E. K. and Ellman, J. A. (1995). *J. Med. Chem.*, **38**, 1427.

20. Hanessian, S. and Xie, F. (1998). *Tetrahedron Lett.*, **38**, 733.

21. Hauske, J. R. and Dorff, P. (1995). *Tetrahedron Lett.*, **36**, 1589.

22. Krchñák, V., Flegelova, Z., Weichsel, A. S., and Lebl, M. (1995). *Tetrahedron Lett.*, **36**, 6193.

23. Barn, D. R., Morphy, J. R., and Rees, D. C. (1996). *Tetrahedron Lett.*, **37**, 3213.

24. Beaver, K. A., Siegmund, A. C., and Spear, K. L. (1996). *Tetrahedron Lett.*, **37**, 1145.

25. Barany, G. and Merrifield, R. B. (1979). *Peptides*, **2**, 1.

26. Hirao, A., Itsuno, S., Hattori, I., Yamaguchi, K., Nakahama, S., and Yamazaki, N. (1983). *J. Chem. Soc. Chem. Commun.*, 25.

27. Barlos, K., Gatos, D., Kallitsis, I., Papaioannou, D., and Sotiriou, P. (1988). *Liebigs Ann. Chem.*, 1079.

28. Wang, S.-S. (1973). *J. Am. Chem. Soc.*, **95**, 1328.

29. Orlowski, R. C., Walter, R., and Winkler, D. (1976). *J. Org. Chem.*, **41**, 3701.

30. van Vlient, A., Rietman, B. H., Karkdijk, S. C. F., Adams, P. J. H. M., and Tesser, G. I. (1994). In *Peptides chemistry, structure and biology* (ed. R. S. Hodges and J. A. Smith), p. 151. ESCOM, Leiden.

31. Mergler, M., Tanner, R., Gosteli, J., and Grogg, P. (1988). *Tetrahedron Lett.*, **29**, 4005.

32. Mergler, M., Nyfeler, R., Tanner, R., Gosteli, J., and Grogg, P. (1988). *Tetrahedron Lett.*, **29**, 4009.

33. Rink, H. (1987). *Tetrahedron Lett.*, **28**, 3787.

34. Albericio, F. and Barany, G. (1991). *Tetrahedron Lett.*, **32**, 1015.

35. Brown, A. R., Rees, D. C., Rankovic, Z., and Morphy, J. R. (1997). *J. Am. Chem. Soc.*, **119**, 3288.

36. Morphy, J. R., Rankovic, Z., and Rees, D. C. (1996). *Tetrahedron Lett.*, **37**, 3209.

37. Mutter, M. and Bellof, D. (1984). *Helv. Chim. Acta*, **67**, 2009.

38. Fehrentz, J.-A., Paris, M., Heitz, A., Velek, J., Liu, C.-F., Winternitz, F., *et al.* (1995). *Tetrahedron Lett.*, **36**, 7871.

39. Atherton, E., Logan, C. J., and Sheppard, R. C. (1981). *J. Chem. Soc. Perkin Trans. 1*, 538.

40. Mizoguchi, T., Shigezane, K., and Takamura, N. (1970). *Chem. Pharm. Bull.*, **18**, 1465.

41. DeGrado, W. F. and Kaiser, E. T. (1980). *J. Org. Chem.*, **45**, 1295.

42. Roussel, P. and Bradley, M. (1997). *Tetrahedron Lett.*, **38**, 4861.

43. Sheehan, J. C. and Umezawa, K. (1973). *J. Org. Chem.*, **38**, 3771.

44. Bellof, D. and Mutter, M. (1985). *Chimia*, **39**, 317.

45. Holmes, C. P. and Jones, D. G. (1995). *J. Org. Chem.*, **60**, 2318.

46. Pátek, M. and Lebl, M. (1991). *Tetrahedron Lett.*, **32**, 3891.

47. Hoffmann, S. and Frank, R. (1994). *Tetrahedron Lett.*, **35**, 7763.

48. Panke, G. and Frank, R. (1998). *Tetrahedron Lett.*, **38**, 17.

49. Backes, B. J., Virgilio, A. A., and Ellman, J. A. (1996). *J. Am. Chem. Soc.*, **118**, 3055.

50. Chenera, B., Finkelstein, J. A., and Veber, D. F. (1995). *J. Am. Chem. Soc.*, **117**, 11999.

51. Boehm, T. L. and Showalter, H. D. H. (1996). *J. Org. Chem.*, **61**, 6498.

52. Colombo, R. (1981). *J. Chem. Soc. Chem. Commun.*, 1012.

53. Schuster, M., Wang, P., Paulson, J. C., and Wong, C.-H. (1994). *J. Am. Chem. Soc.*, **116**, 1135.

54. Seitz, O. and Kunz, H. (1995). *Angew. Chem.* **107**, 901. *Angew. Chem. Int. Ed. Engl.*, **34**, 803.

55. Seitz, O. and Kunz, H. (1997). *J. Org. Chem.*, **62**, 813.

56. Guibé, F., Dangles, O., and Balavoine, G. (1989). *Tetrahedron Lett.*, **30**, 2641.

57. Nicolaou, K. C., *et al.* (1997). *Angew. Chem.*, **109**, 2181.

58. Meienhofer, J., Wakin, M., Heimer, E. P., Lambros, T. S., Makofske, R. C., Chang, C.-D. (19XX). *Int. J. Peptide Protein Res.*, **13**, 35.

Chapter 7

Organic reactions on solid support—an overview

William D. Bennett

Advanced ChemTech, Inc., 5609 Fern Valley Road, Louisville, KY 40228, USA

1 Introduction

Organic synthesis on solid supports offers several advantages:

(a) Separating reaction products from excess reagents, by-products and solvents is reduced to a simple filtration.

(b) Reagents can be added to drive the reaction to completion; hence, the yields of a solid phase synthesis are usually higher than their solution phase equivalent.

The greatest advantage of solid phase synthesis, however, is that it facilitates the rapid and simultaneous synthesis of a large number of compounds suitable for pharmaceuticals screening. The success of this approach, as the following tables attest to it, has generated a tremendous interest in expanding the field of solid phase organic synthesis.

These tables document the rapidly expanding number of traditional solution phase reactions that have been successfully adapted to solid phase synthesis over the past two years. Details about the resins and solvents used and the yields obtained are also included. Peptide and oligonucleotide synthesis reactions are well documented and are therefore not included in these tables unless they illustrate new resins, reagents, or synthetic strategies. Reflecting the growing interest in using polymer-supported reagents for solution phase synthesis and work-up, tables of scavengers, catalysts, and reagents on insoluble supports are also featured.

The tables of anchoring and cleavage reactions illustrate the wide variety of resins and linkers developed for solid phase organic synthesis. The tables of amide bond forming reactions also include reactions to form carbamates, ureas, and guanidines. Additional tables list heterocycle forming reactions, oxidation and reduction reactions, olefin forming reactions, condensations, cycloaddition, and organometallic reactions. A brief survey of the tables on aromatic substitution reactions shows the importance of this reaction in recent solid phase synthesis effort. Likewise, a review of the aliphatic substitution tables shows the important role of the Mitsunobu reaction. The goal of these tables is to assist the solid phase organic chemist in selecting the ideal resin, experimental conditions, and reactions in order to design and implement successful synthetic strategies.

2 Solid phase organic reactions—1998

2.1 Anchoring reactions

Acid attachment	Solvent/Temp.	Yield	Resin	Ref.
	DCM/THF 0 °C	92-95%	hydroxymethyl resin	1

Acid attachment	Solvent/Temp.	Yield	Resin	Ref.
	NMP 110 °C		Rink resin	2

Acid attachment	Solvent/Temp.	Yield	Resin	Ref.
	DCM RT		thiomethyl resin	3

Acid attachment	Solvent/Temp.	Yield	Resin	Ref.
	DMF 80 °C		PhFl chloride resin	4

Acid attachment	Solvent/Temp.	Yield	Resin	Ref.
	DCM 25°C		hydroxymethyl resin	5

Acid attachment	Solvent/Temp.	Yield	Resin	Ref.
	DCM or DME 23 °C		Wang resin	6

Acid attachment	Solvent/Temp.	Yield	Resin	Ref.
	DMF RT		Merrifield resin	7

Acid attachment	Solvent/Temp.	Yield	Resin	Ref.
	DCM RT	>85%	diazo resin	8

Acid attachment	Solvent/Temp.	Yield	Resin	Ref.
	DMA		PEG$_4$-PS	9

Acid attachment	Solvent/Temp.	Yield	Resin	Ref.
	THF 70 °C		Wang resin or Merrifield resin	10

Acid attachment	Solvent/Temp.	Yield	Resin	Ref.
	DCM/Et₃N 0 °C to RT	82%	MeO-PEG	11

Acid attachment	Solvent/Temp.	Yield	Resin	Ref.
	Toluene	96%	MeO-PEG	11

Acid attachment	Solvent/Temp.	Yield	Resin	Ref.
	DCM		9-phenylfluorenyl resin	12

Acid attachment	Solvent/Temp.	Yield	Resin	Ref.
	DCM	>95%	hydroxymethyl polystyrene	13

Alcohol attachment	Solvent/Temp.	Yield	Resin	Ref.
	DCM		polystyrenediisopropylsilyl chloride	14

Alcohol attachment	Solvent/Temp.	Yield	Resin	Ref.
	DCM RT		polystyrene or TentaGel resin	15

Alcohol attachment	Solvent/Temp.	Yield	Resin	Ref.
	NMP 70 °C	96%	carboxypolystyrene	16

Alcohol attachment	Solvent/Temp.	Yield	Resin	Ref.
	DCM RT		sulfonyl chloride resin	17

Aldehyde anchoring	Solvent/Temp.	Yield	Resin	Ref.
	DMF		TentaGel S NH$_2$	18

Aldehyde attachment	Solvent/Temp.	Yield	Resin	Ref.
	CHCl$_3$ 0 °C		1,3-dithiane resin	19

Aldehyde attachment	Solvent/Temp.	Yield	Resin	Ref.
	dichloroethane reflux		TentaGel S OH	20

Alkane attachment	Solvent/Temp.	Yield	Resin	Ref.
	DMF RT	94%	MeO-PEG	21

Amidine attachment	Solvent/Temp.	Yield	Resin	Ref.
	THF/DCM		TentaGel S NH$_2$	22

Amine attachment	Solvent/Temp.	Yield	Resin	Ref.
	THF 25 °C		poly(N-isopropylacrylamide) resin	23

Amine attachment	Solvent/Temp.	Yield	Resin	Ref.
	DMF 25 °C		PS-PEG or MBHA resin	24

Amine attachment	Solvent/Temp.	Yield	Resin	Ref.
	DCM RT		Rink resin	25

Amine attachment	Solvent/Temp.	Yield	Resin	Ref.
	DCM		oxime resin	26

Amine attachment	Solvent/Temp.	Yield	Resin	Ref.
	DMF		4-nitropheoxycarbonate	27

Amine attachment

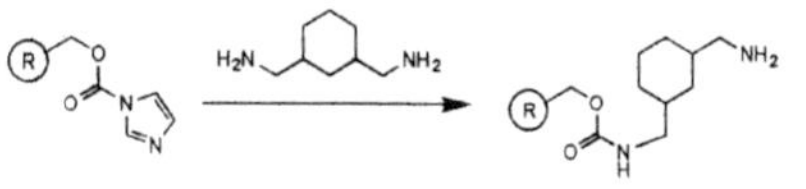

Solvent/Temp.	Yield	Resin	Ref.
NMP 60 °C		imidazolecarbonyl Wang resin	28

Amine attachment

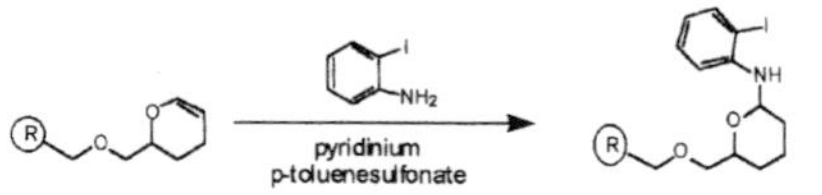

Solvent/Temp.	Yield	Resin	Ref.
Dichloroethane 70 °C		Merrifield resin	29

Amine attachment

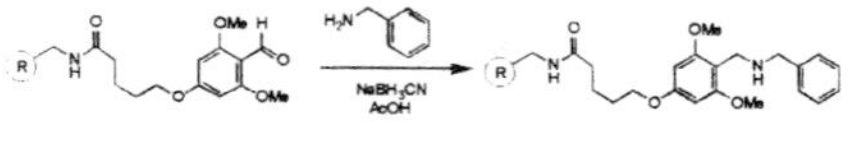

Solvent/Temp.	Yield	Resin	Ref.
DMF/DCM		aminomethyl resin	30

Aryl compound attachment

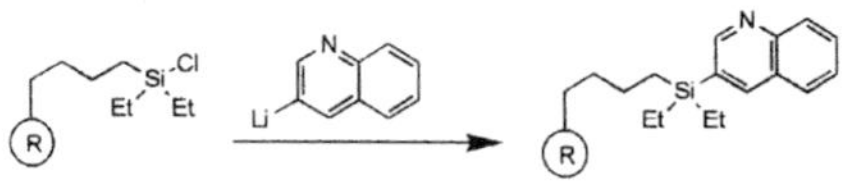

Solvent/Temp.	Yield	Resin	Ref.
THF -78 °C to RT		PS-silane resin	31

Azide attachment

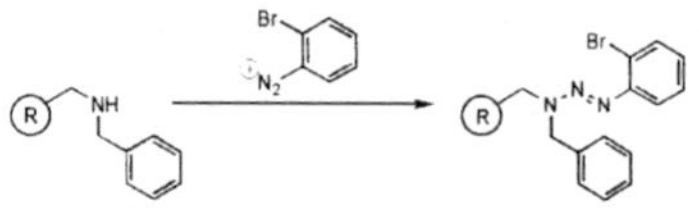

Solvent/Temp.	Yield	Resin	Ref.
THF 0 °C to 25°C		N-benzylaminomethyl resin	32

Glycoside attachment

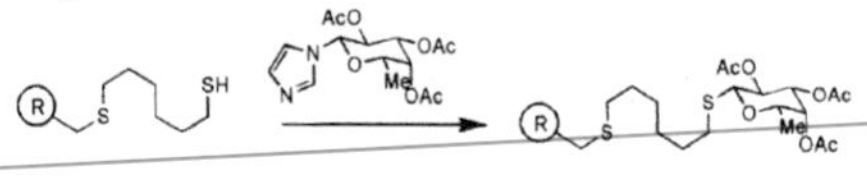

Solvent/Temp.	Yield	Resin	Ref.
DCM		Merrifield resin	33

Indole attachment

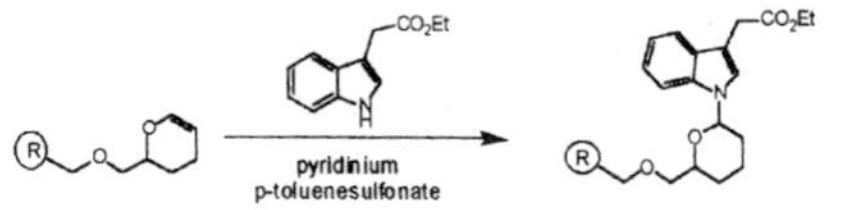

Solvent/Temp.	Yield	Resin	Ref.
Dichloroethane 70 °C		Merrifield resin	29

N-Hydroxyimidazole attachment

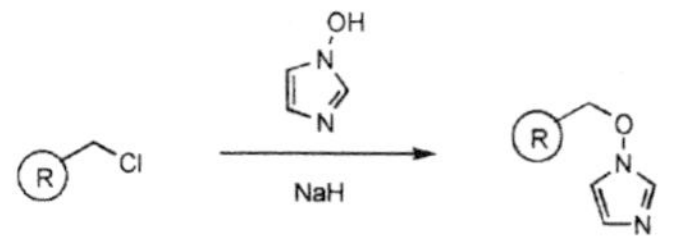

Solvent/Temp.	Yield	Resin	Ref.
DMF 20 °C		Merrifield resin	34

Phenol attachment

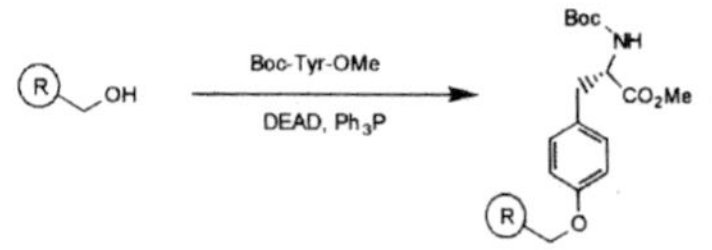

Solvent/Temp.	Yield	Resin	Ref.
NMM	>80%	Wang resin or Merrifield resin	10

Phenol attachment

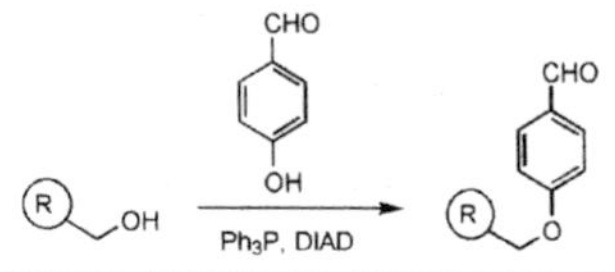

Solvent/Temp.	Yield	Resin	Ref.
NMM 0 °C, then RT		Wang resin	35

143

Reaction		Solvent/Temp.	Yield	Resin	Ref.
Phenol attachment		DMF/THF RT		Wang resin	36
Phosphate ester formation		THF		Wang resin	37
Phosphonic ester attachment		23 °C	70%	Wang resin	6
Phosphorous alkylation		DMF 70 °C		phosphine resin	38
Selenium alkylation		EtOH RT		selenylpolystyrene	39
Thiol attachment		DCM 23 °C	97%	Wang resin	6
Thiol attachment				not specified	40
Thiourea attachment		EtOH/dioxane 85 °C	96%	Merrifield resin	41
Thiourea attachment		THF		imidazole carbonate-Wang resin	42

2.2 Amide bond forming reactions

Acyl azide formation

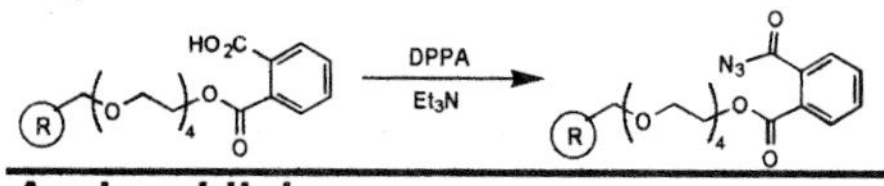

	Solvent/Temp.	Yield	Resin	Ref.
	Toluene RT		PEG4-PS	9

Acyl pyridinium

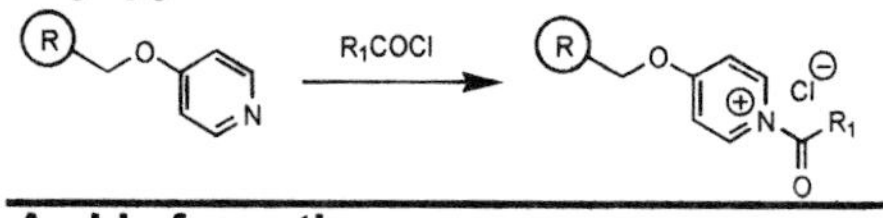

	Solvent/Temp.	Yield	Resin	Ref.
	THF RT		Wang resin	36

Amide formation

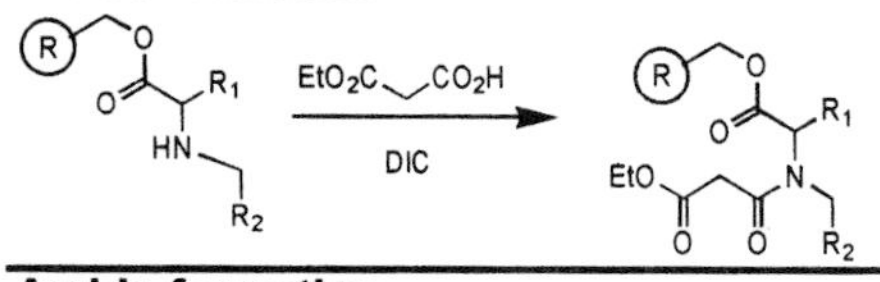

	Solvent/Temp.	Yield	Resin	Ref.
	DMF RT	100%	Wang resin	43

Amide formation

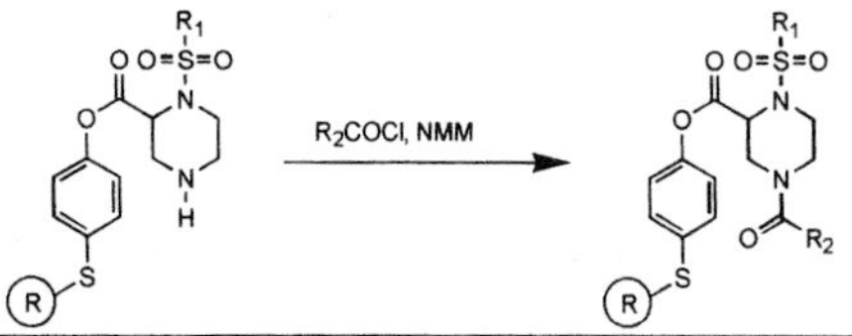

	Solvent/Temp.	Yield	Resin	Ref.
	DCM RT		Merrifield resin	7

Amide formation

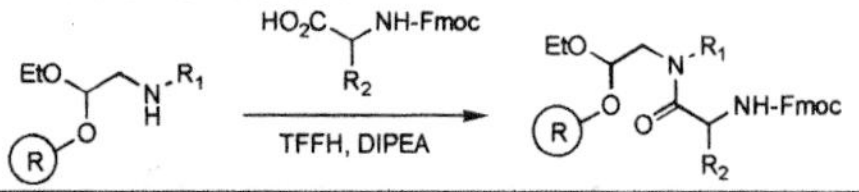

	Solvent/Temp.	Yield	Resin	Ref.
	Dichloroethane		TentaGel S OH	20

Amide formation

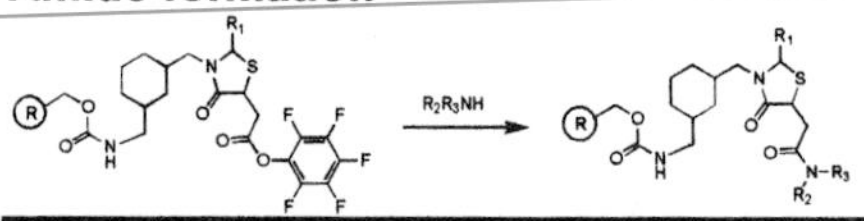

	Solvent/Temp.	Yield	Resin	Ref.
	DMF		imidazolecarbonyl Wang resin	28

Amide formation

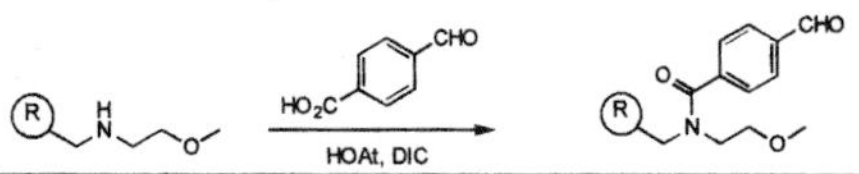

	Solvent/Temp.	Yield	Resin	Ref.
	DMF		AMEBA resin	35

Amide formation

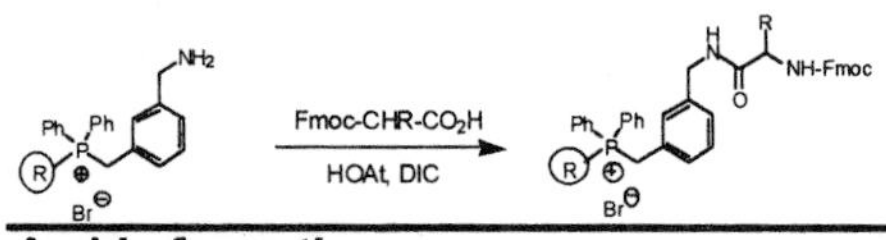

	Solvent/Temp.	Yield	Resin	Ref.
	DCM RT		phosphine resin	38

Amide formation

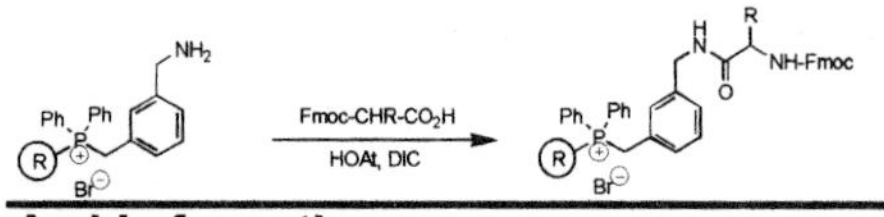

	Solvent/Temp.	Yield	Resin	Ref.
	DCM/DMF RT		phosphine resin	38

Amide formation

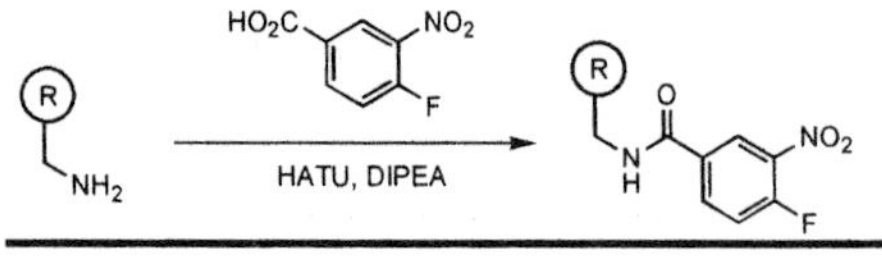

	Solvent/Temp.	Yield	Resin	Ref.
	DMF		Rink resin	44

145

Amide formation

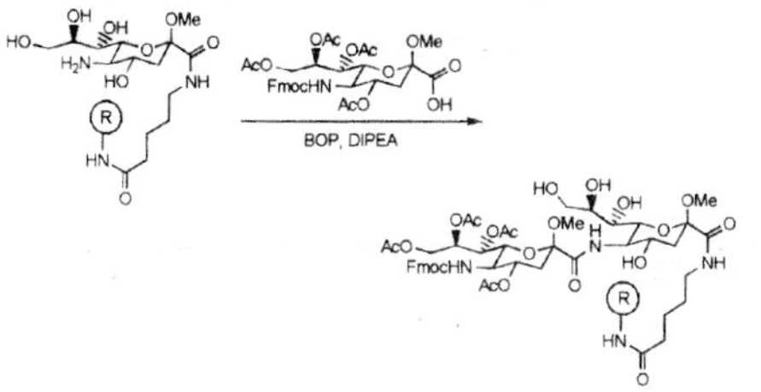

	Solvent/Temp.	Yield	Resin	Ref.
	NMP		Knorr resin	45

BOP, DIPEA

Amide formation

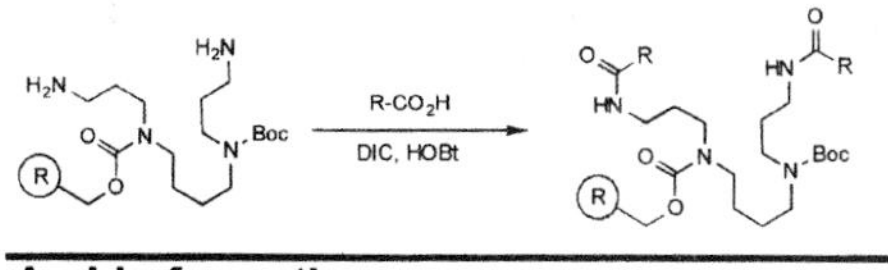

	Solvent/Temp.	Yield	Resin	Ref.
	DCM		HMPA-aminomethyl resin	46

R-CO$_2$H, DIC, HOBt

Amide formation

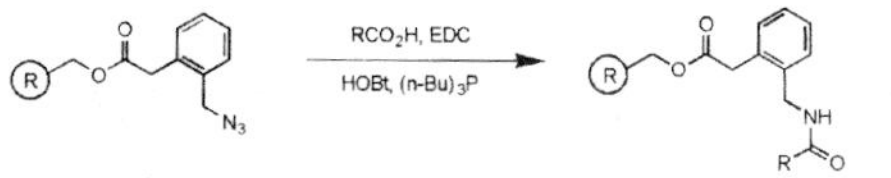

	Solvent/Temp.	Yield	Resin	Ref.
	Dioxane 20 °C	82-100%	Wang resin	47

RCO$_2$H, EDC, HOBt, (n-Bu)$_3$P

Amide formation

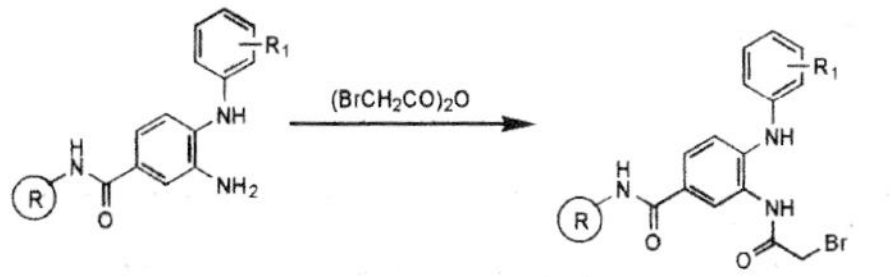

	Solvent/Temp.	Yield	Resin	Ref.
	DMF		PS-PEG-Rink	48

(BrCH$_2$CO)$_2$O

Amide formation

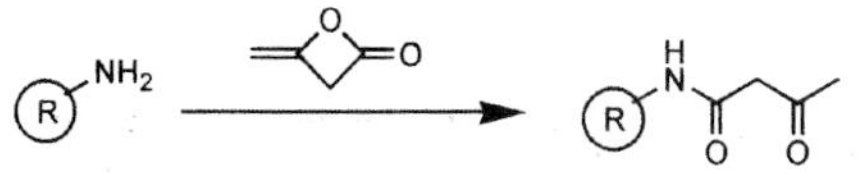

	Solvent/Temp.	Yield	Resin	Ref.
	DCM -15 °C to RT		Rink resin	49

Amide formation

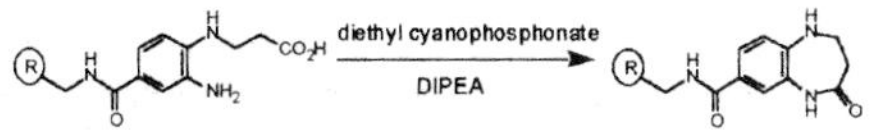

	Solvent/Temp.	Yield	Resin	Ref.
	DMF RT	>90%	ArgoGel-Rink resin	50

diethyl cyanophosphonate, DIPEA

Amide formation

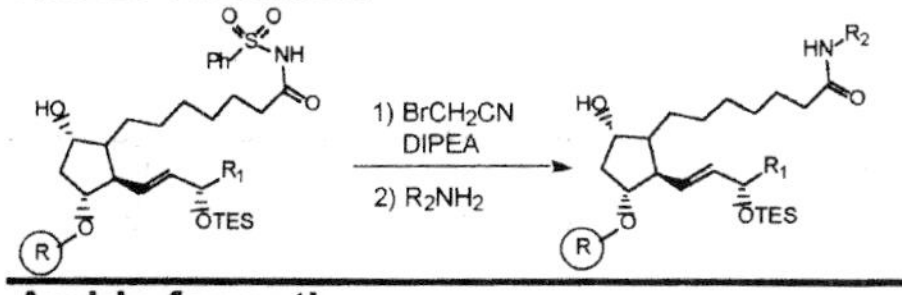

	Solvent/Temp.	Yield	Resin	Ref.
	NMP RT		dibutylsilyl chloride resin	51

1) BrCH$_2$CN, DIPEA 2) R$_2$NH$_2$

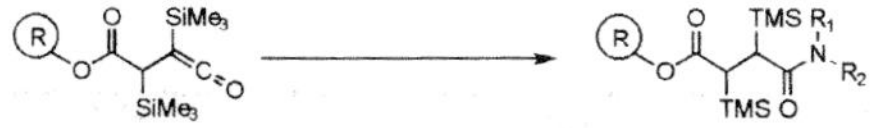

Amide formation

	Solvent/Temp.	Yield	Resin	Ref.
	DCM RT		MeO-PEG	11

Carbamate formation

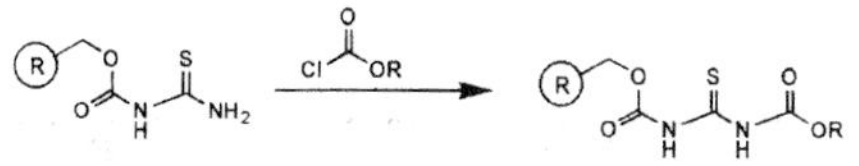

	Solvent/Temp.	Yield	Resin	Ref.
	THF		Wang resin	42

Carboxyazide formation

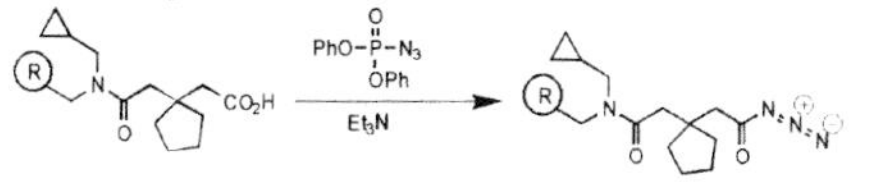

	Solvent/Temp.	Yield	Resin	Ref.
	NMP RT		Rink resin	52

PhO–P–N$_3$, OPh, Et$_3$N

Guanidine formation	Solvent/Temp.	Yield	Resin	Ref.
	DMF		trityl resin	53

Guanidine formation	Solvent/Temp.	Yield	Resin	Ref.
	DMF/DCM RT		Wang resin	54

Sulfonamide formation	Solvent/Temp.	Yield	Resin	Ref.
	DCM		Wang resin	55

Sulfonamide formation	Solvent/Temp.	Yield	Resin	Ref.
	DMF		TentaGel S OH	56

Sulfonamide formation	Solvent/Temp.	Yield	Resin	Ref.
	DCM	87%	2-chlorotrityl resin	57

Urea formation	Solvent/Temp.	Yield	Resin	Ref.
	DMF RT		hydroxymethyl resin	1

Urea formation	Solvent/Temp.	Yield	Resin	Ref.
	DMA		PEG$_4$-PS	9

Urea formation	Solvent/Temp.	Yield	Resin	Ref.
	DMF RT		Wang resin	54

Urea formation	Solvent/Temp.	Yield	Resin	Ref.
	DCM/DMF RT		hydroxymethylene resin	58

Urea formation	Solvent/Temp.	Yield	Resin	Ref.
	DCM/DMF RT		hydroxymethylene resin	58

147

Urea formation	Solvent/Temp.	Yield	Resin	Ref.
	NMP		2-chlorotrityl resin	59

Urea formation	Solvent/Temp.	Yield	Resin	Ref.
	DCM RT		polystyrene	60

Urethane formation	Solvent/Temp.	Yield	Resin	Ref.
	THF 80 °C		Wang resin	61

2.3 Aromatic substitution

C-N cross coupling	Solvent/Temp.	Yield	Resin	Ref.
	Dioxane 70 °C to 80 °C	75% to >95%	TentaGel S RAM	62

C-S cross coupling	Solvent/Temp.	Yield	Resin	Ref.
	DMA 60 °C	90% to 93%	Merrifield crowns	63

Cu catalysed stannane coupling	Solvent/Temp.	Yield	Resin	Ref.
	NMP 100 °C	up to 94%	Merrifield resin	64

Freidel-Crafts alkylation	Solvent/Temp.	Yield	Resin	Ref.
			macroporous polystyrene	65

Heck reaction	Solvent/Temp.	Yield	Resin	Ref.
	DMF 25 °C		2-chlorotrityl resin	66

Heck reaction	Solvent/Temp.	Yield	Resin	Ref.
	DMF 80 °C		N-benzylaminomethyl resin	32

Herocyclic N-oxide

	Solvent/Temp.	Yield	Resin	Ref.
	DCM	66-100%	Wang resin	67
	0 °C –20 °C			

Iodination

	Solvent/Temp.	Yield	Resin	Ref.
	DCM		Rink resin	68
	RT			

Nucleophilic aromatic substitution

	Solvent/Temp.	Yield	Resin	Ref.
	DMSO		PS-PEG-Rink	48
	50 °C			

Nucleophilic aromatic substitution

	Solvent/Temp.	Yield	Resin	Ref.
	DMF		TentaGel S NH$_2$	22

Nucleophilic aromatic substitution

	Solvent/Temp.	Yield	Resin	Ref.
	DMF		Rink resin	44

Nucleophilic aromatic substitution

	Solvent/Temp.	Yield	Resin	Ref.
	acetone/ H$_2$O	>90%	ArgoGel-Rink resin	50
	70 °C			

Nucleophilic aromatic substitution

	Solvent/Temp.	Yield	Resin	Ref.
	NMP	50%	2-chlorotrityl resin	59
	60 °C			

Nucleophilic aromatic substitution

	Solvent/Temp.	Yield	Resin	Ref.
	DIPEA/DMF		Wang resin	61
	RT			

Nucleophilic aromatic substitution

	Solvent/Temp.	Yield	Resin	Ref.
	NMP		Wang resin	69
	110 °C			

149

	Solvent/Temp.	Yield	Resin	Ref.
Nucleophilic aromatic substitution			MeO-PEG	70
Nucleophilic aromatic substitution		99%	MeO-PEG	70
Nucleophilic aromatic substitution	DMF 25 °C	up to 90%	Rink resin	71
Stille reaction	DMF 60 °C		Merrifield resin	72
Stille reaction	DMF 60 °C		Merrifield resin	72
Stille reaction	toluene 100 °C	54%	polystyrene	73
Stille reaction	THF	100%	Merrifield crowns	63
Suzuki coupling	toluene/EtOH 90 °C		Rink resin	74
Suzuki coupling	DME	87% to 95%	glycerol (diol) resin	75

Suzuki coupling	Solvent/Temp.	Yield	Resin	Ref.
	THF/H$_2$O	100%	Merrifield crowns	63

2.4 Condensation reactions

Aldol condensation	Solvent/Temp.	Yield	Resin	Ref.
	Et$_3$N -78 °C		Merrifield resin	76

Aldol condensation	Solvent/Temp.	Yield	Resin	Ref.
	DCM -78 °C		thiomethyl resin	3

Aldol condensation	Solvent/Temp.	Yield	Resin	Ref.
	DMF		aminomethyl microtubes with Knorr linker	77

Aldol condensation	Solvent/Temp.	Yield	Resin	Ref.
	DME RT	64% to 98%	Rink resin	78

Aldol condensation	Solvent/Temp.	Yield	Resin	Ref.
	Et$_2$O -78 °C, 0 °C		diisopropylsilyl resin	79

Claisen condensation	Solvent/Temp.	Yield	Resin	Ref.
	DMF RT	59-69%	Wang resin or Tentagel SH	80

Eneamine formation	Solvent/Temp.	Yield	Resin	Ref.
	DMF RT		Rink resin	49

Eneamine formation	Solvent/Temp.	Yield	Resin	Ref.
	CH(OMe)$_3$	91% to 94%	MeO-PEG	11

151

Eneamine-cyclization/cleavage	Solvent/Temp.	Yield	Resin	Ref.
	Toluene heat		Merrifield resin	81

Guanidine formation	Solvent/Temp.	Yield	Resin	Ref.
	DMF 25 °C		2-chlorotrityl chloride resin	82

Hydrazone formation	Solvent/Temp.	Yield	Resin	Ref.
	DMF RT		hydroxymethyl resin	1

Hydrazone formation	Solvent/Temp.	Yield	Resin	Ref.
	THF RT		Merrifield resin	8

Imine formation	Solvent/Temp.	Yield	Resin	Ref.
			imidazolecarbonyl Wang resin	28

Imine formation	Solvent/Temp.	Yield	Resin	Ref.
	DMF 80 °C		formylpolystyrene	83

Imine formation	Solvent/Temp.	Yield	Resin	Ref.
	60 °C		MeO-PEG	84

Imine formation	Solvent/Temp.	Yield	Resin	Ref.
	DMF		Merrifield resin	85

Imine formation	Solvent/Temp.	Yield	Resin	Ref.
	DCM		polystyrene	86

Imine formation/reduction	Solvent/Temp.	Yield	Resin	Ref.
	CH(OMe)$_3$		aminomethyl microtubes with Knorr linker	77

Reaction	Solvent/Temp.	Yield	Resin	Ref.
Imine formation/reduction H₂N~O~ ; NaBH(OAc)₃, AcOH	DMF		AMEBA resin	35
Imine formation/reduction CH₃CH₂CHO ; NaBH(OAc)₃	DCM/MeOH		Wang resin	69
Imine formation/reduction RNH₂, Me₄NBH(OAc)₃, NaBH₃CN	DCE/MeOH		aminomethyl resin	87
Imine formation/reduction R-CHO, NaBH(OAc)₃, 1% AcOH	DMF		Knorr resin	88
Mukaiyama aldol reaction OTBS / StBu ; BF₃·Et₂O	DCM -78 °C		Merrifield resin	79
Nitrone formation R-CHO ; (MeO)₃CH	DCM RT		Merrifield resin	89
Nitrone formation Ar-CHO	Toluene 80 °C		Rink resin	90
Oxime formation NH₂OH·HCl	Pyridine/EtOH reflux		Merrifield resin	89
Oxime formation NH₂OH·HCl	Pyridine RT		2-chlorotrityl resin	91

2.5 Cycloaddition reactions

[2+2] cycloaddition	Solvent/Temp.	Resin	Ref.
	DCM 23 °C	MeO-PEG	84

[3+2] cycloaddition	Solvent/Temp.	Resin	Ref.
	Toluene 80 °C	Rink resin	90

[3+2] cycloaddition	Solvent/Temp.	Resin	Ref.
	THF 60 °C	Merrifield resin	92

[3+2] cycloaddition	Solvent/Temp.	Resin	Ref.
	Toluene RT	Merrifield resin	89

[3+2] cycloaddition	Solvent/Temp.	Resin	Ref.
	DCM	2-chlorotrityl resin	91

[3+2] cycloaddition	Solvent/Temp.	Resin	Ref.
	THF 60 °C	Merrifield resin	93

Diels-Alder	Solvent/Temp.	Resin	Ref.
	DMF	not specified	94

Diels-Alder	Solvent/Temp.	Resin	Ref.
	DCM -78 °C	polystyrene	95

Diels-Alder	Solvent/Temp.	Resin	Ref.
	Toluene 105 °C	Wang resin	96

Diels-Alder	Solvent/Temp.	Resin	Ref.
	100 °C	piperazinomethyl resin	32

Diels-Alder	Solvent/Temp.	Resin	Ref.
	Toluene 25 °C	Merrifield resin	97

Diels-Alder	Solvent/Temp.	Resin	Ref.
	Toluene reflux	hydroxymethyl polystyrene	13

Diels-Alder	Solvent/Temp.	Resin	Ref.
	DCM	hydroxymethyl polystyrene	13

Diels-Alder	Solvent/Temp.	Resin	Ref.
	DCM	hydroxymethyl polystyrene	13

2.6 Orgamometallic reactions

Addition to acyl pyridinium	Solvent/Temp.	Yield	Resin	Ref.
	THF RT		Wang resin	36

Addition to aldehyde	Solvent/Temp.	Yield	Resin	Ref.
	THF 0 °C to 23 °C	>92%	polystyrene	73

Addition to aldehyde	Solvent/Temp.	Yield	Resin	Ref.
	THF -35 °C	95%	Wang resin	98

Addition to aldehyde		Solvent/Temp.	Yield	Resin	Ref.
		THF -20 °C	85%	MeO-PEG	99

Addition to aldehyde		Solvent/Temp.	Yield	Resin	Ref.
		THF -30 °C to RT		trityl resin	100

Addition to ester		Solvent/Temp.	Yield	Resin	Ref.
				piperazinomethyl resin	101

Addition to imine		Solvent/Temp.	Yield	Resin	Ref.
		THF RT	97%	polystyrenesulfonic acid	102

Addition to ketone		Solvent/Temp.	Yield	Resin	Ref.
		Cyclohexane 65 °C		polystyrene	4

Alkylation		Solvent/Temp.	Yield	Resin	Ref.
		THF -60 °C to 20 °C	92%	Merrifield resin	34

Alkylation		Solvent/Temp.	Yield	Resin	Ref.
		THF -35 °C	98% to 99%	Wang resin	98

Aryl alkylation		Solvent/Temp.	Yield	Resin	Ref.
		Toluene 65 °C		polystyrene	95

Halogenation		Solvent/Temp.	Yield	Resin	Ref.
		THF -60 °C to 20 °C	77% to 93%	Merrifield resin	34

Organolithium coupling	Solvent/Temp.	Yield	Resin	Ref.
	THF -30 °C to RT	72% to 84%	trityl resin	100
Thiolation	Solvent/Temp.	Yield	Resin	Ref.
	THF -60 °C to 20 °C	74% to 90%	Merrifield resin	34
Thiolation	Solvent/Temp.	Yield	Resin	Ref.
	THF -35 °C	87%	Wang resin	98
Transmetalation	Solvent/Temp.	Yield	Resin	Ref.
	THF -78 °C to 23 °C	87%	polystyrene	73
Weinreb amide reaction	Solvent/Temp.	Yield	Resin	Ref.
	THF		Rink resin	103

2.7 Michael addition

C-nucleophile (organocopper reagent)	Solvent/Temp.	Yield	Resin	Ref.
	THF -50 °C		linear polystyrene	104
C-nucleophile	Solvent/Temp.	Yield	Resin	Ref.
	THF -78 °C to -20 °C		dibutylsilyl chloride resin	51
N-nucleophile	Solvent/Temp.	Yield	Resin	Ref.
	DMSO RT		Wang resin	105
N-nucleophile	Solvent/Temp.	Yield	Resin	Ref.
	CH$_3$CN 25 °C	75-78%	Wang resin	106
S-nucleophile	Solvent/Temp.	Yield	Resin	Ref.
			methylsulfinate resin	107

S-nucleophile	Solvent/Temp.	Yield	Resin	Ref.
	EtOH RT °C		thiomethyl resin	107

2.8 Heterocycle forming reactions

3,1-Benzoxazine-4-one formation	Solvent/Temp.	Yield	Resin	Ref.
	THF		Wang resin or Sasrin™ resin	108

Benzimidazole	Solvent/Temp.	Yield	Resin	Ref.
	PhNO$_2$ 130 °C		Wang resin	109

Benzimidazole-2-thione	Solvent/Temp.	Yield	Resin	Ref.
	THF		Rink resin	44

Benzimidazoles	Solvent/Temp.	Yield	Resin	Ref.
	DMF RT		Wang resin or Rink resin	110

Benzopiperazinone formation	Solvent/Temp.	Yield	Resin	Ref.
	DMF 80 °C		Wang resin	61

Bicyclic guanidine	Solvent/Temp.	Yield	Resin	Ref.
	DCM		MBHA resin	111

Dihydrobenzofuran formation	Solvent/Temp.	Yield	Resin	Ref.
	Benzene reflux	95-97%	N-methylaminomethyl resin	112

Dihydrobenzofuran formation	Solvent/Temp.	Yield	Resin	Ref.
	THF RT	45%	TentaGel S PHB	113

Diketopiperazine formation

	Solvent/Temp.	Yield	Resin	Ref.
	DMF 25 °C	up to 100%	PAL resin	114

Hantzsch pyrrole synthesis

	Solvent/Temp.	Yield	Resin	Ref.
	DMF RT		Rink resin	2

Hydantoin formation

	Solvent/Temp.	Yield	Resin	Ref.
	DCM		MBHA resin	115

Hydrobenzofuran formation

	Solvent/Temp.	Yield	Resin	Ref.
	DMF 100 °C	90%	Rink resin	116

Hydrobenzopyran formation

	Solvent/Temp.	Yield	Resin	Ref.
	DMF 100 °C	84%	Rink resin	116

Imidazole formation

	Solvent/Temp.	Yield	Resin	Ref.
	DCM RT	up to 99%	PS-PEG	117

Imidazolidone formation

	Solvent/Temp.	Yield	Resin	Ref.
	DMF RT, 55 °C		Rink acid resin or HMPB-BHA resin	118

Indole formation

	Solvent/Temp.	Yield	Resin	Ref.
	DMF 110 °C	up to 97%	Merrifield resin	29

Indole formation

	Solvent/Temp.	Yield	Resin	Ref.
	DMF/Et$_3$N 80 °C	86% to 96%	Rink resin	119

Indoline formation

	Solvent/Temp.	Yield	Resin	Ref.
	DMF 100 °C	76% to 91%	Rink resin	116

Oxopiperazines

	Solvent/Temp.	Yield	Resin	Ref.
	DMF 60 °C	48%	2-chlorotrityl resin	57

Pyrazolines		Solvent/Temp.	Yield	Resin	Ref.
		EtOH reflux		Rink resin	120

Pyridone formation		Solvent/Temp.	Yield	Resin	Ref.
		Toluene reflux	88% to 92%	MeO-PEG	11

Pyrimidine formation		Solvent/Temp.	Yield	Resin	Ref.
		DMA 100 °C	up to 98%	Rink resin	78

Pyrimidine formation		Solvent/Temp.	Yield	Resin	Ref.
		DMF 0 °C to RT	74%	Merrifield resin	41

Pyrrole formation		Solvent/Temp.	Yield	Resin	Ref.
		DMF/EtOH 60 °C		Rink resin	49

Quinazolinedione formation		Solvent/Temp.	Yield	Resin	Ref.
		NMP 60 °C		2-chlorotrityl resin	59

Tetrahydroquinoline formation		Solvent/Temp.	Yield	Resin	Ref.
		DMF 100 °C	88% to 93%	Rink resin	116

Thiazole		Solvent/Temp.	Yield	Resin	Ref.
		MeOH reflux		Merrifield resin	97

Thiazolidinone		Solvent/Temp.	Yield	Resin	Ref.
		80 °C		imidazolecarbonyl Wang resin	28

2.9 Multi-component reactions

3-Component condensation	Solvent/Temp.	Yield	Resin	Ref.
	DMF/EtOH 70 °C		Rink resin	49

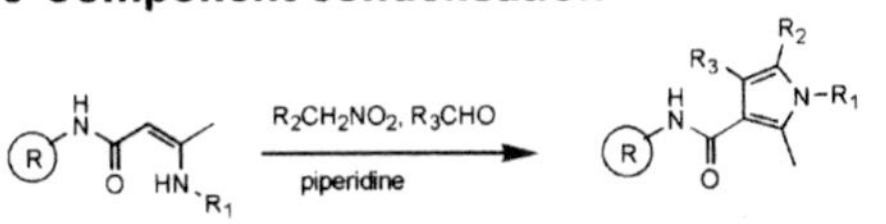

3-Component condesation	Solvent/Temp.	Yield	Resin	Ref.
	DCM/MeOH RT	up to 78%	Rink resin	121

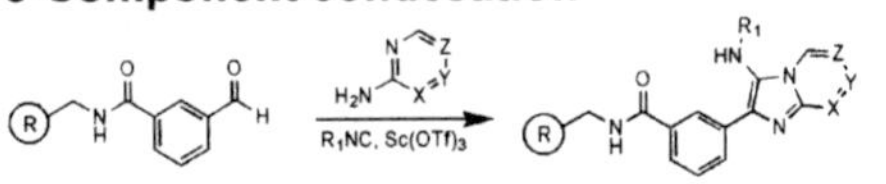

3-Component condensation	Solvent/Temp.	Yield	Resin	Ref.
	CH$_3$CN	60% to 88%	AMEBA resin	35

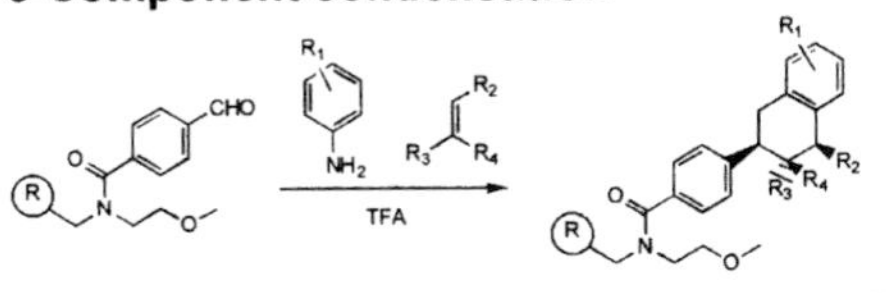

3-Component condensation	Solvent/Temp.	Yield	Resin	Ref.
	CH$_3$CN	61% to 84%	Wang resin	35

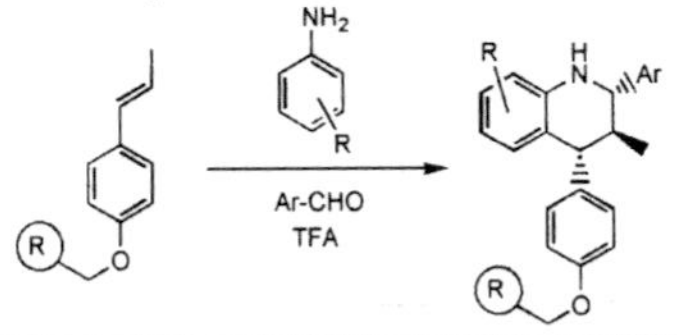

Mannich reaction	Solvent/Temp.	Yield	Resin	Ref.
	Dioxane 23 °C	90% to 100%	Rink resin	119

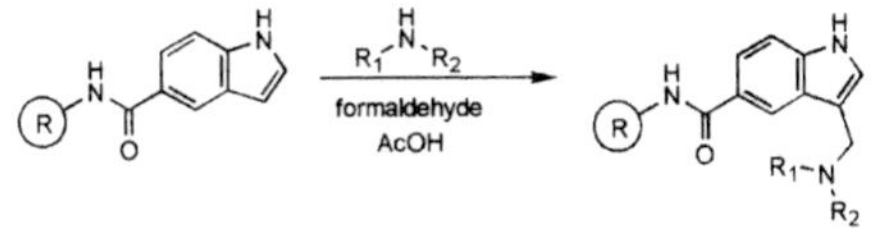

Tsuge Reaction	Solvent/Temp.	Yield	Resin	Ref.
	THF RT	76-92%	Wang resin	122

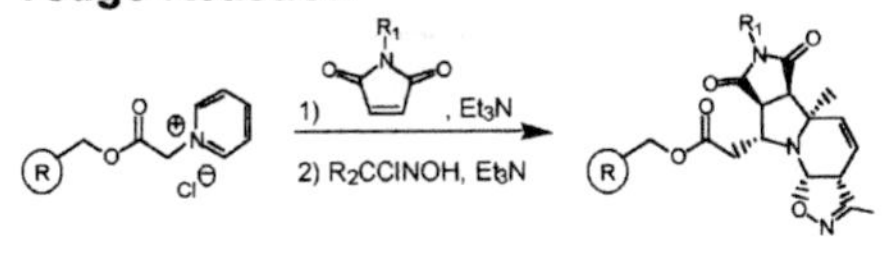

Ugi 4-component	Solvent/Temp.	Yield	Resin	Ref.
	THF/MeOH RT		Wang resin	27

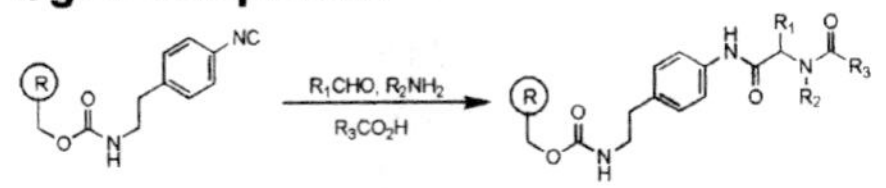

Ugi 4-component	Solvent/Temp.	Yield	Resin	Ref.
	DCM/MeOH RT		Rink resin	103

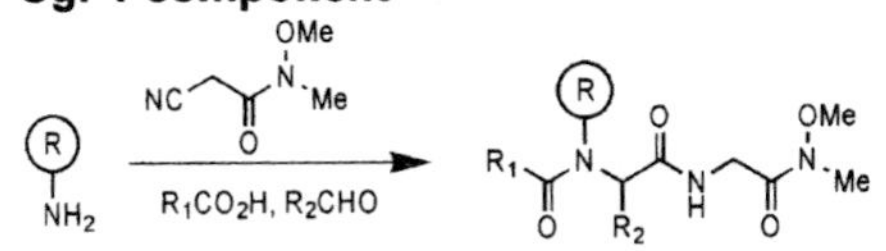

Ugi 4-component	Solvent/Temp.	Yield	Resin	Ref.
	DCM/MeOH RT	up to 95%	Rink resin	71

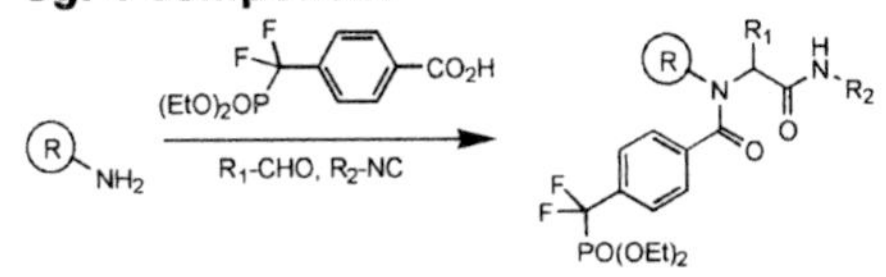

161

2.10 Olefin forming reactions

Dimethylsulfonium metylide		Solvent/Temp.	Yield	Resin	Ref.
		THF 0 °C to RT		Merrifield resin	123

Horner-Emmons		Solvent/Temp.	Yield	Resin	Ref.
		DMF 25 °C	100%	Rink resin	120

Tebbe reagent		Solvent/Temp.	Yield	Resin	Ref.
		Toluene/THF 25 °C		Merrifield resin	97

Wittig		Solvent/Temp.	Yield	Resin	Ref.
		THF 0 °C to RT	>95%	Merrifield resin	5

Wittig		Solvent/Temp.	Yield	Resin	Ref.
		Toluene 50 °C	100%	carboxypolystyrene	16

Wittig		Solvent/Temp.	Yield	Resin	Ref.
		THF -20 °C, then RT		Wang resin	124

Wittig		Solvent/Temp.	Yield	Resin	Ref.
		THF 23 °C		carboxypolystyrene	73

Wittig		Solvent/Temp.	Yield	Resin	Ref.
		DMA 60 °C	71%	Rink resin	78

Wittig		Solvent/Temp.	Yield	Resin	Ref.
		Toluene 80 °C		thiomethyl resin	107

Wittig	Solvent/Temp.	Yield	Resin	Ref.
	THF RT		PS-CH$_2$PPh$_3$Br	125

Wittig-Horner	Solvent/Temp.	Yield	Resin	Ref.
	DMF RT	96%	MBHA resin	126

2.11 Oxidation reactions

Alcohol to aldehyde	Solvent/Temp.	Yield	Resin	Ref.
	DMSO RT		Merrifield resin	76

Alcohol to aldehyde	Solvent/Temp.	Yield	Resin	Ref.
	DCM RT	>95%	Merrifield resin	5

Alcohol to aldehyde	Solvent/Temp.	Yield	Resin	Ref.
	THF 0 °C to 23 °C	>92%	polystyrene	73

Alcohol to aldehyde	Solvent/Temp.	Yield	Resin	Ref.
	DMF		2-chlorotrityl resin	127

Alcohol to ketone	Solvent/Temp.	Yield	Resin	Ref.
	DCM 45 °C		dibutylsilyl chloride resin	51

Alcohol to ketone	Solvent/Temp.	Yield	Resin	Ref.
	DMSO		not specified	94

Epoxidation	Solvent/Temp.	Yield	Resin	Ref.
	DCM		2-pivaloylglycerol resin	128

163

Epoxidation

	Solvent/Temp.	Yield	Resin	Ref.
	DCM 0 °C		diisopropylsilyl chloride resin	14

N-oxide

	Solvent/Temp.	Yield	Resin	Ref.
	DCM 20 °C		Wang resin	67

Olefin to diol

	Solvent/Temp.	Yield	Resin	Ref.
	Acetone/H_2O	53%	piperazinomethyl resin	32

Ozonolysis

	Solvent/Temp.	Yield	Resin	Ref.
	DCM 20 °C		diisopropylsilyl resin	79

Regioselective sulfide to sulfone

	Solvent/Temp.	Yield	Resin	Ref.
	DCM 0 °C to RT	100%	Merrifield resin	41

Sulfide to sulfone

	Solvent/Temp.	Yield	Resin	Ref.
	H_2O RT	91%	MeO-PEG	21

Sulfide to sulfone

	Solvent/Temp.	Yield	Resin	Ref.
	DCM RT		Merrifield resin	129

Sulfone formation

	Solvent/Temp.	Yield	Resin	Ref.
	DCM		thiomethyl resin	107

Wacker oxidation

	Solvent/Temp.	Yield	Resin	Ref.
	H_2O	42%	macroporous polystyrene	65

2.12 Reduction reactions

Amide to amine		Solvent/Temp.	Yield	Resin	Ref.
		DME reflux		Merrifield resin	85

Amide to amine		Solvent/Temp.	Yield	Resin	Ref.
		THF reflux		2-chlorotrityl resin	130

Azide to amine		Solvent/Temp.	Yield	Resin	Ref.
		DMF 50 °C		Rink resin	131

Carboxylic acid to alcohol		Solvent/Temp.	Yield	Resin	Ref.
				Rink resin	132

Ester to alcohol		Solvent/Temp.	Yield	Resin	Ref.
		THF <0 °C	98%	polystyrenesulfonic acid	102

Ester to alcohol		Solvent/Temp.	Yield	Resin	Ref.
		DCM -78 °C		piperazinomethyl resin	32

Hydrostanylation		Solvent/Temp.	Yield	Resin	Ref.
		Toluene 100 °C		polystyrene	73

Hydrostanylation		Solvent/Temp.	Yield	Resin	Ref.
		Toluene 0 °C	90%	vinylpolystyrene	73

Imine to amine		Solvent/Temp.	Yield	Resin	Ref.
		DMA		polystyrene	60

Imine to amine		Solvent/Temp.	Yield	Resin	Ref.
		THF RT		formypolystyrene	83

Ketone reduction		Solvent/Temp.	Yield	Resin	Ref.
		NMP 60 °C	50-61%	polystyrene	133

Ketone to alcohol	Solvent/Temp.	Yield	Resin	Ref.
L-Selectride	THF -78 °C		dibutylsilyl chloride resin	51

Ketone to alcohol	Solvent/Temp.	Yield	Resin	Ref.
	Et$_2$O -78 °C		diisopropylsilyl resin	79

Ketone to alcohol	Solvent/Temp.	Yield	Resin	Ref.
NaBH$_4$	THF/EtOH		Knorr resin	88

Nitro to amine	Solvent/Temp.	Yield	Resin	Ref.
SnCl$_2$ · H$_2$O	DMF		Rink resin	44

Nitro to amine	Solvent/Temp.	Yield	Resin	Ref.
SnCl$_2$·2H$_2$O	NMP 50 °C		PS-PEG-Rink	48

Nitro to amine	Solvent/Temp.	Yield	Resin	Ref.
SnCl$_2$·2H$_2$O NaHCO$_3$	DMF RT	>90%	ArgoGel-Rink resin	50

Olefin reduction	Solvent/Temp.	Yield	Resin	Ref.
	DMF 100 °C	>95%	Wang resin	134

Oxime to hydroxylamine	Solvent/Temp.	Yield	Resin	Ref.
BH$_3$·pyridine	THF RT		Merrifield resin	89

Reductive alkylation	Solvent/Temp.	Yield	Resin	Ref.
R$_2$CHO NaBH(OAc)$_3$ AcOH	DCM		Merrifield resin	81

Reductive alkylation	Solvent/Temp.	Yield	Resin	Ref.
1) Ti(OiPr)$_4$ R$_2$NH$_2$, toluene 2) Na(OAc)$_3$BH AcOH, THF			Merrifield resin	135

2.13 Substitution reactions

4-Fluorobenzoate leaving group	Solvent/Temp.	Yield	Resin	Ref.
	Piperazine 130 °C		Wang resin	69

Allylic substitution	Solvent/Temp.	Yield	Resin	Ref.
	THF		piperazinomethyl resin	32

C-alkylation	Solvent/Temp.	Yield	Resin	Ref.
	THF/hexane		1,3-dithiane resin	19

C-alkylation	Solvent/Temp.	Yield	Resin	Ref.
	Toluene 100 °C	95%	Rink resin	119

C-alkylation	Solvent/Temp.	Yield	Resin	Ref.
	THF 0 °C		bromomethyl polystyrene	86

C-Alkylation	Solvent/Temp.	Yield	Resin	Ref.
	CH$_3$CN reflux	65% to 78%	PEG	136

C-Alkylation	Solvent/Temp.	Yield	Resin	Ref.
	THF RT		2-chlorotrityl resin	48

Chloride formation	Solvent/Temp.	Yield	Resin	Ref.
	DCM		PhFl alcohol resin	4

Chloride formation

	Solvent/Temp.	Yield	Resin	Ref.
	DCM		aminomethyl resin	12

Dialkylation on carbon

	Solvent/Temp.	Yield	Resin	Ref.
	THF −78 °C to RT		Wang resin	132

Enol alkylation

	Solvent/Temp.	Yield	Resin	Ref.
	THF −78 °C to −23 °C		linear polystyrene	104

Epoxide opening

	Solvent/Temp.	Yield	Resin	Ref.
	Dichloroethane 80 °C	100%	Merrifield resin	137

Glycosylation

	Solvent/Temp.	Yield	Resin	Ref.
	DCM		polystyrenediisopropylsilyl chloride	14

Glycosylation

	Solvent/Temp.	Yield	Resin	Ref.
	DCM/ cyclohexane	95%	controlled pore glass	124

Glycosylation

	Solvent/Temp.	Yield	Resin	Ref.
	DCM 30 °C		TentaGel S NH$_2$	138

Mitsunobu bromination

	Solvent/Temp.	Yield	Resin	Ref.
	DCM RT	100%	Merrifield resin	16

Mitsunobu bromination

Solvent/Temp.	Yield	Resin	Ref.
DCM		Rink resin	132

Mitsunobu coupling

Solvent/Temp.	Yield	Resin	Ref.
THF/DCM		trityl chloride resin	139

Mitsunobu esterification

Solvent/Temp.	Yield	Resin	Ref.
THF 0 °C to 23 °C	>76%	polystyrene	73

Mitsunobu reaction

Solvent/Temp.	Yield	Resin	Ref.
DCM 0 °C		Merrifield resin	5

Mitsunobu reaction

Solvent/Temp.	Yield	Resin	Ref.
Toluene 50 °C	93%	Wang resin	16

Mitsunobu reaction

Solvent/Temp.	Yield	Resin	Ref.
NMM RT		selenylpolystyrene	39

Mitsunobu reaction

Solvent/Temp.	Yield	Resin	Ref.
DCM		TentaGel S OH	56

Mitsunobu reaction

Solvent/Temp.	Yield	Resin	Ref.
DCM		TentaGel S OH	140

Mitsunobu reaction

Solvent/Temp.	Yield	Resin	Ref.
THF		Merrifield resin	141

Mitsunobu reaction

Solvent/Temp.	Yield	Resin	Ref.
THF RT	96%	aminomethyl resin	142

169

Reaction	Solvent/Temp.	Yield	Resin	Ref.
N-alkylation (RNH₂)	DMSO		TentaGel S OH	20
N-alkylation (RNH₂)			not specified	40
N-alkylation (1) LiOtBu, THF; 2) R₂X, DMSO)			MBHA resin	143
N-alkylation (R₂NH₂)	DMSO RT		PS-PEG-Rink	48
N-alkylation (XCH₂R₂)	THF/DMF RT	>90%	ArgoGel-Rink resin	50
N-nucleophile (XCH₂R)	DMF 50 °C	>90%	ArgoGel-Rink resin	50
N-nucleophile (H₂N-cyclopropyl)	DMSO 60 °C		Rink resin	52
N-nucleophile	THF		Wang resin	61
N-nucleophile (1) tBuOLi, THF; 2) R₂C≡CCH₂OMs, DMSO)	RT		Wang resin	96
N-nucleophile (TsO-epoxide; LiN(SiMe₃)₂, LiI)	NMP/THF RT		Wang resin	144

N-nucleophile	Solvent/Temp.	Yield	Resin	Ref.
	DMA RT		Merrifield resin	145

N-nucleophile	Solvent/Temp.	Yield	Resin	Ref.
	CH_3CN 70 °C–78 °C		Merrifield resin	146

O-nucleophile	Solvent/Temp.	Yield	Resin	Ref.
	DMA 100 °C		Merrifield resin	9

O-nucleophile	Solvent/Temp.	Yield	Resin	Ref.
	DMSO 80 °C		Wang chloride resin	52

O-nucleophile	Solvent/Temp.	Yield	Resin	Ref.
	DMA 60 °C-70 °C		Merrifield resin	146

P-alkylation	Solvent/Temp.	Yield	Resin	Ref.
	DMF 100 °C		Merrifield resin	146

P-nucleophile	Solvent/Temp.	Yield	Resin	Ref.
	Toluene 40 °C		thiomethyl resin	107

P-nucleophile	Solvent/Temp.	Yield	Resin	Ref.
	Toluene heat		MBHA resin	126

Phosphite formation	Solvent/Temp.	Yield	Resin	Ref.
	CH_3CN RT		controlled pore glass	147

S-nucleophile	Solvent/Temp.	Yield	Resin	Ref.
	DMF		Rink resin	44

Stille coupling	Solvent/Temp.	Yield	Resin	Ref.
	NMP RT or 60 °C		PEG-5000 or Wang resin	148

Sulfonamide alkylation	Solvent/Temp.	Yield	Resin	Ref.
	DMF		TentaGel S NH$_2$	149

Suzuki coupling	Solvent/Temp.	Yield	Resin	Ref.
	THF 65 °C		dibutylsilyl chloride resin	51

2.14 Protection/deprotection reactions

1,2-Diol protection	Solvent/Temp.	Yield	Resin	Ref.
	DCM		Merrifield resin	33

1,3-Diol protection	Solvent/Temp.	Yield	Resin	Ref.
	DCM 20 °C		diisopropylsilyl resin	79

Alcohol deprotection	Solvent/Temp.	Yield	Resin	Ref.
	DMF/MeOH		Merrifield resin	33

Alcohol deprotection	Solvent/Temp.	Yield	Resin	Ref.
	THF 23 °C	94%	polystyrene	73

Alcohol deprotection	Solvent/Temp.	Yield	Resin	Ref.
	MeOH/DMF 25 °C		TentaGel S NH$_2$	138

Aldehyde deprotection	Solvent/Temp.	Yield	Resin	Ref.
	NMP/acetone 50 °C	100%	carboxypolystyrene	16

Amine deprotection	Solvent/Temp.	Yield	Resin	Ref.
	DMF RT		phosphine resin	38

Amine deprotection

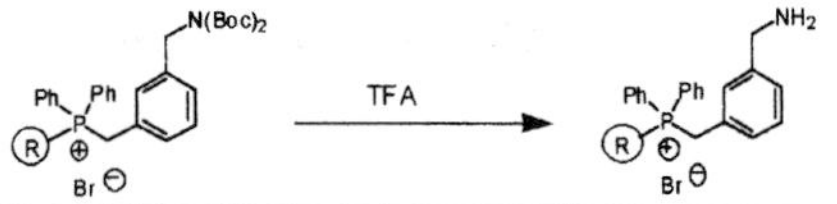

	Solvent/Temp.	Yield	Resin	Ref.
	DCM RT		phosphine resin	38

Amine deprotection

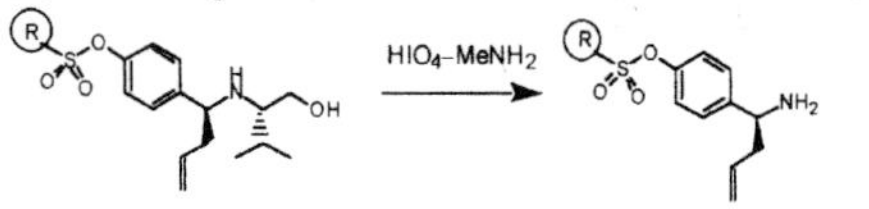

	Solvent/Temp.	Yield	Resin	Ref.
	MeOH/THF RT	95%	polystyrenesulfonic acid	102

Amine deprotection

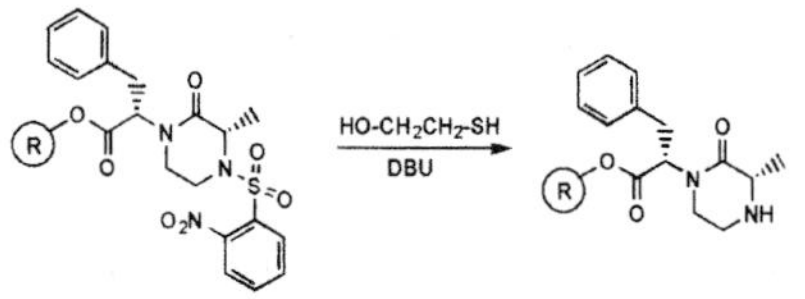

	Solvent/Temp.	Yield	Resin	Ref.
	DMF	85%	2-chlorotrityl resin	57

Carboxylic acid deprotection

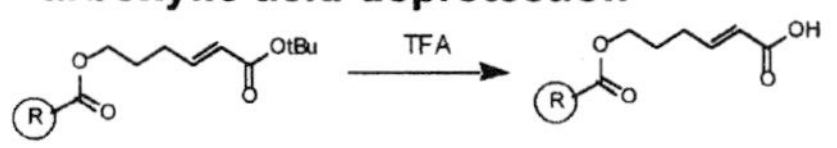

	Solvent/Temp.	Yield	Resin	Ref.
	DCM RT	94%	carboxypolystyrene	16

Carboxylic acid deprotection

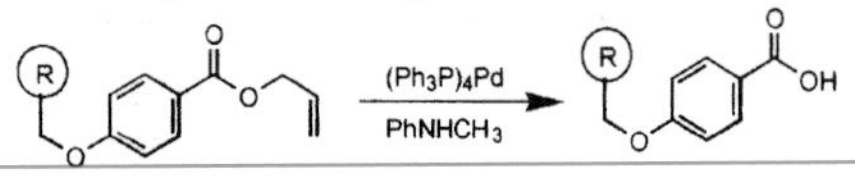

	Solvent/Temp.	Yield	Resin	Ref.
	DMSO/DMF 55 °C		Wang resin	96

Fmoc/Ac removal

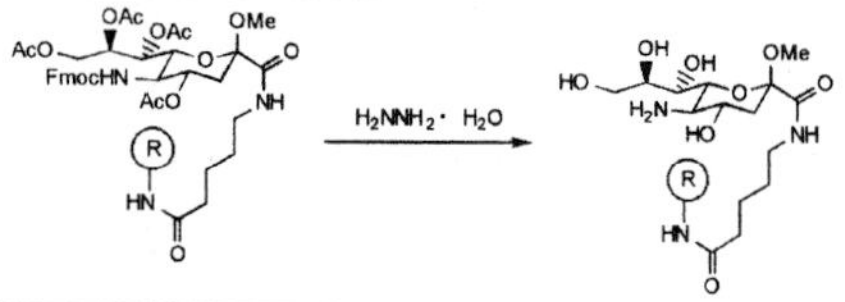

	Solvent/Temp.	Yield	Resin	Ref.
	MeOH		Knorr resin	45

Imine hydrolysis

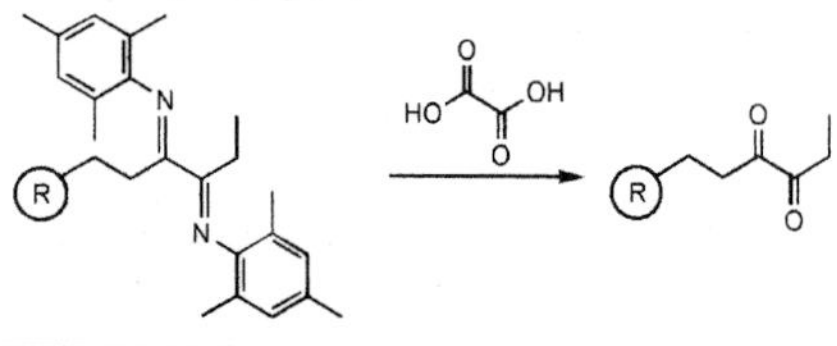

	Solvent/Temp.	Yield	Resin	Ref.
	THF/H$_2$O 70 °C		polystyrene	86

N-deprotection

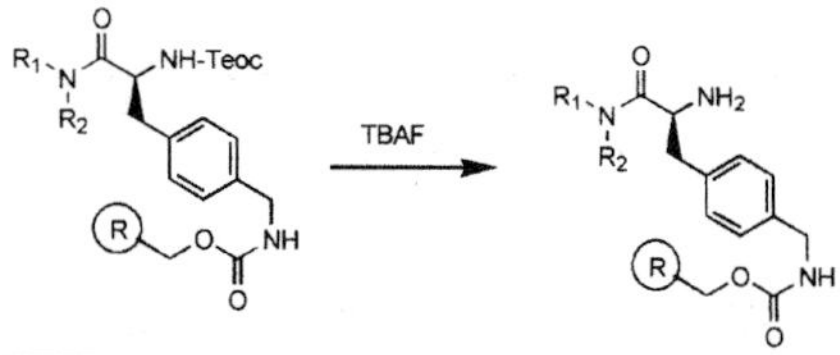

	Solvent/Temp.	Yield	Resin	Ref.
	THF 50 °C		Wang resin	55

N-deprotection

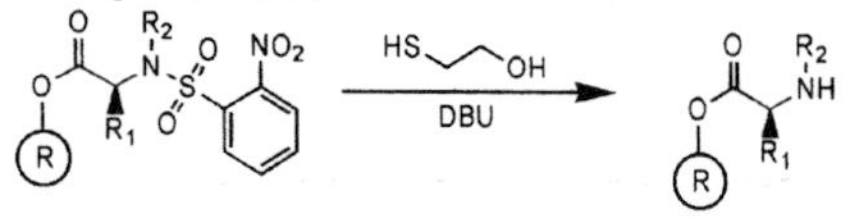

	Solvent/Temp.	Yield	Resin	Ref.
	DMF		TentaGel S OH	56

N-deprotection	Solvent/Temp.	Yield	Resin	Ref.
DBU / HSCH₂CH₂OH	DMF		TentaGel S OH	140

N-deprotection	Solvent/Temp.	Yield	Resin	Ref.
AlCl₃ / anisole	DCM/CH₃NO₂	>97%	controlled pore glass	150

N-deprotection	Solvent/Temp.	Yield	Resin	Ref.
hydrazine	DMF		NovaSyn KR-Knorr	151

O-deprotection	Solvent/Temp.	Yield	Resin	Ref.
HCO₂H	DCM RT		dibutylsilyl chloride resin	51

S-deprotection	Solvent/Temp.	Yield	Resin	Ref.
(i-Bu)₃SiH / TFA	DCM		MBHA resin	143

2.15 Other solid phase reactions

Acid anhydride formation	Solvent/Temp.	Yield	Resin	Ref.
Et₃N	DCM 25 °C		hydroxymethyl resin	5

Acid chloride formation	Solvent/Temp.	Yield	Resin	Ref.
(COCl)₂ / DMF cat.	Dichloroethane RT	100%	carboxypolystyrene	16

Acid fluoride formation	Solvent/Temp.	Yield	Resin	Ref.
CNF	DCM -10 °C	98%	2-chlorotrityl resin	152

Bromination	Solvent/Temp.	Yield	Resin	Ref.
Br₂	DCM 25 °C		Merrifield resin	97

Bromination

	Solvent/Temp.	Yield	Resin	Ref.
	AcOH 60 °C		thiomethyl resin	107

Carbamate formation

	Solvent/Temp.	Yield	Resin	Ref.
	DCM RT		sulfonyl chloride resin	17

Carbene insertion

	Solvent/Temp.	Yield	Resin	Ref.
	DCM 25 °C		Wang resin	153

Carbonate formation

	Solvent/Temp.	Yield	Resin	Ref.
	THF/DCM		TentaGel S NH$_2$	22

Chlorination

	Solvent/Temp.	Yield	Resin	Ref.
	DCM		2-chlorotrityl resin	91

Curtius rearrangement

	Solvent/Temp.	Yield	Resin	Ref.
	Toluene 90 °C		PEG$_4$-PS	9

Curtius rearrangement

	Solvent/Temp.	Yield	Resin	Ref.
	Xylene 90 °C		Rink resin	52

Desilylation

	Solvent/Temp.	Yield	Resin	Ref.
	THF/MeOH		MeO-PEG	11

Deacylation

	Solvent/Temp.	Yield	Resin	Ref.
	DMF 25 °C		Wang resin	153

Diazo formation

	Solvent/Temp.	Yield	Resin	Ref.
	MeOH/THF 90 °C		Merrifield resin	8

Diazo formation

	Solvent/Temp.	Yield	Resin	Ref.
	DMF		Wang resin	153

175

Enol addition to imine

Solvent/Temp.	Yield	Resin	Ref.
DCM RT		Merrifield resin	85

Epoxide opening

Solvent/Temp.	Yield	Resin	Ref.
DCM -78 °C to RT		diisopropylsilyl chloride resin	14

Ester formation

Solvent/Temp.	Yield	Resin	Ref.
DCM or DMF		aminomethyl microtubes with Knorr linker	77

Ester formation

Solvent/Temp.	Yield	Resin	Ref.
DCM		piperazinomethyl resin	32

Guanine formation

Solvent/Temp.	Yield	Resin	Ref.
NMP		Wang resin	42

Halogen-metal exchange

Solvent/Temp.	Yield	Resin	Ref.
-78 °C		piperazinomethyl resin	32

Halogen-metal exchange

Solvent/Temp.	Yield	Resin	Ref.
THF -35 °C		Wang resin	98

Hydroformylation

Solvent/Temp.	Yield	Resin	Ref.
Toluene 40 °C	83%	polystyrene crowns with trityl linker	154

Iodosulfonamide formation

Solvent/Temp.	Yield	Resin	Ref.
DCM 0 °C		diisopropylsilyl resin	155

Isonitrile formation

Solvent/Temp.	Yield	Resin	Ref.
DCM		Wang resin	27

Metalation		Solvent/Temp.	Yield	Resin	Ref.
		THF -30 °C		trityl resin	100

Metalation		Solvent/Temp.	Yield	Resin	Ref.
		THF -60 °C		Merrifield resin	34

Metathesis		Solvent/Temp.	Yield	Resin	Ref.
		Benzene 75 °C		Wang resin	96

Metathesis		Solvent/Temp.	Yield	Resin	Ref.
		DCM reflux	56%	dimethylsilyl resin	156

Olefin hydroxylation		Solvent/Temp.	Yield	Resin	Ref.
1) 9-BBN, THF 2) H$_2$O$_2$, NaOH MeOH/H$_2$O				vinyl polystyrene	123

Organometallic complex		Solvent/Temp.	Yield	Resin	Ref.
		DCM 23 °C		polystyrene	86

Phenol formation		Solvent/Temp.	Yield	Resin	Ref.
		EtOH reflux	up to 85%	Merrifield resin	146

Rearrangement-displacement		Solvent/Temp.	Yield	Resin	Ref.
		DMF -40 °C to 0 °C		diisopropylsilane resin	155

Silyl chloride formation		Solvent/Temp.	Yield	Resin	Ref.
		DCM 25 °C		PS-silane resin	31

Silyl enol ether formation		Solvent/Temp.	Yield	Resin	Ref.
		DCM RT		thiomethyl resin	3

Sulfurization

Solvent/Temp.	Yield	Resin	Ref.
CH$_3$CN		controlled pore glass	147

Titanocene formation

Solvent/Temp.	Yield	Resin	Ref.
		polystyrene	95

Transesterification

Solvent/Temp.	Yield	Resin	Ref.
pyridine/DMF		imidazolecarbonyl Wang resin	28

Transimination

Solvent/Temp.	Yield	Resin	Ref.
DCM RT		polystyrene	60

Transimination

Solvent/Temp.	Yield	Resin	Ref.
DCM		PEG	136

Transmetalation

Solvent/Temp.	Yield	Resin	Ref.
50 °C		polystyrene	39

Trichloroacetaimidate formation

Solvent/Temp.	Yield	Resin	Ref.
DCM 0 °C then 25 °C	96%	Wang resin	6

2.16 Cleavage from supports

Acidolysis

Solvent/Temp.	Yield	Resin	Ref.
	81%	9-phenylfluorenyl resin	12

Acyl transfer

Solvent/Temp.	Yield	Resin	Ref.
DMF	up to 97%	aminomethyl resin	157

Acylation/cleavage

	Solvent/Temp.	Yield	Resin	Ref.
	DCM RT	up to 80%	Wang bromide resin	158

Addition/cyclization/release

	Solvent/Temp.	Yield	Resin	Ref.
	THF RT		thiomethyl resin	107

Amide hydrolysis

	Solvent/Temp.	Yield	Resin	Ref.
	EtOH/H$_2$O 100 °C		Tentagel S NH$_2$	138

Cleavage from dibutylsilyl resin

	Solvent/Temp.	Yield	Resin	Ref.
	THF RT		dibutylsilyl chloride resin	51

Cyclization/release

	Solvent/Temp.	Yield	Resin	Ref.
	EtOH 85 °C	100%	Wang resin	43

Cyclization/release

	Solvent/Temp.	Yield	Resin	Ref.
	THF 60 °C		Merrifield resin	92

Cyclization/release

	Solvent/Temp.	Yield	Resin	Ref.
	THF 40 °C		thiomethyl resin	3

Cyclization/release

	Solvent/Temp.	Yield	Resin	Ref.
	RT to 60 °C		TentaGel S OH	20

179

Cyclization/release	Solvent/Temp.	Yield	Resin	Ref.
	DIPEA RT		polystyrene	60

Cyclization/release	Solvent/Temp.	Yield	Resin	Ref.
	toluene 100 °C	54%	polystyrene	73

Cyclization/release	Solvent/Temp.	Yield	Resin	Ref.
	DMF 90 °C		Merrifield resin	93

Cyclization/release	Solvent/Temp.	Yield	Resin	Ref.
	DCM RT		sulfonyl chloride resin	17

Cyclization/release	Solvent/Temp.	Yield	Resin	Ref.
	CH$_3$CN 60 °C	72-81%	PEG$_4$-PS	9

Cyclization/release	Solvent/Temp.	Yield	Resin	Ref.
	toluene 80 °C		Wang resin	80

Cyclization/release	Solvent/Temp.	Yield	Resin	Ref.
	THF	up to 91%	Wang resin	159

Cyclization/release	Solvent/Temp.	Yield	Resin	Ref.
	DCM		oxime resin	160

Cyclization/release	Solvent/Temp.	Yield	Resin	Ref.
	THF RT		hydroxymethyl resin	161

Desulfonylation

Solvent/Temp.	Yield	Resin	Ref.
MeOH/DMF RT	97%	MeO-PEG	21

Displacement of amine

Solvent/Temp.	Yield	Resin	Ref.
		Merrifield resin	86
50 °C			

Disulfide cleavage

Solvent/Temp.	Yield	Resin	Ref.
		not specified	40

Elimination

Solvent/Temp.	Yield	Resin	Ref.
DCM	80% to 100%	indole resin	87

Eneamine-cyclization/cleavage

Solvent/Temp.	Yield	Resin	Ref.
Toluene heat		Merrifield resin	81

Enzymeatic cleavage

Solvent/Temp.	Yield	Resin	Ref.
0.1 M potassium phosphate buffer 25 °C	up to 50%	TentaGel resin	15

Epoxide opening/cyclization/cleavage

Solvent/Temp.	Yield	Resin	Ref.
THF RT		Wang resin	144

Fluoride

Solvent/Temp.	Yield	Resin	Ref.
THF		Merrifield resin	76

Hofmann elimination

Solvent/Temp.	Yield	Resin	Ref.
DMF		REM resin	162

Hydrolysis

Solvent/Temp.	Yield	Resin	Ref.
DCM RT		Merrifield resin	85

Lewis acid	Solvent/Temp.	Yield	Resin	Ref.
1) SnCl₄ 2) H₂O → ROH	CHCl₃ RT	70-95%	Merrifield resin	137

Nucleophilic aromatic substitution	Solvent/Temp.	Yield	Resin	Ref.
R-NH₂	Dioxane RT	up to 93%	Merrifield resin	41

Nucleophilic displacement	Solvent/Temp.	Yield	Resin	Ref.
Sc(OTf)₂ Ac₂O	DCM		PEG-PS	163

Nucleophilic displacement	Solvent/Temp.	Yield	Resin	Ref.
R₃NH₂	Pyridine, DIPEA RT		Merrifield resin	7

Nucleophilic displacement	Solvent/Temp.	Yield	Resin	Ref.
Et₂NH	Toluene 75 °C	98%	oxime resin	26

Nucleophilic displacement	Solvent/Temp.	Yield	Resin	Ref.
R₄NH₂	Pyridine RT		Merrifield resin	135

Nucleophilic displacement	Solvent/Temp.	Yield	Resin	Ref.
Et₃N	CH₃CN 60 °C	79% to 98%	Merrifield resin	129

Nucleophilic displacement	Solvent/Temp.	Yield	Resin	Ref.
R₁-NH-R₂	CH₃CN 60 °C	up to 95%	aryl sulfonate resin	164

Oxidation/displacement	Solvent	Yield	Resin	Ref.
H-Nuc DBU or pyridine Cu(OAc)₂	RT	89% to >95%	TentaGel S OH with hydrazide linker	165

Oxidative cleavage

	Solvent/Temp.	Yield	Resin	Ref.
	THF 0 °C to RT		1,3-dithiane resin	19

Oxidative cleavage

	Solvent/Temp.	Yield	Resin	Ref.
	Benzene RT		Merrifield resin	85

Oxidative cleavage

	Solvent/Temp.	Yield	Resin	Ref.
	DCM/H$_2$O		Merrifield resin	89

Ozonolysis

	Solvent/Temp.	Yield	Resin	Ref.
	DCM -78 °C		polystyrene	125

Ozonolysis

	Solvent/Temp.	Yield	Resin	Ref.
	DCM -78 °C	up to 100%	MBHA resin	126

Phosphate ester hydrolysis

	Solvent/Temp.	Yield	Resin	Ref.
	DCM	up to 98%	Wang resin	37

Photolytic cleavage

	Solvent/Temp.	Yield	Resin	Ref.
	H$_2$O/MeOH		TentaGel S NH$_2$	166

Photolytic cleavage

	Solvent/Temp.	Yield	Resin	Ref.
	THF	65%	2-pivaloylglycerol resin	128

Photolytic cleavage

	Solvent/Temp.	Yield	Resin	Ref.
	THF	74%	2-pivaloylglycerol resin	128

Ring-closing metathesis	Solvent/Temp.	Yield	Resin	Ref.
	Toluene 50 °C	86%	Merrifield resin	16

Safety-catch	Solvent/Temp.	Yield	Resin	Ref.
			Wang resin	27

Selective cleavage	Solvent/Temp.	Yield	Resin	Ref.
	Dichloroethane 83 °C	78% to 96%	chloroacetylpolystyrene	167

Selective cleavage	Solvent/Temp.	Yield	Resin	Ref.
	Dichloroethane reflux	80% to 100%	PAM resin or Merrifield resin	168

Silyl ether cleavage	Solvent/Temp.	Yield	Resin	Ref.
	THF 40 °C		diisopropylsilyl resin	14

Traceless	Solvent/Temp.	Yield	Resin	Ref.
	THF		PS-silane resin	31

Traceless	Solvent/Temp.	Yield	Resin	Ref.
	DCM 25 °C	80	PS-silane resin	31

Traceless	Solvent/Temp.	Yield	Resin	Ref.
	dioxane/H_2O RT		phosphine resin	38

Traceless	Solvent/Temp.	Yield	Resin	Ref.
	THF 50 °C		piperazinomethyl resin	32

Traceless	Solvent/Temp.	Yield	Resin	Ref.
	Toluene 90 °C		selenylpolystyrene	39

2.17 Polymer-supported reagents

Acylsulfonamides

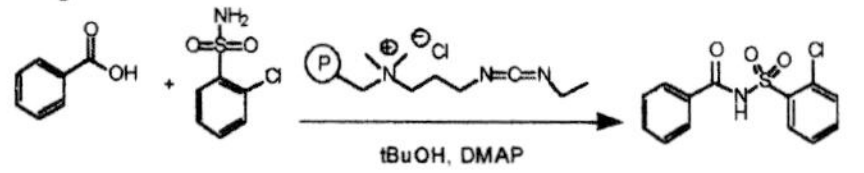

P = Polystyrene

	Solvent/Temp.	Yield	Resin	Ref.
	Dichloroethane RT	62-81%	PS-EDC	169

Aldol condensation

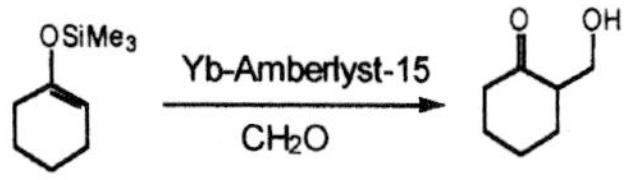

	Solvent/Temp.	Yield	Resin	Ref.
	THF/H$_2$O	95%	Yb-Amberlyst-15	170

Alkyl halide reduction

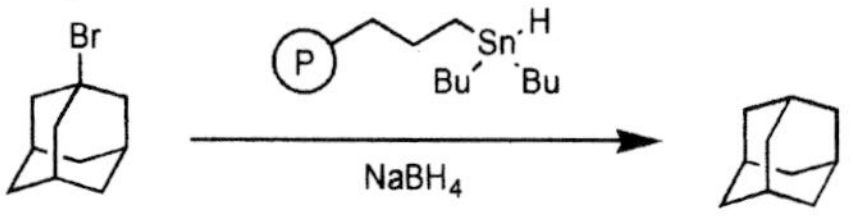

P = Polystyrene

	Solvent/Temp.	Yield	Resin	Ref.
	EtOH 65 °C	93-97%	PS-dibutyltin hydride	171

Amide formation

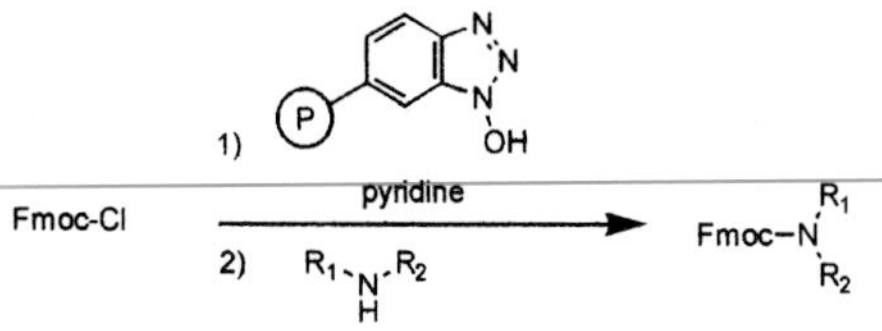

P = Polystyrene

	Solvent/Temp.	Yield	Resin	Ref.
	DCM 25 °C	up to 99%	polymer supported HOBt	172

Amide formation

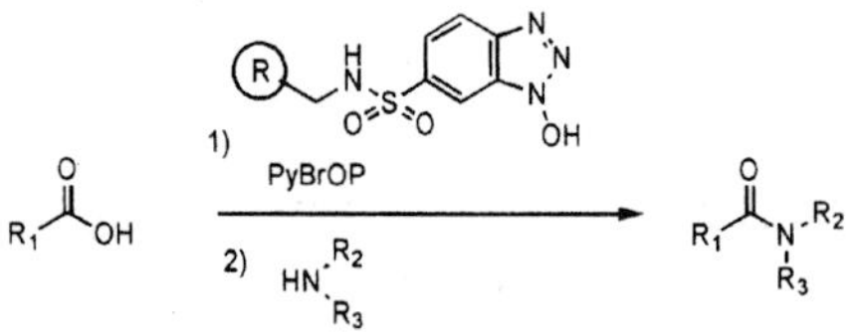

	Solvent/Temp.	Yield	Resin	Ref.
	DMF	50% to 87%	polymer supported HOBt	173

Amide formation

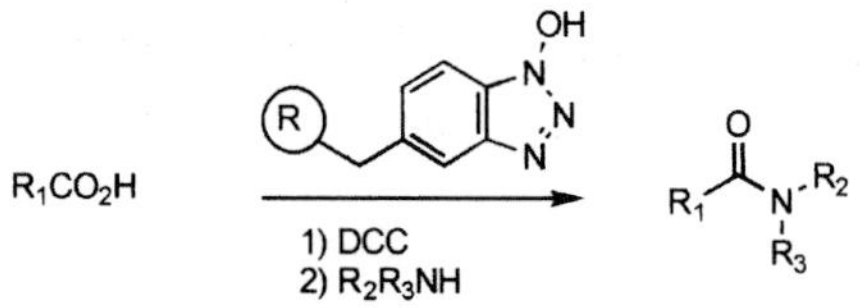

	Solvent/Temp.	Yield	Resin	Ref.
	DCM RT	72% to 99%	polymer supported HOBt	174

Asymmetric hydrogenation

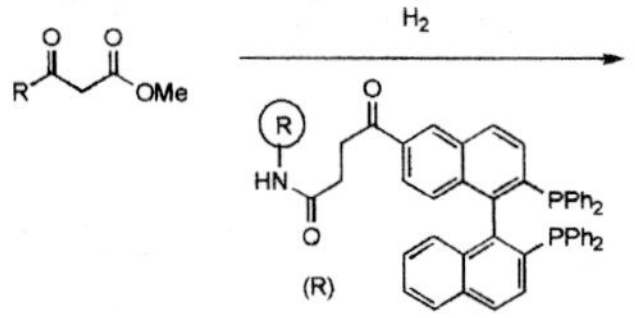

	Solvent/Temp.	Yield	Resin	Ref.
	THF/MeOH 70 °C	99%	aminomethyl resin supported BINAP Ru catalyst	175

Benzoxanone synthesis

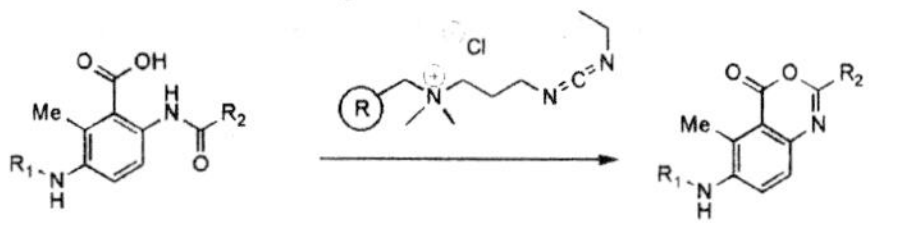

	Solvent/Temp.	Yield	Resin	Ref.
	DMF RT	up to 85%	polymer supported EDC	176

Benzoxazine synthesis

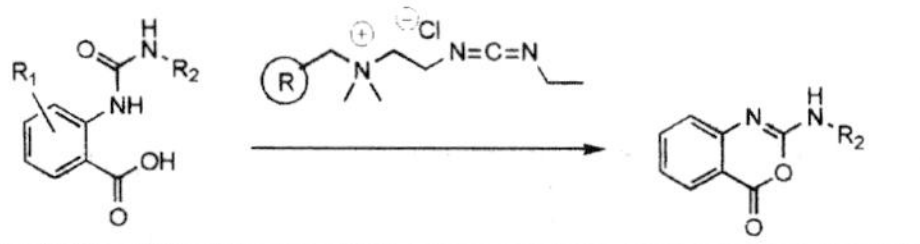

	Solvent/Temp.	Yield	Resin	Ref.
	DMF	73% to 94%	polymer supported EDC	177

Boc deprotection

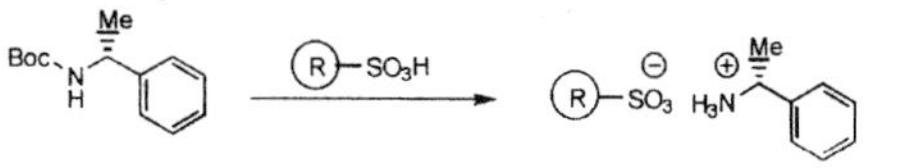

	Solvent/Temp.	Yield	Resin	Ref.
	DCM 25 °C	93%	Amberlyst 15	178

Bromination

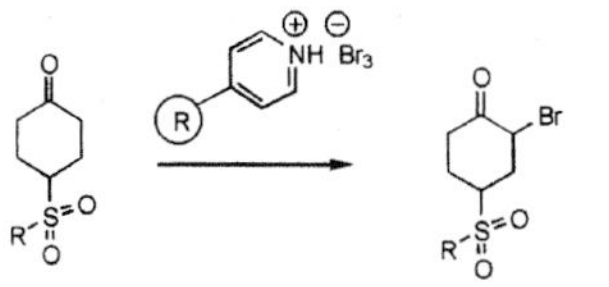

	Solvent/Temp.	Yield	Resin	Ref.
	Toluene	Up to 85%	pyridinium tribromide resin	179

Coupling to amines

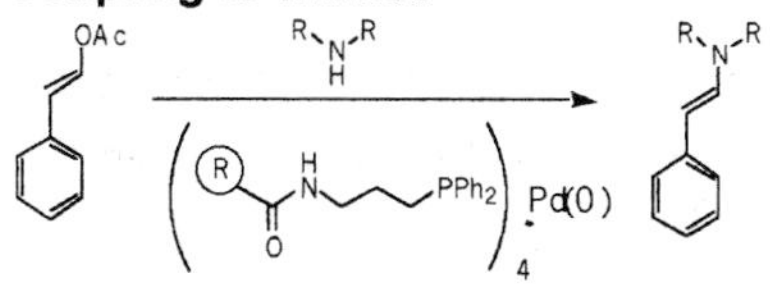

	Solvent/Temp.	Yield	Resin	Ref.
	EtOH/H$_2$O/ heptane 70 °C	73% to 100%	poly(N-isopropylacrylamide) resin	23

Diels-Alder

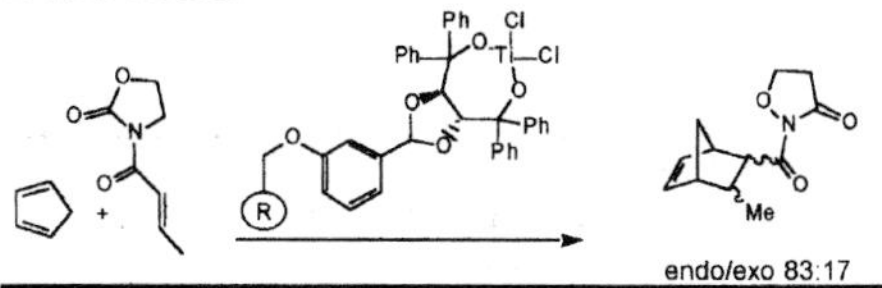

endo/exo 83:17

	Solvent/Temp.	Yield	Resin	Ref.
	Toluene RT	99%	Merrifield-Ti-TADDOL resin	180

Elimination

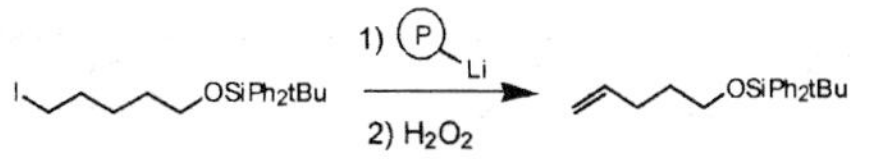

(P) = Polystyrene

	Solvent/Temp.	Yield	Resin	Ref.
	THF 23 °C	78%	PS-selenium	181

Ethylene polymerization

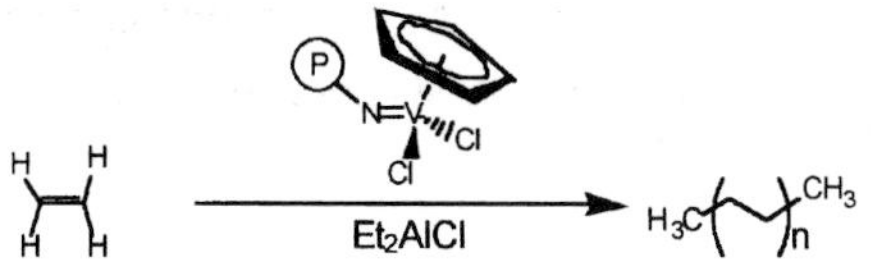

(P) = Polystyrene

	Solvent/Temp.	Yield	Resin	Ref.
	Chlorobenzene 25 °C		aminopolystyrene-vanadium catalyst	182

Hydroformylation

	Solvent/Temp.	Yield	Resin	Ref.
	Toluene 80 °C	100 %	polystyrene-Rh$_2$(OAc)$_2$	183

Hydrogenation

	Solvent/Temp.	Yield	Resin	Ref.
	EtOH/H$_2$O/ heptane 70°C		poly(N-isopropylacrylamide) resin	23

Imine reduction

	Solvent/Temp.	Yield	Resin	Ref.
	MeOH		Amberlite IRA-400 borohydride resin	184

Michael addition

	Solvent/Temp.	Yield	Resin	Ref.
	EtOH 20 °C	up to 99%	silica-phenylate	185

Michael addition

	Solvent/Temp.	Yield	Resin	Ref.
	CH$_3$CN RT	92% to 97%	microencapsulated Sc(OTf)$_3$	186

Mitsunobu reaction

	Solvent/Temp.	Yield	Resin	Ref.
	DCM 25 °C	69-81%	PS-PPh$_2$	187

Monothioacetalization

	Solvent/Temp.	Yield	Resin	Ref.
	CH$_3$CN 60 °C	up to 86%	polymethacrylate pi acid	188

Ⓟ = Polystyrene/ethyleneglycol dimethacrylate

N-alkylation

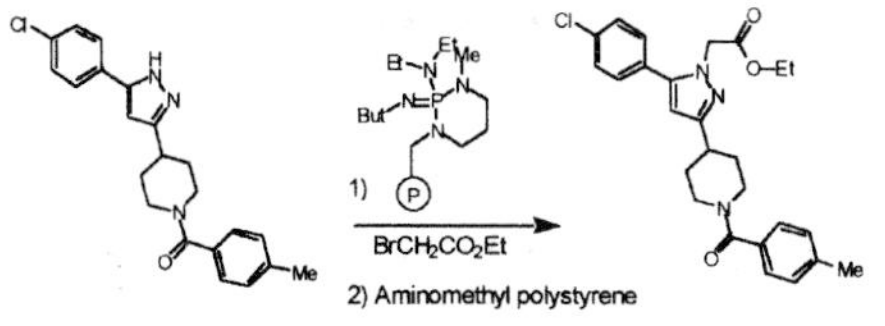

		Solvent/Temp.	Yield	Resin	Ref.
		CH$_3$CN RT	93- 99%	PBEMP	189

1) (P)
BrCH$_2$CO$_2$Et
2) Aminomethyl polystyrene

(P) = polystyrene

N-hydroxysuccinimide ester formation

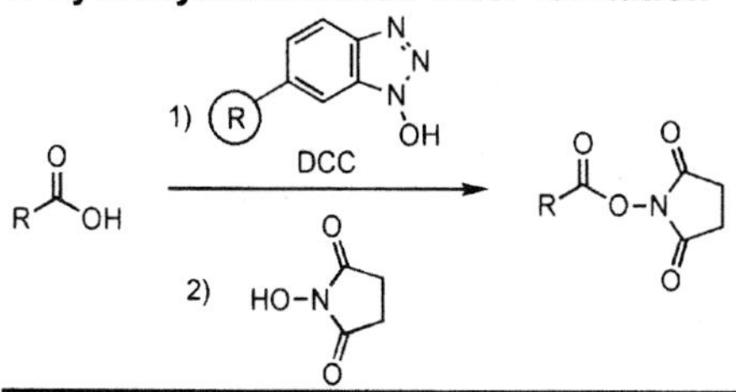

		Solvent/Temp.	Yield	Resin	Ref.
		DCM RT	up to 100%	polystyrene-HOBt	190

1) (R) DCC

2) HO-N

Olefin hydroxylation

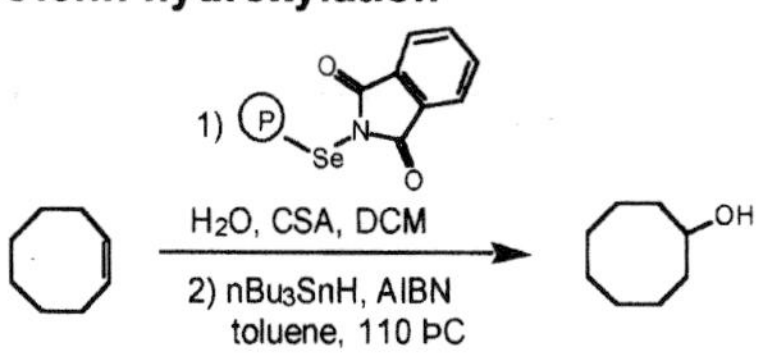

		Solvent/Temp.	Yield	Resin	Ref.
		DCM	80- 82%	PS-selenium	181

1) (P)-Se

H$_2$O, CSA, DCM

2) nBu$_3$SnH, AIBN
toluene, 110 ºC

(P) = Polystyrene

Olefin oxidation to diol

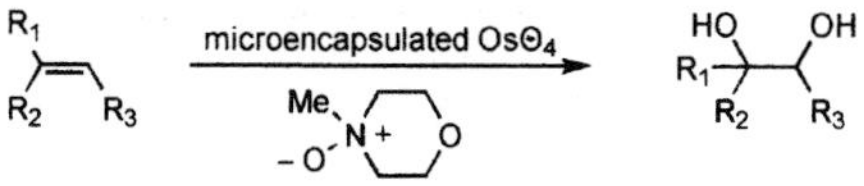

		Solvent/Temp.	Yield	Resin	Ref.
	microencapsulated OsO$_4$	H$_2$O/acetone/CH$_3$CN RT	63% to 89%	microencapsulated OsO$_4$	191

R$_1$ R$_2$ R$_3$ → HO OH R$_1$ R$_2$ R$_3$

Organolithium reactions

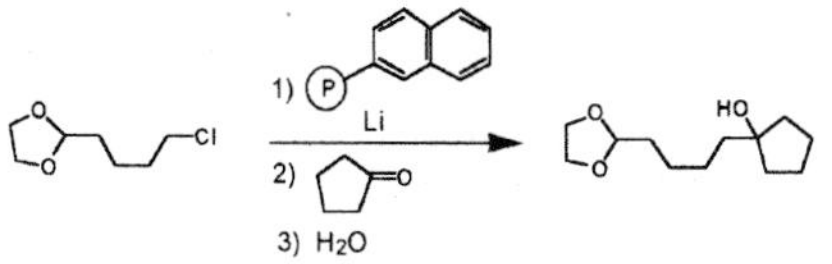

		Solvent/Temp.	Yield	Resin	Ref.
		THF -90 °C to -78 °C	up to 98%	PS-PVN	192

1) (P) Li
2)
3) H$_2$O

(P) = Polystyrene

Organozinc addition to aldehyde

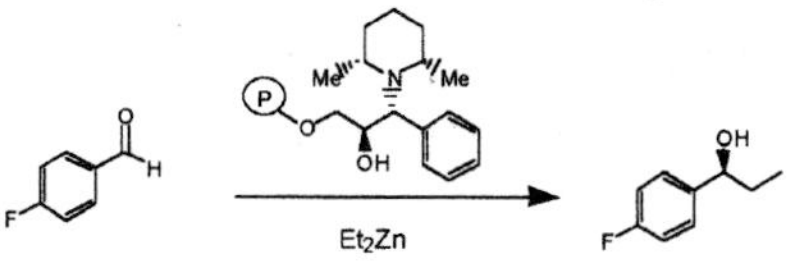

		Solvent/Temp.	Yield	Resin	Ref.
		Toluene 0 °C	to >99%	aminoalcohol-2- chlorotrityl resin	193

Et$_2$Zn

(P) = 2-chlorotrityl resin

Oxidation

		Solvent/Temp.	Yield	Resin	Ref.
		CHCl$_3$ 30 °C	73- 82%	PMM-isoxazolinium Cr(VI)	194

Me—◯—CH$_2$OH → Me—◯—CHO

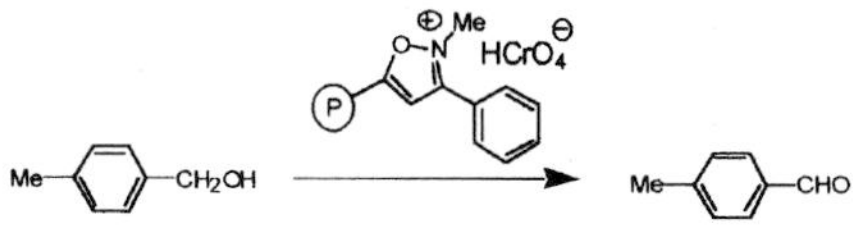

(P) = poly(methyl methacrylate)
with 2% ethyleglycol demethacrylate

Oxidation

	Solvent/Temp.	Yield	Resin	Ref.
Me$_2$N$\smile\frown$OH $\xrightarrow{\;P\text{—}NMe_3\,RuO_4\;}$ Me$_2$N$\smile\frown$CHO	Toluene 75 °C	>95%	PS-perruthenate	195

(P) = Polystyrene

Oxidation

	Solvent/Temp.	Yield	Resin	Ref.
(reaction scheme) $\xrightarrow{P\text{—}NMe_3\,RuO_4}$ (nitrone) $\xrightarrow[CHCl_3\;60\,°C]{CO_2Me}$ (isoxazolidine)	DCM RT	55-91%	PS-perruthenate	196

(P) = Polystyrene

Oxidation

	Solvent/Temp.	Yield	Resin	Ref.
(stilbene) + P—O$_2$C–C$_6$H$_4$–C(O)CF$_3$ $\xrightarrow[2KHSO_3\cdot KHSO_4\cdot K_2SO_4]{EDTA,\;NaHCO_3}$ (epoxide)	CH$_3$CN	76-97%	TentaGel-phenoxy-trifluoromethyldioxiran	197

(P) = TentaGel

Pd(0) catalyst

	Solvent/Temp.	Yield	Resin	Ref.
(allylic acetate) + (diketone) $\xrightarrow[K_2CO_3]{Pd\text{-}PEP\text{-}MOP}$ (product)	H$_2$O 25 °C		PEG-PS	198

Pd-PEP-MOP = (structure)

Reduction

	Solvent/Temp.	Yield	Resin	Ref.
(biphenyl-CHO) $\xrightarrow[\substack{RhCl(PPh_3)_3\\2)\,HF/pyridine}]{1)\;P\text{—}SiEt_2H}$ (biphenyl-CH$_2$OH)	NMP 60 °C	up to 84%	PS-trialkysilane	133

(P) = Polystyrene

Reductive amination

	Solvent/Temp.	Yield	Resin	Ref.
(ketone + amine) $\xrightarrow{R\text{—}NMe_3\,CNBH_3}$ (product)	MeOH	Up to 100%	polymer supported cyanoborohydride	179

S-alkylation

	Solvent/Temp.	Yield	Resin	Ref.
(bromide + thiol) $\xrightarrow{R\text{—tetramethylguanidine}}$ (cyclic product)			tetramethylguanidine resin	40

Sharpless asymmetric epoxidation

	Solvent/Temp.	Yield	Resin	Ref.
R$\smile$OH $\xrightarrow[\substack{Ti(OiPr)_4\\TBHP}]{polytartrate\;ester}$ R$\smile$epoxide$\smile$OH	DCM -20 °C	up to 57%	polytartrate ester	199

189

Substitution

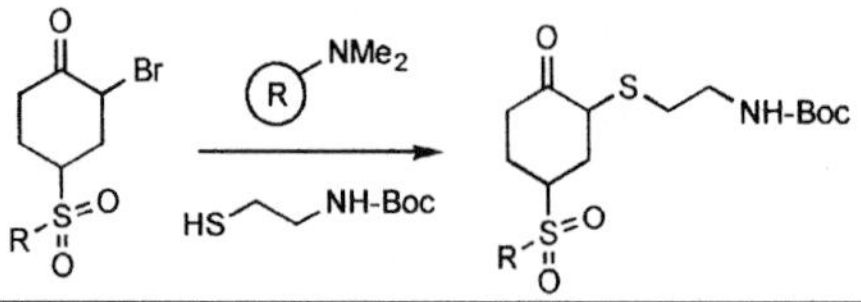

	Solvent/Temp.	Yield	Resin	Ref.
	THF	Up to 100%	Amberlyst A21	179

Sulfonamide formation

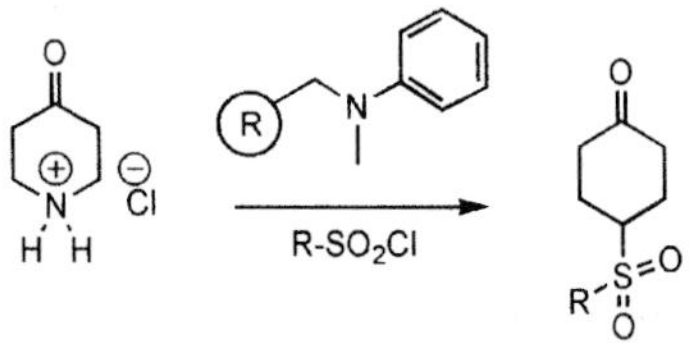

	Solvent/Temp.	Yield	Resin	Ref.
	DCM	Up to 94%	PolyDMAP	179

Suzuki coupling

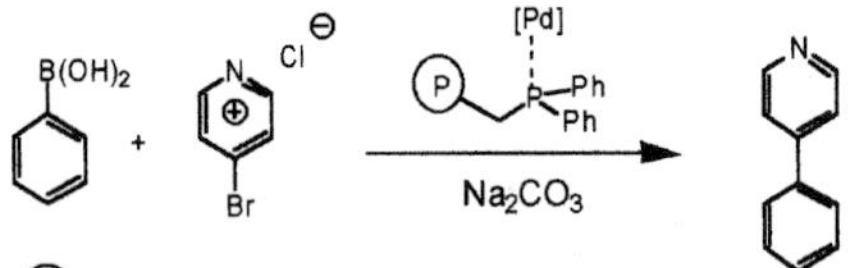

	Solvent/Temp.	Yield	Resin	Ref.
	Toluene/EtOH/H$_2$O 80 °C	up to 98%	PS-PPh$_2$/Pd	200

P = Polystyrene

Swern oxidation

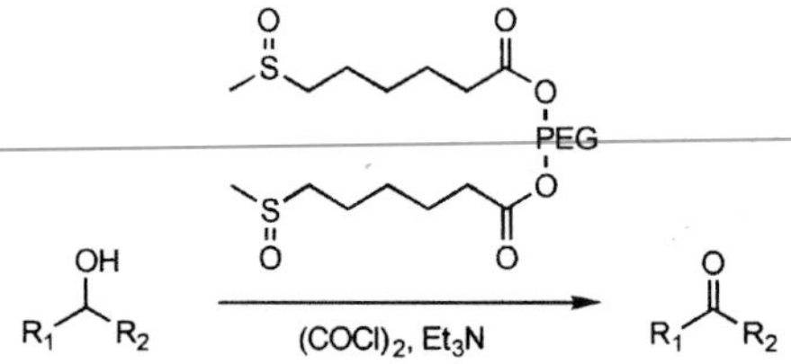

	Solvent/Temp.	Yield	Resin	Ref.
	DCM -50 °C to -60 °C	91% to 99%	PEG-bis[6-methylsulfinyl)hexanoate	201

THP ether hydrolysis

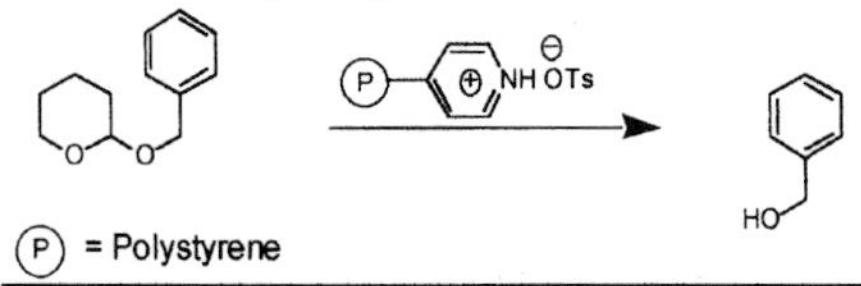

	Solvent/Temp.	Yield	Resin	Ref.
	EtOH/THF 75 °C	86-100%	polyvinylpyridinium p-toluenesulfonate	202

P = Polystyrene

Transfer hydrogenation

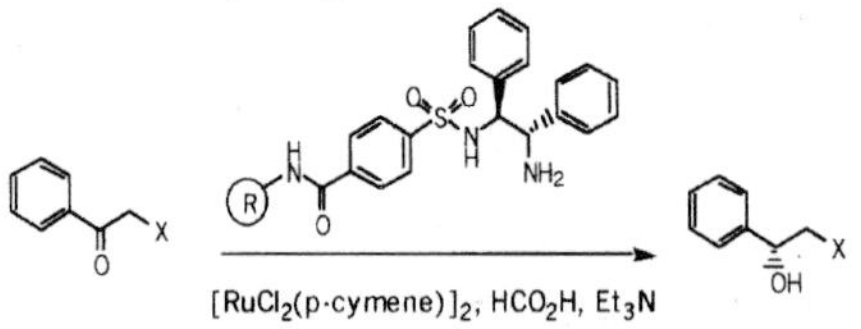

	Solvent/Temp.	Yield	Resin	Ref.
	30°C	90% to 95%	Ru catalyst on TentaGel S NH$_2$	203

Transfer hydrogenation

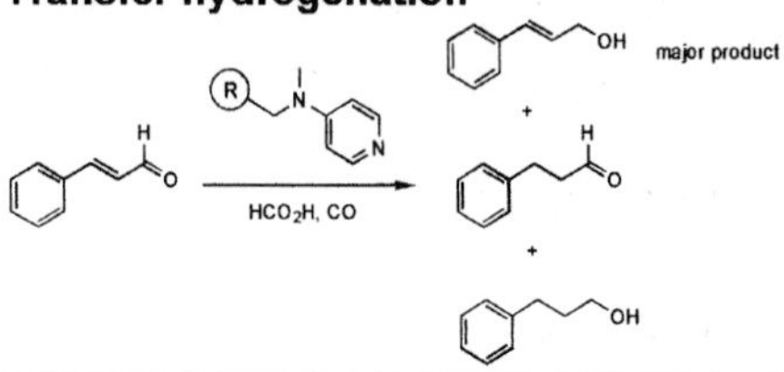

	Solvent/Temp.	Yield	Resin	Ref.
	THF 30 °C	86%	POLYDMAP	204

2.18 Scavenger resins

2-Thioethylamine resin		Ref.
	thioacetic acid, ketone scavenger	205

Amberlyst 15		Ref.
	base scavenger	179
	base scavenger	169

Aminomethyl resin		Ref.
	alkyl halide scavenger	189
	acid scavenger	36
	acid scavenger	179

Bis(aminoethyl)aminomethyl resin		Ref.
	acid scavenger	206

Dimethylamino resin		Ref.
	acid scavenger	162

Formylpolystyrene		Ref.
	primary amine scavenger	184

Isatoic anhydride resin		Ref.
	amine scavenger	145

Isocyanate resin		Ref.
	amine scavenger	184

Morpholine resin		Ref.
	acid scavenger	184
	acid scavenger	207

POLYDMAP		Ref.
	acid scavenger	176

Polystyrene-linked 4-hydroxybenzaldehyde		Ref.
	primary amine scavenger	184

Sulfonic acid resin/calcium sulfonate resin		Ref.
	tetrabutyl-ammonium fluoride scavenger	176

Thiourea resin		Ref.
	alpha-bromoketone scavenger	208

Tris(2-aminoethyl)amine resin		Ref.
	acid chloride scavenger	184
	acid scavenger	207
	ketone and imine scavengeer	184
	thioacetic acid, ketone scavenger	205

References

1. Wilson, L. J., Li, M., and Portlock, D. E. (1998). *Tetrahedron Lett.*, **39**, 5135.
2. Trautwein, A. W., Süßmuth, R. D., and Jung, G. (1998). *Bioorg. Med. Chem. Lett.*, **8**, 2381.
3. Kobayashi, S., Wakabayashi, T., and Yasuda, M. J. (1998). *Org. Chem.*, **63**, 4868.
4. Bleicher, K. H. and Wareing, J. R. (1998). *Tetrahedron Lett.*, **39**, 4587.
5. Nicolaou, K. C., Winssinger, N., Vourloumis, D., Ohshima, T., Kim, S., Pfefferkorn, J., *et al.* (1998). *J. Am. Chem. Soc.*, **120**, 10814.
6. Phoon, C. W., Oliver, S. F., and Abell, C. (1998). *Tetrahedron Lett.*, **39**, 7959.
7. Breitenbucher, J. G., Johnson, C. R., Haight, M., and Phelan, J. C. (1998). *Tetrahedron Lett.*, **39**, 1295.
8. Bhalay, G. and Dunstan, A. R. (1998). *Tetrahedron Lett.*, **39**, 7803.
9. Shao, H., Colucci, M., Tong, S., Zhang, H., and Castelhano, A. L. (1998). *Tetrahedron Lett.*, **39**, 7235.
10. Winkler, J. D. and McCoull, W. (1998). *Tetrahedron Lett.*, **39**, 4935.
11. Far, A. R. and Tidwell, T. T. (1998). *J. Org. Chem.*, **63**, 8636.
12. Henkel, B. and Bayer, E. (1998). *Tetrahedron Lett.*, **39**, 9401.
13. Winkler, J. D. and Kwak, Y.-S. (1998). *J. Org. Chem.*, **63**, 8634.

14. Zheng, C., Seeberger, P. H., and Danishefsky, S. J. (1998). *J. Org. Chem.*, **63**, 1126.

15. Böhm, G., Dowden, J., Rice, D. C., Burgess, I., Pilard, J.-F., Guibert, B., *et al.* (1998). *Tetrahedron Lett.*, **39**, 3819.

16. Veerman, J. J. N., van Maarseveen, J. H., Visser, G. M., Kruse, C. G., Schoemaker, H. E., Hiemstra, H., *et al.* (1998). *Eur. J. Org. Chem.*, 2583.

17. ten Holte, P., Thijs, L., and Zwanenburg, B. (1998). *Tetrahedron Lett.*, **39**, 7407.

18. Metz, W. A., Jones, W. D., Ciske, F. L., and Peet, N. P. (1998). *Bioorg. Med. Chem. Lett.*, **8**, 2399.

19. Bertini, V., Lucchesini, F., Pocci, M., and De Munno, A. (1998). *Tetrahedron Lett.*, **39**, 9263.

20. Vojkovský, T., Weichsel, A., and Pátek, M. (1998). *J. Org. Chem.*, **63**, 3162.

21. Zhao, X. and Janda, K. D. (1998). *Bioorg. Med. Chem. Lett.*, **8**, 2439.

22. Mohan, R., Yun, W., Buckman, B. O., Liang, A., Trinh, L., and Morrissey, M. M. (1998). *Bioorg. Med. Chem. Lett.*, **8**, 1877.

23. Bergbreiter, D. E., Liu, Y.-S., and Osburn, P. L. (1998). *J. Am. Chem. Soc.*, **120**, 4250.

24. Jensen, K. J., Alsina, J., Songster, M. F., Vágner, J., Albericio, F., and Barany, G. (1998). *J. Am. Chem. Soc.*, **120**, 5441.

25. Tommasi, R. A., Natermet, P. G., Shapiro, M. J., Chin, J., Brill, W. K. D., and Ang, K. (1998). *Tetrahedron Lett.*, **39**, 5477.

26. Scialdone, M. A., Shuey, S. W., Soper, P., Hamuro, Y., and Burns, D. M. J. (1998). *Org. Chem.*, **63**, 4802.

27. Hulme, C., Peng, J., Morton, G., Salvino, J. M., Herpin, T., and Labaudiniere, R. (1998). *Tetrahedron Lett.*, **39**, 7227.

28. Munson, M. C., Cook, A. W., Josey, J. A., and Rao, C. (1998). *Tetrahedron Lett.*, **39**, 7223.

29. Smith, A. L., Stevenson, G. I., Swain, C. J., and Castro, J. L. (1998). *Tetrahedron Lett.*, **39**, 8317.

30. Bui, C. T., Rasoul, F. A., Ercole, F., Pham, Y., and Maeji, N. J. (1998). *Tetrahedron Lett.*, **39**, 9279.

31. Hu, Y., Porco, J. A., Jr., Labadie, J. W., Gooding, O. W., and Trost, B. M. (1998). *J. Org. Chem.*, **63**, 4518.

32. Bräse, S., Enders, D., Köbberling, J., and Avemaria, F. (1998). *Angew. Chem. Int. Ed. Engl.*, **37**, 3413.

33. Deben, I., Goorden, J., van Doornum, E., Ovaa, H., and Kellenbach, E. (1998). *Eur. J. Org. Chem.*, 697.

34. Havez, S., Begtrup, M., and Vedsø, P. (1998). *J. Org. Chem.*, **63**, 7418.

35. Kiselyov, A. S., Smith, L., II, Virgilio, A. A., and Armstrong, R. W. (1998). *Tetrahedron*, **54**, 7987.

36. Chen, C. and Munoz, B. (1998). *Tetrahedron Lett.*, **39**, 6781.

37. Metcalf, C. A., III, Vu, C. B., Sundaramoorthi, R., Jacobsen, V. A., Laborde, E. A., Green, J., *et al.* (1998). *Tetrahedron Lett.*, **39**, 3435.

38. Hughes, I. J. (1998). *Med. Chem.*, **41**, 3804.

39. Ruhland, T., Andersen, K., and Pendersen, H. (1998). *J. Org. Chem.*, **63**, 9204.

40. Souers, A. J., Virgilio, A. A., Schürer, S. S., Ellman, J. A., Kogan, T. P., West, H. E., *et al.* (1998). *Bioorg. Med. Chem. Lett.*, **8**, 2297.

41. Masquelin, T., Sprenger, D., Baer, R., Gerber, F., and Mercadal, Y. (1998). *Helv. Chim. Acta*, **81**, 646.

42. Josey, J. A., Tarlton, C. A., and Payne, C. E. (1998). *Tetrahedron Lett.*, **39**, 5899.

43. Matthews, J. and Rivero, R. A. (1998). *J. Org. Chem.*, **63**, 4808.

44. Lee, J., Gauthier, D., and Rivero, R. A. (1998). *Tetrahedron Lett.*, **39**, 201.

45. Szabo, L., Smith, B. L., McReynolds, K. D., Parrill, A. L., Morris, E. R., and Gervay, J. (1998). *J. Org. Chem.*, **63**, 1074.

46. Page, P., Burrage, S., Baldock, L., and Bradley, M. (1998). *Bioorg. Med. Chem. Lett.*, **8**, 1751.

47. Tang, Z. and Pellatier, J. C. (1998). *Tetrahedron Lett.*, **39**, 4773.

48. Tumelty, D., Schwarz, M. K., and Needels, M. C. (1998). *Tetrahedron Lett.*, **39**, 7467.

49. Trautwein, A. W. and Jung, G. (1998). *Tetrahedron Lett.*, **39**, 8263.

50. Schwarz, M. K., Tumelty, D., and Gallop, M. A. (1998). *Tetrahedron Lett.*, **39**, 8397.

51. Thompson, L. A., Moore, F. L., Moon, Y.-C., and Ellman, J. A. (1998). *J. Org. Chem.*, **63**, 2066.

52. Richter, L. S. and Andersen, S. (1998). *Tetrahedron Lett.*, **39**, 8747.

53. Linkletter, B. A. and Bruice, T. C. (1998). *Bioorg. Med. Chem. Lett.*, **8**, 1285.

54. Lin, P. and Ganesan, A. (1998). *Tetrahedron Lett.*, **39**, 9789.

55. Kim, S. W., Hong, C. Y., Lee, K., Lee, E. J., and Koh, J. S. (1998). *Bioorg. Med. Chem. Lett.*, **8**, 735.

56. Boeijen, A., Kruijtzer, J. A. W., and Liskamp, R. M. J. (1998). *Bioorg. Med. Chem. Lett.*, **8**, 2375.

57. Mohamed, N., Bhatt, U., and Just, G. (1998). *Tetrahedron Lett.*, **39**, 8213.

58. Nieuwenhuijzen, J. W., Conti, P. G. M., Ottenheijm, H. C. J., and Linders, J. T. M. (1998). *Tetrahedron Lett.*, **39**, 7811.

59. Gordeev, M. F., Luehr, G. W., Hui, H. C., Gordon, E. M., and Patel, D. V. (1998). *Tetrahedron*, **54**, 15879.

60. Lee, S. H., Chung, S.-H., and Lee, Y.-S. (1998). *Tetrahedron Lett.*, **39**, 9469.

61. Morales, G. A., Corbett, J. W., and DeGrado, W. F. (1998). *J. Org. Chem.*, **63**, 1172.

62. Willoughby, C. A. and Chapman, K. T. (1998). *Tetrahedron Lett.*, **37**, 7181.

63. Wendeborn, S., Beaudegnies, R., Ang, K. H., and Maeji, N. J. (1999). *Biotech. Bioeng. (Combin. Chem.)*, **61**, 89.

64. Kang, S.-K., Kim, J.-S., Yoon, S.-K., Lim, K.-H., and Yoon, S. S. (1998). *Tetrahedron Lett.*, **39**, 3011.

65. Hori, M., Gravert, D. J., Wentworth, P., Jr., and Janda, K. D. (1998). *Bioorg. Med. Chem. Lett.*, **8**, 2363.

66. Khan, S. I. and Grinstaff, M. W. (1998). *Tetrahedron Lett.*, **39**, 8031.

67. Hoemann, M. Z., Melikian-Badalian, A., Kumaravel, G., and Hauske, J. R. (1998). *Tetrahedron Lett.*, **39**, 4749.

68. Arsequell, G., Espuña, G., Valencia, G., Baruenga, J., Carlón, R. P., and González, J. M. (1998). *Tetrahedron Lett.*, **39**, 7393.

69. Yamamoto, Y., Ajito, K., and Ohtsuka, Y. (1998). *Chem. Lett.*, 379.

70. Pan, P.-C. and Sun, C.-M. (1998). *Tetrahedron Lett.*, **39**, 9505.

71. Feng, Y., Wang, Z., Jin, S., and Burgess, K. J. (1998). *Am. Chem. Soc.*, **120**, 10768.

72. Chamoin, S., Houldsworth, S., and Snieckus, V. (1998). *Tetrahedron Lett.*, **39**, 4175.

73. Nicolaou, K. C., Winssinger, N., Pastor, J., and Murphy, F. (1998). *Angew. Chem. Int. Ed. Engl.*, **37**, 2534.

74. Lago, M. A., Nguyen, T. T., and Bhatnagar, P. (1998). *Tetrahedron Lett.*, **39**, 3885.

75. Chamoin, S., Houldsworth, S., Kruse, C. G., Bakker, W. I., and Snieckus, V. (1998). *Tetrahedron Lett.*, **39**, 4179.

76. Reggelin, M., Brenig, V., and Welcker, R. (1998). *Tetrahedron Lett.*, **39**, 4801.

77. Shi, S., Xiao, X.-Y., and Czarnik, A. W. (1998). *Biotech. Bioeng. (Combin. Chem.)*, **61**, 7.

78. Marzinzik, A. L. and Felder, E. R. (1998). *J. Org. Chem.*, **63**, 723.

79. Gennari, C., Ceccarelli, S., Piarulli, U., Aboutayab, K., Donghi, M., and Paterson, I. (1998). *Tetrahedron*, **54**, 14999.

80. Sim, M. M., Lee, C. L., and Ganesan, A. (1998). *Tetrahedron Lett.*, **39**, 6399.

81. Li, W.-R. and Peng, S.-Z. (1998). *Tetrahedron Lett.*, **39**, 7373.

82. Schneider, S. E., Bishop, P. A., Salazar, M. A., Bishop, O. A., and Anslyn, E. V. (1998). *Tetrahedron*, **54**, 15063.

83. Adrian, F. M., Altava, B., Burguete, M. I., Luis, S. V., Salvador, R. V., and García-España, E. (1998). *Tetrahedron*, **54**, 3581.

84. Molteni, V., Annunziata, R., Cinquini, M., Cozzi, F., and Benaglia, M. (1998). *Tetrahedron Lett.*, **39**, 1257.

85. Kobayashi, S. and Aoki, Y. (1998). *Tetrahedron Lett.*, **39**, 7345.

86. Boussie, T. R., Coutard, C., Turner, H., Murphy, V., and Powers, T. S. (1998). *Angew. Chem. Int. Ed. Engl.*, **37**, 3272.

87. Estep, K. G., Neipp, C. E., Stramiello, L. M. S., Adam, M. D., Allen, M. P., Robison, S., *et al.* (1998). *J. Org. Chem.*, **63**, 5300.

88. Nizi, E., Botta, M., Corelli, F., Manetti, F., Messina, F., and Maga, G. (1998). *Tetrahedron Lett.*, **39**, 3307.

89. Kobayashi, S. and Akiyama, R. (1998). *Tetrahedron Lett.*, **39**, 9211.

90. Haap, W. J., Kaiser, D., Walk, T. B., and Jung, G. (1998). *Tetrahedron*, **54**, 3705.

91. Shankar, B. B., Yang, D. Y., Girton, S., and Ganguly, A. K. (1998). *Tetrahedron Lett.*, **39**, 2447.

92. Park, K.-H., Abbate, E., Najdi, S., Olmstead, M. M., and Kurth, M. J. (1998). *Chem. Commun.*, 1679.

93. Park, K.-H., Olmstead, M. M., and Kurth, M. J. (1998). *J. Org. Chem.*, **63**, 6579.

94. Ogbu, C. O., Qabar, M. N., Boatman, P. D., Urban, J., Meara, J. P., Ferguson, M. D., *et al.* (1998). *Bioorg. Med. Chem. Lett.*, **8**, 2321.

95. Barrett, A. G. M. and de Miguel, Y. R. (1998). *Chem. Commun.*, 2079.

96. Heerding, D. A., Takata, D. T., Kwon, C., Huffman, W. F., and Samanen, J. (1998). *Tetrahedron Lett.*, **39**, 6815.

97. Ball, C. P., Barrett, A. G. M., Commerçon, A., Compère, D., Kuhn, C., Roberts, R. S., *et al.* (1998). *Chem. Commun.*, 2019.

98. Boymond, L., Rottländer, M., Cahiez, G., and Knochel, P. (1998). *Angew. Chem. Int. Ed. Engl.*, **37**, 1701.

99. Blaskovich, M. A. and Kahn, M. (1998). *Synthesis*, 965.

100. Li, Z. and Ganesan, A. (1998). *Synlett*, 405.

101. Khim, S.-K. and Nuss, J. M. (1999). *Tetrahedron Lett.*, **40**, 1827.

102. Itsuno, S., El-Shehawy, A. A., Abdelaal, M. Y., and Ito, K. (1998). *New J. Chem.*, 775.

103. Kim, S. W., Bauer, S. M., and Armstrong, R. W. (1998). *Tetrahedron Lett.*, **39**, 6993.

104. Chen, S. and Janda, K. D. (1998). *Tetrahedron Lett.*, **39**, 3943.

105. Hamper, B. C., Kolodziej, S. A., Scates, A. M., Smith, R. G., and Cortez, E. (1998). *J. Org. Chem.*, **63**, 708.

106. Barbaste, M., Rolland-Fulcrand, V., Roumestant, M.-L., Viallefont, P., and Martinez, J. (1998). *Tetrahedron Lett.*, **39**, 6287.

107. Barco, A., Benetti, S., De Risi, C., Marchetti, P., Pollini, G. P., and Zanirato, V. (1998). *Tetrahedron Lett.*, **39**, 7591.

108. Gordeev, M. F. (1998). *Biotech. Bioeng. (Combin. Chem.)*, **61**, 13.

109. Sun, Q. and Yan, B. (1998). *Bioorg. Med. Chem. Lett.*, **8**, 361.

110. Mayer, J. P., Lewis, G. S., McGee, C., and Bankaitis-Davis, D. (1998). *Tetrahedron Lett.*, **39**, 6655.

111. Ostresh, J. M., Schoner, C. C., Hamashin, V. T., Nefzi, A., Meyer, J.-P., and Houghten, R. A. (1998). *J. Org. Chem.*, **63**, 8622.

112. Berteina, S. and De Mesmaeker, A. (1998). *Tetrahedron Lett.*, **39**, 5759.

113. Du, X. and Armstrong, R. W. (1998). *Tetrahedron Lett.*, **39**, 2281.

114. del Fresno, M., Alsina, J., Royo, M., Barany, G., and Albericio, F. (1998). *Tetrahedron Lett.*, **39**, 2639.

115. Nefzi, A., Ostresh, J. M., Giulianotti, M., and Houghten, R. A. (1998). *Tetrahedron Lett.*, **39**, 8199.

116. Wang, Y. and Huang, T.-N. (1998). *Tetrahedron Lett.*, **39**, 9605.

117. Bilodeau, M. T. and Cunningham, A. M. (1998). *J. Org. Chem.*, **63**, 2800.

118. Goff, D. A. (1998). *Tetrahedron Lett.*, **39**, 1477.

119. Zhang, H.-C., Brumfield, K. K., Jaroskova, L., and Maryanoff, B. E. (1998). *Tetrahedron Lett.*, **39**, 4449.

120. Lyngsø, L. O. and Nielsen, J. (1998). *Tetrahedron Lett.*, **39**, 5845.

121. Blackburn, C. (1998). *Tetrahedron Lett.*, **39**, 5469.

122. Bicknell, A. J., Hird, N. W., and Readshaw, S. A. (1998). *Tetrahedron Lett.*, **39**, 5869.

123. Sylvain, C., Wagner, A., and Mioskowski, C. (1998). *Tetrahedron Lett.*, **39**, 9679.

124. Adinolfi, M., Barone, G., De Napoli, L., Iadonisi, A., and Piccialli, G. (1998). *Tetrahedron Lett.*, **39**, 1953.

125. Hall, B. J. and Sutherland, J. D. (1998). *Tetrahedron Lett.*, **39**, 6593.

126. Paris, M., Heitz, A., Guerlavais, V., Cristau, M., Fehrentz, J.-A., and Martinez, J. (1998). *Tetrahedron Lett.*, **39**, 7287.

127. Yan, B., and Sun, Q. (1998). *J. Org. Chem.*, **63**, 55.

128. Peukert, S. and Giese, B. (1998). *J. Org. Chem.*, **63**, 9045.

129. Dressman, B. A., Singh, U., and Kaldor, S. W. (1998). *Tetrahedron Lett.*, **39**, 3631.

130. Karigiannis, G., Mamos, P., Balayiannis, G., Katsoulis, I., and Papaioannou, D. (1998). *Tetrahedron Lett.*, **39**, 5117.

131. Long, D. D., Smith, M. D., Marquess, D. G., Claridge, T. D. W., and Fleet, G. W. J. (1998). *Tetrahedron Lett.*, **39**, 9293.

132. Bhandari, A., Jones, D. G., Schullek, J. R., Vo, K., Schunk, C. A., Tamanaha, L. L., *et al.* (1998). *Bioorg. Med. Chem. Lett.*, **8**, 2303.

133. Hu, Y. and Porco, J. A., Jr. (1998). *Tetrahedron Lett.*, **39**, 2711.

134. Lacombe, P., Castagner, B., Gareau, Y., and Ruel, R. (1998). *Tetrahedron Lett.*, **39**, 6785.

135. Breitenbucher, J. G. and Hui, H. C. (1998). *Tetrahedron Lett.*, **39**, 8207.

136. Sauvagnat, B., Lamaty, F., Lazaro, R., and Martinez, J. (1998). *Tetrahedron Lett.*, **39**, 821.

137. Stones, D., Miller, D. J., Beaton, M. W., Rutherford, T. J., and Gani, D. (1998). *Tetrahedron Lett.*, **39**, 4875.

138. Kanemitsu, T., Kanie, O., and Wong, C.-H. (1998). *Angew. Chem. Int. Ed. Engl.*, **37**, 3415.

139. Collins, J. L., Blanchard, S. G., Boswell, G. E., Charifson, P. S., Cobb, J. E., Henke, B. R., *et al.* (1998). *J. Med. Chem.*, **41**, 5037.

140. Reichwein, J. F. and Liskamp, R. M. J. (1998). *Tetrahedron Lett.*, **39**, 1243.

141. Dodd, D. S. and Wallace, O. B. (1998). *Tetrahedron Lett.*, **39**, 5701.

142. Aronov, A. M. and Gelb, M. H. (1998). *Tetrahedron Lett.*, **39**, 4947.

143. Nefzi, A., Dooley, C. T., Ostresh, J. M., and Houghten, R. A. (1998). *Bioorg. Med. Chem. Lett.*, **8**, 2273.

144. Buchstaller, H.-P. (1998). *Tetrahedron*, **54**, 3465.

145. Coppola, G. M. (1998). *Tetrahedron Lett.*, **39**, 8233.

146. Katritzky, A. R., Belyakov, S. A., Fang, Y., and Kiely, J. S. (1998). *Tetrahedron Lett.*, **39**, 8051.

147. Iyer, R. P., Guo, M.-J., Yu, D., and Agrawal, S. (1998). *Tetrahedron Lett.*, **39**, 2491.

148. Blaskovich, M. A. and Kahn, M. (1998). *J. Org. Chem.*, **63**, 1119.

149. Gennari, C., Longari, C., Ressel, S., Salom, B., Piarulli, U., Ceccarelli, S., *et al.* (1998). *Eur. J. Org. Chem.*, 2437.

150. James, K. D. and Ellington, A. D. (1998). *Tetrahedron Lett.*, **39**, 175.

151. Kellam, B., Bycroft, B. W., Chan, W. C., and Chhabra, S. R. (1998). *Tetrahedron*, **54**, 6817.
152. Karygiannis, G., Athanassopoulos, C., Mamos, P., Karamonos, N., Papaioannou, D., and Francis, G. W. (1998). *Acta Chem. Scand.*, **52**, 1144.
153. Cano, M., Camps, F., and Joglar, J. (1998). *Tetrahedron Lett.*, **39**, 9819.
154. Takahashi, T., Ebata, S., and Doi, T. (1998). *Tetrahedron Lett.*, **39**, 1369.
155. Zheng, C., Seeberger, P. H., and Danishefsky, S. J. (1998). *Angew. Chem. Int. Ed. Engl.*, **37**, 786.
156. Schuster, M. and Blechert, S. (1998). *Tetrahedron Lett.*, **39**, 2295.
157. Kaljuste, K. and Tam, J. P. (1998). *Tetrahedron Lett.*, **39**, 9327.
158. Miller, M. W., Vice, S. F., and McCombie, S. W. (1998). *Tetrahedron Lett.*, **39**, 3429.
159. Kulkarni, B. A. and Ganesan, A. (1998). *Tetrahedron Lett.*, **39**, 4369.
160. Smith, R. A., Bobko, M. A., and Lee, W. (1998). *Bioorg. Med. Chem. Lett.*, **8**, 2369.
161. van Loevezijn, A., van Maarseveen, J. H., Stegman, K., Visser, G. M., and Koomen, G.-J. (1998). *Tetrahedron Lett.*, **39**, 4737.
162. Ouyang, X., Armstrong, R. W., and Murphy, M. M. (1998). *J. Org. Chem.*, **63**, 1027.
163. Mehta, S. and Whitfield, D. (1998). *Tetrahedron Lett.*, **39**, 5907.
164. Rueter, J. K., Nortey, S. O., Baxter, E. W., Leo, G. C., and Reitz, A. B. (1998). *Tetrahedron Lett.*, **39**, 975.
165. Millington, C. R., Quarrell, R., and Lowe, G. (1998). *Tetrahedron Lett.*, **40**, 7201.
166. Sternson, S. M. and Schreiber, S. L. (1998). *Tetrahedron Lett.*, **39**, 7451.
167. Furlán, R. L. E., Mata, E. G., and Mascararetti, O. A. (1998). *J. Chem. Soc. Perkin Trans.*, **1**, 355.
168. Furlán, R. L. E., Mata, E. G., Mascararetti, O. A., Peña, C., and Coba, M. P. (1998). *Tetrahedron*, **54**, 13023.
169. Sturino, C. F. and Labelle, M. (1998). *Tetrahedron Lett.*, **39**, 5891.
170. Yu, L., Chen, D., Li, J., and Wang, P. G. (1997). *J. Org. Chem.*, **62**, 3575.
171. Dumartin, G., Pourcel, M., Delmond, B., Donard, O., and Pereyre, M. (1998). *Tetrahedron Lett.*, **39**, 4663.
172. Dendrinos, K. G. and Kalivretenos, A. G. (1998). *J. Chem. Soc. Perkin Trans.*, **1**, 1463.
173. Pop, I. E., Dèprez, B. P., and Tartar, A. L. (1997). *J. Org. Chem.*, **62**, 2594.
174. Dendrinos, K., Jeong, J., Huang, W., and Kalivretenos, A. G. (1998). *Chem. Commun.*, 499.
175. Bayston, D. J., Fraser, J. L., Ashton, M. R., Baxter, A. D., Polywka, M. E. C., and Moses, E. (1998). *J. Org. Chem.*, **63**, 3137.
176. Parlow, J. J. and Flynn, D. L. (1999). *Tetrahedron*, **54**, 4013.
177. Buckman, B. O., Morrissey, M. M., and Mohan, R. (1998). *Tetrahedron Lett.*, **39**, 1487.
178. Liu, Y.-S., Zhao, C., Bergbreiter, D. E., and Romo, D. (1998). *J. Org. Chem.*, **63**, 3471.
179. Habermann, J., Ley, S. V., and Scott, J. S. (1998). *J. Chem. Soc. Perkin Trans.*, **1**, 3127.
180. Altava, B., Burguete, M. I., Escuder, B., Luis, S. V., Salvador, R. V., Fraile, J. M., *et al.* (1997). *J. Org. Chem.*, **62**, 3126.
181. Nicolaou, K. C., Pastor, J., Barluengà, S., and Winssinger, N. (1998). *Chem. Commun.*, 1947.
182. Chan, M. C. W., Chew, K. C., Dalby, C. I., Gibson, V. C., Kohlmann, A., Little, I. R., *et al.* (1998). *Chem. Commun.*, 1673.
183. Andersen, J.-A. M., Karodia, N., Miller, D. J., Stones, D., and Gani, D. (1998). *Tetrahedron Lett.*, **39**, 7815.
184. Creswell, M. W., Bolton, G. L., Hodges, J. C., and Meppen, M. (1998). *Tetrahedron*, **54**, 3983.
185. Macquarrie, D. J. (1998). *Tetrahedron Lett.*, **39**, 4125.

186. Kobayashi, S. and Nagayama, S. (1998). *J. Am. Chem. Soc.*, **120**, 2985.

187. Tunoori, A. R., Dutta, D., and Georg, G. I. (1998). *Tetrahedron Lett.*, **39**, 8751.

188. Masaki, Y., Tanaka, N., and Miura, T. (1998). *Tetrahedron Lett.*, **39**, 5799.

189. Xu, W., Mohan, R., and Morrissey, M. M. (1998). *Bioorg. Med. Chem. Lett.*, **8**, 1089.

190. Dendrinos, K. G. and Kalivretenos, A. G. (1998). *Tetrahedron Lett.*, **39**, 1321.

191. Nagayama, S., Endo, M., and Kobayashi, S. (1998). *J. Org. Chem.*, **63**, 6094.

192. Gómez, C., Ruiz, S., and Yus, M. (1998). *Tetrahedron Lett.*, **39**, 1397.

193. Vidal-Ferran, A., Bampos, N., Moyano, A., Pericàs, M. A., Riera, A., and Sanders, J. K. M. (1998). *J. Org. Chem.*, **63**, 6309.

194. Abraham, S., Rajan, P. K., and Sreekumar, K. (1997). *Indian J. Chem.*, **36B**, 769.

195. Hinzen, B., Lenz, R., and Ley, S. V. (1998). *Synthesis*, 977.

196. Hinzen, B. and Ley, S. V. (1998). *J. Chem. Soc. Perkin Trans.*, **1**, 1.

197. Boehlow, T. R., Buxton, P. C., Grocock, E. L., Marples, B. A., and Waddington, V. L. (1998). *Tetrahedron Lett.*, **39**, 1839.

198. Uozumi, Y., Danjo, H., and Hayashi, T. (1998). *Tetrahedron Lett.*, **39**, 8303.

199. Karjalainen, J. K., Hormi, O. E. O., and Sherrington, D. C. (1998). *Tetrahedron: Asymmetry*, **9**, 2019.

200. Fenger, I. and Le Drian, C. (1998). *Tetrahedron Lett.*, **39**, 4287.

201. Harris, J. M., Liu, Y., Chai, S., Andrews, M. D., and Vederas, J. C. (1998). *J. Org. Chem.*, **63**, 2407.

202. Li, Z. and Ganesan, A. (1998). *Synth. Commun.*, **28**, 3209.

203. Bayston, D. J., Travers, C. B., and Polywka, M. E. C. (1998). *Tetrahedron: Asymmetry*, **9**, 2015.

204. Mizugaki, T., Kanayama, Y., Ebitani, K., and Kaneda, K. (1998). *J. Org. Chem.*, **63**, 2378.

205. Ault-Justus, S. E., Hodges, J. C., and Wilson, M. W. (1998). *Biotech. Bioeng. (Combin. Chem.)*, **61**, 17.

206. Starkey, G. W., Parlow, J. J., and Flynn, D. L. (1998). *Bioorg. Med. Chem. Lett.*, **8**, 2385.

207. Blackburn, C., Guan, B., Fleming, P., Shiosaki, K., and Tsai, S. (1998). *Tetrahedron Lett.*, **39**, 3635.

208. Warmus, J. S., Ryder, T. R., Hodges, J. C., Kennedy, R. M., and Brady, K. D. (1998). *Bioorg. Med. Chem. Lett.*, **8**, 2309.

3 Solid phase organic reactions—1999

3.1 Anchoring reactions

Acid attachment	Solvent/Temp.	Yield	Resin	Ref.
	THF RT	98%	Wang resin	1

Acid attachment	Solvent/Temp.	Yield	Resin	Ref.
	DMF RT		Merrifield resin	2

198

Acid attachment

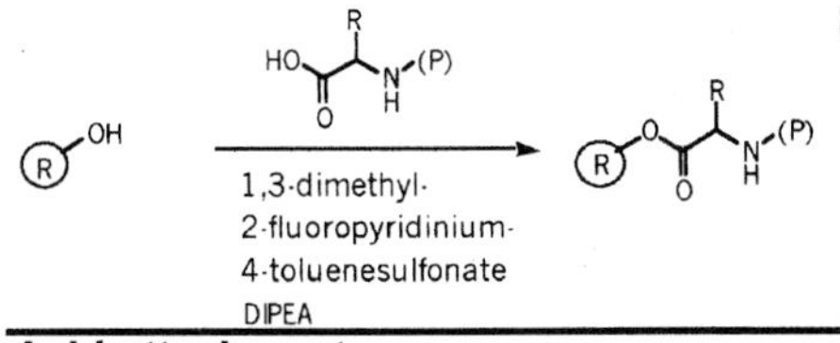

Solvent/Temp.	Yield	Resin	Ref.
DCM		TentaGel S OH	3

1,3-dimethyl-
2-fluoropyridinium-
4-toluenesulfonate
DIPEA

Acid attachment

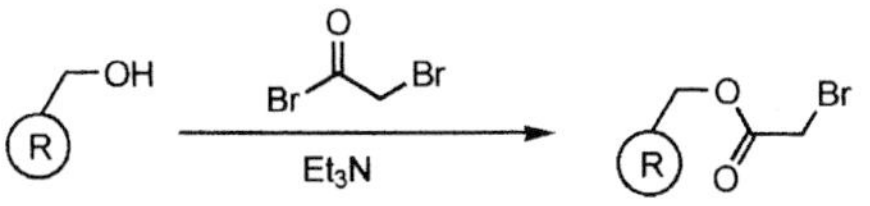

Solvent/Temp.	Yield	Resin	Ref.
DCM -20° C, then 20°C	95%	PEG	4

Et₃N

Acid attachment

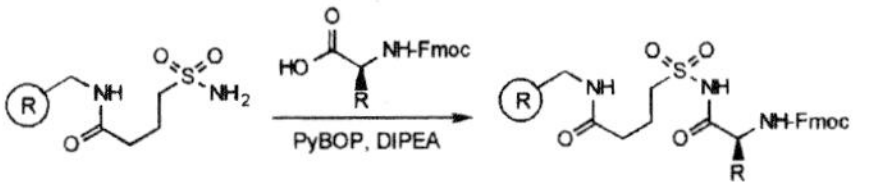

Solvent/Temp.	Yield	Resin	Ref.
CHCl₃ -20° C	up to 95%	Aliphatic safety-catch resin	5

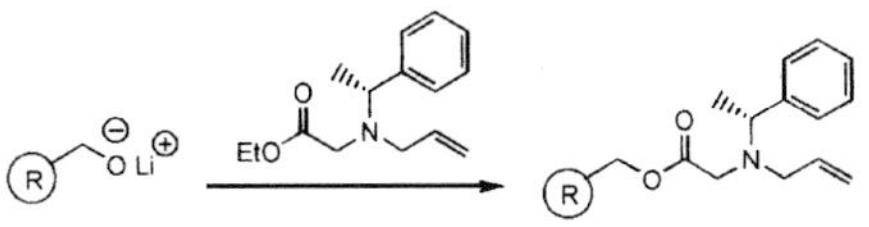

PyBOP, DIPEA

Acid attachment

Solvent/Temp.	Yield	Resin	Ref.
THF -78° C, then RT		Wang resin	6

Acid attachment

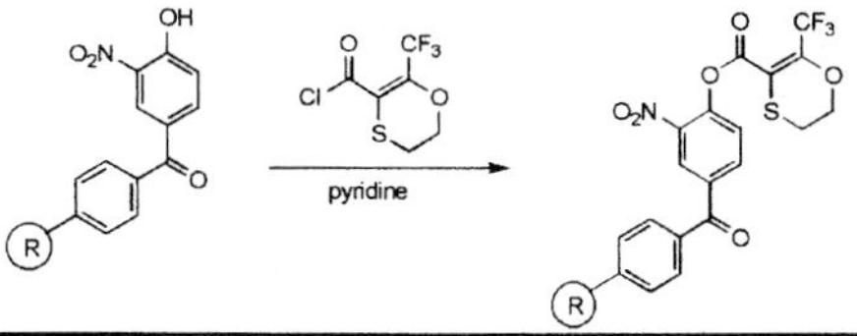

Solvent/Temp.	Yield	Resin	Ref.
DCM RT		4-hydroxy-3-nitro-benzophenone resin	7

pyridine

Acid attachment

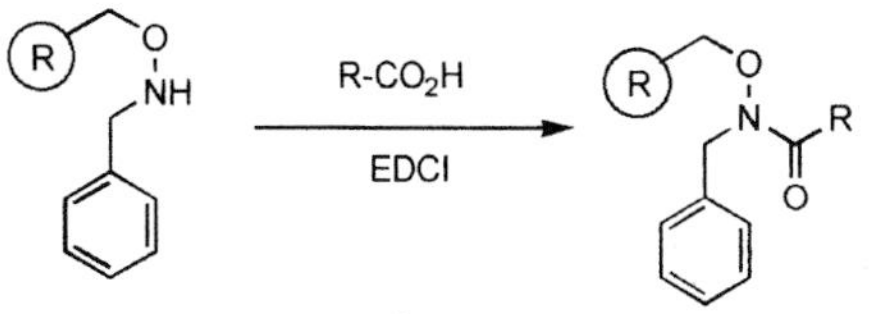

Solvent/Temp.	Yield	Resin	Ref.
DMF		hydroxylamine Wang resin	8

R-CO₂H

EDCI

Acid attachment

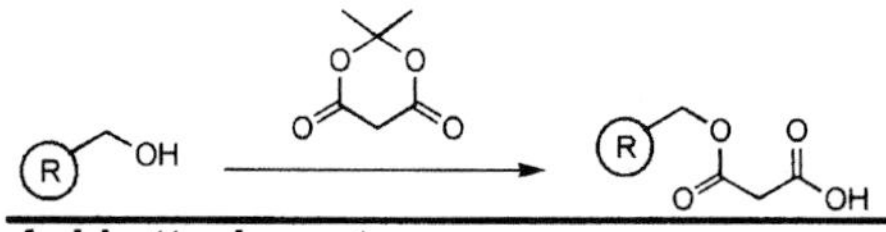

Solvent/Temp.	Yield	Resin	Ref.
		Wang resin	9

Acid attachment

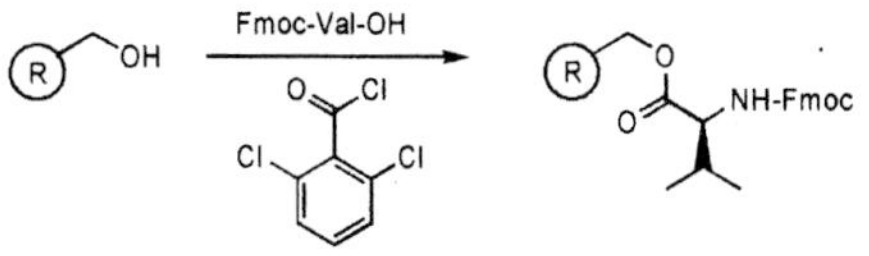

Solvent/Temp.	Yield	Resin	Ref.
NMP/pyridine	80% to 90%	Merrifield resin, high loading	10

Fmoc-Val-OH

Alcohol attachment

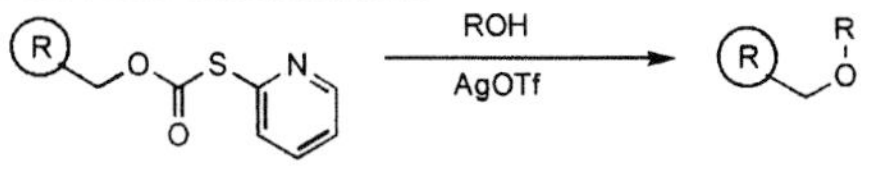

Solvent/Temp.	Yield	Resin	Ref.
DCM		2-pyridinyl thiocarbonate resin	11

ROH

AgOTf

199

Alcohol attachment	Solvent/Temp.	Yield	Resin	Ref.
	Dichloroethane		Merrifield resin with dihydropyran linker	12

Alcohol attachment	Solvent/Temp.	Yield	Resin	Ref.
	DMF 25° C		Merrifield resin	13

Alcohol attachment	Solvent/Temp.	Yield	Resin	Ref.
	DCM 0° C	68% to 92%	Merrifield resin	14

Alcohol attachment	Solvent/Temp.	Yield	Resin	Ref.
	DCM	90%	thiopyridine carbonyl Wang resin	15

Alcohol attachment	Solvent/Temp.	Yield	Resin	Ref.
	DCM		polymeric diphenyl diazomethane	16

Alcohol attachment	Solvent/Temp.	Yield	Resin	Ref.
	DCM RT		diethylsilyl chloride resin	17

Amidine attachment	Solvent/Temp.	Yield	Resin	Ref.
	dioxane/DMF/DCM		HMPA p-nitrophenylcarbonate resin	18

Amine attachment	Solvent/Temp.	Yield	Resin	Ref.
	CH₃CN/DCM	90%	hydroxymethyl resin	19

Amine attachment	Solvent/Temp.	Yield	Resin	Ref.
	DMF		Wang resin	20

Amine attachment	Solvent/Temp.	Yield	Resin	Ref.
	DMF 25° C		TentaGel S NH₂	21

200

Amine attachment	Solvent/Temp.	Yield	Resin	Ref.
	DMF		AMEBA resin	22

Amine attachment	Solvent/Temp.	Yield	Resin	Ref.
	DCM RT		4-benzyloxytrityl chloride resin	23

Amine attachment	Solvent/Temp.	Yield	Resin	Ref.
	NMP RT		methoxybenzhydrylchloride resin	24

Amine attachment	Solvent/Temp.	Yield	Resin	Ref.
	THF RT		benzylsulfonylethoxy carbonyl resin	25

Amine attachment	Solvent/Temp.	Yield	Resin	Ref.
	THF -10° C to RT		Merrifield resin with azide linker	26

Aniline attachment	Solvent/Temp.	Yield	Resin	Ref.
	DMF RT	up to 100%	Wang p-nitrocarbonate resin	27

Carbamate attachment	Solvent/Temp.	Yield	Resin	Ref.
	Toluene 100° C	90% or greater	Wang resin	28

Carbonyl attachment	Solvent/Temp.	Yield	Resin	Ref.
	DCM 25° C		PS-DES-OTf	29

Carbonyl attachment	Solvent/Temp.	Yield	Resin	Ref.
	DCM 25° C		PS-DES-OTf	30

Carbonyl attachment	Solvent/Temp.	Yield	Resin	Ref.
	Benzene reflux		hydroxymethyl resin	31

Diazo attachment	Solvent/Temp.	Yield	Resin	Ref.
	THF 0° C to 25° C	78%	piperazinylmethyl polystyrene	32

Diol attachment	Solvent/Temp.	Yield	Resin	Ref.
	CH$_3$CN 85° C		Wang aldehyde resin	33

Diol attachment	Solvent/Temp.	Yield	Resin	Ref.
	DCM 20° C		Wang aldehyde resin	34

Imidazole attachment	Solvent/Temp.	Yield	Resin	Ref.
	DCM	83%	trityl chloride resin	35

Imidazole attachment	Solvent/Temp.	Yield	Resin	Ref.
	DCM		Trityl type resins	36

Nitroacetic acid attachment	Solvent/Temp.	Yield	Resin	Ref.
	THF 0° C		hydroxymethyl polystyrene	37

Oxylate attachment	Solvent/Temp.	Yield	Resin	Ref.
	DCM 0° C to 25° C		hydroxymethyl polystyrene	38

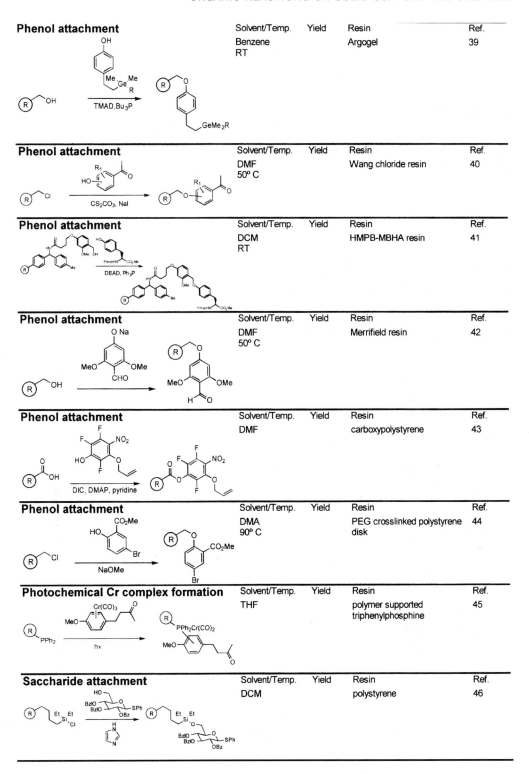

Phenol attachment

	Solvent/Temp.	Yield	Resin	Ref.
	Benzene RT		Argogel	39

Phenol attachment

	Solvent/Temp.	Yield	Resin	Ref.
	DMF 50° C		Wang chloride resin	40

Phenol attachment

	Solvent/Temp.	Yield	Resin	Ref.
	DCM RT		HMPB-MBHA resin	41

Phenol attachment

	Solvent/Temp.	Yield	Resin	Ref.
	DMF 50° C		Merrifield resin	42

Phenol attachment

	Solvent/Temp.	Yield	Resin	Ref.
	DMF		carboxypolystyrene	43

Phenol attachment

	Solvent/Temp.	Yield	Resin	Ref.
	DMA 90° C		PEG crosslinked polystyrene disk	44

Photochemical Cr complex formation

	Solvent/Temp.	Yield	Resin	Ref.
	THF		polymer supported triphenylphosphine	45

Saccharide attachment

	Solvent/Temp.	Yield	Resin	Ref.
	DCM		polystyrene	46

Thiol attachment		Solvent/Temp.	Yield	Resin	Ref.
		DMF		Merrifield resin	47

3.2 Amide bond forming reactions

Acyl hydrazide formation		Solvent/Temp.	Yield	Resin	Ref.
	hydrazine hydrate	1,3-dimethyl-2-imidazolidinone 90° C		carboxypolystyrene	48

Acyl hydrazide formation		Solvent/Temp.	Yield	Resin	Ref.
	DIC, HOBt, Et₃N	DCM		aminomethyl resin or TentaGel S NH₂	49

Acylisothiocyanate formation		Solvent/Temp.	Yield	Resin	Ref.
	Bu₄NNCS	THF/dichloroethane		carboxypolystyrene	50

Amide formation		Solvent/Temp.	Yield	Resin	Ref.
	PhCOCl	Pyridine 25° C		TentaGel S NH₂	21

Amide formation		Solvent/Temp.	Yield	Resin	Ref.
	EEDQ	DCM		polymethacrylate resin	51

Amide formation		Solvent/Temp.	Yield	Resin	Ref.
		Dichloroethane		Wang resin	52

Amide formation		Solvent/Temp.	Yield	Resin	Ref.
	1) NaH 2) RX	DMF/collidine		MBHA resin	53

Amide formation		Solvent/Temp.	Yield	Resin	Ref.
		DMF RT		macrocrowns with Knorr linker	54

Amide formation		Solvent/Temp.	Yield	Resin	Ref.
		DMF RT		Rink resin	55

Amide formation		Solvent/Temp.	Yield	Resin	Ref.
		DMF RT		Rink resin	55

Amide formation		Solvent/Temp.	Yield	Resin	Ref.
		DMF	100%	Wang resin	56

Amide formation		Solvent/Temp.	Yield	Resin	Ref.
		THF/pyridine		methoxybenzhydryl chloride resin	24

Amide formation		Solvent/Temp.	Yield	Resin	Ref.
		Dioxane 80° C		Wang resin, high loading	10

Amide formation		Solvent/Temp.	Yield	Resin	Ref.
		DCM		PS-TTEGDA	57

Amide formation		Solvent/Temp.	Yield	Resin	Ref.
		Pyridine 120° C		Merrifield resin	58

Amide formation		Solvent/Temp.	Yield	Resin	Ref.
		THF 60° C		2-chlorotrityl resin	59

Carbamate formation		Solvent/Temp.	Yield	Resin	Ref.
		DMF RT		Wang resin	60

Carbamate formation		Solvent/Temp.	Yield	Resin	Ref.
		DCM RT	78% to 89%	not specified	61

Carbamoyl chloride formation		Solvent/Temp.	Yield	Resin	Ref.
		DCM		Wang resin	62

Guanidine formation		Solvent/Temp.	Yield	Resin	Ref.
		CHCl$_3$ or DMF		carboxypolystyrene	50

Guanidine formation		Solvent/Temp.	Yield	Resin	Ref.
		DCM RT	85%	MeO-PEG	63

Semicarbazide formation		Solvent/Temp.	Yield	Resin	Ref.
		DCM		dibenzosuberyl resin	64

Sulfonamide formation		Solvent/Temp.	Yield	Resin	Ref.
		DCM/pyridine	Up to 97%	Sasrin resin	65

206

Sulfonamide formation

Solvent/Temp.	Yield	Resin	Ref.
DCM RT	84%	MeO-PEG	66

Sulfonamide formation

Solvent/Temp.	Yield	Resin	Ref.
DCM		Wang resin	62

Thiourea formation

Solvent/Temp.	Yield	Resin	Ref.
DMF		carboxypolystyrene	50

Thiourea formation

Solvent/Temp.	Yield	Resin	Ref.
DIPEA		oxime resin	67

Urea formation

Solvent/Temp.	Yield	Resin	Ref.
DCM RT		Wang resin	68

Urea formation

Solvent/Temp.	Yield	Resin	Ref.
DCM		MBHA resin	69

Urea formation

Solvent/Temp.	Yield	Resin	Ref.
DCM microwaves		Wang resin	70

Urea formation

Solvent/Temp.	Yield	Resin	Ref.
DCM		Wang resin	62

Urea formation

Solvent/Temp.	Yield	Resin	Ref.
THF		Merrifield resin	71

3.3 Aromatic substitution

Aryl coupling

Solvent/Temp.	Yield	Resin	Ref.
DMF 80° C		piperazinylmethyl polystyrene	32

Arylzinc coupling

Solvent/Temp.	Yield	Resin	Ref.
THF	85%	Merifield resin	72

C-N cross coupling

Solvent/Temp.	Yield	Resin	Ref.
pyridine/NMP 80° C	64%	PS-PEG-PAL	73

Friedel-Crafts acylation

Solvent/Temp.	Yield	Resin	Ref.
-10° C		polystyrene	16

Friedel-Crafts acylation

Solvent/Temp.	Yield	Resin	Ref.
DCM RT		polystyrene	24

Friedel-Crafts alkylation

Solvent/Temp.	Yield	Resin	Ref.
Nitrobenzene 70° C		polystyrene	74

Friedel-Crafts alkylation

Solvent/Temp.	Yield	Resin	Ref.
Dichloroethane RT		polystyrene	75

Friedel-Crafts alkylation

Solvent/Temp.	Yield	Resin	Ref.
THF 50° C		PS-TTEGDA	57

Heck coupling

Solvent/Temp.	Yield	Resin	Ref.
DMF 80° C		piperazinylmethyl polystyrene	32

Heck coupling

	Solvent/Temp.	Yield	Resin	Ref.
	DMA	83-85%	aminomethyl resin or	49
	100° C		TentaGel S NH$_2$	

Heck coupling

	Solvent/Temp.	Yield	Resin	Ref.
	DMF	82% to	MeO-PEG	76
	145° C	98%		

Heck coupling

	Solvent/Temp.	Yield	Resin	Ref.
	DMF	up to	Wang resin	77
	50° C to 100° C	100%		

Heck coupling

	Solvent/Temp.	Yield	Resin	Ref.
	DMF		benzylaminomethyl	78
	80° C		polystyrene	

Iodination

	Solvent/Temp.	Yield	Resin	Ref.
	Nitrobenzene		linear polystyrene	79
	90° C			

Nucleophilic aromatic substitution

	Solvent/Temp.	Yield	Resin	Ref.
	DMF		AMEBA resin	22
	RT			

Nucleophilic aromatic substitution

	Solvent/Temp.	Yield	Resin	Ref.
	i-PrOH/DMF/		hydroxymethyl polystyrene	38
	HCl			
	RT			

Nucleophilic aromatic substitution

	Solvent/Temp.	Yield	Resin	Ref.
	DMSO		Wang resin	52
	RT			

Nucleophilic aromatic substitution

	Solvent/Temp.	Yield	Resin	Ref.
	Bu$_3$N		Merrifield resin	42
	105° C -110° C			

209

Nucleophilic aromatic substitution	Solvent/Temp.	Yield	Resin	Ref.
	DMF RT		REM resin	80

Nucleophilic aromatic substitution	Solvent/Temp.	Yield	Resin	Ref.
	DMF RT	71% to 96%	Rink resin	81

Nucleophilic aromatic substitution	Solvent/Temp.	Yield	Resin	Ref.
			MBHA resin	82

Nucleophilic aromatic substitution	Solvent/Temp.	Yield	Resin	Ref.
	2-ethoxyethanol		polystyrene	74

Nucleophilic aromatic substitution	Solvent/Temp.	Yield	Resin	Ref.
	DMF 20° C		Wang resin	83

Nucleophilic aromatic substitution	Solvent/Temp.	Yield	Resin	Ref.
	DMF RT		Rink resin	55

Nucleophilic aromatic substitution	Solvent/Temp.	Yield	Resin	Ref.
	DCM RT	99%	MeO-PEG	15

Nucleophilic aromatic substitution	Solvent/Temp.	Yield	Resin	Ref.
	DMF RT	76% to 79%	ArgoGel-Rink resin	84

Nucleophilic aromatic substitution	Solvent/Temp.	Yield	Resin	Ref.
	THF 55° C		oxime resin	67

Nucleophilic aromatic substitution	Solvent/Temp.	Yield	Resin	Ref.
	20° C		Wang resin	83

Nucleophilic aromatic substitution	Solvent/Temp.	Yield	Resin	Ref.
	DMSO 60° C		ArgoGel MB	85

Nucleophilic aromatic substitution	Solvent/Temp.	Yield	Resin	Ref.
	DMF RT		Wang resin	86

Nucleophilic aromatic substitution	Solvent/Temp.	Yield	Resin	Ref.
	NMP		Wang resin	87

Sonogashira coupling	Solvent/Temp.	Yield	Resin	Ref.
	Et$_2$NH/THF		dihydropyranyl resin	88

Sonogashira coupling	Solvent/Temp.	Yield	Resin	Ref.
	dioxane/Et$_3$N RT	up to 93%	aminomethyl resin or TentaGel S NH$_2$	49

211

Stille coupling	Solvent/Temp.	Yield	Resin	Ref.
	DMF 100° C		REM resin	89

Stille coupling	Solvent/Temp.	Yield	Resin	Ref.
	Dioxane 60° C	79-90%	aminomethyl resin or TentaGel S NH$_2$	49

Stille coupling	Solvent/Temp.	Yield	Resin	Ref.
	NMP 90° C	80% to >95%	Merrifield resin	90

Stille coupling	Solvent/Temp.	Yield	Resin	Ref.
	DMF 90° C		polystyrene sulfonylhydrazine	91

Suzuki coupling	Solvent/Temp.	Yield	Resin	Ref.
	DMF 85° C		polystyrene	92

Suzuki coupling	Solvent/Temp.	Yield	Resin	Ref.
	DMF 80° C to 90° C	up to 93%	aminomethyl resin or TentaGel S NH$_2$	49

Suzuki coupling	Solvent/Temp.	Yield	Resin	Ref.
	DMF 110° C	up to 89%	PEG	93

Suzuki coupling	Solvent/Temp.	Yield	Resin	Ref.
	DMF 80° C	99%	non-crosslinked Merrifield resin	94

Suzuki coupling

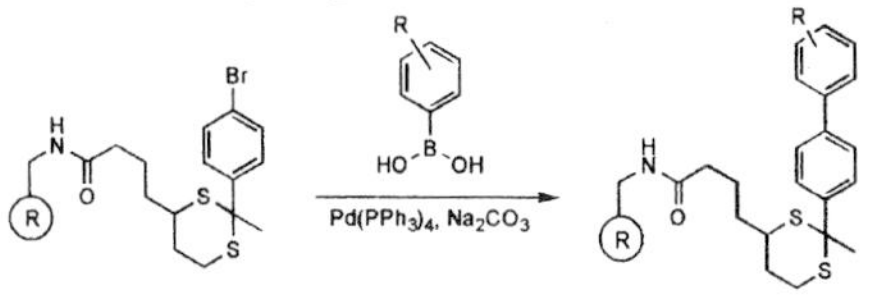

	Solvent/Temp.	Yield	Resin	Ref.
	H_2O/DMF/DME 80° C	28% to 34%	aminomethyl resin	95

Pd(PPh$_3$)$_4$, Na$_2$CO$_3$

Suzuki coupling

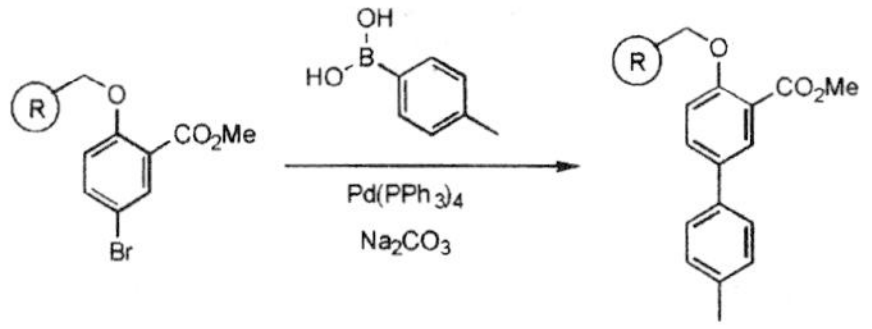

	Solvent/Temp.	Yield	Resin	Ref.
	DME 90° C		PEG crosslinked polystyrene	44

Pd(PPh$_3$)$_4$
Na$_2$CO$_3$

3.4 Condensation reactions

Acyl hydrozone formation

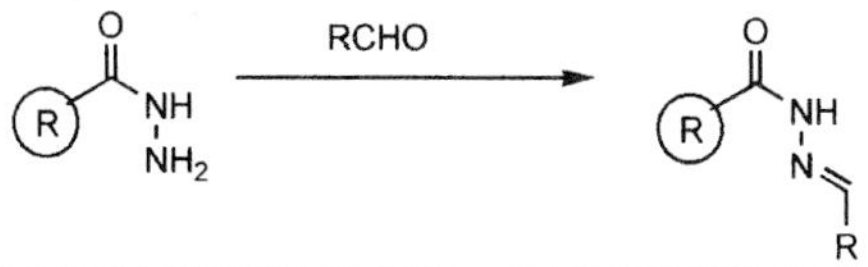

	Solvent/Temp.	Yield	Resin	Ref.
RCHO	DMF-AcOH 50° C		carboxypolystyrene	48

Baylis-Hillman reaction

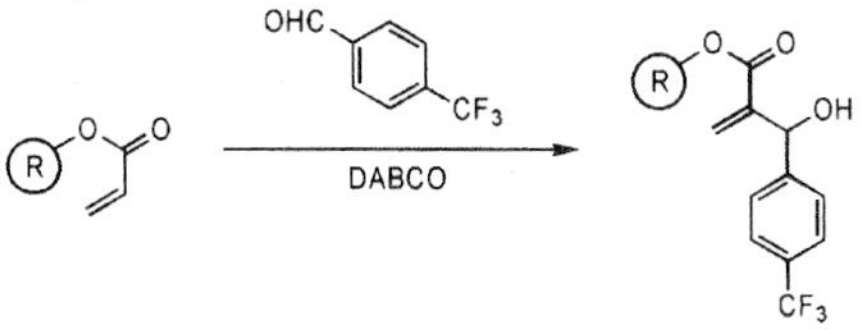

	Solvent/Temp.	Yield	Resin	Ref.
DABCO	CHCl$_3$/DMSO RT	85%	2-Cl-Trt resin	96

Claisen-Schmidt condensation

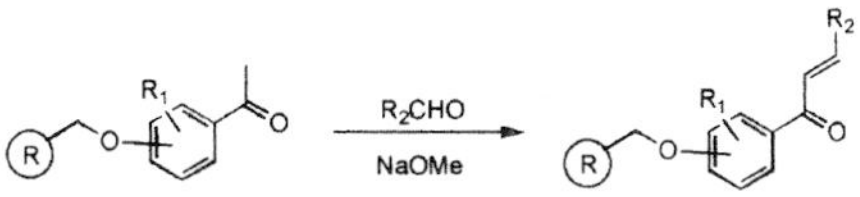

	Solvent/Temp.	Yield	Resin	Ref.
R$_2$CHO, NaOMe	trimethyl ortho-formate/MeOH		Wang resin	40

Condensation/reduction

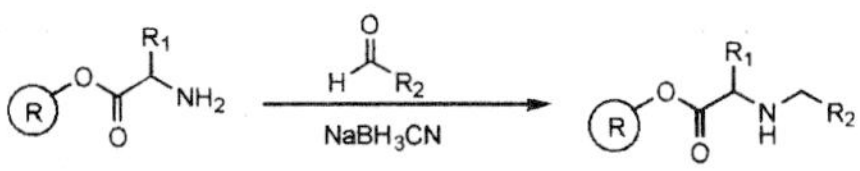

	Solvent/Temp.	Yield	Resin	Ref.
NaBH$_3$CN	trimethyl ortho-formate/MeOH		TentaGel S OH	3

Condensation/reduction

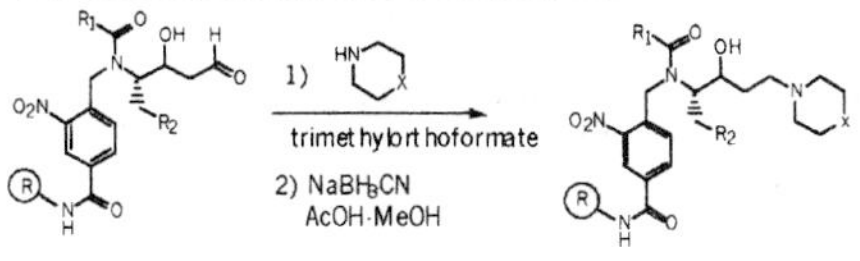

	Solvent/Temp.	Yield	Resin	Ref.
1) trimethyl orthoformate 2) NaBH$_3$CN AcOH·MeOH	25° C		TentaGel S NH$_2$	21

Condensation/reduction

	Solvent/Temp.	Yield	Resin	Ref.
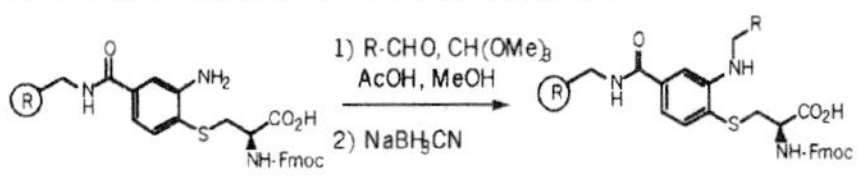 1) R-CHO, CH(OMe)$_3$ AcOH, MeOH 2) NaBH$_3$CN	trimethyl ortho-formate/DMF/THF 45° C -50° C	up to 73%	ArgoGel-Rink resin	82

Hydrazone formation

	Solvent/Temp.	Yield	Resin	Ref.
	n-butanol reflux		polystyrene	16

Imine formation

	Solvent/Temp.	Yield	Resin	Ref.
	Toluene RT		trityl resin	97

Imine formation

	Solvent/Temp.	Yield	Resin	Ref.
	timethylothofor mate/methanol		ArgoGel-MB-CHO	98

Imine formation

	Solvent/Temp.	Yield	Resin	Ref.
	DCM		not specified	61

Imine formation

	Solvent/Temp.	Yield	Resin	Ref.
	THF		Merrifield resin	71

Imine formation

	Solvent/Temp.	Yield	Resin	Ref.
	Trimethylotho-formate		Wang resin	86

Imine formation

	Solvent/Temp.	Yield	Resin	Ref.
	DCM		Sasrin resin, high loading	10

Imine formation

	Solvent/Temp.	Yield	Resin	Ref.
	Dichloroethane		carboxyethyl polystyrene	99

Imine formation/reduction

	Solvent/Temp.	Yield	Resin	Ref.
	DCM/AcOH RT		aminomethyl resin	49

Imine formation/reduction

Solvent/Temp.	Yield	Resin	Ref.
DMF		MBHA resin	82

Imine formation/reduction

Solvent/Temp.	Yield	Resin	Ref.
DMF RT		Rink resin	55

Imine formation/reduction

Solvent/Temp.	Yield	Resin	Ref.
DMF/EtOH		Wang resin	100

Imine formation/reduction

Solvent/Temp.	Yield	Resin	Ref.
DMF		Wang resin	86

Imine formation/reduction

Solvent/Temp.	Yield	Resin	Ref.
DCM		Wang resin, high loading	10

Imine formation/reduction

Solvent/Temp.	Yield	Resin	Ref.
THF 50° C		ArgoGel-MB-CHO	85

Knoevenagel condensation

Solvent/Temp.	Yield	Resin	Ref.
Benzene reflux		Merrifield resin	42

Knoevenagel condensation

Solvent/Temp.	Yield	Resin	Ref.
Toluene reflux	79% to 89%	Wang resin	9

215

Knoevenagel condensation

	Solvent/Temp.	Yield	Resin	Ref.
	Pyridine RT	up to 40%	Wang resin	101

Semicarbazone formation

	Solvent/Temp.	Yield	Resin	Ref.
			dibenzosuberyl resin	64

Sulfonylhydrazone formation

	Solvent/Temp.	Yield	Resin	Ref.
	AcOH/THF 50° C		polystyrene sulfonylhydrazine	91

3.5 Cycloaddition reactions

[2+2] Cycloaddition

	Solvent/Temp.	Yield	Resin	Ref.
	DCM RT		MeO-PEG	102

[2+2] Cycloaddition

	Solvent/Temp.	Yield	Resin	Ref.
	DCM RT	89% to 98%	not specified	61

[2+2] Cycloaddition

	Solvent/Temp.	Yield	Resin	Ref.
	DCM 0° C		Sasrin resin, high loading	10

[3+2] Cycloaddition

	Solvent/Temp.	Yield	Resin	Ref.
	Toluene 100° C		aminomethylpolystyrne grafted PTFE tubes	103

[3+2] Cycloaddition

	Solvent/Temp.	Yield	Resin	Ref.
	THF		Merrifield resin	71

Diels-Alder	Solvent/Temp.	Yield	Resin	Ref.
	DCM 25° C	up to 100%	diethylsilane resin	29

Diels-Alder	Solvent/Temp.	Yield	Resin	Ref.
	Xylene 145° C		aminomethylpolystyrne grafted PTFE tubes	103

Diels-Alder	Solvent/Temp.	Yield	Resin	Ref.
	DMF or dioxane		Rink resin	104

3.6 Organometallic reactions

Addition to aldehydes	Solvent/Temp.	Yield	Resin	Ref.
	THF/NMP		not specified	105

Addition to imines	Solvent/Temp.	Yield	Resin	Ref.
	Benzene RT	85% to 95%	trityl resin	97

Allylindium addition to aldehydes	Solvent/Temp.	Yield	Resin	Ref.
	THF/H_2O 25° C		TentaGel S NH_2	21

C-P coupling	Solvent/Temp.	Yield	Resin	Ref.
	THF -78° C to 25° C		Merrifield resin	13

C-P coupling	Solvent/Temp.	Yield	Resin	Ref.
	THF 60° C		Merrifield resin	13

217

Grignard addition to aldehyde	Solvent/Temp.	Yield	Resin	Ref.
	THF -78° C to -10° C	85%	non-crosslinked Merrifield resin	95

Grignard coupling	Solvent/Temp.	Yield	Resin	Ref.
	Toluene reflux		polystyrene	74

Grignard coupling	Solvent/Temp.	Yield	Resin	Ref.
	THF -10° C		Diethylsilylpolystyrene	17

Iodination	Solvent/Temp.	Yield	Resin	Ref.
	THF 0° C		Wang resin	6

Li-dithiane addition to aldehydes	Solvent/Temp.	Yield	Resin	Ref.
			Merrifield resin	2

Organozinc addition to aldehyde	Solvent/Temp.	Yield	Resin	Ref.
	THF 0° C to RT	89%	Merrifield resin	72

3.7 Michael addition

C-nucleophile	Solvent/Temp.	Yield	Resin	Ref.
	DMSO 70° C		Wang resin	40

C-nucleophile	Solvent/Temp.	Yield	Resin	Ref.
	THF -20° C	61%	Merrifield resin	72

C-nucleophile

	Solvent/Temp.	Yield	Resin	Ref.
	THF RT	75%	2-Cl-Trt resin	96

C-nucleophile

	Solvent/Temp.	Yield	Resin	Ref.
			Wang resin	15

N-nucleophile

	Solvent/Temp.	Yield	Resin	Ref.
	DMF 80° C		REM resin	80

O-nucleophile

	Solvent/Temp.	Yield	Resin	Ref.
			polyvinylphenol	106

S-nucleophile

	Solvent/Temp.	Yield	Resin	Ref.
	THF	100%	aminomethyl resin	107

S-nucleophile

	Solvent/Temp.	Yield	Resin	Ref.
	pyridine/DMF RT, 55° C	96%	thiomethyl polystyrene	108

Tandem Michael addition/substitution

	Solvent/Temp.	Yield	Resin	Ref.
	THF -60° C to -20° C, 0° C		hydroxymethyl resin	31

219

3.8 Heterocycle forming reactions

1,5-Benzodiazepin-2-one formation	Solvent/Temp.	Yield	Resin	Ref.
DIC, HOBt	DMF RT		Rink resin	81

Amino-zinc enolate cyclization	Solvent/Temp.	Yield	Resin	Ref.
1) LDA 2) ZnBr$_2$	THF -78° C, then RT		Wang resin	6

Azepane formation	Solvent/Temp.	Yield	Resin	Ref.
	DCM/MeOH RT		Rink resin	109

Benzimidazole formation	Solvent/Temp.	Yield	Resin	Ref.
R$_3$-CHO	NMP RT, then 50° C		ArgoGel-MB	85

Benzimidazolone formation	Solvent/Temp.	Yield	Resin	Ref.
Cl$_3$C-O-C(O)-O-CCl$_3$	DCM RT		MeO-PEG	15

Benzimidazothione formation	Solvent/Temp.	Yield	Resin	Ref.
Cl-C(S)-Cl Et$_3$N	DCM RT		MeO-PEG	110

220

Benzisoxazole formation

	Solvent/Temp.	Yield	Resin	Ref.
	NMP 20° C	66%	Wang resin	83

Benzothiazepinone formation

	Solvent/Temp.	Yield	Resin	Ref.
	Benzene RT	up to 70%	ArgoGel-Rink resin	83

Benzothiazepinone formation

	Solvent/Temp.	Yield	Resin	Ref.
	DMF		MBHA	82

Dihydropyrimidinone formation

	Solvent/Temp.	Yield	Resin	Ref.
	DMA 70° C		Wang resin	9

Hydantoin formation

	Solvent/Temp.	Yield	Resin	Ref.
	CHCl$_3$ 70° C		Wang resin	111

Hydroxybenzotriazole formation

	Solvent/Temp.	Yield	Resin	Ref.
	EtOH		polystyrene	74

Hydroxyindole formation

	Solvent/Temp.	Yield	Resin	Ref.
	NMP 20° C	up to 74%	Wang resin	83

Imidazolinone formation

	Solvent/Temp.	Yield	Resin	Ref.
	DCM		Sasrin resin	65

221

Imidazopyridine formation

	Solvent/Temp.	Yield	Resin	Ref.
	Nitrobenzene 180° C	93%	PEG	94

Oxadiazole formation

	Solvent/Temp.	Yield	Resin	Ref.
	DCM/EtOH	90%	TentaGel S OH	112

Perhydrodiazepinone formation

	Solvent/Temp.	Yield	Resin	Ref.
	THF 0° C, then 20° C	up to 89%	PEG	4

Pyrazolone formation

	Solvent/Temp.	Yield	Resin	Ref.
	THF RT	80%	2-Cl-Trt resin	96

Pyridine formation

	Solvent/Temp.	Yield	Resin	Ref.
	DMF 100° C		Wang resin	40

Pyrimidine formation

	Solvent/Temp.	Yield	Resin	Ref.
	DMF RT		thiourea resin	113

Pyrimido[4,5-d]pyrimidine formation

	Solvent/Temp.	Yield	Resin	Ref.
	120° C to 200° C		thiourea resin	113

Pyrimido[4,5-d]pyrimidine formation

	Solvent/Temp.	Yield	Resin	Ref.
	Diphenyl ether 259° C		thiourea resin	113

222

Pyrrolidine formation	Solvent/Temp.	Yield	Resin	Ref.
	DCM RT		benzylsulfonylethoxy-carbonyl resin	25

Quinazolinone formation	Solvent/Temp.	Yield	Resin	Ref.
	Dioxane 110° C		hydroxymethyl polystyrene	38

Quinolone formation	Solvent/Temp.	Yield	Resin	Ref.
	Dowtherm 260° C		Merrifield resin	58

Ring closing metathesis	Solvent/Temp.	Yield	Resin	Ref.
	DCM reflux	85% to 89%	trityl resin	97

Tandem cyclization/substitution	Solvent/Temp.	Yield	Resin	Ref.
	DMSO 80° C		Wang resin	52

Tetrahydrofuran formation	Solvent/Temp.	Yield	Resin	Ref.
	Benzene reflux	up to 93%	Merrifield resin	14

Tetrahydropyrazine-2-one formation	Solvent/Temp.	Yield	Resin	Ref.
	CH$_3$CN/DMF		ArgoGel-MB-CHO	98

Thiazole formation	Solvent/Temp.	Yield	Resin	Ref.
	DMF 70° C	up to 97%	Rink resin	114

3.9 Multi-component reactions

3-Component condensation

	Solvent/Temp.	Yield	Resin	Ref.
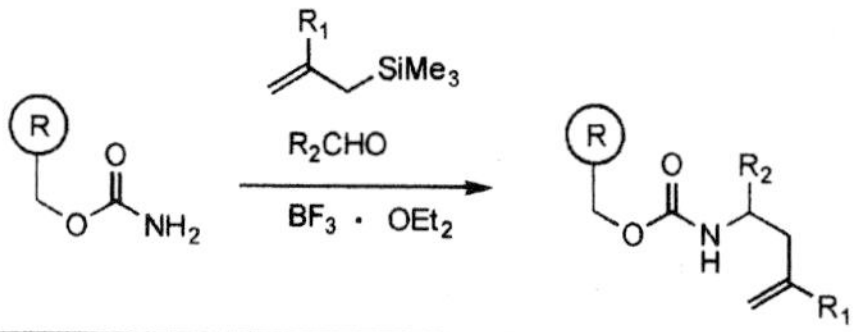	CH_3CN 0° C to 20° C		Wang resin	60

R_2CHO

$BF_3 \cdot OEt_2$

3-Component coupling

	Solvent/Temp.	Yield	Resin	Ref.
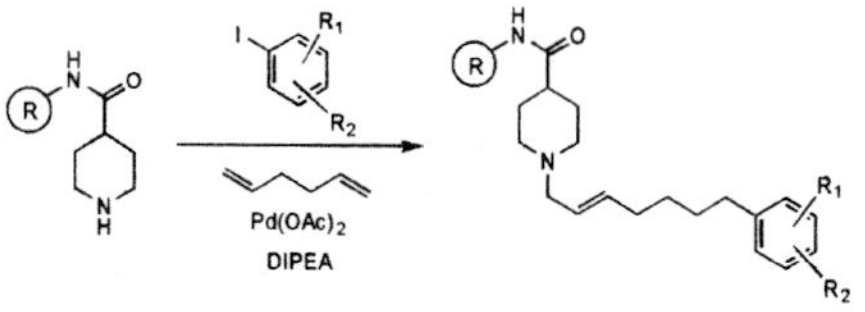	DMF 100° C	up to 86%	Rink resin	115

$Pd(OAc)_2$
DIPEA

Mannich reaction

	Solvent/Temp.	Yield	Resin	Ref.
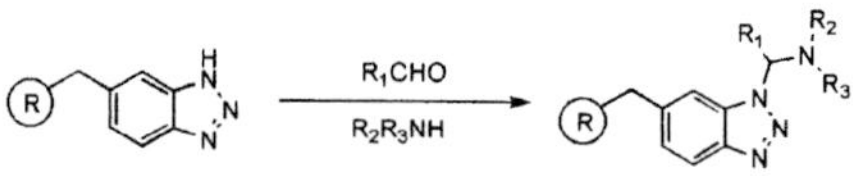	THF, EtOH or DMF reflux or 70° C		polystyrene	74

R_1CHO

R_2R_3NH

Ugi reaction

	Solvent/Temp.	Yield	Resin	Ref.
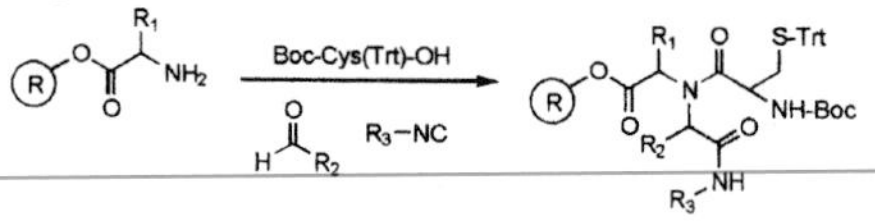	DCM/MeOH		TentaGel S OH	3

Boc-Cys(Trt)-OH

$R_3{-}NC$

Ugi reaction

	Solvent/Temp.	Yield	Resin	Ref.
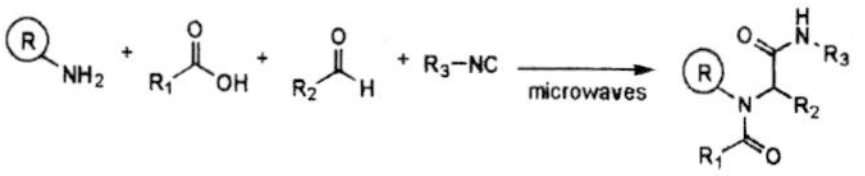	DCM/MeOH	up to 96%	TentaGel S Ram Fmoc	116

R_1 , R_2 , $+ R_3{-}NC$ microwaves

Ugi reaction

	Solvent/Temp.	Yield	Resin	Ref.
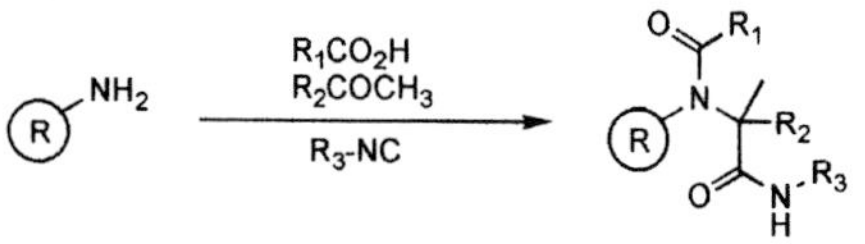	DCM RT	up to 71%	Rink resin	117

R_1CO_2H
R_2COCH_3
R_3-NC

Ugi reaction

	Solvent/Temp.	Yield	Resin	Ref.
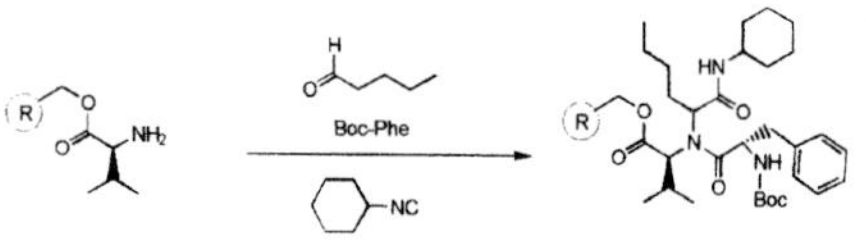	DCM/MeOH	79%	Merrifield resin, high loading	10

Boc-Phe

{-}NC

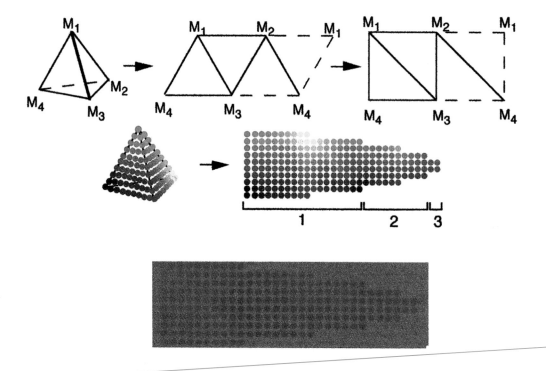

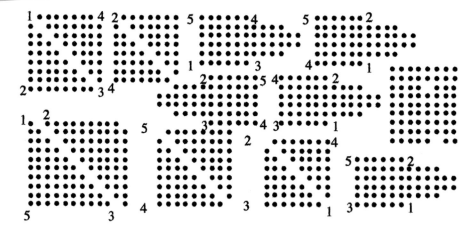

Plate 3. *Top.* Unfolding of a four-component discrete composition map into two dimensions. The four elements are represented as cyan, magenta, yellow, and black, and mixtures of these colours. Dashed lines represent redundant binaries that are eliminated in the two-dimensional map. At the resolutions shown, there are ten compositions along each binary edge. The three concentric shells in the resulting 220-spot array are indicated by numbered brackets.
Middle. Scanned image of a printed quaternary array. Actual size: 3.5 cm × 11 cm. Catalyst spots are 2mm in diameter.
Bottom. Discrete map containing all single elements, binary, tenary, quaternary, and pentanary compositions at the same resolution. The 70 pentanary spots form the rectangle at the far right.

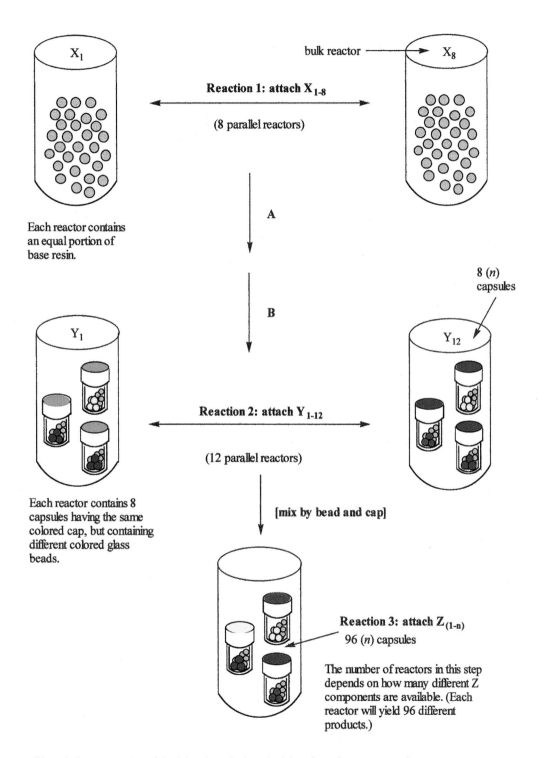

Plate 2. Representation of the 'visual tagging' methodology for a three-step reaction.

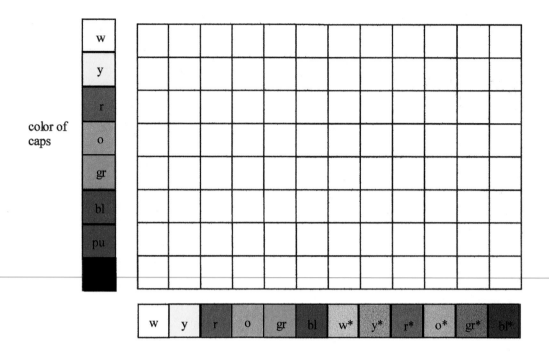

color of caps

w
y
r
o
gr
bl
pu

| w | y | r | o | gr | bl | w* | y* | r* | o* | gr* | bl* |

color of glass beads

Plate 1. An 8 × 12 colour coded array employed for identifying a 96-component library.

3.10 Olefin forming reactions

Horner-Emmons	Solvent/Temp.	Yield	Resin	Ref.
	cyclohexane/ THF	up to 82%	Wang resin	8

Horner-Wadsworth-Emmons	Solvent/Temp.	Yield	Resin	Ref.
	THF		Rink resin	104

Tandem cyclization/P-alkylation/Wittig	Solvent/Temp.	Yield	Resin	Ref.
	NMP	up to 43%	Wang resin	52

Wadsworth-Emmons	Solvent/Temp.	Yield	Resin	Ref.
			MeO-PEG	76

Wittig	Solvent/Temp.	Yield	Resin	Ref.
	THF RT		hydroxymethyl resin	31

Wittig	Solvent/Temp.	Yield	Resin	Ref.
			Wang resin	15

Wittig	Solvent/Temp.	Yield	Resin	Ref.
	THF reflux		MBHA resin	118

Wittig-Horner	Solvent/Temp.	Yield	Resin	Ref.
	THF -78° C to -40° C		diethylsilylpolystyrene	17

3.11 Oxidation reactions

Alcohol to aldehyde

	Solvent/Temp.	Yield	Resin	Ref.
iodobenzoic acid oxide	DMSO 25° C		TentaGel S NH$_2$	21

Alcohol to aldehyde

	Solvent/Temp.	Yield	Resin	Ref.
SO$_3$-pyridine, Et$_3$N	anhydrous DMSO		aminomethylpolystyrene-PEG	119

Baeyer-Villiger oxidation

	Solvent/Temp.	Yield	Resin	Ref.
m-CPBA	DCM reflux		hydroxymethy resin	31

Davis oxidation

	Solvent/Temp.	Yield	Resin	Ref.
1) KHMDS 2) PhSO$_2$			Wang resin	15

Diol to aldehyde

	Solvent/Temp.	Yield	Resin	Ref.
NaIO$_4$	H$_2$O 25° C		TentaGel S NH$_2$	21

Hydrazone to diazo

	Solvent/Temp.	Yield	Resin	Ref.
CH$_3$CO$_3$H	DCM		polystyrene	16

Iodine to hypervalent iodine

	Solvent/Temp.	Yield	Resin	Ref.
H$_3$C-C(O)O-OH	40° C	100%	iodopolystyrene	120

Iodine to hypervalent iodine

	Solvent/Temp.	Yield	Resin	Ref.
CH$_3$CO$_3$H	DCM RT	96%	linear iodopolystyrene	79

Olefin to diol

	Solvent/Temp.	Yield	Resin	Ref.
OsO$_4$ morpholine N-oxide	acetone/H$_2$O 25° C		TentaGel S NH$_2$	21

Pyrimidinone formation

	Solvent/Temp.	Yield	Resin	Ref.
Cerric Ammonium Nitrate	DMA RT	up to 99%	Wang resin	9

Sufide oxidation

	Solvent/Temp.	Yield	Resin	Ref.
Oxone	Dioxane/H_2O		thiourea resin	113

Sufide oxidation

	Solvent/Temp.	Yield	Resin	Ref.
m-CPBA	DCM RT		benzylthioethanol resin	25

Swern oxidation

	Solvent/Temp.	Yield	Resin	Ref.
Swern Oxidation			Wang resin	15

3.12 Reduction reactions

Acetal reduction

	Solvent/Temp.	Yield	Resin	Ref.
DIBAL-H	DCM -70° C to -20° C		Wang-CHO resin	34

Aldehyde to alcohol

	Solvent/Temp.	Yield	Resin	Ref.
$NaBH_4$	THF RT		Merrifield resin	42

Amide to amine

	Solvent/Temp.	Yield	Resin	Ref.
1) BH_3, 55-65 C 2) I_2, DIPEA, AcOH	THF	66% to 86%	Wang resin	121

Amide to amine

	Solvent/Temp.	Yield	Resin	Ref.
BH_3 $B(OH)_3$, BMe_3	THF 65° C		MBHA resin	122

Azide reduction

	Solvent/Temp.	Yield	Resin	Ref.
Ph_3P	H_2O/THF		Wang resin	100

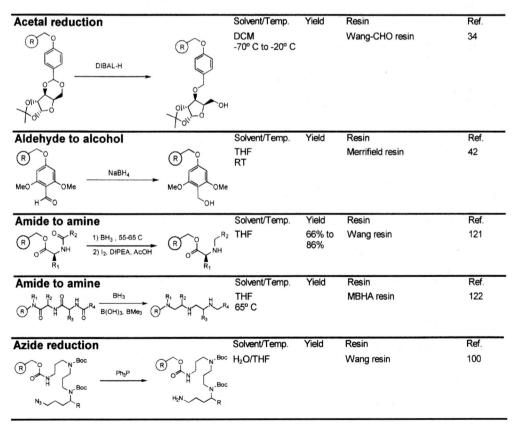

Azide reduction/amide formation	Solvent/Temp.	Yield	Resin	Ref.
1) prionic acid, EDC,HOBt, dioxane 2) Bu$_3$P, toluene — RT	RT		aminomethyl resin	123

Diester to diol	Solvent/Temp.	Yield	Resin	Ref.
LiAlH$_4$	THF 30° C to 60° C		Merrifield resin	124

Disulfide reduction	Solvent/Temp.	Yield	Resin	Ref.
DTT/Et$_3$N	THF/MeOH 20° C		trityl resin	125

Epoxide to alcohol	Solvent/Temp.	Yield	Resin	Ref.
LiAlH$_4$	No solvent	60%	Merrifield resin	74

Ester to alcohol	Solvent/Temp.	Yield	Resin	Ref.
DIBAL-H			Wang resin	16

Hydroxybenzotriazole reduction	Solvent/Temp.	Yield	Resin	Ref.
PCl$_3$ or SmI$_2$	CHCl$_3$ or THF reflux		polystyrene	74

Imidazolinone reduction	Solvent/Temp.	Yield	Resin	Ref.
NaBH$_3$CN, AcOH Sonication	MeOH		Sasrin Resin	65

Imine reduction	Solvent/Temp.	Yield	Resin	Ref.
BH$_3$-pyridine HOAc			Argogel-MB-CHO	98

Imine reduction	Solvent/Temp.	Yield	Resin	Ref.
NaCNBH$_3$ MeOH, AcOH	THF		Merrifield resin	71

Imine reduction	Solvent/Temp.	Yield	Resin	Ref.
NaBH(OAC)$_3$ AcOH	DMF		Wang resin	86

Imine reduction

	Solvent/Temp.	Yield	Resin	Ref.
	dichloroethane		carboxyethyl polystyrene	99

Isoxazolidide reduction

	Solvent/Temp.	Yield	Resin	Ref.
	THF 0° C	96%	non-crosslinked Merrifield resin	95

Ketone to alcohol

	Solvent/Temp.	Yield	Resin	Ref.
			polymer supported Cr complex	45

Ketone to alcohol

	Solvent/Temp.	Yield	Resin	Ref.
	THF 60° C		polystyrene	24

Nitrile to amine

	Solvent/Temp.	Yield	Resin	Ref.
	Diglyme 150° C		polyacrylonitrile	126

Nitro to amine

	Solvent/Temp.	Yield	Resin	Ref.
	DMF RT		Rink resin	81

Nitro to amine

	Solvent/Temp.	Yield	Resin	Ref.
	DMF RT		AMEBA resin	22

Nitro to amine

	Solvent/Temp.	Yield	Resin	Ref.
	DMF RT	up to 100%	Wang p-nitrocarbonate resin	27

Nitro to amine

	Solvent/Temp.	Yield	Resin	Ref.
	NMP		Wang resin	52

Nitro to amine	Solvent/Temp.	Yield	Resin	Ref.
	DMF RT	76% to 84%	ArgoGel-Rink resin	84

Nitro to amine	Solvent/Temp.	Yield	Resin	Ref.
	DMF RT	97%	Rink resin	127

Nitro to amine	Solvent/Temp.	Yield	Resin	Ref.
	DMF		MBHA resin	82

Nitro to amine	Solvent/Temp.	Yield	Resin	Ref.
	DMF RT		Rink resin	57

Nitro to amine	Solvent/Temp.	Yield	Resin	Ref.
	NMP RT		Argogel-MB	85

Olefin reduction	Solvent/Temp.	Yield	Resin	Ref.
	THF/H$_2$O 75° C		Merrifield resin	42

Reduction of primary azides	Solvent/Temp.	Yield	Resin	Ref.
	THF		Merrifield resin with dihydropyran linker	12

Reduction of secondary azides	Solvent/Temp.	Yield	Resin	Ref.
	THF		Merrifield resin with dihydropyran linker	12

Weinreb amide to aldehyde	Solvent/Temp.	Yield	Resin	Ref.
LiAlH$_4$ or LiAlH(Ot-Bu)$_3$	THF 0° C	90%	MBHA resin	118

3.13 Substitution reactions

Allylic Mitsunobu reaction	Solvent/Temp.	Yield	Resin	Ref.
DIAD, Ph$_3$P	THF RT	89%	2-Cl-Trt resin	96

Allylic substitution	Solvent/Temp.	Yield	Resin	Ref.
BEMP	DMF RT	76%	2-Cl-Trt resin	96

Amine alkylation	Solvent/Temp.	Yield	Resin	Ref.
R$_2$-Br, LiOtBu	THF/DMSO RT	up to 100%	Wang p-nitrocarbonate resin	27

C-Nucleophile	Solvent/Temp.	Yield	Resin	Ref.
R-X, BEMP or BTPP	DCM -78° C	61% to 100%	Wang resin	128

C-Nucleophile	Solvent/Temp.	Yield	Resin	Ref.
NaH	THF 60° C		Merrifield resin	124

C-Nucleophile	Solvent/Temp.	Yield	Resin	Ref.
1) KHMDS 2) R-I	THF -78° C		Wang resin	15

C-Nucleophile	Solvent/Temp.	Yield	Resin	Ref.
R-NH$_2$	60° C		Merrifield resin	99

Chloride formation	Solvent/Temp.	Yield	Resin	Ref.
CH$_3$SO$_2$Cl, Collidine, LiCl	DMF 0° C to RT		Wang resin	40

231

Chloride formation

Solvent/Temp.	Yield	Resin	Ref.
DCM RT		Merrifield resin	23

Chloride formation

Solvent/Temp.	Yield	Resin	Ref.
Toluene 60° C		polystyrene	24

COOH nucleophile

Solvent/Temp.	Yield	Resin	Ref.
Toluene 80° C		Wang resin	20

Diazo displacement

Solvent/Temp.	Yield	Resin	Ref.
DCM RT		aminomethyl crowns	129

Heteroatom nucleophile

Solvent/Temp.	Yield	Resin	Ref.
NMP 70° C	up to 95%	Rink resin	114

Mitsunobu reaction

Solvent/Temp.	Yield	Resin	Ref.
THF/DCM		HMPA carbonate resin	18

Mitsunobu reaction

Solvent/Temp.	Yield	Resin	Ref.
THF		Merrifield resin	47

Mitsunobu reaction

Solvent/Temp.	Yield	Resin	Ref.
THF -15° C, then RT		carboxypolystyrene	43

Mitsunobu reaction

Solvent/Temp.	Yield	Resin	Ref.
DCM/THF RT	37% to 59%	aminomethyl resin	95

Mitsunobu reaction

Solvent/Temp.	Yield	Resin	Ref.
THF		Rink resin	104

Mitsunobu reaction		Solvent/Temp.	Yield	Resin	Ref.
		DCM		Sasrin resin	65

Mitsunobu reaction		Solvent/Temp.	Yield	Resin	Ref.
		DCM		Wang resin	62

Mitsunobu reaction (modified)		Solvent/Temp.	Yield	Resin	Ref.
		DCM RT		Merrifield resin	42

Mitsunobu reaction (modified)		Solvent/Temp.	Yield	Resin	Ref.
		DCM RT		Merrifield resin	42

N-Alkylation		Solvent/Temp.	Yield	Resin	Ref.
		DMF 60° C		REM resin	89

N-Alkylation		Solvent/Temp.	Yield	Resin	Ref.
		Acetone 55° C	up to 97%	Rink resin	81

N-Alkylation		Solvent/Temp.	Yield	Resin	Ref.
		CH_3CN	87%	PEG	4

N-Alkylation	Solvent/Temp.	Yield	Resin	Ref.
	NMP		Aliphatic safety-catch resin	5

N-Alkylation	Solvent/Temp.	Yield	Resin	Ref.
	Toluene		hydroxylamine Wang resin	8

N-Alkylation	Solvent/Temp.	Yield	Resin	Ref.
1) NaH 2) RX	DMF	77% to >95%	Wang resin	28

N-Alkylation	Solvent/Temp.	Yield	Resin	Ref.
Cs_2CO_3	DMF		macrocrowns with Knorr linker	54

N-Alkylation	Solvent/Temp.	Yield	Resin	Ref.
DMAP	DCM		MeO-PEG	66

N-Alkylation	Solvent/Temp.	Yield	Resin	Ref.
R_2-X LiOtBu	DMSO		MBHA resin	122

N-Alkylation	Solvent/Temp.	Yield	Resin	Ref.
	DMSO	65% to 100%	Wang resin	130

N-Alkylation	Solvent/Temp.	Yield	Resin	Ref.
DIPEA	DMSO		MBHA resin	53

N-Alkylation	Solvent/Temp.	Yield	Resin	Ref.
NaI	NMP 80° C		methoxybenzhydryl chloride resin	24

N-Alkylation

	Solvent/Temp.	Yield	Resin	Ref.
	DMF 110° C to 120° C		PS-TTEGDA	57

N-Alkylation

	Solvent/Temp.	Yield	Resin	Ref.
	DCM RT		PS-TTEGDA	57

Organoborane coupling

	Solvent/Temp.	Yield	Resin	Ref.
	THF RT	up to 99%	Wang resin	131

S-Alkylation

	Solvent/Temp.	Yield	Resin	Ref.
	DMF		Wang resin	6

S-Alkylation

	Solvent/Temp.	Yield	Resin	Ref.
	DCM RT		MeO-PEG	110

S-Alkylation

	Solvent/Temp.	Yield	Resin	Ref.
	THF 20° C		trityl resin	125

S-Alkylation

	Solvent/Temp.	Yield	Resin	Ref.
	DMF 25° C		Knorr-MBHA resin	132

S-Alkylation

	Solvent/Temp.	Yield	Resin	Ref.
	DMF 60° C to RT		Merrifield resin	25

235

Sonogashira coupling		Solvent/Temp. DMF	Yield	Resin controlled pore glass	Ref. 133

Tosyl displacement		Solvent/Temp. NMP	Yield	Resin Merrifield resin with dihydropyran linker	Ref. 12

3.14 Protection/deprotection reactions

Acetylene deprotection		Solvent/Temp. THF	Yield	Resin dihydropyranyl resin	Ref. 88

Acid deprotection		Solvent/Temp. DCM	Yield	Resin REM resin	Ref. 89

Acid deprotection		Solvent/Temp. N-methylaniline	Yield	Resin Wang resin	Ref. 56

Alcohol deprotection		Solvent/Temp.	Yield	Resin	Ref. 98

Alcohol deprotection		Solvent/Temp.	Yield	Resin	Ref. 61

Alcohol protection

	Solvent/Temp.	Yield	Resin	Ref.
			Wang resin	15

Amine deprotection

	Solvent/Temp.	Yield	Resin	Ref.
	DCM RT		hydroxymethyl resin	19

Amine deprotection

	Solvent/Temp.	Yield	Resin	Ref.
	iPrOH/DCM 40° C		trityl resin	64

Amine deprotection

	Solvent/Temp.	Yield	Resin	Ref.
	DMF RT		benzyloxytrityl chloride resin	23

Amine deprotection

	Solvent/Temp.	Yield	Resin	Ref.
	MeOH		Sasrin resin	65

Amine deprotection

	Solvent/Temp.	Yield	Resin	Ref.
	DCM RT		oxime resin	67

Amine deprotection

	Solvent/Temp.	Yield	Resin	Ref.
	DMF		Wang resin	62

237

Amine deprotection		Solvent/Temp.	Yield	Resin	Ref.
		H₂O/THF		Wang resin	100

Amine deprotection		Solvent/Temp.	Yield	Resin	Ref.
		EtOH reflux		PS-TTEGDA	57

Carboxylic acid deprotection		Solvent/Temp.	Yield	Resin	Ref.
				Wang resin	134

Diol deprotection		Solvent/Temp.	Yield	Resin	Ref.
				amino PEGA resin	135

Ketone deprotection		Solvent/Temp.	Yield	Resin	Ref.
		THF RT		Merrifield resin	2

Lactam deprotection		Solvent/Temp.	Yield	Resin	Ref.
		DMF		2-chlorotrityl resin	59

N-Protection		Solvent/Temp.	Yield	Resin	Ref.
		DCM RT		hydroxylamine Wang resin	8

N-Protection		Solvent/Temp.	Yield	Resin	Ref.
		DCM/DMF		MBHA resin	122

O-Deprotection

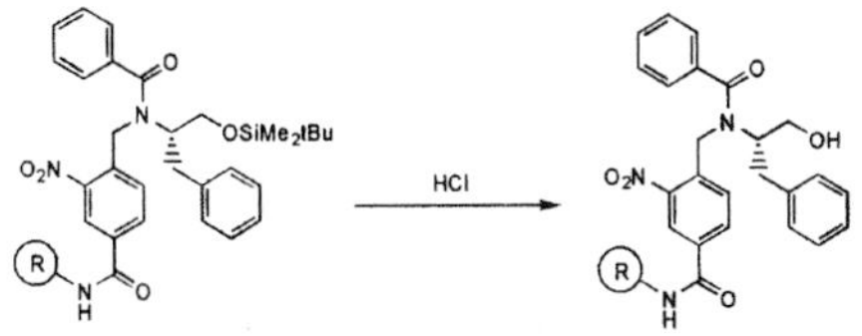

	Solvent/Temp.	Yield	Resin	Ref.
	MeOH 25° C		TentaGel S NH$_2$	21

Phenol deprotection

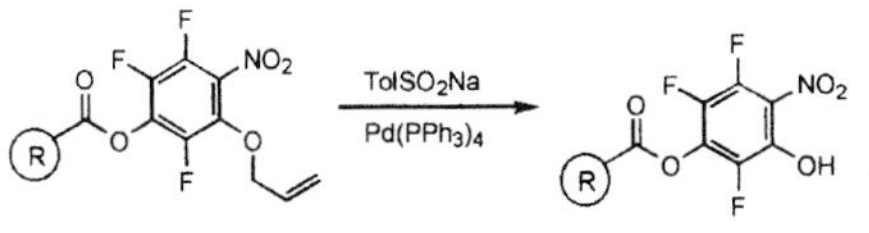

	Solvent/Temp.	Yield	Resin	Ref.
	MeOH/THF		carboxypolystyrene	43

Phenol deprotection

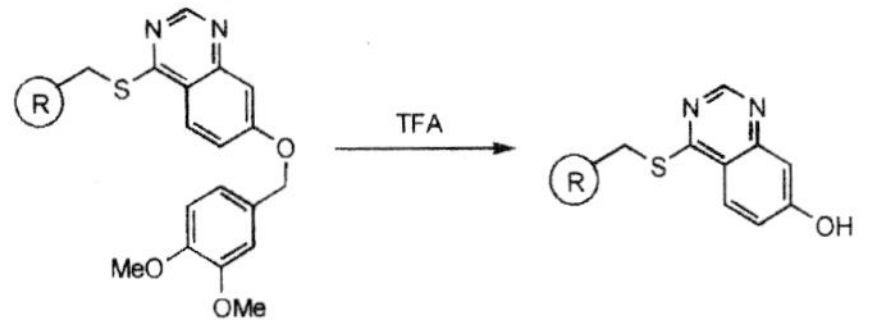

	Solvent/Temp.	Yield	Resin	Ref.
	DCM RT		Merrifield resin	47

Saponification

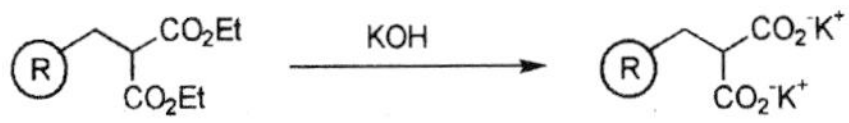

	Solvent/Temp.	Yield	Resin	Ref.
	THF		Merrifield resin	99

TPH ether hydrolysis

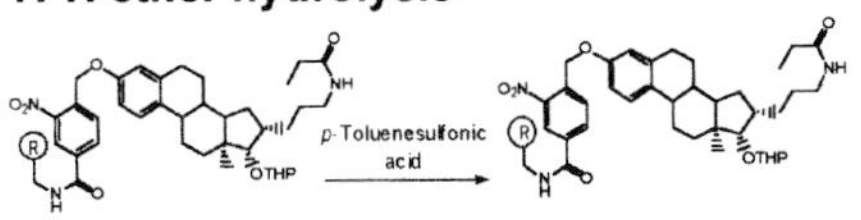

	Solvent/Temp.	Yield	Resin	Ref.
	1-butanol/ dichloroethane RT		aminomethyl resin	123

TPH ether hydrolysis

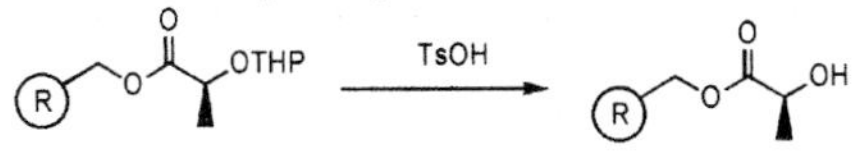

	Solvent/Temp.	Yield	Resin	Ref.
	MeOH/DCM		Wang resin	136

3.15 Other solid phase reactions

Acetal formation

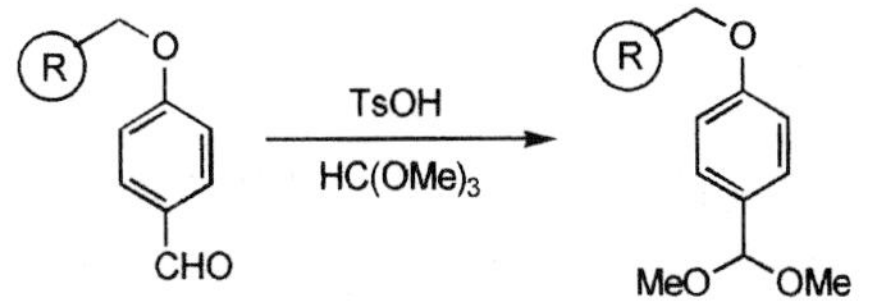

	Solvent/Temp.	Yield	Resin	Ref.
	RT		Wang-CHO resin	34

Acid chloride formation

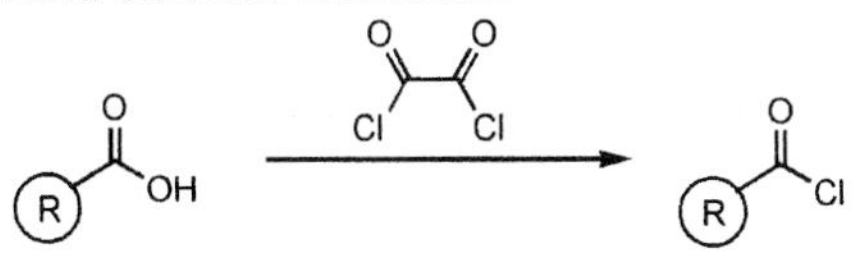

	Solvent/Temp.	Yield	Resin	Ref.
	DMF/dichloro- ethane		carboxypolystyrene	50

Acid chloride formation

	Solvent/Temp.	Yield	Resin	Ref.
	Benzene reflux		carboxypolystyrene	137

Addition to hydrazone

	Solvent/Temp.	Yield	Resin	Ref.
	DCM RT		carboxypolystyrene	48

Allylsilane addition

	Solvent/Temp.	Yield	Resin	Ref.
	DCM	70%	carboxyethylpolystyrene	99

Azide formation

	Solvent/Temp.	Yield	Resin	Ref.
	THF -10° C		Merrifield resin	26

Carbonate formation

	Solvent/Temp.	Yield	Resin	Ref.
	CH$_3$CN/DCM RT	100%	hydroxymethyl polystyrene	19

Carbonate formation

	Solvent/Temp.	Yield	Resin	Ref.
	DCM RT	80% to 93%	not specified	61

Carbonate formation

	Solvent/Temp.	Yield	Resin	Ref.
	DCM RT		benzenesulfonylethanol resin	25

Chlorination

	Solvent/Temp.	Yield	Resin	Ref.
	DMF reflux		hydroxymethyl polystyrene	38

Chlorosilane formation

	Solvent/Temp.	Yield	Resin	Ref.
	DCM		diethylsilane resin	46

Diazotization

	Solvent/Temp.	Yield	Resin	Ref.
	H_2O		aminomethyl crowns	129

Decarboxylation

	Solvent/Temp.	Yield	Resin	Ref.
	THF		Merrifield resin	99

Enamine formation

	Solvent/Temp.	Yield	Resin	Ref.
	DMF		Merrifield resin	58

Ester formation

	Solvent/Temp.	Yield	Resin	Ref.
	DCM 20° C		REM resin	89

Ester formation

	Solvent/Temp.	Yield	Resin	Ref.
			polymer supported Cr complex	45

Ester formation

	Solvent/Temp.	Yield	Resin	Ref.
	THF 60° C		carboxypolystyrene	48

Ester formation

	Solvent/Temp.	Yield	Resin	Ref.
	DMF		macrocrowns with Knorr linker	54

Ester formation

	Solvent/Temp.	Yield	Resin	Ref.
	Et_3N		MeO-PEG	66

241

Ester formation	Solvent/Temp.	Yield	Resin	Ref.
	DCM RT		aminomethylpolystyrne grafted PTFE tubes	103

Ester formation	Solvent/Temp.	Yield	Resin	Ref.
	DCM RT		aminomethyl polystyrene	123

Ester formation	Solvent/Temp.	Yield	Resin	Ref.
	THF		Wang resin	136

Ethylene oxide polymerization	Solvent/Temp.	Yield	Resin	Ref.
	THF		Merrifield resin	124

Glycosylation	Solvent/Temp.	Yield	Resin	Ref.
	Et$_2$O 25° C	62%	aminomethyl macroporous polystyrene	138

Halogen-metal exchange	Solvent/Temp.	Yield	Resin	Ref.
	THF/NMP -40° C		not specified	105

Hydroboration	Solvent/Temp.	Yield	Resin	Ref.
	THF 0° C to RT		vinylpolystyrene	92

Iodination	Solvent/Temp.	Yield	Resin	Ref.
			polystyrene	122

Ireland-Claisen rearrangement	Solvent/Temp.	Yield	Resin	Ref.
	THF 50° C	up to 60%	PS-DES	30

Isocyanate formation

	Solvent/Temp.	Yield	Resin	Ref.
	DCM		MBHA resin	69

Isonitrile formation

	Solvent/Temp.	Yield	Resin	Ref.
	DCM		dibenzosuberyl resin	64

Ketone formation

	Solvent/Temp.	Yield	Resin	Ref.
	Toluene RT	77 to 92%	Wang resin	139

Lithium-halogen exchange

	Solvent/Temp.	Yield	Resin	Ref.
	THF -78° C		Merrifield resin	13

Mixed anhydride formation

	Solvent/Temp.	Yield	Resin	Ref.
	DCM -20° C	100%	Wang resin	56

Nitrile hydrolysis

	Solvent/Temp.	Yield	Resin	Ref.
	H_2O RT		thiourea resin	113

Nitrile hydrolysis

	Solvent/Temp.	Yield	Resin	Ref.
	H_2O 50° C		thiourea resin	113

Radical addition

	Solvent/Temp.	Yield	Resin	Ref.
	DCM/H_2O RT	up to 60 %	Wang resin	1

Radical addition

	Solvent/Temp.	Yield	Resin	Ref.
	DCM 20° C	up to 83%	Wang resin or TentaGel S OH	140

Scandium complex formation

	Solvent/Temp.	Yield	Resin	Ref.
	THF RT		polyaminoethylene	126

Silane activation	Solvent/Temp.	Yield	Resin	Ref.
	DCM		diethylsilane resin	29

Silane activation	Solvent/Temp.	Yield	Resin	Ref.
	DCM 25° C		diethylsilane resin	30

Silane activation	Solvent/Temp.	Yield	Resin	Ref.
	DCM RT		diethylsilane resin	17

Sodium thiolate formation	Solvent/Temp.	Yield	Resin	Ref.
	THF 20° C		trityl resin	125

Thioacylation	Solvent/Temp.	Yield	Resin	Ref.
	THF 90° C		TentaGel S NH$_2$	141

Thioamide formation	Solvent/Temp.	Yield	Resin	Ref.
	THF/H$_2$O 70° C		Rink resin	114

Thiocarbonate formation	Solvent/Temp.	Yield	Resin	Ref.
	DCM		Wang resin	11

Triflate leaving group	Solvent/Temp.	Yield	Resin	Ref.
	THF -70° C to 20° C		Wang resin	34

3.16 Cleavage reactions

Cleavage/coupling	Solvent/Temp.	Yield	Resin	Ref.
	MeOH 40° C	up to 91%	piperazinylmethyl polystyrene	32

Cyclization/release	Solvent/Temp.	Yield	Resin	Ref.
	Dichloroethane 60° C		polystyrene sulfonylhydrazine	91

Cyclization/release	Solvent/Temp.	Yield	Resin	Ref.
	MeOH		TentaGel S OH	3

Cyclization/release	Solvent/Temp.	Yield	Resin	Ref.
	DCM RT	up to 100%	Wang p-nitrocarbonate resin	27

Cyclization/release	Solvent/Temp.	Yield	Resin	Ref.
	TFA/5 N HCl 55° C		oxime resin	67

Cyclization/release	Solvent/Temp.	Yield	Resin	Ref.
	H_2O/EtOH 50° C		TentaGel S NH_2	141

Cyclization/release	Solvent/Temp.	Yield	Resin	Ref.
	DCM 20° C		Wang resin	34

Cyclization/release	Solvent/Temp.	Yield	Resin	Ref.
	THF 60° C		Merrifield resin	71

Cyclization/release

Solvent/Temp.	Yield	Resin	Ref.
dioxane/DCM		Wang resin, high loading	10

Cyclization/release

Solvent/Temp.	Yield	Resin	Ref.
THF heat	up to 95%	carboxyethylpolystyrene	99

Decarboxylation/cleavage

Solvent/Temp.	Yield	Resin	Ref.
DCM		Wang resin	139

Decarboxylative traceless linking

Solvent/Temp.	Yield	Resin	Ref.
MeCN/dioxane 75° C	64% to 69%	hydroxymethyl polystyrene	38

Desilyation

Solvent/Temp.	Yield	Resin	Ref.
AcOH/THF	73%	Wang resin	134

Diol oxidation

Solvent/Temp.	Yield	Resin	Ref.
H₂O/AcOH		amino PEGA resin	135

Electrophilic ipso-degermylation

Solvent/Temp.	Yield	Resin	Ref.
THF 67° C		germanyl resin	39

246

Electrophilic ipso-degermylation		Solvent/Temp.	Yield	Resin	Ref.
		RT		germanyl resin	39

Electrophilic ipso-degermylation		Solvent/Temp.	Yield	Resin	Ref.
		DCM RT		germanyl resin	39

Electrophilic ipso-degermylation		Solvent/Temp.	Yield	Resin	Ref.
		DCM RT		germanyl resin	39

Elimination		Solvent/Temp.	Yield	Resin	Ref.
		THF/MeOH RT		benzylsulfonylethoxy-carbonyl resin	25

Ester/amide exchange		Solvent/Temp.	Yield	Resin	Ref.
		DCM 20° C		Wang resin	142

Guanidine release		Solvent/Temp.	Yield	Resin	Ref.
		CHCl₃/MeOH		carboxypolystyrene	50

247

Hofmann elimination

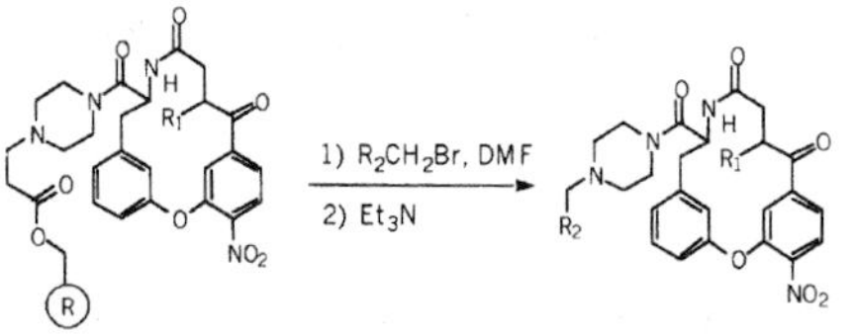

	Solvent/Temp.	Yield	Resin	Ref.
	RT		REM resin	80

1) R$_2$CH$_2$Br, DMF
2) Et$_3$N

Hydrolysis

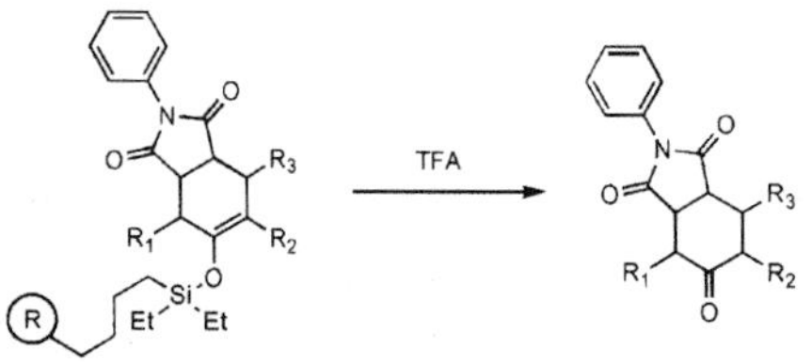

	Solvent/Temp.	Yield	Resin	Ref.
	DCM		diethylsilane resin	29

TFA

Hydrolysis

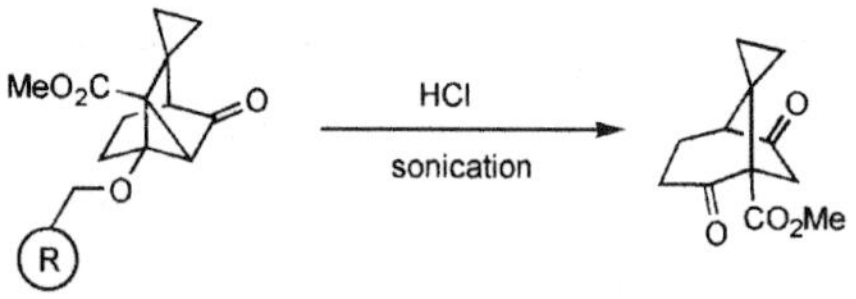

	Solvent/Temp.	Yield	Resin	Ref.
	DCM		hydroxymethyl resin	31

HCl
sonication

Hydrolysis

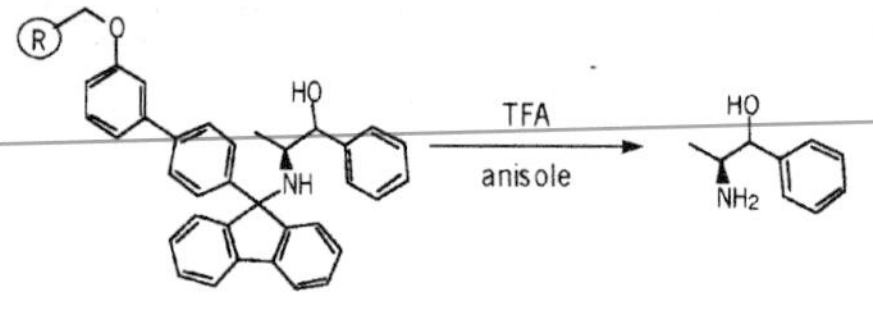

	Solvent/Temp.	Yield	Resin	Ref.
	DCM		non-crosslinked Merrifield resin	95

TFA
anisole

Hydrolysis

	Solvent/Temp.	Yield	Resin	Ref.
	DCM		benzhydryl resin	16

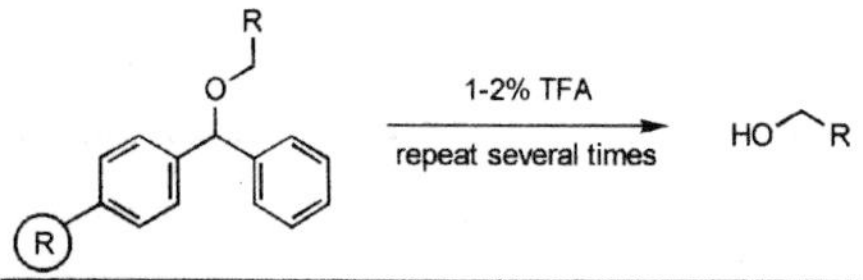

1-2% TFA
repeat several times

Hydrolysis

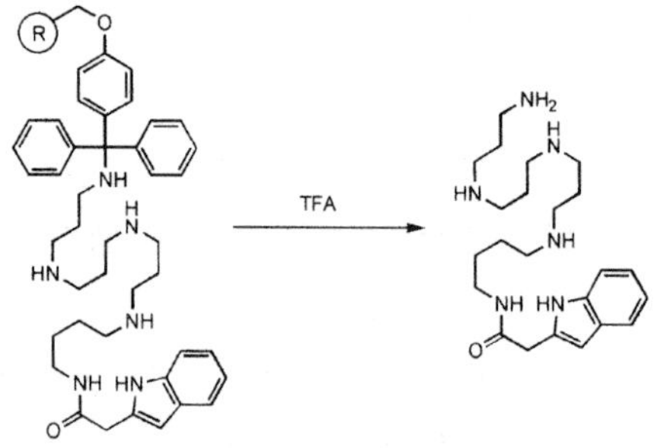

	Solvent/Temp.	Yield	Resin	Ref.
	DCM		benzyloxytrityl chloride resin	23

TFA

Hydrolysis

Solvent/Temp.	Yield	Resin	Ref.
DCM RT		methoxybenzhydryl resin	24

Hydrolysis

Solvent/Temp.	Yield	Resin	Ref.
H_2O		dibenzosuberyl resin	64

Ligand exchange

Solvent/Temp.	Yield	Resin	Ref.
Pyridine		polymer supported Cr complex	45

Linker reduction/cyclization

Solvent/Temp.	Yield	Resin	Ref.
THF/H_2O RT		aminomethyl resin/quinone linker	143

Linker oxidation

Solvent/Temp.	Yield	Resin	Ref.
DCM/H_2O	91%	aminomethyl macroporous resin	138

Nucleophilic aromatic substitution

Solvent/Temp.	Yield	Resin	Ref.
DMSO 100° C	up to 78%	Merrifield resin	47

Nucleophilic aromatic substitution

Solvent/Temp.	Yield	Resin	Ref.
DMF 60° C	80% to 90%	thiourea resin	113

Nucleophilic displacement

Solvent/Temp.	Yield	Resin	Ref.
Dioxane 90° C	84%	Aliphatic safety-catch resin	5

249

Nucleophilic displacement	Solvent/Temp.	Yield	Resin	Ref.
	THF	up to 96%	aliphatic safety-catch resin	5

Nucleophilic displacement	Solvent/Temp.	Yield	Resin	Ref.
	Et$_2$O/THF	68%	hydroxylamine Wang resin	8

Oxidation	Solvent/Temp.	Yield	Resin	Ref.
	Acetone RT		Merrifield resin	14

Oxidative cleavage	Solvent/Temp.	Yield	Resin	Ref.
	RT	55%	TentaGel S NH$_2$	49

Oxidative cleavage	Solvent/Temp.	Yield	Resin	Ref.
	n-Propylamine RT	93%	aminomethyl resin	49

Oxidative cleavage	Solvent/Temp.	Yield	Resin	Ref.
	DCM/EtOH/H$_2$O RT		thioacetal resin	95

Oxidative release	Solvent/Temp.	Yield	Resin	Ref.
			polymer supported Cr complex	45

Photolysis	Solvent/Temp.	Yield	Resin	Ref.
	THF/MeOH	up to 97%	Merrifield resin	2

250

Photolysis		Solvent/Temp.	Yield	Resin	Ref.
		MeOH 50° C		TentaGel S NH$_2$	21

Photolysis		Solvent/Temp.	Yield	Resin	Ref.
		MeOH RT		aminomethyl resin	123

Photolysis		Solvent/Temp.	Yield	Resin	Ref.
		TFE/DCM	up to 83%	PS-TTEGDA	57

Raney nickel desulfization		Solvent/Temp.	Yield	Resin	Ref.
		EtOH	83% to 98%	thiourea resin	113

Reductive cleavage		Solvent/Temp.	Yield	Resin	Ref.
		THF 0° C	67% to 91%	hydroxylamine Wang resin	8

Resin reduction		Solvent/Temp.	Yield	Resin	Ref.
		DCM/Et$_3$SiH	up to 100%	trityl type resins	36

Richter reaction		Solvent/Temp.	Yield	Resin	Ref.
		acetone/H$_2$O	up to 95%	benzylaminomethyl-polystyrene	78

Saponification		Solvent/Temp.	Yield	Resin	Ref.
		MeOH/THF	up to 94%	carboxypolystyrene	43

Semicarbazone hydrolysis		Solvent/Temp.	Yield	Resin	Ref.
		THF 65° C		BHA resin	144

Siloxane cleavage

	Solvent/Temp.	Yield	Resin	Ref.
	THF RT		diethylsilyl resin	17

Transesterification

	Solvent/Temp.	Yield	Resin	Ref.
	MeOH/dichloroethane 55° C		PS-DES	30

Transesterification

	Solvent/Temp.	Yield	Resin	Ref.
	MeOH	up to 93%	PEG	94

Transesterification

	Solvent/Temp.	Yield	Resin	Ref.
	MeOH RT		MeO-PEG	63

Triazine cleavage

	Solvent/Temp.	Yield	Resin	Ref.
	DCM RT		Merrifield resin	26

3.17 Resin-supported reagents

1,3,4-Oxadiazole formation

	Solvent/Temp.	Yield	Resin	Ref.
	THF	70% to 98%	PEG supported Burgess reagent	145

3-Component condensation

	Solvent/Temp.	Yield	Resin	Ref.
	CH_3CN	up to 68%	polyaniline-Co(II)(OAc)$_2$	146

3-Component condensation

	Solvent/Temp.	Yield	Resin	Ref.
	DCM/CH_3CN 60° C	65% to 100%	PA-Sc-TAD	126

Acetophenones to acyloins

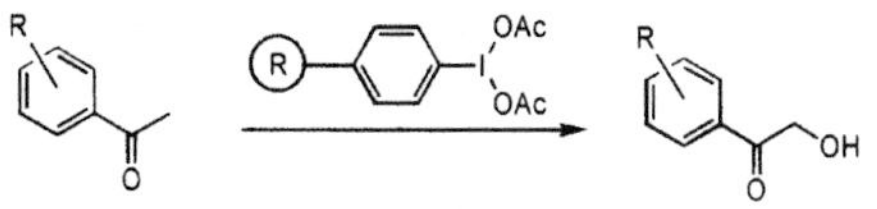

Solvent/Temp.	Yield	Resin	Ref.
DCM/CH$_3$CN	100%	polymer supported phenyliodine(III) diacetate	120

Acid bromide formation

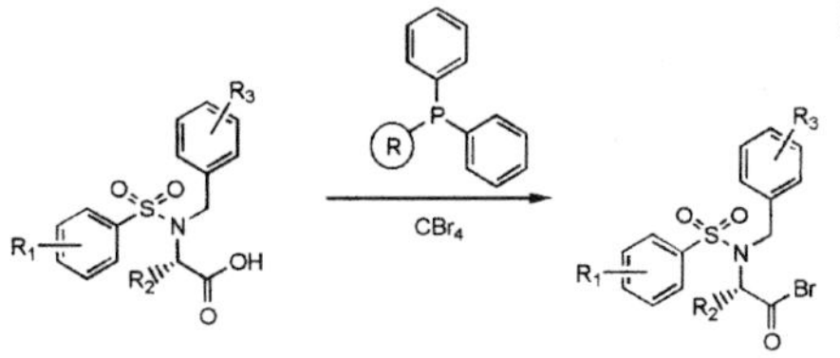

Solvent/Temp.	Yield	Resin	Ref.
DCM		polystyrene-PPh$_2$	147

Allyl coupling

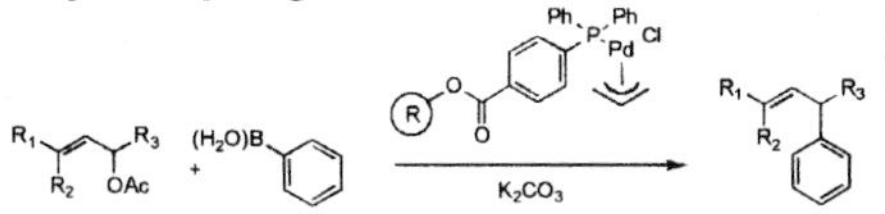

Solvent/Temp.	Yield	Resin	Ref.
H$_2$O 25° C	up to 99%	Pd-PEG-PS	148

Amide formation

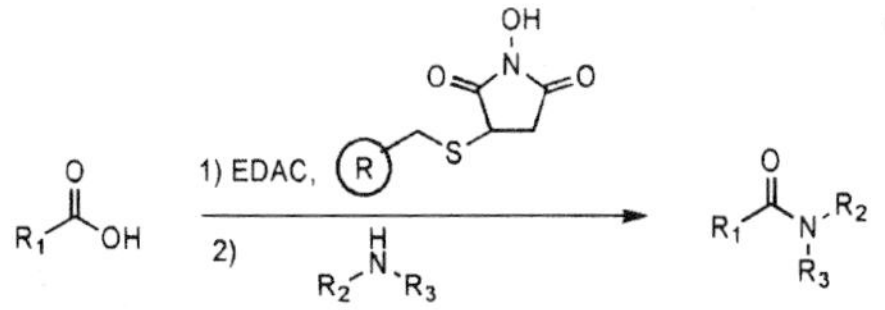

Solvent/Temp.	Yield	Resin	Ref.
CHCl$_3$/DMF RT	89% to 99%	N-hydroxysuccinimide resin	108

Amide formation

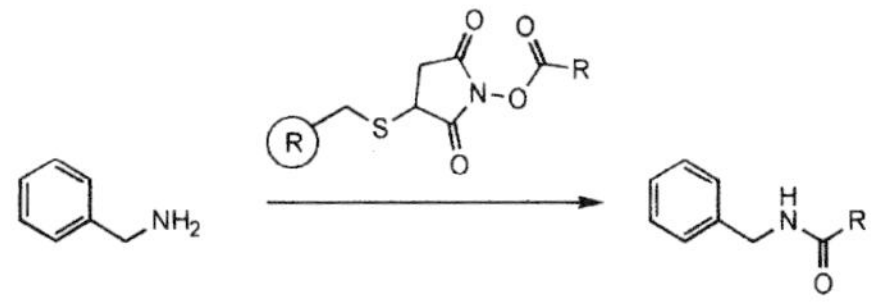

Solvent/Temp.	Yield	Resin	Ref.
DMF	87% to 98%	N-hydroxysuccinimide resin	149

Amide formation

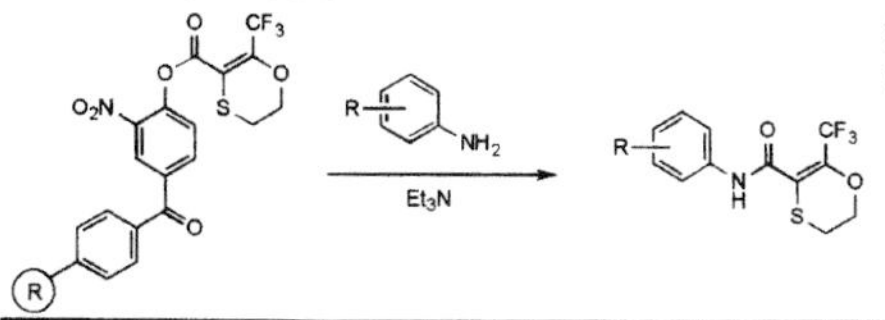

Solvent/Temp.	Yield	Resin	Ref.
CH$_3$CN reflux	up to 100%	4-hydroxy-3-nitro-benzophenone resin	7

Amine deprotection

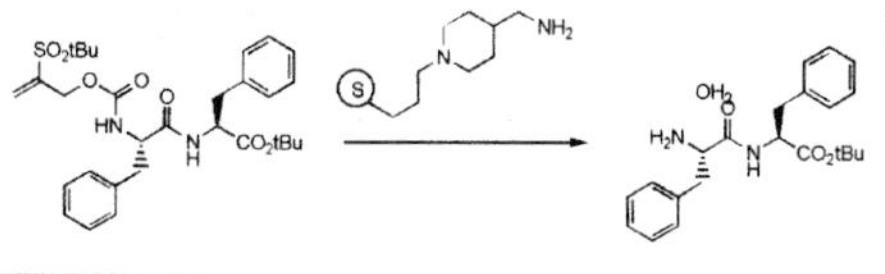

Solvent/Temp.	Yield	Resin	Ref.
DCM		aminomethylpiperidinyl silica gel	150

Amine deprotection

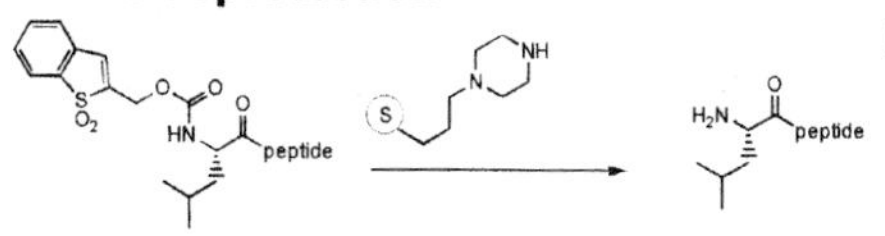

Solvent/Temp.	Yield	Resin	Ref.
DCM		piperazinyl silica gel	151

Benzyl alcohols to benzaldehydes

	Solvent/Temp.	Yield	Resin	Ref.
		100%	polymer supported phenyliodine(III) diacetate	120

Carbonylation

	Solvent/Temp.	Yield	Resin	Ref.
	190° C		pyrrolidinylpyridine-polystyrene	152
CO, MeI				

α-Chlorination of ketones

	Solvent/Temp.	Yield	Resin	Ref.
	THF -78° C to RT	60%	chloro-p-toluenesulfonyl resin	153

Cleavage from REM resin

	Solvent/Temp.	Yield	Resin	Ref.
	DCM RT		N-ethylpiperazine Wang resin	154

Disulfide formation

	Solvent/Temp.	Yield	Resin	Ref.
	DCM RT	89% to 93%	polymer supported phenyliodine(III) diacetate	79

R-SH ⟶ RS-SR

Fluorescent labelling

	Solvent/Temp.	Yield	Resin	Ref.
	THF 23° C	81% to 95%	polymer supported OSu ester	107

R_1 = 1-pyrenebutyric acid, 6-carboxyfluorescein diacetate

Grignard quench

	Solvent/Temp.	Yield	Resin	Ref.
	THF		macroporous polystyrene-sufonic acid	91

Hydrazone oxidation

	Solvent/Temp.	Yield	Resin	Ref.
	DCM RT	78%	polymer supported phenyliodine(III) diacetate	89

Iminocoumarin formation	Solvent/Temp.	Yield	Resin	Ref.
	Cyclohexane reflux	up to 92%	Amberlyte IRA 900	155

Lithium-halogen exchange	Solvent/Temp.	Yield	Resin	Ref.
	THF -78° C	up tp 99%	PS-PVBP	156

Lithium-halogen exchange	Solvent/Temp.	Yield	Resin	Ref.
	THF -78° C	up to 92%	PS-PVN	156

Organozinc addition to aldehydes	Solvent/Temp.	Yield	Resin	Ref.
	Toluene -20° C	94%	polymer supported chiral catalyst	157

Organozinc addition to aldehydes	Solvent/Temp.	Yield	Resin	Ref.
	Toluene -20° C		dendritically crosslinked polystyrene bound Ti-TA	158

Phenol alkylation	Solvent/Temp.	Yield	Resin	Ref.
	MeOH reflux	80% to 85%	polymer-supported phenoxides	159

Quinols to quinones	Solvent/Temp.	Yield	Resin	Ref.
	DCM RT	100%	polymer-supported (diacetoxyiodo)benzene	120

Quinone formation	Solvent/Temp.	Yield	Resin	Ref.
	DCM RT	63%	polymer-supported phenyliodine(III) diacetate	79

255

Re complex formation	Solvent/Temp.	Yield	Resin	Ref.
	CH_3CN 80° C	60% to 71%	polymer-supported cyclo-pentadienylphosphazine	160

Sulfonamide alkylation	Solvent/Temp.	Yield	Resin	Ref.
	DCM	89% to 100%	polystyrene-BEMP	147

Sulfoxide formation	Solvent/Temp.	Yield	Resin	Ref.
	DCM reflux	65% to 80%	polymer-supported phenyliodine(III) diacetate	79

Suzuki coupling	Solvent/Temp.	Yield	Resin	Ref.
	H_2O 25° C	66% to 91%	Pd-PEG-PS	148

Suzuki coupling	Solvent/Temp.	Yield	Resin	Ref.
	iPrOH/H_2O 80° C	up to 99%	Deloxan resin-Pd	161

Tyrosine to spirocyclic dienone	Solvent/Temp.	Yield	Resin	Ref.
	DCM/CH_3CN 60° C	75% to 96%	polymer-supported phenyliodine(III) diacetate	120

3.18 Scavenger resins

AG 50W-X2		Ref.
	organometal quench	162

Amberlyst A-21		Ref.
	acid scavenger	163
	scavenge acyl-transfer by-products	164

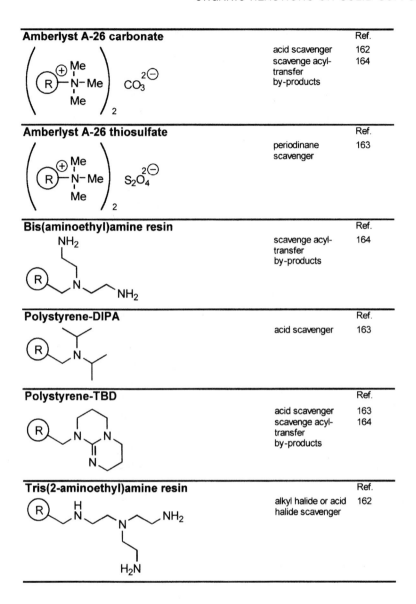

		Ref.
Amberlyst A-26 carbonate	acid scavenger	162
	scavenge acyl-transfer by-products	164
Amberlyst A-26 thiosulfate	periodinane scavenger	163
Bis(aminoethyl)amine resin	scavenge acyl-transfer by-products	164
Polystyrene-DIPA	acid scavenger	163
Polystyrene-TBD	acid scavenger	163
	scavenge acyl-transfer by-products	164
Tris(2-aminoethyl)amine resin	alkyl halide or acid halide scavenger	162

References

1. Yim, A.-M., Vidal, Y., Viallefont, P., and Martinez, J. (1999). *Tetrahedron Lett.*, **40**, 4535.
2. Lee, H. B. and Balasubramanian, S. (1999). *J. Org. Chem.*, **64**, 3454.
3. Szardenings, A. K., Antonenko, V., Campbell, D. A., DeFrancisco, N., Ida, S., Shi, L., *et al.* (1999). *J. Med. Chem.*, **42**, 1348.
4. Nouvet, A., Binard, M., Lamaty, F., Martinez, J., and Lazaro, R. (1999). *Tetrahedron*, **55**, 4685.
5. Backes, B. J. and Ellman, J. A. (1999). *J. Org. Chem.*, **64**, 2322.
6. Karoyan, P., Triolo, A., Nannicini, R., Giannotti, D., Altamura, M., Chassaing, G., *et al.* (1999). *Tetrahedron Lett.*, **40**, 71.

7. Hahn, H.-G., Chang, K. H., Nam, K. D., Bae, S. Y., and Mah, H. (1998). *Heterocycles*, **48**, 2253.

8. Salvino, J. M., Mervic, M., Mason, H. J., Kiesow, T., Teager, D., Airey, J., *et al.* (1999). *J. Org. Chem.*, **64**, 1823.

9. Hamper, B. C., Gan, K. Z., and Owen, T. J. (1999). *Tetrahedron Lett.*, **40**, 4973.

10. Raillard, S. P., Ji, G., Mann, A. D., and Baer, A. (1999). *Org. Process Res. Dev.*, **3**, 177.

11. Hanessian, S. and Huynh, H. K. (1999). *Tetrahedron Lett.*, **40**, 671.

12. Zhou, J., Termin, A., Wayland, M., and Tarby, C. M. (1999). *Tetrahedron Lett.*, **40**, 2729.

13. Crabtree, R. H. (1999). *Chemtech*, 21.

14. Watanabe, Y., Ishikawa, S., Takao, G., and Toru, T. (1999). *Tetrahedron Lett.*, **40**, 3411.

15. Hanessian, S., Ma, J., and Wang, W. (1999). *Tetrahedron Lett.*, **40**, 4631.

16. Mergler, M., Dick, F., Gosteli, J., and Nyfeler, R. (1999). *Tetrahedron Lett.*, **40**, 4663.

17. Doi, T., Hijikuro, I., and Takahashi, T. (1999). *J. Am. Chem. Soc.*, **121**, 6749.

18. Roussel, P., Bradley, M., Kane, P., Bailey, C., Arnold, R., and Cross, A. (1999). *Tetrahedron*, **55**, 6219.

19. Adang, A. E. P., Peters, C. A. M., Gerritsma, S., de Zwart, E., and Veeneman, G. (1999). *Bioorg. Med. Chem. Lett.*, **9**, 1227.

20. Yu, K.-L., Civiello, R., Roberts, D. G. M., Seiler, S. M., and Meanwell, N. A. (1999). *Bioorg. Med. Chem. Lett.*, **9**, 663.

21. Cavallaro, C. L., Herpin, T., McGuinness, B. F., Shimshock, Y. C., and Dolle, R. E. (1999). *Tetrahedron Lett.*, **40**, 2711.

22. Ouyang, X., Tamayo, N., and Kiselyov, A. S. (1999). *Tetrahedron*, **55**, 2827.

23. Hidai, Y., Kan, T., and Fukuyama, T. (1999). *Tetrahedron Lett.*, **40**, 4711.

24. Brown, D. S., Revill, J. M., and Shute, R. E. (1998). *Tetrahedron Lett.*, **39**, 8533.

25. Veerman, J. J. N., Rutjes, F. P. J. T., vanMaarseveen, J. H., and Hiemstra, H. (1999). *Tetrahedron Lett.*, **40**, 6079.

26. Bräse, S., Köbberling, J., Enders, D., Lazny, R., Wang, M., and Brandtner, S. (1999). *Tetrahedron Lett.*, **40**, 2105.

27. Huang, W. and Scarborough, R. M. (1999). *Tetrahedron Lett.*, **40**, 2665.

28. Sunami, S., Sagara, T., Ohkubo, M., and Morishima, H. (1999). *Tetrahedron Lett.*, **40**, 1721.

29. Smith, E. M. (1999). *Tetrahedron Lett.*, **40**, 3285.

30. Hu, Y. and Porco, J. A., Jr. (1999). *Tetrahedron Lett.*, **40**, 3289.

31. Gutke, H.-J. and Spitzner, D. (1999). *Tetrahedron*, **55**, 3931.

32. Bräse, S. and Schroen, M. (1999). *Angew. Chem. Int. Ed. Engl.*, **38**, 1071.

33. Hanessian, S. and Huynh, H. K. (1999). *Synlett*, 102.

34. Furman, B., Thürmer, R., Kaluza, Z., Voelter, W., and Chmielewski, M. (1999). *Tetrahedron Lett.*, **40**, 5909.

35. Sabatino, G., Chelli, M., Mazzucco, S., Ginanneschi, M., and Papini, A. M. (1999). *Tetrahedron Lett.*, **40**, 809.

36. Eleftheriou, S., Gatos, D., Panagopolous, A., Stathopoulos, S., and Barlos, K. (1999). *Tetrahedron Lett.*, **40**, 2825.

37. Sylvain, C., Wagner, A., and Mioskowski, C. (1999). *Tetrahedron Lett.*, **40**, 875.

38. Cobb, J. M., Fiorini, M. T., Goddard, C. R., Theoclitou, M.-E., and Abell, C. (1999). *Tetrahedron Lett.*, **40**, 1045.

39. Spivey, A. C., Diaper, C. M., and Rudge, A. (1999). *J. Chem. Commun.*, 835.

40. Chiu, C., Tang, Z., and Ellingboe, J. W. (1999). *J. Comb. Chem.*, **1**, 73.

41. Cabrele, C., Langer, M., and Beck-Sickinger, A. G. (1999). *J. Org. Chem.*, **64**, 4353.

42. Brummond, K. M. and Lu, J. (1999). *J. Org. Chem.*, **64**, 1723.

43. Barber, A. M., Hardcastle, I. R., Rowlands, M. G., Nutley, B. P., Marriott, J. H., and Jarman, M. (1999). *Bioorg. Med. Chem. Lett.*, **9**, 623.

44. Hind, N., Hughs, I., Hunter, D., Morrison, M. G. J. T., Sherrington, D. C., and Stevenson, L. (1999). *Tetrahedron*, **55**, 9575.

45. Gibson, S. E., Hales, N. J., and Peplow, M. A. (1999). *Tetrahedron Lett.*, **40**, 1417.

46. Doi, T., Sugiki, M., Yamada, H., Takahashi, T., and Porco, J. A., Jr. (1999). *Tetrahedron Lett.*, **40**, 2141.

47. Hennequin, L. F. and Piva-Le Blanc, S. (1999). *Tetrahedron Lett.*, **40**, 3881.

48. Kobayashi, S., Furuta, T., Sugita, K., Okitsu, O., and Oyamada, H. (1999). *Tetrahedron Lett.*, **40**, 1341.

49. Stieber, F., Grether, U., and Waldmann, H. (1999). *Angew. Chem. Int. Ed. Engl.*, **38**, 1073.

50. Wilson, L. J., Klopfenstein, S., and Li, M. (1999). *Tetrahedron Lett.*, **40**, 3999.

51. Lewandoski, K., Murer, P., Svec, F., and Fréchet, J. M. J. (1999). *J. Comb. Chem.*, **1**, 105.

52. Zaragoza, F. and Stephensen, H. (1999). *J. Org. Chem.*, **64**, 2555.

53. Lohse, A., Jensen, K. B., and Bols, M. (1999). *Tetrahedron Lett.*, **40**, 3033.

54. Linn, J. A., Gerritz, S. W., Handlon, A. L., Hyman, C. E., and Heyer, D. (1999). *Tetrahedron Lett.*, **40**, 2227.

55. Ouyang, X. and Kiselyov, A. S. (1999). *Tetrahedron*, **55**, 8295.

56. Lee, D., Adams, J. L., Brandt, M., DeWolf, W. E., Jr., Keller, P. M., and Levy, M. A. (1999). *Bioorg. Med. Chem. Lett.*, **9**, 1667.

57. Kumar, K. S. and Pillai, V. N. R. (1999). *Tetrahedron*, **55**, 10437.

58. Srivastava, S. K., Haq, W., Murthy, P. K., and Chauhan, P. M. S. (1999). *Bioorg. Med. Chem. Lett.*, **9**, 1885.

59. Khim, S.-K. and Nuss, J. M. (1999). *Tetrahedron Lett.*, **40**, 1827.

60. Meester, W. J. N., Rutjes, F. P. J. T., Hermkens, P. H. H., and Hiemstra, H. (1999). *Tetrahedron Lett.*, **40**, 1601.

61. Singh, R. and Nuss, J. M. (1999). *Tetrahedron Lett.*, **40**, 1249.

62. Sciscinski, J. J., Barker, M. D., Murray, P. J., and Jarvie, E. M. (1998). *Bioorg. Med. Chem. Lett.*, **8**, 3609.

63. Ho, K.-C. and Sun, C.-M. (1999). *Bioorg. Med. Chem. Lett.*, **9**, 1517.

64. Patterson, J. A. and Ramage, R. (1999). *Tetrahedron Lett.*, **40**, 6121.

65. Barlaam, B., Kiza, P., and Berriot, J. (1999). *Tetrahedron*, **55**, 7221.

66. Shey, J.-Y. and Sun, C.-M. (1999). *Bioorg. Med. Chem. Lett.*, **9**, 519.

67. Lepore, S. D. and Wiley, M. R. (1999). *J. Org. Chem.*, **64**, 4547.

68. Chong, P. Y. and Petillo, P. A. (1999). *Tetrahedron Lett.*, **40**, 4501.

69. Limal, D., Semetey, V., Dalbon, P., Jolivet, M., and Briand, J.-P. (1999). *Tetrahedron Lett.*, **40**, 2749.

70. Yu, A.-M., Zhang, Z.-P., Yang, H. Z., Zhang, C. X., and Liu, Z. (1999). *Synth. Commun.*, **29**, 1595.

71. Park, K.-H. and Kurth, M. J. (1999). *Tetrahedron Lett.*, **40**, 5841.

72. Kondo, Y., Komine, T., Fujinami, M., Uchiyama, M., and Sakamoto, T. J. (1999). *Comb. Chem.*, **1**, 123.

73. Combs, A. P., Saubern, S., Rafalski, M., and Lam, P. Y. S. (1999). *Tetrahedron Lett.*, **40**, 1623.

74. Schiemann, K. and Showalter, H. D. H. (1999). *J. Org. Chem.*, **64**, 4972.

75. Scott, R. H., Barnes, C., Gerhard, U., and Balasubramanian, S. (1999). *Chem. Commun.*, 1331.

76. Blettner, C. G., König, W. A., Stenzel, W., and Schotten, T. (1999). *Tetrahedron Lett.*, **40**, 2101.

77. Shaughnessy, K. H., Kim, P., and Hartwig, J. F. (1999). *J. Am. Chem. Soc.*, **121**, 2123.

78. Bräse, S., Dahmen, S., and Heuts, J. (1999). *Tetrahedron Lett.*, **40**, 6201.

79. Wang, G.-P. and Chen, Z.-C. (1999). *Synth. Commun.*, **29**, 2859.

80. Kiselyov, A. S., Eisenberg, S., and Luo, Y. (1999). *Tetrahedron Lett.*, **40**, 2465.

81. Lee, J., Gauthier, D., and Rivero, R. A. (1999). *J. Org. Chem.*, **64**, 3060.

82. Nefzi, A., Ong, N. A., Giulianotti, M. A., Ostresh, J. M., and Houghten, R. A. (1999). *Tetrahedron Lett.*, **40**, 4939.

83. Stephensen, H. and Zaragoza, F. (1999). *Tetrahedron Lett.*, **40**, 5799.

84. Schwarz, M. K., Tumelty, D., and Gallop, M. A. (1999). *J. Org. Chem.*, **64**, 2219.

85. Tumelty, D., Schwarz, M. K., Cao, K., and Needels, M. C. (1999). *Tetrahedron Lett.*, **40**, 6185.

86. Ouyang, X. and Kiselyov, A. S. (1999). *Tetrahedron Lett.*, **40**, 5827.

87. Fotsch, C., Kumaravel, G., Sharma, S. K., Wu, A. D., Gounarides, J. S., Nirmala, N. R., et al. (1999). *Bioorg. Med. Chem. Lett.*, **9**, 2125.

88. Huang, S. and Tour, J. M. (1999). *J. Am. Chem. Soc.*, **121**, 4908.

89. Cottney, J., Rankovic, Z., and Morphy, J. R. (1999). *Bioorg. Med. Chem. Lett.*, **9**, 1323.

90. Brody, M. S. and Finn, M. G. (1999). *Tetrahedron Lett.*, **40**, 415.

91. Hu, Y., Baudart, S., and Porco, J. A., Jr. (1999). *J. Org. Chem.*, **64**, 1049.

92. Vanier, C., Wagner, A., and Mioskowski, C. (1999). *Tetrahedron Lett.*, **40**, 4335.

93. Blettner, C. G., König, W. A., Stenzel, W., and Schotten, T. (1999). *Synlett*, 307.

94. Gosselin, F., Van Betsbrugge, J., Hatam, M., and Lubell, W. D. (1999). *J. Org. Chem.*, **64**, 2486.

95. Huwe, C. M. and Künzer, H. (1999). *Tetrahedron Lett.*, **40**, 683.

96. Richter, H., Walk, T., Holtzel, A., and Jung, G. (1999). *J. Org. Chem.*, **64**, 1362.

97. Pernerstorfer, J., Schuster, M., and Blechert, S. (1999). *Synthesis*, 138.

98. Kung, P.-P. and Swayze, E. (1999). *Tetrahedron Lett.*, **40**, 5651.

99. Brown, R. C. D. and Fisher, M. (1999). *Chem. Commun.*, 1547.

100. Carrington, S., Renault, J., Tomasi, S., Corbel, J.-C., Uriac, P., and Blagbrough, I. S. (1999). *Chem. Commun.*, 1341.

101. Watson, B. T. and Christiansen, G. E. (1998). *Tetrahedron Lett.*, **39**, 6087.

102. Benaglia, M., Cinquini, M., and Cozzi, F. (1999). *Tetrahedron Lett.*, **40**, 2019.

103. Zhao, C., Shi, S., Mir, D., Hurst, D., Li, R., Xiao, X.-Y., et al. (1999). *J. Comb. Chem.*, **1**, 91.

104. Boldi, A. M., Johnson, C. R., and Eissa, H. O. (1999). *Tetrahedron Lett.*, **40**, 619.

105. Rottländer, M., Boymond, L., Cahiez, G., and Knochel, P. (1999). *J. Org. Chem.*, **64**, 1080.

106. Wilhelm, P. and Neuenschwander, M. (1999). *Helv. Chim. Acta*, **82**, 338.

107. Katoh, M. and Sodeoka, M. (1999). *Bioorg. Med. Chem. Lett.*, **9**, 881.

108. Adamczyk, M., Fishpaugh, J. R., and Mattingly, P. G. (1999). *Tetrahedron Lett.*, **40**, 463.

109. Gauzy, L., Le Merrer, Y., Depezay, J.-C., Clerc, F., and Mignani, S. (1999). *Tetrahedron Lett.*, **40**, 6005.

110. Yeh, C. M. and Sun, C.-M. (1999). *Synlett*, 810.

111. Chong, P. Y. and Petillo, P. A. (1999). *Tetrahedron Lett.*, **40**, 2493.

112. Liang, G.-B. and Qian, X. (1999). *Bioorg. Med. Chem. Lett.*, **9**, 2101.

113. Srivastava, S. K., Haq, W., and Chauhan, P. M. S. (1999). *Comb. Chem. High Throughput Screening*, **2**, 33.

114. Goff, D. and Fernandez, J. (1999). *Tetrahedron Lett.*, **40**, 423.

115. Wang, Y. and Huang, T.-N. (1999). *Tetrahedron Lett.*, **40**, 5837.

116. Hoel, A. M. L. and Nielsen, J. (1999). *Tetrahedron Lett.*, **40**, 3941.

117. Kim, S. W., Shin, Y. S., and Ro, S. (1998). *Bioorg. Med. Chem. Lett.*, **8**, 1665.

118. Paris, M., Douat, C., Heitz, A., Gibbons, W., Martinez, J., and Fehrentz, J.-A. (1999). *Tetrahedron Lett.*, **40**, 5179.

119. Page, P., Bradley, M., Walters, I., and Teague, S. (1999). *J. Org. Chem.*, **64**, 794.

120. Ley, S. V., Thomas, A. W., and Finch, H. (1999). *J. Chem. Soc. Perkin Trans.*, **1**, 669.

121. Hall, D. G., Laplante, C., Manku, S., and Nagendran, J. (1999). *J. Org. Chem.*, **64**, 698.

122. Nefzi, A., Ostresh, J. M., and Houghten, R. A. (1999). *Tetrahedron*, **55**, 335.

123. Tremblay, M. R. and Poirier, D. (1999). *Tetrahedron Lett.*, **40**, 1277.

124. Gooding, O. W., Baudart, S., Deegan, T. L., Heisler, K., Labadie, J. W., Newcomb, W. S., *et al.* (1999). *J. Comb. Chem.*, **1**, 113.

125. Hummel, G. and Hindsgaul, O. (1999). *Angew. Chem. Int. Ed. Engl.*, **38**, 1782.

126. Kobayashi, S. (1999). *Chem. Soc. Rev.*, **28**, 1.

127. Hari, A. and Miller, B. L. (1999). *Tetrahedron Lett.*, **40**, 245.

128. O'Donnell, M. J., Delgado, F., and Pottorf, R. S. (1999). *Tetrahedron*, **55**, 6347.

129. Bui, C. T., Maeji, N. J., Rasoul, F., and Bray, A. M. (1999). *Tetrahedron Lett.*, **40**, 5383.

130. Cosquer, A., Pichereau, V., Le Mée, D., Le Roch, M., Renault, J., Carboni, B., *et al.* (1999). *Bioorg. Med. Chem. Lett.*, **9**, 49.

131. O'Donnell, M. J., Delgado, F., Drew, M. D., Pottorf, R. S., Zhou, C., and Scott, W. L. (1999). *Tetrahedron Lett.*, **40**, 5831.

132. Feng, Y., Pattarawarapan, M., Wang, Z., and Burgess, K. (1999). *Org. Lett.*, **1**, 121.

133. Khan, S. I. and Grinstaff, M. W. (1999). *J. Am. Chem. Soc.*, **121**, 4704.

134. Nakamura, K., Hanai, N., Kanno, M., Kobayashi, A., Ohnishi, Y., Ito, Y., *et al.* (1999). *Tetrahedron Lett.*, **40**, 515.

135. Fruchart, J.-S., Gras-Masse, H., and Melnyk, O. (1999). *Tetrahedron Lett.*, **40**, 6225.

136. Kuisle, O., Quiñoá, E., and Riguera, R. (1999). *Tetrahedron Lett.*, **40**, 1203.

137. Katritzky, A. R., Belyakov, S. A., Strah, S., Cage, B., and Dalal, N. S. (1999). *Tetrahedron Lett.*, **40**, 407.

138. Fukase, K., Nakai, Y., Egusa, K., Porco, J. A., Jr., and Kusumoto, S. (1999). *Synlett*, 1074.

139. Kim, D. Y. and Suh, K. H. (1999). *Synth. Commun.*, **29**, 1271.

140. Miyabe, H., Fujishima, Y., and Naito, T. (1999). *J. Org. Chem.*, **64**, 2174.

141. Albert, R., Knecht, H., Andersen, E., Hungerford, V., Schreier, M. H., and Papageorgiou, C. (1998). *Bioorg. Med .Chem. Lett.*, **8**, 2203.

142. Barn, D. R., Bom, A., Cottney, J., Caulfield, W. L., and Morphy, J. R. (1999). *Bioorg. Med. Chem. Lett.*, **9**, 1329.

143. Zheng, A., Shan, D., and Wang, B. (1999). *J. Org. Chem.*, **64**, 156.

144. Poupart, M.-A., Fazal, G., Goulet, S., and Mar, L. T. (1999). *J. Org. Chem.*, **64**, 1356.

145. Brain, C. T., Paul, J. M., Loong, Y., and Oakley, P. J. (1999). *Tetrahedron Lett.*, **40**, 3275.

146. Prabhakaran, E. N. and Iqbal, J. (1999). *J. Org. Chem.*, **64**, 3339.

147. Caldarelli, M., Habermann, J., and Ley, S. V. (1999). *Bioorg. Med. Chem. Lett.*, **9**, 2049.

148. Uozumi, Y., Danjo, H., and Hayashi, T. (1999). *J. Org. Chem.*, **64**, 3384.

149. Adamczyk, M., Fishpaugh, J. R., and Mattingly, P. G. (1999). *Bioorg. Med. Chem. Lett.*, **9**, 217.

150. Carpino, L. A. and Philbin, M. (1999). *J. Org. Chem.*, **64**, 4315.

151. Carpino, L. A., Ismail, G., Truran, G. A., Mansour, E. M. E., Iguchi, S., Ionescu, D., *et al.* (1999). *J. Org. Chem.*, **64**, 4324.

152. Terekhova, G. V., Kolesnichenko, N. V., Batov, A. E., Alieva, E. D., Trukhmanova, N. I., Slivinskii, E. V., *et al.* (1999). *Russ. Chem. Bull.*, **48**, 818.

153. Brummond, K. M. and Gesenberg, K. D. (1999). *Tetrahedron Lett.*, **40**, 2231.

154. Yamamoto, Y., Tanabe, K., and Okonogi, T. (1999). *Chem. Lett.*, 103.

155. Mhiri, C., Gharbi, R. E., and Le Bigot, Y. (1999). *Synth. Commun.*, **29**, 3385.

156. Gómez, C., Ruiz, S., and Yus, M. (1999). *Tetrahedron*, **55**, 7017.

157. Sung, D. W. L., Hodge, P., and Stratford, P. W. (1999). *J. Chem. Soc. Perkin Trans.*, **1**, 1463.

158. Sellner, H. and Seebach, D. (1999). *Angew. Chem. Int. Ed. Engl.*, **38**, 1918.

159. Damle, S. V., Patil, P. N., and Salunkhe, M. M. (1999). *Synth. Commun.*, **29**, 1639.

160. Minutolo, F. and Katzenellenbogen, J. A. (1999). *Angew. Chem. Int. Ed. Engl.*, **38**, 1617.
161. Zhang, T. Y. and Allen, M. J. (1999). *Tetrahedron Lett.*, **40**, 5813.
162. Bussolari, J. C., Rehborn, D. C., and Combs, D. W. (1999). *Tetrahedron Lett.*, **40**, 1241.
163. Parlow, J. J., Case, B. L., and South, M. S. (1999). *Tetrahedron*, **55**, 6785.
164. Weidner, J. J., Parlow, J. J., and Flynn, D. L. (1999). *Tetrahedron Lett.*, **40**, 239.

Chapter 8

Analytical methods in combinatorial chemistry

Bing Yan, Yen-ho Chu,[†] Michael Shapiro,[†]
Ramsay Richmond,[§] Jefferson Chin,[†] Lina Liu,[†]
and Zhiguang Yu[†]

Axys Advanced Technologies, 385 Oyster Point Blvd., South San Francisco, CA 94080, USA.

[†]Novartis Pharmaceuticals Corporation, 556 Morris Avenue, Summit, NJ 07901, USA.

[†]Department of Chemistry, Ohio State University, Columbus, Ohio 43210, USA.

[§]Core Technologies Area, Novartis Pharma AG, CH 4002, Basel, Switzerland.

1 Introduction

This chapter describes an array of techniques successfully applied to the optimization of solid phase organic synthesis (SPOS) (1, 2), quality control (QC) of combinatorial libraries, and in the selection of active lead compounds. Since low quality of the final library is largely attributable to poorly developed chemistry at the optimization stage, investing more time and effort in reaction optimization is crucial in combinatorial synthesis. QC data gives confidence on both the molecular diversity generated and the subsequent screening results. While the screening of pure compounds is a well-developed routine operation, the screening of compound mixtures is more difficult and this will be covered at the end of this chapter. Various analytical techniques are described below in the order encountered in a combinatorial synthesis process.

2 Analytical methods in reaction optimization

Organic reactions on solid supports such as polystyrene (PS), polystyrene-polyethylene glycol (PS-PEG), and modified polymeric surfaces often require conditions different from those used in solution. Before a compound library can be made (usually in weeks), a time-consuming (likely in months) solid phase reaction optimization on a small number of individual compounds is required.

The process of cleaving and analysing the product in order to assess the progress of a solid phase synthetic scheme is destructive and time-consuming. Furthermore, some synthetic intermediates may not be stable to the cleavage conditions. Therefore, on-support analytical methods including FTIR gel phase and magic angle spinning (MAS) NMR, and spectrophotometric methods have been developed.

2.1 FTIR methods

The most widely used on-support method for routine monitoring of SPOS is FTIR (1–3). In particular, single bead FTIR microspectroscopy (4, 5) (see *Protocol 1*), which performs a non-destructive analysis on a single bead, is advantageous in both sensitivity and speed. This analysis offers a wide range of information required for reaction optimization such as qualitative 'yes-or-no' answers, quantitative percentage of conversion, kinetics, and comparison or selection of resin supports and reaction conditions.

Figure 1 shows IR spectra taken at various times on PS and PS-PEG resins for the reaction depicted in *Scheme 1* (6). The disappearance of bands at 3480 (PS-PEG) and 3450 cm^{-1} (PS), and the formation of the product carbonyl bands at 1732/1716 cm^{-1} indicate the completion of the reaction. Integration of these bands against time can be fitted to a pseudo first order rate equation with a rate constant of 2.2×10^{-4} s^{-1} (data not shown). In this reaction, the rates on PS and PS-PEG resins are similar.

Protocol 1

Single bead FTIR measurement

1 The required instrument is a FTIR spectrometer coupled with a microscope accessory, which is equipped with an IR objective and a liquid nitrogen-cooled mercury–cadmium–telluride (MCT) detector.

2 Place resin beads (five to ten beads) in the centre of the diamond window and sandwich the sample with another diamond window. For fragile beads (such as PS-PEG resins), go directly to step 5.

3 Screw the cap down carefully to tighten the sandwiched diamond windows.

4 Unscrew the cap and remove the top diamond window and place the one with the flattened beads on the microscope stage.

5 Centre one bead on the cross hairs of the reticle under the microscope, and refocus to a sharp image.

6 Collect the sample spectrum, remove the bead and collect the diamond cell background spectrum.

A low-cost beam condenser (7) (see *Protocol 2*) can be used to obtain similar information from the analysis of 50–100 beads. The reaction of succinimidyl 6-(*N*-(7-nitrobenz-2-oxa-1,3-diazo-4-yl)amino)hexanoate with Wang resin (*Scheme 2*) produces compound **3** (*Figure 2*). The disappearance of the starting material bands at 3577 and 3458 cm^{-1} with time coincides with the formation of the

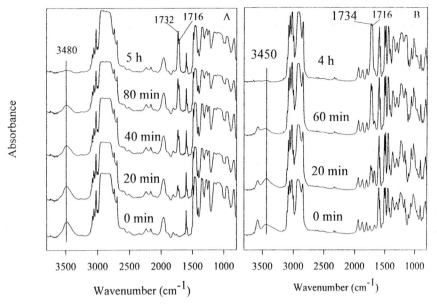

Figure 1 IR spectra taken from a single bead (A) or a single flattened bead (B) at various times during the reaction depicted in *Scheme 1* on TentaGel (A) or PS-based (B) resins.

product bands at 1732, 1580, and 3352 cm^{-1}. The completion of the reaction can be concluded with confidence by observing the disappearance of the hydroxyl bands at 3577 and 3458 cm^{-1} (*Figure 2*). The area integration of both the disappearing bands at 3577 and 3458 cm^{-1} and the emerging band at 1732 cm^{-1} were plotted against time and the time course was fitted to a pseudo first order rate equation yielding a rate constant of 3.8×10^{-4} s^{-1} (data not shown).

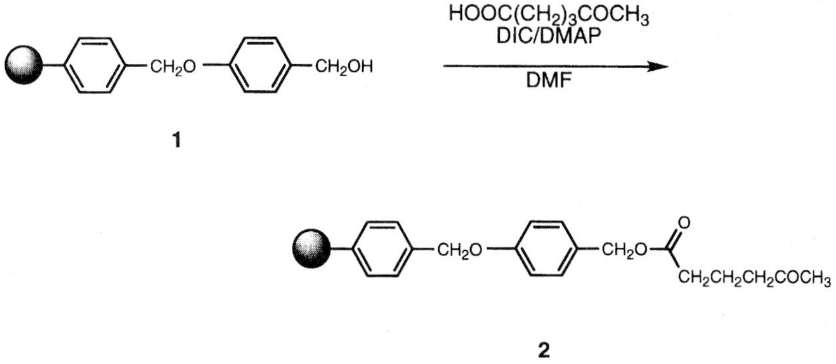

Scheme 1

With these methods, rapid analysis of trace amounts of sample reveals qualitative, quantitative, and other reaction optimization-related information directly on the solid support (3). Many other FTIR methods have also been proven useful for solid phase samples. These methods include DRIFT (8), photoacoustic (9), micro- (10) and macro-ATR (11) methods.

Protocol 2

Beam condenser FTIR measurement

1 The required instrument is a FTIR spectrometer equipped with a beam condenser accessory with a diamond window.

2 Place resin beads (50–100 beads) on a diamond window and sandwich the sample with another diamond window.

3 Screw the cap down to tighten diamond windows.

4 Place the diamond cell in the beam condenser mount, cover the sample compartment, and collect a sample spectrum.

5 Take the diamond windows out and clean them. Place them back in the mount, cover the sample compartment, and collect a background spectrum.

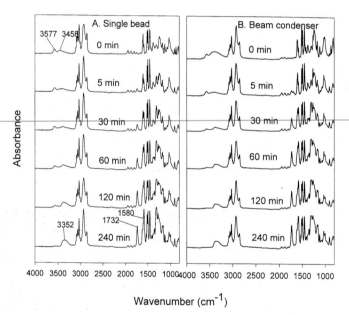

Figure 2 IR spectra of the reaction in *Scheme 2* obtained by single bead FTIR (A) and beam condenser FTIR at various time intervals (B).

2.2 Gel phase NMR

The largest obstacle to NMR analysis of compounds on a resin is the broad line in solids NMR spectra, particularly in proton spectra. The line broadening within the NMR data primarily arise from the restricted mobility of the pendant compound, the heterogeneity of the sample (magnetic susceptibility), and dipolar coupling. The mobility of the pendant compound is dependent upon the resin swelling and the tether length. The polymer matrix causes a magnetically

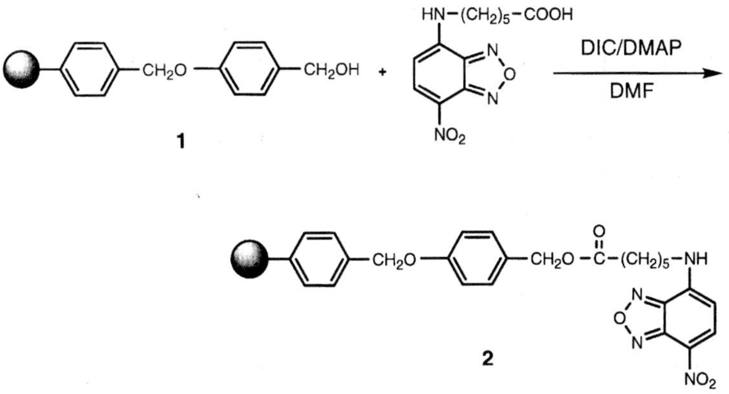

Scheme 2

inhomogeneous environment throughout the sample making it difficult to shim, which in turn causes linewidths broadening.

Gel phase NMR (see *Protocol 3*) usually refers to experiments on the solvent swollen resins using a standard liquid NMR probe. The narrow chemical shift range and the broad linewidths that are observed within the proton spectrum complicate ^{1}H NMR data interpretation. However, the utility of gel phase NMR is found with nuclei such as ^{13}C, ^{19}F, ^{31}P that have wider chemical shift ranges, and therefore, linewidths are less important. Because gel phase NMR utilizes a conventional liquids probe, hardware requirements are minimal and implementation into an analytical laboratory can be readily accomplished.

Protocol 3

General procedure for gel phase NMR

General considerations for resin analysis by gel phase NMR

1 Choice of resin. The quality of the NMR analysis is dependent upon the mobility of the compounds of interest. The degree of mobility depends upon the length of the tether that attaches the solid phase matrix to the cleavable linker. Data collected from PS-PEG resin, which has a long tether, shows better linewidths than the Merrifield resin.

2 Choice of solvent. The swelling of the resin varies based upon the solvent used. General purpose solvents for swelling of resins are DMF-D$_7$ or CD$_2$Cl$_2$. The majority of the resins used will shrink in the presence of alcohols.

3 The resin should be handled as a dry solid. The tendency for the resin to stick to glass increases as the resin becomes solvated.

Protocol 3 continued

Method

1   Transfer dry resin (40–100 mg) into a NMR tube.

2   Slowly add deuterated solvent (0.4–0.7 ml), such that the solvated resin will be within the probe's coil region. An exact quantity of resin and solvent needed depends upon the spectrometer and probe used.

3   Acquire NMR data.

2.3 Magic angle spinning (MAS) NMR

Proton spectra are crucial for the monitoring of reactions and in structural elucidation. MAS NMR (see *Protocol 4*) has been applied towards the analysis of resin samples to reduce the NMR linewidths. The magic angle (54.7 from the Z-axis, see *Figure 3*) spinning at 2–5 kHz allows magnetic susceptibility-induced line broadening terms to average out, resulting in narrower linewidths. This technique requires the use of a cross-polarization (CP)-MAS (solid state probe), or a dedicated high resolution (HR)-MAS probe. A pneumatics system is also required in order to spin the sample at a typical MAS speed (2–5 kHz). A cylindrical rotor (*Figure 3*) is used to contain the sample. Spherical inserts for the rotors have been designed to reduce the required sample volume and provide a liquid tight seal.

Protocol 4

General procedure for MAS NMR spectroscopy

General considerations

1   Spectra of the unmodified resin should be obtained as reference data.

2   The choice of spinning speed depends upon the compounds of interest and resin used, and is chosen so that spinning side bands do not obscure the area of interest. Alternatively, data can be collected with variable spinning speeds allowing for elimination of side bands.

3   The 90° pulsewidth should be checked with each sample.

4   The CPMG spinlock time (generally 70 msec) should be varied as the attenuation of the resin resonance may change.

5   MAS NMR will not distinguish whether the proton resonance from compounds are attached to the resin or entrapped within the resin matrix and leach out upon solvation.

Protocol 4 continued

6 Solution spectra of the solvent for the swelling of the resin should be acquired.

7 Different solvents should be tried if the initial results are inconclusive.

Method

1 Transfer the dry resin (1–2 mg) into a MAS rotor that contains a spherical insert.

2 Add the deuterated solvent, typically DMF-d_7 or CD_2Cl_2 (10–20 μl).

3 Seal the rotor.

4 Shim the MAS probe. Shimming a MAS probe differs significantly from the high resolution solution probes. Since the MAS probe's coils are tilted, the X or Y shims have a greater impact on shimming and the Z shims influence is diminished. The position of the stator within the magnet contributes also to the shimming. If properly aligned within the magnet, shimming the probe can be accomplished with only one of the linear shims (X or Y) while the others can be set to zero.

5 Acquire spectra.

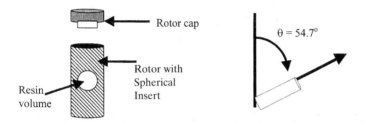

Figure 3 Diagram of the MAS rotor with spherical insert (*left*). Pictorial representation of a rotor spinning at the magic angle (*right*).

Proton spectra obtained from resin samples are generally dominated by resonances that arise from the solid support. The reduction of the resin resonance can be accomplished using solvent suppression techniques. Alternatively, the Carr–Purcell–Meiboom–Gill (CPMG) spinlock $90° - (\tau - 180° - \tau)_n$ can be used for the attenuation of the resin resonance (*Figure 4*) (12).

The acquisition of a simple 1D proton spectrum may not provide sufficient information because the solid phase matrix may still obscure spectral regions. Although methods have been developed to further decrease the linewidths, coupling information is still lost. The reintroduction of the coupling information can be achieved by the use of the 2D J-resolved (13) experiment (*Figure 5*). For data collection, the F2 dimension was set up with a 4K by 64 data matrix with 32–64 scans per increment. The data was processed with an unshifted sine multiplication in both dimensions. The processed data should

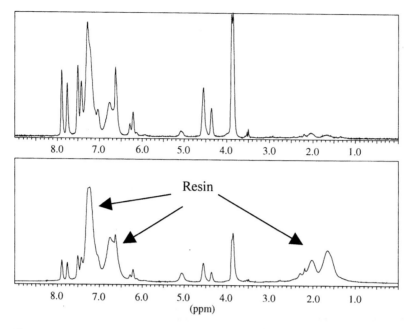

Figure 4 1D proton spectra of Fmoc-amide Rink resin (1.5 mg) in CD_2Cl_2: 4000 Hz. Top spectrum: CPMG experiment, 70 msec echo time. Bottom spectrum: single pulse.

not be tilted to retain the coupling information within the F2 dimension projection. The spin echo correlated spectroscopy (SECSY) experiment has also been revisited as a technique that yields spin system connectivity as well as a resolution enhanced chemical shift projection of the proton spectrum (14). The traditional 2D experiments that are used for structure elucidation of small molecules can also be applied to resin-bound compounds. Total correlation spectroscopy (TOCSY) experiment has been used for spin system connectivity, and heteronuclear single quantum correlation (HSQC) and heteronuclear multiple bond correlation (HMBC) for ^1H–^{13}C information. Nuclear Overhauser effect spectroscopy (NOESY) experiment has been applied to structural conformation determination (15, 16). Experimental parameters as well as data processing are similar to those used in solution experiments. Incorporation of gradient technology with the HR MAS probe has allowed the experiments to be collected with reduced acquisition time and spectral noise.

2.4 Spectrophotometric methods

In solid phase peptide synthesis (SPPS) research, spectrophotometric methods for quantitation of primary and secondary amine groups have been developed to monitor amide bond couplings. SPOS, however, involves many diverse organic reactions and deals with various functional groups. Quantitative methods are, thus, needed to guide the reaction optimization effort. One

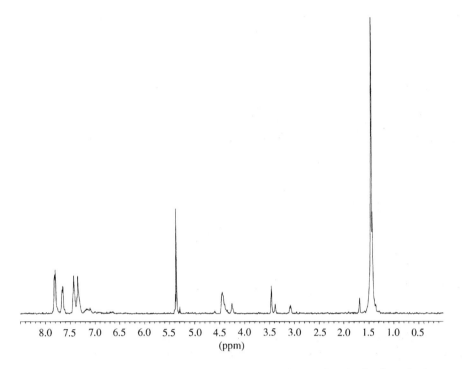

Figure 5 Fmoc-Lys (Boc) Wang resin 1.5 mg in CD_2Cl_2: 5000 Hz. F2 projection from the J-resolved experiment.

approach is to use molecular markers that react specifically with a particular functional group. Their disappearance from the supernatant solution can be monitored by UV-Vis or fluorescence spectroscopy. Based on this principle, methods for determination of absolute amounts of aldehyde and ketone (17) (see *Protocol 5*), hydroxyl and carboxyl groups (18) have been developed. These analyses require 5–10 mg of resin and usually take 1–2 h.

Protocol 5

Quantitative determination of aldehyde and ketone groups on resins

1   Add 3–11 mg (1.8–6.5 μmol) of weighed beads (polystyrene control and aldehyde beads) in separate 1 ml Supelco filtration tubes.

2   Swirl the sealed tubes with 0.5 ml of DMF on an orbital shaker for 20 min. Drain the resin on a visiprep vacuum manifold.

3   Add 200 μl dansylhydrazine in DMF to each tube (approx. twofold molar ratio relative to the loading of the resin).

Protocol 5 continued

4   Rotate the mixtures (control and sample) on a Glas-Col Laboratory rotator (32 r.p.m.) at room temperature for 30 min.

5   Add 5 μl of supernatant from each reaction tube to 2 ml DMF in a cuvette for UV-Vis or fluorescence measurements.

6   The loading of aldehyde groups on the resin can be calculated based on the consumed amount of dansylhydrazine in the supernatant after subtracting the amount of non-covalently associated dye molecules in the control beads.

3 Quality control of libraries from parallel synthesis

The size of combinatorial libraries is often too large to allow the characterization of every library member. It may be argued that only an examination of the active compounds is necessary since an exhaustive characterization is not only expensive, but also associated with an excessive amount of data. However, without the analytical data, it is impossible to determine whether the lack of active compounds means that the tested compounds were not active or the synthesis failed in the first place. Therefore, it is important to confirm the desired randomness and diversity of a library. Although biological assays can be carried out directly on beads, solution assays are more popular for screening discrete compound libraries. When a solution assay is the desired approach, the characterization of library members should be performed after cleavage from the solid support. Therefore, the sample for QC analysis is in a form similar to libraries made by high-throughput solution phase synthesis.

3.1 MS analysis

Discrete compound libraries obtained from parallel synthesis are well suited for high-throughput MS analysis, in contrast to split and mix libraries. First, multiple parallel synthesis should generate a single end-product with a known molecular weight (MW) in a specific rack well. Secondly, these end-products are delivered for analysis in solution and in standardized cassettes suitable for automated laboratory handling, e.g. 96-well plates. Thirdly, sensitivity is rarely a problem for MS from the amount of compounds generated by combinatorial synthesis.

The new generation of atmospheric pressure ionization LC/MS interfaces developed in the early 1990s, such as electrospray (ESI), proved to be timely technologies for rapidly providing molecular weight information for diverse samples. The only challenge is the daunting high-throughput needed (e.g. 50 000 to 100 000 analyses per year) and the resulting 'data interpretation fatigue' problem.

Automated open-access flow injection analysis (FIA) MS system (see *Protocol 6*) has been developed for the QC of combinatorial libraries (19). Using this system, an analysis of compounds from a 96-well plate can be completed in three hours.

Three tasks are automatically accomplished in each run: the confirmation of the presence of the expected compound, the detection of any inter-sample carry-over, and the automated reporting of results to the chemist's bench. Enhancements to the FIA-MS system (20) were specifically designed to help the combinatorial chemist improve the end-product purity by pin-pointing the synthetic inefficiencies. A 3D spectral map of the entire combinatorial chemistry rack assists in reaction optimization by colour highlighting contamination ions. The size of some large libraries originally deterred full characterization (in the mid 1990s) in a reasonable time period. In this situation, analysis of a statistically significant subset of the combinatorial library by FIA-MS is done prior to biological testing. However the recent development of sub-minute duty cycles for FIA-MS can now support complete or near complete analytical surveys of even large libraries.

Protocol 6

High-throughput flow injection analysis (FIA) MS

1 Generally a single quadrupole mass spectrometer with an electrospray interface is recommended.

2 Multi 96-well rack autosamplers can be used for overnight and weekend analysis. Normally 2–10 μl of sample are introduced into the FIA solvent stream by the autosampler.

3 Microsoft Excel files describing each rack are deposited by the combinatorial chemist in a directory that the MS operator can access. Script files (e.g. PERL) then automatically build up the analysis methods and set up the MS parameters, e.g. chemist's name, rack's name, well locations of compounds with their formulae and expected monoisotopic molecular weight, and adduct m/z positions.

4 Typical MS operating conditions are electrospray ionization positive ion mode; Finnigan heated capillary temperature, 220 °C; conversion dynode at 15 kV; electron multiplier at 0.9 kV, a collision-induced dissociation offset of 0.0 V, and a spray voltage of 4.5 kV. This maximizes the collection of ion current in predictable gas–adduct m/z positions rather than in unpredictable fragmentation positions.

5 The FIA solvent is usually CH_3CN/H_2O mixture (e.g. 70:30, v/v), with a flow rate of 0.05–0.10 ml/min. Samples are also usually dissolved in this solvent mixture.

6 ESI fused silica capillary of dimensions 100 μm i.d. rather than 50 μm i.d. should be used to avoid clogging.

7 A scan range of 160–1560 m/z is recommended, with centroid data acquisition. This range is wide enough to include any gas dimers for purity estimation. This range avoids three occasional but highly ESI responding contaminants, i.e. [DMSO gas dimer + H]$^+$, [triethylamine + H]$^+$, [triethylamine + H + CH_3CN]$^+$, and [N-ethyldiisopropylamine + H]$^+$.

Protocol 6 continued

8 A program processes the data and attributes the ion current to the expected end-product with purity estimate. The Rack Viewer Visual Basic application on the chemists' workbench PC accesses these results with colour codes. With respect to the success or failure of the synthesis, a green suggests OK, amber possibly OK, and red not OK.

3.2 MS-guided purification

Although the parallel synthesis (one compound per well) strategy eliminates the deconvolution steps necessary when pooled libraries are screened, parallel synthesis products are usually screened as crude mixtures because purification slows down the process of lead discovery. However, screening crude products can produce misleading results due to active side-products, unreacted material, or synergistic interaction of multiple components. Data from 'active' wells must be re-evaluated after resynthesis and purification, and data from 'inactive' wells is unreliable. Although a total purification approach may not be necessary for lead finding, screening pure compounds is the only sure route to get immediate, reliable structure–activity relationships. Therefore, it is necessary to purify individual products obtained from parallel synthesis.

The rapid separations can be performed at an elevated flow rate of 5 ml/min and a short column (10 mm × 5 mm C-18) (21). While use of a shorter HPLC column length sacrifices theoretical plates, the vast majority of separations involving crude reaction mixtures do not require high efficiency. The reduction in column length with an increased flow allows for sharply increased throughput. An automated parallel analytical/preparative HPLC/MS workstation has been developed to increase the analysis speed for characterizing and purifying combinatorial libraries (22). The system incorporates two columns operated in parallel for both LC/MS analysis and preparative LC/MS purification. This system has a multiple sprayer IonSpray interface to support flows from multiple columns and a valving configuration, which is under complete software control for delivering the crude samples to the two HPLC columns from a single auto-sampler. By incorporating fast chromatography and mass-based fraction collection on the system, more than 200 compounds can be characterized per instrument per day, and greater than 200 compounds can be purified overnight.

3.3 High-throughput NMR

Although HT-MS has had a major impact on the characterization of large volumes of compounds from combinatorial libraries, NMR has not been used in a similar fashion. The reasons are:

(a) The samples are normally dissolved in a protic solvent which generates large background signals that complicate the NMR spectra.

(b) The volumes of solute are often too small for conventional NMR spectroscopy (< 500 μl). The extreme case is a 'one-bead one-compound' library (see Chapters. 1–3).

(c) The samples are often in a 96-well microtitre plate format which could not be used directly by regular NMR probes.

(d) Conventional NMR analysis is too slow.

The development of the HPLC-NMR probe, the solvent suppression methods, and the automated liquid handling systems have made the first generation high-throughput NMR (HT-NMR) possible. HT-NMR (23) or tubeless NMR is an extension of HPLC-NMR, being assembled by the removal of the HPLC system and the incorporation of a liquid handling system. The liquid handling system allows the samples to be taken from and returned to a 96-well plate. The deuterated solvent is no longer required with the use of solvent suppression methods. The samples are transferred via capillary tubing to the NMR spectrometer and then pumped back out after data acquisition. The sample analysis takes 2–3 min. However, the generation of a large number of spectra has made the data interpretation a major bottleneck.

4 Analysis of compounds from pooled libraries

Combinatorial synthesis using split and pool protocol (24–26) produces one-bead one-compound libraries (Chapters 1–3). Although discrete compounds are actually made at the single bead level, the analysis of a mixture is encountered when the library is to be analysed as a pool. Characterization of a pooled library is not trivial. Any destructive analytical method will not be useful since theoretically each bead carries one unique compound. The huge number of compounds in the final library made by this method precludes the possibility of full characterization. The focus has to be on the synthetic route optimization and the characterization of small size trial libraries.

In combinatorial synthesis, the formation of by-products during the synthesis cannot be completely avoided. Major reasons for the occurrence of by-products, for example, are incomplete coupling or deprotection steps and undesirable oxidation by oxygen in air. Since the reaction is applied not to a single compound but to an entire population of different molecules, there is always the potential of encountering some selectivity toward the generation of certain molecules at the expense of others. Impurities or missing components in library mixtures used in biological assays can lead to misinterpretation of the results. Therefore, the quality control of pooled libraries must accomplish two tasks:

(a) To confirm whether all of the expected compounds were produced during the combinatorial synthesis.

(b) To identify by-products. So far, analytical methods are highly limited for such analysis and only small size libraries can be analysed by current methods.

4.1 **MS analysis**

ESI-MS and matrix-assisted laser desorption–time of flight mass spectrometry (MALDI-TOF MS) produce essentially fragmentation-free spectra with a pseudo-molecular ion peak for every component and have been used for mixture analysis. For example, ESI-MS was used to determine if the combination of a xanthene acid chloride core with a set of amines results in the formation of a mixture in which all of the expected compounds are present above a certain concentration threshold. Direct observation of 65 341 compounds in the final library was out of question. Therefore, smaller yet representative trial libraries were constructed. In a library of 55 expected compounds, 43 (78%) were detected. Missing were predominantly those compounds which contained the arginine-*p*-benzyl-amide building block (27). With the MS data in hand, one can be confident that the synthetic methodology produced highly diverse libraries of well-defined composition, and one is able to put faith in the results of subsequent screening assays.

For a medium size library, the severe overlap in the MS spectrum precludes distinction between individual members. It is useful to use a computer simulated spectrum as a comparison. However, the ionization efficiency of components in a mixture is not predictable and the ionization of some may be partially or completely suppressed, even though the same compounds when analysed alone give intense molecular ions. Qualitatively, the presence of side-products can be immediately confirmed if there are peaks outside the envelope formed by the peaks which originate from the lowest and the highest MW molecular ions in a MS spectrum.

4.2 **Other methods**

4.2.1 HPLC-NMR

This method couples the separation power of chromatography with the structural information from NMR. The requirements for HPLC-NMR are a HPLC system, a LC-NMR probe, and the sampling unit that directs the effluent to either the probe, storage cells, or to solvent waste. The HPLC system consists of the mobile phase pumps and a HPLC detector (typically an UV or diode array detector). The LC-NMR probe is available in a variety of configurations depending upon the user's primary work. This technique can only analyse very small mixtures limited by HPLC separation capacity. The utility of HPLC-NMR for routine analysis of combinatorial libraries still needs to be established (see Section 3.3).

4.2.2 LC-MS

The combination of HPLC with MS divides a complicated mixture library into many small mixture groups and identification of compounds becomes easier. Different LC/MS interfaces are now available to cope with a variety of compound types and solvent systems. Similar to HPLC-NMR, only small mixtures can be analysed due to the limited separation power of HPLC. LC-MS is widely used in analysing and guiding the purification of discrete compound libraries and small mixture libraries (28) (also see Section 3.2.).

4.2.3 CE-MS

Capillary electrophoresis is a rapid, high resolution method for separating compounds with different electrophoretic mobilities. Small diameter fused capillary has a bare silica inner wall, which is negatively charged due to the silanol groups. An excess of positive charges in the electrophoretic buffer will form a layer and as a high voltage is applied to the capillary, electro-osmotic flow (EOF) drive the analytes toward the negative cathode, regardless of their charge. Charged species are separated based on their m/z differences. CE is mostly suitable for compounds containing groups with diverse acid/base properties. Application of CE before mass spectrometric detection provides fast and efficient separation of library components for their subsequent characterization by MS. In CE analysis, only nanolitres of sample are injected for each analysis. This can be a tremendous advantage when extremely small amounts of sample are available as is typical with combinatorial chemistry. In general, CE is a complementary technique to HPLC.

4.2.4 High resolution MS

In a combinatorial mixture, some components may have nearly identical masses that cannot be distinguished by conventional mass spectrometers. For example, the substitution of lysine for glutamine in a peptide yields a molecular weight difference of 0.036 mass units. ESI Fourier transform mass spectrometry (FTMS) can be used to address such common challenges in characterizing these complex mixtures. For example, a phenylalanine-substituted octapeptide ion at m/z 964 had a measured mass (963.5346) that was in error only by 17 ppm (29). The utility of Fourier transform ion cyclotron resonance (FTICR) mass spectrometry for analysis of selected small peptide combinatorial libraries containing 10^2 to 10^4 individual components was demonstrated (30). The advantages of the FTICR technique over conventional mass spectrometric methods are ultra-high mass resolution, high mass accuracy, multi-stage tandem mass spectrometry (MS^n), and the ability to trap ions for extended periods of time.

4.2.5 Tandem MS

Another way to get information on the composition of a mixture is to use tandem MS. A tandem mass spectrometer consists of an ion source, two analysers separated by a fragmentation region, and an ion detector. The parent ion undergoes collisionally activated dissociation (CAD) through collisions with neutral gas molecules in the fragmentation region to yield various daughter ions. Subsequent mass analysis of the daughter ions by the second mass analyser permits identification of the separated components. The ionization of the mixture produces ions characteristic of the individual components. This analysis can distinguish components with the same molecular mass by looking at their fragmentation pattern. The parent ion and neutral loss scans can be used to screen for a class of compounds that fragments to yield a common sub-structure or for a class of compounds characterized by a common fragmentation pathway. The various tandem MS techniques have been used in the analysis of small mixture libraries (27).

4.2.6 High-throughput quantitation

Quantitation of trace amounts of pure or impure solution samples is a challenge. The commonly used UV detector for HPLC systems is not quantitative. The chemiluminescent nitrogen detection (CLND) (31, 32) and the evaporative light scattering detection (ELSD) (33) in the flow injection mode or coupled with HPLC are better quantitative methods for samples from combinatorial chemistry. Using both analysis modes, the CLND produced a linear response from 25–6400 pmol of nitrogen for a set of diverse compounds. Over the entire linear range, the absolute response exhibits an average error of approximately ± 10% among the compounds. In addition, the response was independent of mobile phase composition. These results demonstrate that the CLND can be used in both modes for rapid and accurate quantitation down to low picomole levels, using a single external standard. The ELSD detector's response is also independent of the sample chromophore, which makes it a better method for the QC of combinatorial chemistry samples compared to UV detection. Furthermore, the ELSD exhibits a nearly equivalent response to compounds of similar structural class. A rapid quantitation of compound libraries may be carried out with the use of a single external standard. The quantitation errors for amino acids by normal phase HPLC with ELSD is on average ± 10%.

5 The screening of pooled combinatorial libraries

Combinatorial synthesis can produce large compound libraries. High-throughput screening methods must be combined with the strength of synthesis so that large pools of compounds can be searched rapidly to find lead compounds for a variety of biological targets. In-solution assays are the major methods for the selection of lead compounds from discrete libraries. The success of these assays relies on the screening of well-defined pure components because of the ambiguity resulting from the screening of mixtures. The focus of this section will be on methods used to screen pooled combinatorial libraries rather than the routine screening of pure compounds.

5.1 On-bead screening

5.1.1 MALDI-MS for identifying lead compounds

Screening a large combinatorial library can be achieved by carrying out an assay directly on bead-bound compound library with a tagging (34) or encoding scheme. One of the encoding methods is partial chain termination. Partial chain termination synthesis followed by MALDI-MS decoding was first developed by Youngquist *et al.* to search for peptide ligands against an anti-gp 120 monoclonal antibody (35). This method has been applied for a variety of receptors, such as searching epitopes for Lyme diagnosis (36). After synthesis, each bead contains a major peptide product (complete sequence, ~ 50 pmol) and some sequentially terminated shorter peptides (5–10 pmol). The solid phase peptide library can be screened by enzyme-linked immunosorbent assay (ELISA, see *Protocol 7*), which can detect specific interactions between the antibody and the resin-bound compounds. Once the positive beads are identified, they are rinsed, and the peptides released from each resin bead individually, and then submitted to MALDI-MS analysis (*Protocol 8*, *Figure 6*).

Protocol 7

Enzyme-linked immunosorbent assay for identifying ligands

Reagents

- 1 × PBS: 2.5 mM Tris, 13.7 mM NaCl, 0.27 mM KCl pH 7.9
- 2 × PBS: 16 mM Na_2HPO_4, 3 mM KH_2PO_4, 247 mM NaCl, 5.4 mM KCl pH 7.4
- 1 × PBST: 2 × PBS diluted onefold with 0.1% Tween 20 in water
- Blocking buffer: 2 × PBS with 0.05% Tween 20 and 1% gelatin

A. Control ELISA experiments to exclude false positives

1 Wash resin twice with water (deionized and distilled) and trice with 1 × PBST buffer.

2 Block the resin with 0.1% gelatin in 2 × PBS buffer pH 7.2 for 2 h with agitation; then drain.

3 Incubate resin with diluted streptavidin–alkaline phosphatase conjugate (such as 1:2500 dilution) in the blocking buffer for 2 h with agitation.

4 Wash with 1 × PBST trice, 2 × PBS twice, and 1 × TBS once to gradually bring the pH to 8.00.

5 Add a standard substrate solution of BCIP/NBT (5-bromo-4-chloro-3-indolyl-phosphate/nitro-blue tetrazolium) to resin and incubate with agitation for 30 min. Any coloured beads are false positive. They can be picked up under a microscope and should be discarded.

6 Regenerate the library by washing with 6.0 M guanidine–HCl pH 1.0 three times.

B. Identify real ligands from regenerated library

1 The screening procedure is the same, except that the target monoclonal antibody is incubated with the beads immediately after the blocking step. Coloured beads are positive lead compounds and need to be decoded by MALDI-MS.

Protocol 8

Decoding the ligands by single bead analysis using matrix-assisted laser desorption and ionization mass spectrometry

1 Prepare cyanogen bromide solution (100 mg/ml in 70% formic acid). Cyanogen bromide solution (20 μl) is used to cleave peptides on a single bead in an Eppendorf tube. The reaction proceeds for 16–24 h in the dark with agitation.

2 Freeze and lyophilize the residue in the tube to remove excess cyanogen bromide. Dissolve the peptide in 20 μl of 0.1% (v/v) aqueous TFA.

3 Place small amount of α-cyano-4-hydroxy-cinnamic acid in 50:50 0.1% aqueous TFA/ acetonitrile, vortex, and centrifuge to make a saturated matrix solution (dihydroxy-benzene can also be used as the matrix). However, this matrix requires higher laser energy and gives rise to a spectrum of lower resolution. Dihydroxybenzene is well suited for analysis of small molecules due to the clean background at low m/z.

4 Calibrate mass spectrometer using at least two standards with known molecular weights covering the mass range of interest, such as substance P, leucine-enkephalin, and renin. The standard solutions (~ 1.0 g/litre) are mixed with the matrix solution in 1:1 volume ratio (0.5 ml: 0.5 ml) on a sample plate and air dried. After desorption using a 337 nm pulsed N_2 laser and extraction at 25 kV, spectra from 50–100 pulses were summed, and the centroids of the unsmoothed data were determined at 50% peak height. Input the molecular weights of the standards. Use computer software to calibrate the instrument.

5 Mix the unknown bead with the matrix solution (0.5 ml: 0.5 ml) on the sample plate. Record the mass spectrum. The terminated peptides will show a mass ladder pattern (see *Figure 6*). The molecular weight differences reveal the amino acid sequence at each randomized position.

5.1.2 Fluorescence-activated bead sorting

Protein binders can be identified as fluorescent beads by reacting the bead-bound split library with a fluorescence labelled target protein. These beads are isolated by a fluorescence-activated cell sorting instrument (FACS) or fluorescence-activated bead sorting instrument (FABS). Ligands that bind to mAbD32.39 (37) and SH2 domains (38) have been identified using this approach.

5.1.3 Chemical tagging

After active compounds are selected from the on-bead assay of combinatorial libraries, the identity of these compounds must be revealed by various decoding procedures (34). Chemical codes, which are added to the synthetic beads to record the synthetic history during the library synthesis, are 'read' by methods such as GC (48) and HPLC (49). Tagging is an important strategy to deal with the

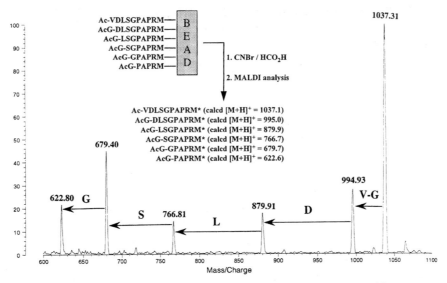

Figure 6 MALDI mass spectrum of peptides (M*, homoserine lactone) cleaved from a single bead. This bead was recognized as H9724 and contained the peptide sequence Ac-VDLSGPAPRM. Note that only 5% of the peptides from a single bead were used for sequence decoding.

'finding a needle in the haystack' problem. However, chemical tagging requires nearly doubled synthetic steps and the extra caution for mutual interference between the tag and the synthesis. Radiofrequency tagging (50, 51) does not have these problems (Chapter 4). It will become more useful if the tag can be made smaller and less expensive.

5.1.4 Structural studies on a single bead

If structural elucidation can be done from single bead analysis, the tedious chemical tagging and decoding processes would not be necessary. Structural information of compounds on a single bead can be obtained by using MS imaging time of flight secondary ion mass spectrometry (TOF-SIMS) (52). Single bead analysis with several other methods have also been demonstrated and reviewed (53). However, the feasibility of a complete structural elucidation of an unknown compound based on single bead analysis is still to be proved.

5.2 Solution affinity selection methods

Screening methods for solid phase attached compounds may suffer from the interference of the support. Immobilized compounds may have different affinity characteristics compared to their native forms in solution. An ideal approach should allow both receptors and ligands to interact in solution and preserve completely their native binding properties. Affinity selection approaches have been developed in which the binding ligands are first physically separated from the rest of the compounds and their structures identified. However, in this kind of affinity selection experiments, the number of compounds in the mixture is limited by the solubility of the compounds and the possible denaturation of the protein (i.e. target receptor) when organic compounds exceed a certain concentration.

5.2.1 Pulsed ultrafiltration MS

Coupling of a pulsed ultrafiltration process to the MS has been shown to be an effective method for affinity selection MS (39). An ultrafiltration chamber and a membrane (such as an Amicon 10 000 Da cut-off membrane) can be used for the separation of binding complexes. During ultrafiltration, ligand–receptor complexes remain in solution in the ultrafiltration chamber while unbound library compounds are washed and filtered through the membrane. Next, the ligand–receptor complexes can be disrupted by changing the mobile phase (pH, temperature, or exposure to other organic solvents). Mass spectra are obtained using quadrupole mass spectrometry. The mobile phase should be completely volatile, and of low ionic strength for compatibility with ESI-MS.

5.2.2 Size exclusion chromatography MS

The receptor of interest is incubated in solution with the compound library under receptor-limiting conditions, allowing only compounds with the highest affinities to bind to the receptor. The receptor/ligand complex is separated rapidly from the unbound library compounds by size exclusion chromatography. The sample is desalted and concentrated on-line in the injector loop using a peptide reversed phase cartridge, and an aliquot of the receptor/ligand complex introduced into the ion source of an electrospray triple-quadrupole mass spectrometer. Under the conditions used for elution off of the reversed phase cartridge, the receptor/ligand complexes dissociate readily. The m/z range of the compound library is scanned to determine the molecular weight of the ligands, which had bound to the receptor in the affinity selection experiment. As a control, the selection experiment is repeated in the presence of a natural receptor ligand competitor with high binding affinity. The compounds that bind specifically at the binding site are hence identified readily since they are observed in the selection experiment in the absence of the competitor, but are not observed in the control experiment. Any ligands that are observed both in the absence and presence of the competitor are assumed to bind through general hydrophobic interactions at sites distinct from the receptor binding site. A high affinity binder was identified by screening a library of 600 compounds by this method (40).

5.2.3 Affinity capillary electrophoresis MS

Coupling of capillary electrophoresis to mass spectrometry provides an approach for screening solution mixture libraries. This method integrates candidate selection and structure analysis into a single experiment. The pioneering work was done using vancomycin as a model target (41). Peptide libraries containing 100 compounds in the form of Fmoc-DDXX (X = ten different D-amino acids) were constructed by split synthesis. The library solution was injected directly into a capillary plugged with vancomycin. The experiments were designed to allow vancomycin to migrate away from the mass spectrometer while the whole library migrates towards it. Ligands bound tightly to the target were retained and separated from non-specific species in the mixture. Three ligands, Fmoc-DDFA, Fmoc-DDYA, and Fmoc-DDHA, were identified to have stronger binding affinities to vancomycin than Fmoc-DDAA. When a larger library needs to be assayed, an affinity chromatography procedure can be incorporated prior to the CE-MS analysis to pre-concentrate active components, remove a large number of non-specific components, and simplify the MS results. Furthermore, CE-MS can directly measure binding constants for interactions between the receptor and the ligand, and can be applied to a variety of libraries of small organic compounds (42).

5.2.4 Fourier transform ion cyclotron resonance (FTICR) MS

A mixture of carbonic anhydrase II (CA II, 2.5 μM) and a peptide library (0.5 μM for each inhibitor, 289 compounds) in a 10 mM NH_4OAc solution pH 7.0 was analysed using ESI-FTICR (43). The resulting mass spectra allow the dissociated ligands to be identified based upon their different molecular weights while their ion intensities reflect the relative binding affinities. Several inhibitors identified from this library were resynthesized and their binding affinities to CA II in solution were studied using a fluorescence binding assay. The binding constants of the seven inhibitors in solution correlated well with the relative intensities of the ions dissociated from the CA II–inhibitor complexes. This result provides support that vapour phase and solution phase ordering of binding constants correlate and that the MS data can be used to infer the order of binding affinities in solution.

This approach should be most effective for the identification of the tight binding ligands, since non-specific (random) association induced during or by the ESI process may limit the quality of information for weaker binding ligands, particularly in larger libraries. In principle, the ESI-FTICR methods should be most useful for determining the relative affinities of a library of compounds to a protein when:

(a) The binding study can be carried out using volatile buffers such as NH_4OAc, Tris acetate, and ammonium bicarbonate.

(b) The protein sample is moderately pure.

(c) All the library members will be charged after dissociation of complexes in the gas phase.

(d) The library members have different masses.

5.2.5 NMR methods

Various NMR methods have been developed to select active compounds from small mixtures of model compounds. Diffusion-ordered 2D NMR spectroscopy (DOSY) (44), diffusion encoded spectroscopy (DECODE) (45), diffusion-assisted nuclear Overhauser effect (NOE) pumping (46), and structure–activity relation-ship (SAR) by NMR (47) are some examples. SAR by NMR in particular has demonstrated great utility in lead optimization process.

6 Concluding remarks

Solid phase organic reaction optimization is fundamentally important for the success of combinatorial synthesis. Optimal reaction conditions should be selected for all reactions in both a non-library format and a small trial library format. In cases where the QC of the final library members cannot be done, the selection of the best synthesis route is the only way to assure a good quality of the final library. Although the 'cleave-and-analyse' method can always be used when no other method is available, on-support analysis is the preferred way to monitor reactions and assist the reaction optimization. The value of FTIR, NMR, and spectrophotometric methods in the reaction optimization stage have been proven in routine applications. New methods, especially quantitative methods, need to be developed to provide more analytical support at this stage.

Both qualitative and quantitative analysis, at least for a statistically selected set of combinatorial library members is crucial for confirming the chemical diversity desired to give confidence on bioassay results. It is evident from above that MS is the major technique in the quality control of soluble discrete libraries while HT-NMR is still in its infancy. Automated spectral interpretation capability needs to be developed before HT-NMR can have a major impact on combinatorial chemistry. Speed and sensitivity are the main advantages of the MS techniques, while quantitation is a major drawback. The best choices for efficient high-throughput quantitation are the use of quantitative detectors such as chemiluminescence nitrogen-specific detection (CLND) and evaporative light scattering detection (ELSD) in FIA or on-line format.

Mixture analysis is still at the very early stage. Comparison of mass distribu-tion obtained from MS measurements with that from computer simulation gives a very crude approximation for small libraries since the ionization efficiency of diverse compounds is unknown. This opens a field for future research and development. Analytical chemists have many good research oppor-tunities in the broad area of combinatorial chemistry (54). In general, the applica-tion of serial analytical methods to deal with products made from parallel synthesis is definitely a mismatch and breakthroughs in parallel qualitative and quantitative analysis are badly needed.

The current preferred screening method is the high-throughput solution phase screening of pure compounds. It should be noted, however, that the purity standard for lead optimization should be higher than that for finding leads. The screening of small mixtures using various affinity selection methods may be used in secondary assay or confirmation studies. The screening of very large mixture libraries using on-bead assays coupled with tagging schemes is a potentially powerful method to accommodate high capacity synthesis. Both of these mixture screening methods need more research and development to become the widely accepted.

References

1. Leznoff, C. C. (1974). *Chem. Soc. Rev.*, **3**, 65.
2. Crowley, J. I. and Rapoport, H. (1976). *Acc. Chem. Res.*, **9**, 135.
3. Yan, B. (1998). *Acc. Chem. Res.*, **31**, 621.
4. Yan, B. and Kumaravel, G. (1996). *Tetrahedron*, **52**, 843.
5. Yan, B., Kumaravel, G., Anjaria, H., Wu, A., Petter, R. C., Jewell, C. F., *et al.* (1995). *J. Org. Chem.*, **60**, 5736.
6. Li, W. and Yan, B. (1998). *J. Org. Chem.*, **63**, 4092.
7. Yan, B., Gremlich, H.-U., Moss, S., Coppola, G. M., Sun, Q., and Liu, L. (1999). *J. Comb. Chem.*, **1**, 46.
8. Chan, T. Y., Chen, R., Sofia, M. J., Smith, B. C., and Glennon, D. (1997). *Tetrahedron Lett.*, **38**, 2821.
9. Gosselin, F., Di Renzo, M., Ellis, T. H., and Lubell, W. D. (1996). *J. Org. Chem.*, **61**, 7980.
10. Yan, B., Fell, J. B., and Kumaravel, G. (1996). *J. Org. Chem.*, **61**, 7467.
11. Gremlich, H.-U. and Berets, S. L. (1996). *Appl. Spectrosc.*, **50**, 532.
12. Garigipati, R. S., Adams, B., Adam, J. L., and Sarkar, S. K. (1996). *J. Org. Chem.*, **61**, 2911.
13. Shapiro, M. J., Chin, J., Marti, R. E., and Jarosinski, M. A. (1997). *Tetrahedron Lett.*, **38**, 1333.
14. Chin, J., Fell, B., Pochapsky, S., Shapiro, M. J., and Wareing, J. R. (1998). *J. Org. Chem.*, **63**, 1309.
15. Dhalluin, C. F., Boutilion, C., Tartar, A. L., and Lippens, G. (1997). *J. Am. Chem. Soc.*, **119**, 10494.
16. Jelinsk, R., Valente, K. G., and Opella, S. J. (1997). *J. Magn. Reson.*, **125**, 185.
17. Yan, B. and Li, W. (1997). *J. Org. Chem.*, **62**, 9354.
18. Yan, B. and Liu, L., Astor, C. A., and Tang, Q. (1999) *Anal. Chem.* **71**, 4564.
19. Hegy, G., Gorlach, E., Richmond, R., and Bitsch, F. (1996). *Rapid Commun. Mass Spectrom.*, **10**, 1894.
20. Gorlach, E., Richmond, R., and Lewis, I. (1998). *Anal. Chem.*, **70**, 3227.
21. Kiplinger, J. P., Cole, R. O., Robinson, S., Roskamp, E. J., Ware, R. S., Connell, H. J. O., *et al.* (1998). *Rapid Commun. Mass Spectrom.*, **12**, 658.
22. Zeng, L. and Kassel, D. B. (1998). *Anal. Chem.*, **70**, 4380.
23. Keifer, P. A. (1998). *Drugs of the Future*, **23**, 301.
24. Furka, A., Sebestyen, F., Asgedom, M., and Dibo, G. (1991). *Int. J. Peptide Protein Res.*, **37**, 487.
25. Lam, K. S., Salmon, S. E., Hersh, E. M., Hruby, V. J., Kazmierski, W. M., and Knapp, R. J. (1991). *Nature*, **354**, 82.
26. Houghten, R. A., Pinilla, C., Blondelie, S. E., Appel, J. R., Dooley, C. T., and Cuervo, J. H. (1991). *Nature*, **354**, 84.
27. Dunayevskiy, Y., Vouros, P., Carell, T., Wintner, E. A., and Rebek, J. Jr. (1995). *Anal. Chem.*, **67**, 2906.
28. Metzger, J. W., Wiesmuller, K.-H., Gnau, V., Brunjes, J., and Jung, G. (1993). *Angew. Chem. Int. Ed. Engl.*, **32**, 894.
29. Winger, B. E. and Campans, J. E. (1996). *Rapid Commun. Mass Spectrom.*, **10**, 1811.
30. Nawrocki, J. P., Wigger, M., Watson, C. H., Hayes, T. W., Senko, M. W., Benner, S. A., *et al.* (1996). *Rapid Communic. Mass Spectrom.*, **10**, 1860.
31. Fitch, W. L., Szardenings, A. K., and Fujinari, E. M. (1997). *Tetrahedron Lett.*, **38**, 1689.

32. Taylor, E. W., Qian, M. G., and Dollinger, G. D. (1998). *Anal. Chem.*, **70**, 3339.

33. Kibbey, C. E. (1995). *Mol. Div.*, **1**, 247.

34. Czarnik, A. W. (1997). *Proc. Natl. Acad. Sci. USA*, **94**, 12738.

35. Youngquist, R. S., Fuentes, G. R., Lacey, M. P., and Keough, T. (1995). *J. Am. Chem. Soc.*, **117**, 3900.

36. Yu, Z., Tu, J., and Chu, Y.-H. (1997). *Anal. Chem.*, **69**, 4515.

37. Needels, M. C., Jones, D. G., Tare, E. H., Heinkel, G. L., Kochersperger, L. M., Dower, W. J., *et al.* (1993). *Proc. Natl. Acad. Sci. USA*, **90**, 10700.

38. Muller, K., Gombert, F. O., Manning, U., Grossmuller, F., Graff, P., Zaegel, H., *et al.* (1996). *J. Biol. Chem.*, **271**, 16500.

39. van Breeman, R. B., Huang, C. R., Nikolic, D., Woodbury, C. P., Zhao, Y. Z., and Venton, D. L. (1997). *Anal. Chem.*, **69**, 2159.

40. Kaur, S., McGuire, L., Tang, D., Dollinger, G., and Huebner, V. (1997). *J. Protein Chem.*, **16**, 505.

41. Chu, Y.-H., Dunayevskiy, Y. M., Kirby, D. P., Vouros, P., and Karger, B. L. (1996). *J. Am. Chem. Soc.*, **118**, 7827.

42. Dunayevskiy, Y. M., Lyubarskaya, Y. V., Chu, Y.-H., Vouros, P., and Karger, B. L. (1998). *J. Med. Chem.*, **41**, 1201.

43. Gao, J., Cheng, X., Chen, R., Sigal, G. B., Bruce, J. E., Schwartz, B. L., *et al.* (1996). *J. Med. Chem.*, **39**, 1949.

44. Morris, K. F. and Johnson, C. S. Jr. (1992). *J. Am. Chem. Soc.*, **114**, 3139.

45. Lin, M., Shapiro, M. J., and Wareing, J. (1997). *J. Am. Chem. Soc.*, **119**, 5249.

46. Chen, A. and Shapiro, M. J. (1998). *J. Am. Chem. Soc.*, **120**, 10258.

47. Hajduk, P. J., Sheppard, G., Nettesheim, D. G., Olejniczak, E. T., Shuker, S. B., Meadows, R. P., *et al.* (1997). *J. Am. Chem. Soc.*, **119**, 5818.

48. Ohlmeyer, M. H. J., Swanson, R. N., Dillard, L. W., Reader, J. C., Asouline, G., Kobayashi, R., *et al.* (1993). *Proc. Natl. Acad. Sci. USA*, **90**, 10922.

49. Ni, Z.-J., Maclean, D., Holmes, C. P., Murphy, M. M., Ruhland, B., Jacobs, J. B., *et al.* (1996). *J. Med. Chem.*, **39**, 1601.

50. Moran, E. J., Sarshar, S., Cargill, J. F., Shahbaz, M. M., Lio, A., Mjalli, A. M. M., *et al.* (1995). *J. Am. Chem. Soc.*, **117**, 10787.

51. Nicolaou, K. C., Xiao, X. Y., Parandoosh, Z., Senyei, A., and Nova, M. (1995). *Angew. Chem. Int. Ed. Engl.*, **34**, 2289.

52. Brummel, C. L., Vickeman, J. C., Hemling, M. E., Roberts, G. D., Johnson, W., Weinstock, J., *et al.* (1996). *Anal. Chem.*, **68**, 237.

53. Swali, V. and Bradley, M. (1997). *Anal. Commun.*, **34**, 15H.

54. Czarnik, A. W. (1998). *Anal. Chem.*, **70**, 378A.

Chapter 9

Multi-component reactions (MCRs) of isocyanides and their chemical libraries

Ivar Ugi and Alexander Dömling[†]

Institute of Organic Chemistry and Biochemistry, Technische Universität München, Lichtenbergstraße 4, D-85747 Garching, Germany.
[†]MORPHOCHEM AG, Grmunder Str, 37–37a, D-81379 München, Germany.

1 Introduction

In principle, all chemical reactions correspond to equilibria between one or two reactants and products. In practice, however, preparative chemical reactions proceed irreversibly and, in the absence of competing reactions, extremely high yields of pure products can be obtained (1–3). If more than two ($n \geq 2$) reactants are required for the synthesis of a product, usually several (at least $n - 1$) preparative steps must be carried out. The more preparative steps there are, the more isolation and characterization work must be achieved, and the lower the yields are after each step of the synthetic effort. In contrast with this approach, many products can also be prepared in a one-step, one-pot multi-component reaction (MCR) of three or more starting materials. MCRs are defined as a collection of subreactions taking place simultaneously and/or sequentially, each involving two reacting molecules.

One can distinguish three types of MCRs. In type I MCRs, the reactants, intermediates, and final products are involved in equilibrating subreactions, whereas in those of type II they are siphoned towards the end-product because of the irreversible nature of the last step. MCRs of type III are sequences of irreversible subreactions that proceed from reactants to product (2). The equilibrating MCRs of type I form mixtures of all participating molecules, and the yields of their products are usually low (5). Under optimal conditions, MCRs of type II can lead to very high yields of rather pure final products. Most MCRs of this type are Ugi four-component reactions (U-4CRs) or their derivatives where the final irreversible formation of the product results from heterocycle ring closure or transition of divalent carbon atoms (C^{II}) of isocyanides to the corresponding tetravalent carbon (C^{IV}) (*Figure 1*) (4, 6, 15). Preparative MCRs of type III can only be carried out in exceptional cases (7), whereas in living cells the majority of organic

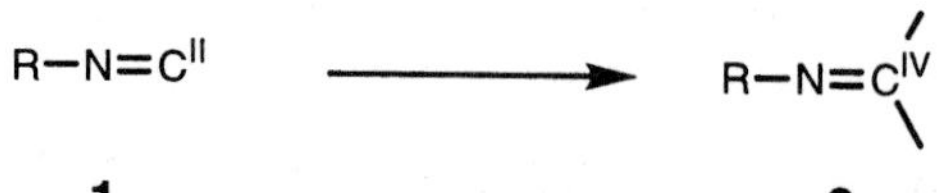

Figure 1 The driving force of isocyanide reactions. The interconversion of divalent carbon atom (C^{II}) of isocyanide (**1**) to tetravalent carbon atom (C^{IV}) of its product (**2**).

compounds result from MCRs of this type (2). In this chapter, only MCRs of type II will be covered.

Because of their ability to combine several building blocks to generate a multitude of highly functionalized products MCRs have had tremendous impact on medicinal and combinatorial chemistry (1–3, 8, 61, 63), an aspect that will be highlighted throughout this chapter.

2 Isocyanide chemistry

There are several naturally occurring isocyanides with biological activities that were synthesized by Barton (34) and Baldwin (35). Others, isolated by Scheuer (36) and Miyaoka (37) from marine sources, have shown medicinal activity such as the anti-malaria Kalihinol-A that was isolated from a marine sponge (37).

Isocyanides were introduced in 1859 when Lieke produced allylisocyanide from allyliodide and silver cyanide (16), and for more than a decade afterward they were investigated by Gautier (17). Even today this approach remains the method of choice for the preparation of isocyano-sugars (17a). At about the same time, Hofmann introduced an alternative method of producing alkyl- and aryl-isocyanides from primary amines and chloroform in concentrated alkaline solutions (18). Since then, and for over a century, no improved method was reported probably as a result of the poor yield and evil odour associated with isocyanide chemistry (4).

A new era of isocyanide chemistry began in 1956 when Hagedorn and Tönjes described the preparation of *0,0'-dimethyl-Xanthocillin*™ **4** by dehydration of **3** with phenylsulfonyl chloride in pyridine (*Figure 2*) (27–29). Other dehydration methods of formylamines in the presence of trialkyl amines, pyridine, or quinoline, as bases, and $COCl_2$ (30), its dimer (31), or its trimer (32) as dehydrating agents produced aliphatic isocyanides in good yields (*Protocol 1*). $POCl_3$ and potassium *tert*-butoxide promote the formation of aryl-isocyanides reasonably

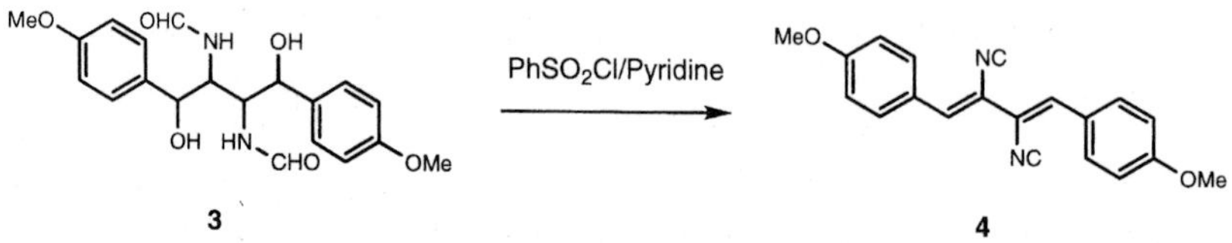

Figure 2 Synthesis of the naturally occurring antibiotic Xanthocyillin.

well (4) and later it was found that $POCl_3$ and diisopropylamine can also promote the formation of alkyl- and aryl-isocyanides in good yields (33) (*Protocol 2*).

Protocol 1

Preparation of alkyl-isocyanides from *N*-alkylformamides using diphosgene (31)

Equipment and reagents

- Rotavapor
- Stirring plate
- Cooling bath
- Standard glassware for organic synthesis (e.g. round-bottom flasks, condensers, distillation apparatus)

- Chemicals and solvents: *N*-alkylformamide, triethylamine, methylene chloride, diphosgene
- $NaHCO_3$ aqueous solution (7.5%, w/v)
- Molecular sieves (4 Å)

Method

1 Add diphosgene (6.03 ml, 9.69 g, 0.05 mol) in methylene chloride (20 ml) at 0 °C to a solution of *N*-alkylformamide (0.10 mol) and triethylamine (29.0 ml, 21.3 g, 0.21 mol) in methylene chloride (100 ml).

2 Stir for 1 h at 0 °C and then raise the temperature to 20 °C. Wash the organic layer with an aqueous solution of $NaHCO_3$ (7.5%, w/v) and dry it over 4 Å molecular sieves.

3 Evaporate the solvent (rotavapor) and purify the residual crude isocyanide by distillation or recrystallization.

Protocol 2

Preparation of isocyanides from the corresponding formamides using $POCl_3$ and diisopropylamine (33)

Equipment and reagents

- Stirring plate
- Rotavap
- Standard glassware for organic synthesis (e.g. round-bottom flasks, condensers, distillation apparatus)

- Chemicals and solvents: *N*-alkyl- or aryl-formamide, diisopropylamine, methylene chloride, phosphoryloxychloride, sodium sulfate
- Na_2CO_3 aqueous solution (20%, w/v)

Method

1. Dissolve (or suspend) the formamide (0.1 mol) in methylene chloride (100 ml) and then add diisopropylamine (0.27 mol) and phosphoryl oxychloride (0.11 mol) drop-

wise while stirring at 0°C. Maintain the stirring for an additional 1 h. Stir the sparingly soluble formamides for 8 h at 20°C.

2. Slowly, add an aqueous solution of sodium carbonate (20%, w/v) so as not to exceed a reaction temperature of 25–30°C.

3. After stirring for 1 h at 20°C, add more water (100 ml) and methylene chloride (50 ml), recover the organic layer and wash it with water (3 × 50 ml), dry it over sodium sulfate, and then evaporate the solvent (rotavap).

4. The residue is either distilled at low pressure or recrystallized to yield the desired isocyanides.

3 MCR chemistry

In 1838 the preparative chemistry of MCRs began when Laurent and Gerhardt (9) converted bitter almond oil (containing benzaldehyde **5**) and ammonia **7** into a crystalline product **10** in the presence of hydrogen cyanide **6** (*Figure 3*). Its primary product **10** reacts once more with benzaldehyde **5** to form **12** (5). α-Aminocyanide **10** is also the product of the Strecker 3CR (S-3CR) (10). Generally the yield of such reactions is relatively low, and often result in by-products like the rather toxic compound **13**. In 1929 Bergs and Bucherer introduced the BB-4CR (5, 11) by adding CO_2 to S-3CR which led to the formation of the hydantoin derivative **11** in high yields. Currently α-amino acids of type **14** are generated through acidic hydrolysis of hydantoins such as **11** rather than through the hydrolysis of the aminocyanide **10**.

Hantzsch MCRs (H-MCR, $2 \leq M \leq 4$) (13), which are similar to Mannich's 3CR (14), generate five and six-membered heterocyclic rings. Hellmann and Opitz description of such reactions is now also referred to as the HO-3CR class of reactions (2, 5). In 1959 Passerini 3CR (P-3CR) (4, 19, 20), which was the first MCR involving isocyanides was combined with HO-3CR (24) in what was termed Ugi four-component reaction (U-4CR) (*Figure 3*) (1, 20a, 25). For more than three decades after this significant development MCR chemistry remained latent. Although in 1961 (22), and again in 1971 (4a), chemical libraries prepared using U-4CR were reported, it was not until the advent of combinatorial chemistry, in the late 1980s, that MCRs underwent an almost instantaneous renaissance. Because of their tolerance towards a wide range of reactants, even the sterically most hindered (*Figure 4*), the full potential of U-4CRs in the design of combinatorial chemical libraries for the discovery of new pharmaceuticals was rapidly recognized (1–3, 8, 39, 62, 63).

Although various types of reactants can be used in U-4CR, primary amines, carboxylic acids, isocyanides, and aldehydes (or ketones) were the most widely studied. The products are usually derivatives of α-amino acid amides such as **23** (*Figure 5*, *Protocol 3*). As a result, many natural products like bicyclomycin, dimethyldysidenin, willardiin, nikkomycin, sinefugin, polyoxins, plumbemycin,

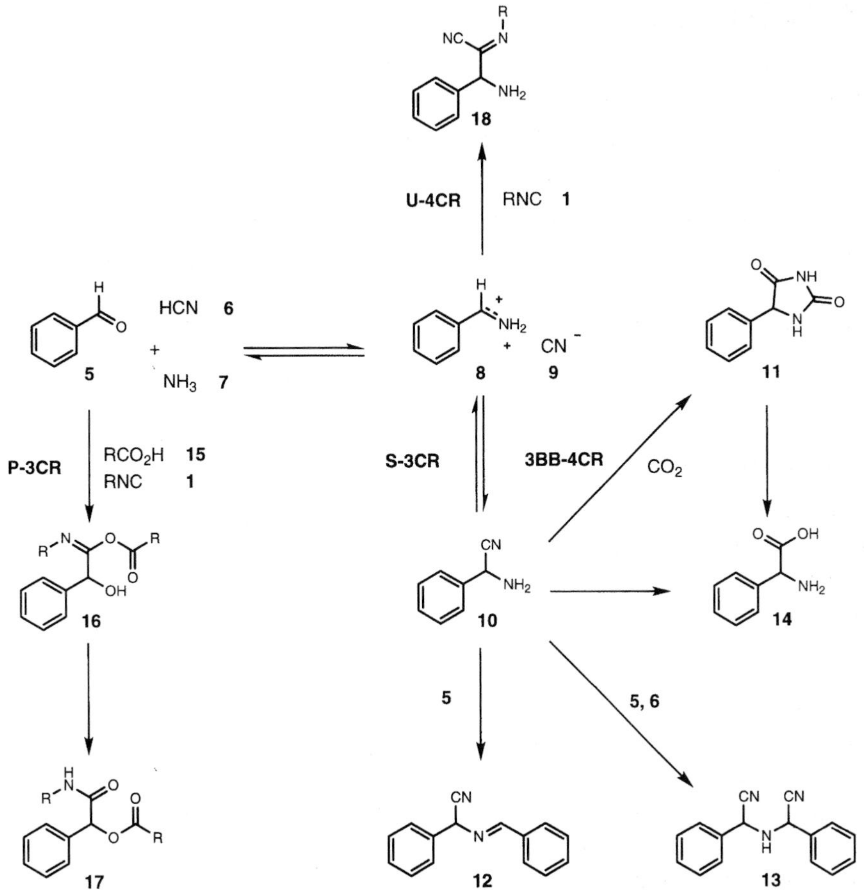

Figure 3 Equilibrium MCRs of type I, the Strecker reaction (S-3CR), and an MCR of type II with an irreversible final step to the product, the Bucherer-Bergs reaction (BB-4CR).

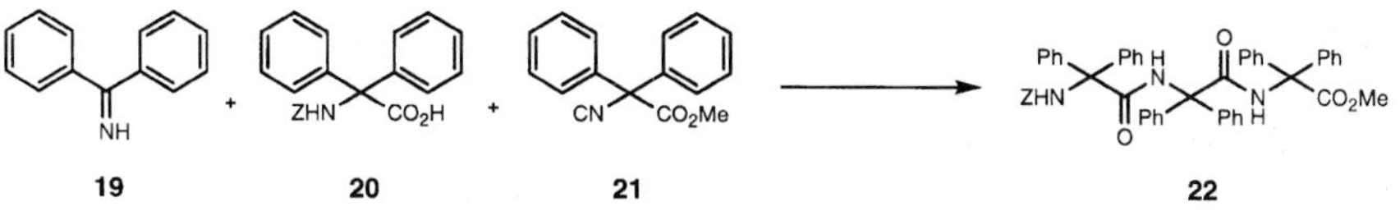

Figure 4 Synthesis of a sterically crowded peptidic product by an U-4CR.

glutathione, eloidisine, and related pharmaceutical compounds were made by such U-4CRs (3). For instance, Merck's new HIV protease inhibitor Crixivan™ (MK 639, *Figure 5*) **29** was synthesized using U-4CR (40). Also, the local anaesthetic *Xylocain*™ **34**, one of the world's best-selling pharmaceuticals (38), was prepared in high yield in a one-pot 4CR from reactants **30–33** (*Figure 5*) (4, 21, 44, 70). Note that in the latter reaction water acts as one of the four reaction components by taking on the role of the acid.

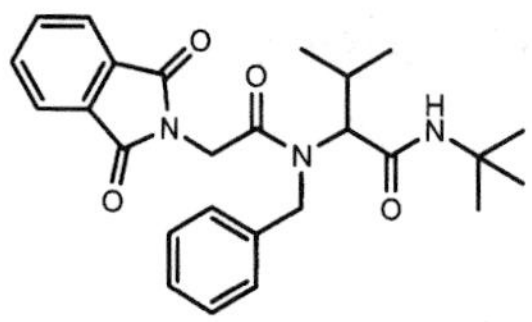

23

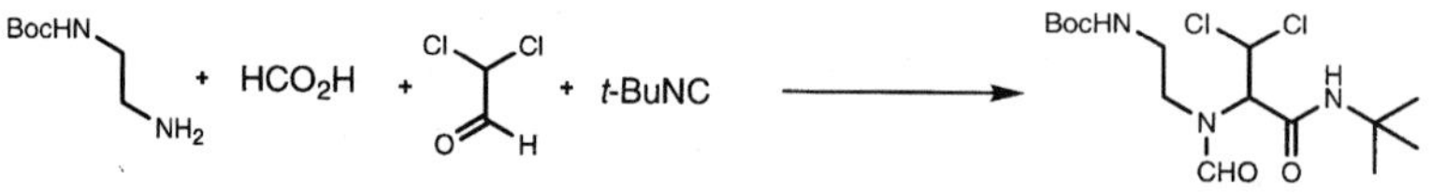

24 25 26 27 28

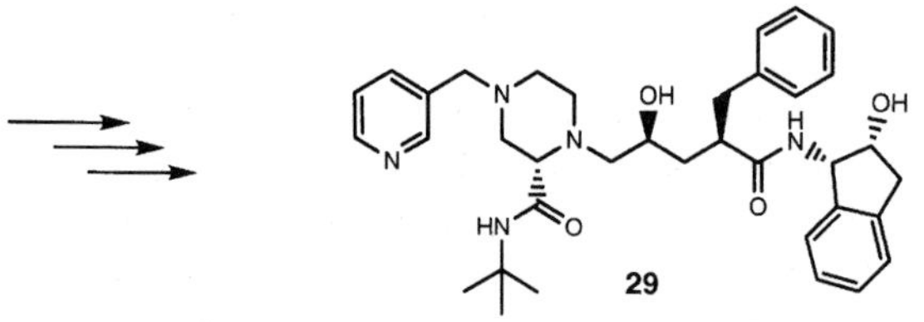

29

Crixivan (MK639)

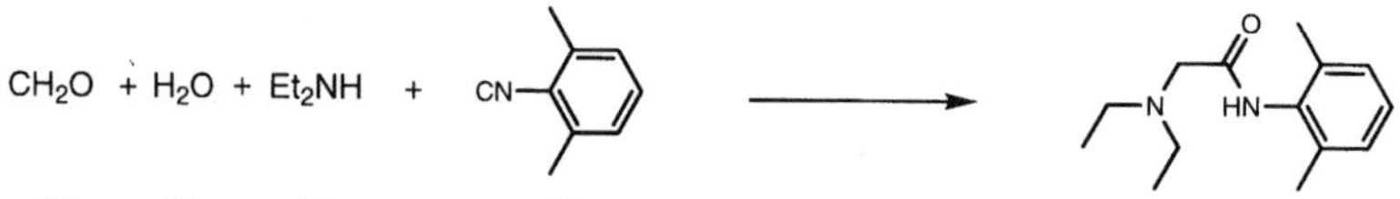

30 31 32 33 34

Xylocain

Figure 5 One-pot synthesis of substituted dipeptide **23**, synthesis of Merck's HIV protease inhibitor *Crixivan* **29** via an U-4CR, and one-pot synthesis of *Xylocain™* **34** by an U-MCR. All three reactions involve a primary amine, an acid, an aldehyde, and an isocyanide.

Protocol 3

N,N-phthalyl-glycil-*N'*-benzyl-valine-*tert*-butylamide 23 (40)

Equipment and reagents

- Stirring plate
- Standard glassware for organic synthesis (e.g. round-bottom flasks, condensers, distillation apparatus)
- Organic chemicals and solvents: phtalyl-glycine, benzylamine, isobutyraldehyde, *tert*-butylisocyanide, methanol

Method

1. Mix phthalyl-glycine (3.08 g), benzylamine (1.61 g), isobutyraldehyde (1.08 g), and *tert*-butylisocyanide (1.25 g) in methanol (25 ml).
2. Isolate the crystalline compound formed after 2 h (m.p. 145–155 °C) and recrystallize **23** (92% yield, m.p. 165–167 °C), from methanol/benzene (1:1).

4 Stereoselective U-4CRs and their applications in the synthesis of α-amino acids, peptides, and related compounds

Usually the P-3CR has no or rather low stereoselectivity (41) while, in contrast, stereoselective U-4CRs can be accomplished. Early mechanistic studies of the U-4CR established that the yield and stereoselectivity of this reaction depend greatly on the ratio of reactants as well as on other reaction conditions (4b). Indeed, the study of isobutyraldehyde-(S)-α-phenylethylimine, benzoic acid, and *tert*-butylisocyanide as model reactants (4, 12) revealed that this reaction starts with an equilibrating 3CR, which is then followed by an irreversible reaction between the final intermediate and the isocyanide. By carefully choosing the ratio and chirality of the reactants it was therefore possible to drive the reaction towards the final desired diastereomer.

U-4CR is particularly suitable for the stereoselective synthesis of α-amino acids, peptides, and their unnatural derivatives (24, 43). This process requires the introduction and mild and selective removal of a chiral accessory on the primary amine (R₂NH₂, *Figure 6*) (24, 44, 48). As a result, a great variety of natural products such as α-amino acids, peptides (44, 45) (*Protocols 4* and *5*), β-lactams (46, 47) (*Protocol 6*), and alkaloids (3) were prepared through this route.

The first chiral amines tested in stereoselective U-4CR were derivatives of α-ferrocenyl alkylamines (4, 44), but the yield and stereoselectivity were poor. A more promising class of chiral amines was introduced by Kunz (49), who found that 1-amino-2,3,4,6-tetra-O-pivaloyl-β-galactopyranosyl **42** as well as its derivatives in the presence of ZnCl₂ can form α-amino acids in high yield and stereoselectivity. These amines, however, did not perform well in the synthesis of peptide derivatives. In addition, the cleavage of the pyranosyl residue (chiral

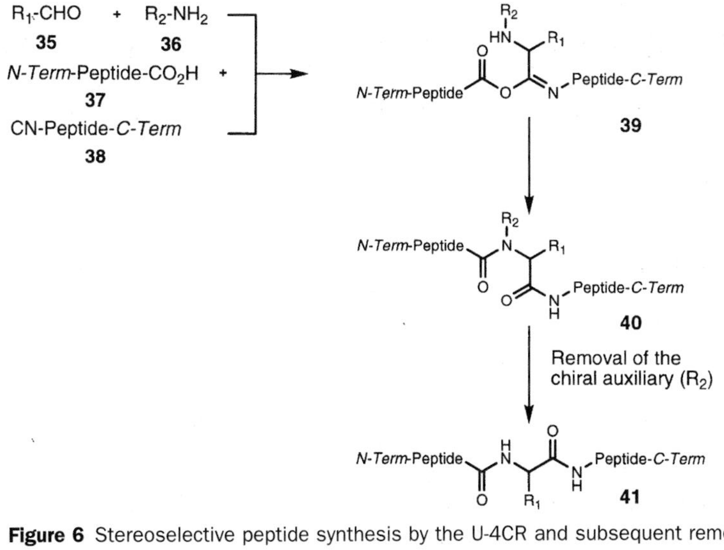

Figure 6 Stereoselective peptide synthesis by the U-4CR and subsequent removal of the chiral auxiliary.

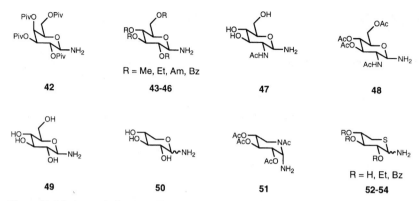

Figure 7 1-Amino carbohydrate chiral auxiliaries investigated in the stereoselective synthesis of peptides by U-4CR.

auxiliary) required hot concentrated hydrochloric acid. Later on Goebel and Ugi (50) explored various 2,3,4,6-tetra-O-alkyl-β-glucopyranosyl-1-amines **43–46** and 1-amino-2-acetamino-2-desoxy-3,4,6-tri-O-acetyl-β-D-glucopyranose **47** (51) but the corresponding U-4CR suffered from the same limitations. Of the 15 different products of the stereoselective U-4CR with chiral amine **48** eight were obtained in 79–90% yield and seven in 5–50% yield, but *all* of them were > 99% optically pure. For example, in the case of **48**, isobutyraldehyde, formic acid, and *tert*-butylisocyanide yielded 90% of the desired chiral isomer. Other amines such as 1-amino-5-acetamino-5-desoxy-2,3,4-O-triacetyl-α-D-xylopiperidose **51** (52), amino sugars **49** and **50** (53, 54), and the sulfur containing derivatives of xylose **52–54** (55) not only promoted stereoselective U-4CR but their sugar accessory was cleaved under relatively mild conditions (*Figure 7*).

Protocol 4

Preparation of α-amino acid and peptide derivatives by the stereoselective U-4CR using amino sugar 48 as a chiral auxiliary (*Figure 7*)

Equipment and reagents

- Stirring plate
- Cooling bath
- Silica thin-layer chromatography plates
- Celite
- Rotavapor
- High vacuum pump
- Standard glassware for organic synthesis (e.g. round-bottom flasks, condensers, distillation apparatus)
- Aqueous solution of NaHCO$_3$ (7.5%, w/v)

- Chemicals and solvents: 1-amino-2-acetamino-2-desoxy-3,4,6-tri-O-acetyl-β-D-glucopyranose **48**, ZnCl$_2$-etherate, methanol (or tetrahydrofuran), isocyanide, aldehyde, carboxylic acid, chloroform, ethanol, methylene chloride, acetic acid, sodium sulfate
- Aqueous solution of tartaric acid (3.5%, w/v)
- Molecular sieves (4 Å)
- Argon source

Protocol 4 continued

Method

1. Combine 1-amino-2-acetamino-2-desoxy-3,4,6-tri-O-acetyl-β-D-glucopyranose **48** (10 mmol), an aldehyde (10 mmol), 4 Å molecular sieves (2.5 g), and $ZnCl_2$-etherate (11 mmol) in methanol or tetrahydrofuran (90 ml) at 20°C.

2. Stir and monitor the reaction by thin-layer chromatography (eluent: chloroform/ethanol) until the Schiff base is formed.

3. Add the isocyanide (10.4 mmol) and the carboxylic acid (10.4 mmol) at a suitable temperature (between 20°C and −75°C, depending on the reactivity of the isocyanide).

4. Stir and monitor the reaction by thin-layer chromatography (eluent: chloroform/ethanol) until the starting materials are all consumed.

5. Filter the resulting mixture on a layer of celite (1 cm), rinse the celite with methylene chloride (20 ml), then remove the solvents under reduced pressure (rotavapor).

6. Dissolve the residue obtained in methylene chloride (100 ml) and stir with an aqueous solution of $NaHCO_3$ (7.5%, w/v, 15 ml). Excess isocyanide is removed by treatment with acetic acid (20 ml) for 16 h. If *tert*-butylisocyanide is to be removed, aqueous tartaric acid solution (3.5%, w/v) should be used.

7. Wash the organic layer with an aqueous solution of $NaHCO_3$ (7.5%, w/v, 250 ml), and dry it over sodium sulfate.

8. Evaporate the solvent (rotavapor) and dry the residual solid under vacuum.

Protocol 5

Procedure for the synthesis of peptide derivative 55 (*Figure 8*) (73)

Equipment and reagents

- Stirring plate
- Standard glassware for organic synthesis (e.g. round-bottom flasks, condensers, distillation apparatus)
- Aqueous solution of $NaHCO_3$ (7.5%, w/v)
- Chemicals and solvents: proline, isobutyraldehyde, *tert*-butylisocyanide, dry methanol, methylene chloride, magnesium sulfate
- Brine

Method

1. Suspend 1 mmol each of proline, isobutyraldehyde, and *tert*-butylisocyanide in dry methanol (5 ml) and stir at 20°C for 24 h.

2. Remove the solvent (rotavapor), dissolve the resulting product **55** in methylene chloride, and wash with water, aqueous solution of $NaHCO_3$ (7.5%, w/v), then brine.

3. Dry the organic phase over magnesium sulfate, filter, and remove the solvent (rotavapor). **55** is obtained in 88% yield.

Protocol 6

General procedure for the preparation of β-lactam 56 (*Figure 8*) (22b, 47, 56, 72)

Equipment and reagents

- Standard glassware for organic synthesis (e.g. round-bottom flasks, condensers, distillation apparatus)
- Stirring plate
- Organic chemicals and solvents: β-alanine, aldehyde, isocyanide, dry methanol
- Silica gel for column chromatography

Method

1. Suspend 1 mmol each of β-alanine, aldehyde, and isocyanide in dry methanol (5 ml) and stir at 20 °C for 24 h.

2. Remove the solvent (rotavapor) and purify the resulting mixture by silica gel column chromatography. **56** is obtained in 90% yield.

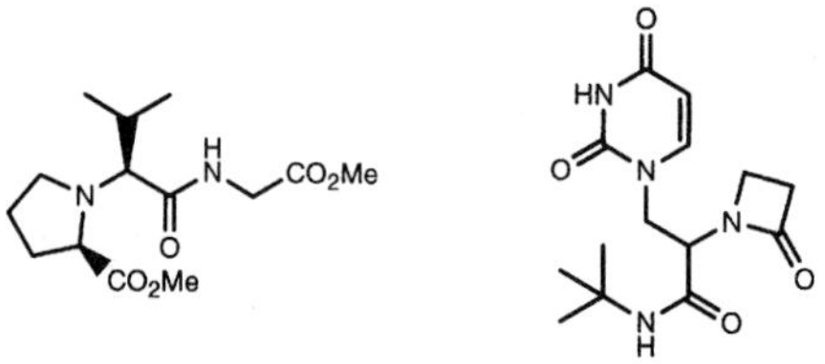

| 55 | 56 |

Figure 8 Compound **55** and **56** prepared according to *Protocols 5* and *6*, respectively.

5 Multi-component reactions of five and more reactants

Since methanol and carbon dioxide are in equilibrium with their monomethyl carbonate, they were used instead of carboxylic acids in the U-4CR, and thus products could be formed from five instead of four components (4). It was also noted rather early that the three- or four-components of the Asinger reaction, A-MCR (57), can be combined with isocyanides and hydrogen cyanide, from which bicyclic products of five-components can result (58). In 1993 Dömling and Ugi introduced the one-pot synthesis of thiazolidine derivatives from a seven-component reaction (59, 60). These 7CR are actually a combination (26) of Asinger 4CR, the equilibrium of methanol and CO_2, and the U-4CR (61). The complexity of MCRs increases further when multi-functional reactants are used. In the last few years in particular, this aspect of MCRs resulted in a great variety of highly functionalized heterocyclic compounds as illustrated in *Figure 9* and *Protocols 7* and *8* (3). The combination of levulinic acid, valine methylester, and methyl isocyanide in a U-4CR results in the formation of the monocyclic compound **60** (64) which was then treated with potassium *tert*-butoxide to generate the bicyclic system **61** (*Figure 9, Protocol 7*). In a U-4CR of a special bifunctional isocyanide **69** (67), thiocarboxylic acid, a primary amine and an aldehyde, highly

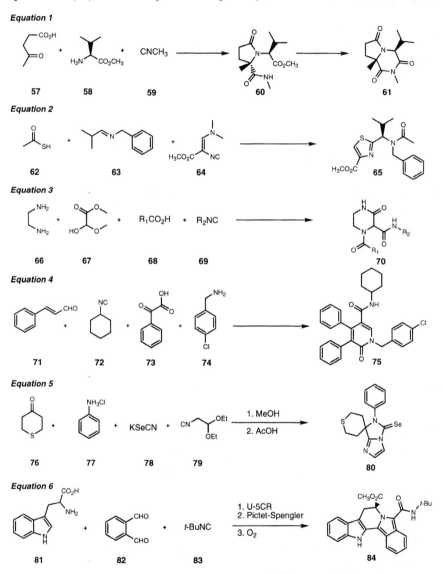

Figure 9 *Equation 1:* reaction of a bifunctional β-ketoacid **57**, isocyanide **59**, and an α-amino acid methylester **58** to form a lactam. The lactam is cyclized subsequently to form the bicyclic diketopiperazine derivative **61**. *Equation 2:* first synthesis of a thiazol derivative by U-4CR. *Equation 3:* ketopiperazine **70** synthesis by the reaction of glyoxal with diamines, isocyanides, and carboxylic acids. *Equation 4:* synthesis of the dihydropyridinone derivative **75** by a tandem U-MCR/cyclization reaction. *Equation 5:* synthesis of bicyclic imidazole derivative by MCR and subsequent ring-closure. *Equation 6:* alkaloid-like polycycle **84** synthesis by a one-pot U-MCR, Pictet-Spengler reaction, and final oxidation.

substituted thiazoles **70** (66) was obtained (*Protocol 8*). This represents the first highly versatile multi-component thiazole synthesis (68).

Rossen (65) described the preparation of piperazine-2-carbonamides from ethylene diamine, chloracetaldehyde, a carboxylic acid, and an isocyanide in a 34–67% yield. Around the same time Zychlinski and Ugi (66) reported a very high yield synthesis of the 2-ketopiperazine **70** from reactants **66–69**. More recently, in 1997, Bossio prepared the pyridone derivative **75** (69) and the bicyclic system **80** (70) by a U-4CR and subsequent ring closure. Finally, The polyheterocyclic product **84** was prepared by a one-pot combination of U-4CR, Pictet-Spengler reaction, and air oxidation (71).

Protocol 7

Synthesis of bicyclic 1,3-diketopiperazine derivative 61 from valine methylester 58, levulinic acid 57, and methyl isocyanide 59 (*Figure 9*) (64)

Equipment and reagents

- Rotavapor
- Stirring/heating plate
- Standard glassware for organic synthesis (round-bottom flasks, condensers, distillation apparatus)
- Silica gel for flash chromatography

- Chemicals and solvents: levulinic acid, valine methylester hydrochloride, triethylamine, methanol, methyl isocyanide, ethylacetate, methylene chloride, hexane, potassium *tert*-butoxide, tetrahydrofuran, sodium sulfate

Method

1. Add levulinic acid (5 mmol) in methanol (5 ml) to a mixture of valine methylester hydrochloride (5 mmol) and triethylamine in methanol (50 ml) and stir for 2 h to form the imine.

2. Add the methyl isocyanide (5 mmol) in methanol (5 ml) and stir for 24 h.

3. Remove the solvent (rotavapor) and add ethylacetate (50 ml).

4. Remove the insolublized material, evaporate the solvent (rotavapor), then re-dissolve in methylene chloride (50 ml).

5. Wash this solution with distilled water, dry the organic layer on sodium sulfate, remove the solvent (rotavapor), then dry the residual material under vacuum.

6. Purify the desired diastereoisomer **60** (71% yield, m.p. 122 °C) by silica gel flash chromatography (eluent: hexane/ethylacetate, 3:1).

7. Combine **60** (3 mmol) and potassium *tert*-butoxide (5 mmol) in 50 ml tetrahydrofuran and reflux for 48 h.

8. Remove the solvent (rotavapor), add methylene chloride (50 ml), wash with water

Protocol 7 continued

(2 × 25 ml), dry the organic layer over sodium sulfate, filter, and then evaporate the solvent (rotavapor).

9. Purify the desired diastereoisomer **61** (25% yield) by silica gel flash chromatography (eluent: hexane/ethylacetate, 3:1).

Protocol 8

The preparation of a thiazole derivative by a one-pot reaction of an U-4CR and a secondary reaction (*Figure 9*)

Equipment and reagents

- Stirring/heating plate
- Rotavapor
- Standard glassware for organic synthesis (e.g. round-bottom flasks, condensers, distillation apparatus)

- Organic chemicals and solvents: dry methanol, benzylamine isobutyraldehyde thioacetic acid, isocyanide
- Silica gel for column chromatography

Method

1. Mix benzylamine (1 mmol, 1.07 g) and isobutyraldehyde (1 mmol, 0.72 g) in dry methanol (1 ml) at 20°C for 2 h to form the Schiff base **68**.

2. Add thioacetic acid **67** (1 mmol, 0.62 g) and isocyanide **69** (1 mmol, 1.42 g) and stir for 12 h at 20°C.

3. Remove the solvent (rotavapor) and purify the residual material by silica gel chromatography. The resulting product **70** is obtained in 95% yield (3.13 g) as a colourless oil.

References

1. (a) Ugi, I. (1995). *Proc. Estonian Acad. Sci. Chem.*, **44**, 237. (b) (1998). **47**, 107.
2. Ugi, I. (1997). *J. Prakt. Chem.*, **339**, 499.
3. Dömling, A. (1998). *Comb. Chem. High Throughput Screening*, **1**, 1.
4. Ugi, I. (1971). *Isonitrile chemistry*. Academic Press, New York. (a) p. 149 (b) p. 161.
5. Hellmann, H. and Opitz, G. (1960). In *α-Aminoalkylierung*. Verlag Chemie, Weinheim.
6. Balkenhohl, F., Buschen-Hünnefeld, C. V., Lanshy, A., and Zechel, C. (1996). *Angew. Chem.*, **108**, 3436. (1996). *Angew. Chem. Int. Ed. Engl.*, **35**, 2288.
7. Chattopadhyaya, J., Dömling, A., Lorenz, K., Ugi, I., and Werner, B. (1997). *Nucleosides Nucleotides*, **16**, 843.
8. (a) Weber, K., Waltbaum, S., Broger, C., and Gubernator, K. (1995). *Angew. Chem.*, **107**, 2452. (1995). *Angew. Chem. Int. Ed. Engl.*, **34**, 2280. (b) Lacke, O. and Weber, L. (1996). *Chimia*, **50**, 445. (c) Weber, L. (1998). *Curr. Opin. Chem. Biol.*, **2**, 381. Groebke, K., Weber, L., and Mehlin, F. (1998). *Synlett*, **6**, 661. (d) Weber, L. (1998). In *Lecture at the ANALYTICA 98*. Conference at Munich, June 1998.

9. Laurent, A. and Gerhardt, C. F. (1838). *Ann. Chim. Phys.*, **66**, 181. (1838). *Liebigs Ann. Chem.*, **28**, 265.

10. Strecker, A. (1850). *Ann. Chem.*, **75**, 27.

11. Bergs, H. (1929). *Ger. Pat.*, 566094. (1933). *Chem. Abstr.*, **27**, 1001. Bucherer, H. T. and Steiner, W. (1934). *J. Prakt. Chem.*, **140**, 291. Bucherer, H. T. (1934). *J. Prakt. Chem.*, **141**, 5.

12. Ugi, I. and Kaufhold, G. (1967). *Liebigs Ann. Chem.*, **709**, 11. Ugi, I. (1969). *Rec. Chem. Progr.*, **30**, 289.

13. Hantzsch, A. (1882). *Justus Liebigs Ann.*, **215**, 1. Eisener , U. and Kuthan, J. (1972). *Chem. Rev.*, **72**, 1.

14. Mannich, C. and Krötsche, I. (1912). *Arch. Pharm.*, **250**, 647. Blick, F. F. (1942). In *Organic reaction* (ed. R. Adams), Vol. 1, p. 303. John Wiley and Sons, New York.

15. Oliveri-Mandala, E. and Alagnna, B. (1910). *Gazz. Chim. Ital.*, **40 II**, 441.

16. Lieke, W. (1859). *Justus Liebigs Ann. Chem.*, **112**, 316.

17. Gautier, A. (1867). *Justus Liebigs Ann. Chem.*, **142**, 289. (1869). *Ann. Chim. (Paris)*, (4) **17**, 103. (a) Boullanger, P. and Descotes, G. (1976). *Tetrahedron Lett.*, **38**, 4327.

18. Hofmann, A. W. (1870). *Ber. Dtsch. Chem. Ges.*, **3**, 653. Weber, W. P., Gokel, G. W., and Ugi, I. (1972). *Angew. Chem.*, **84**, 587. (1972). *Angew. Chem. Int. Ed. Engl.*, **11**, 530.

19. Passerini, M. (1921). *Gazz. Chim. Ital.*, **51 II**, 126. *ibid.* (1921). **51**, 181. *ibid.* (1926). **56**, 826. Passerini, M. and Ragni, G. (1931). *ibid.* **61**, 964.

20. Ugi, I., Lohberger, S., and Karl, R. (1991). In *Comprehensive organic synthesis: selectivity for synthetic efficiency* (ed. B. M. Trost and C. H. Heathcock), Vol. 2, Chap. 4.6, p. 1083. Pergamon, Oxford. (a) p. 1090.

21. Ugi, I., Meyr, R., Fetzer, U., and Steinbrückner, C. (1959). *Angew. Chem.*, **71**, 386.

22. Ugi, I. and Steinbrückner, C. (1961). *Chem. Ber.*, **94**, (a) 734. (b) 2802.

23. Ugi, I. (1962). *Angew. Chem.*, **74**, 9. (1962). *Angew. Chem. Int. Ed. Engl.*, **1**, 8.

24. Ugi, I. (1962). *Angew. Chem.*, **74**, 9. (1962). *Angew. Chem. Int. Ed. Engl.*, **1**, 8.

25. Opitz, G. and Merz, W. (1962). *Justus Liebigs Ann. Chem.*, **659**, 163. Sjöberg, K. (1963). *Sv. Kem. Tidskr.*, **75**, 493. (a) McFarland, W. (1963). *J. Org. Chem.*, **28**, 2179. (b) Zinner, G. and Kleigel, W. (1966). *Arch. Pharm.*, **299**, 746.

26. Mac Lane, S. W. and Birkhoff, G. (1967). *Algebra*. MacMillan Company, New York (p. 3: Given sets of R and S have the intersection R Ç S with the common elements R and S).

27. Rothe, W. (1950). *Pharmazie*, **5**, 190. Griesebach, H. and Achenbach, H. (1965). *Z. Naturforsch.*, **B20**, 137.

28. Hagedorn, I. and Tönjes, H. (1956). *Pharmazie*, **11**, 409. (1957). **12**, 567. Hagedorn, I., Eholzer, U., and Lüttringhaus, A. (1969). *Chem. Ber.*, **93**, 1584.

29. Hertler, W. R. and Corey, E. J. (1958). *J. Org. Chem.*, **23**, 1221.

30. Ugi, I. and Meyr, R. (1958). *Angew. Chem.*, **70**, 702. Ugi, I. and Meyr, R. (1960). *Chem. Ber.*, **93**, 239.

31. Skorna, G. and Ugi, I. (1977). *Angew. Chem.*, **89**, 267. (1977). *Angew. Chem. Int. Ed. Engl.*, **16**, 259.

32. Eckert, H. and Forster, B. (1987). *Angew. Chem.*, **99**, 922. (1987). *Angew. Chem. Int. Ed. Engl.*, **26**, 1221.

33. Obrecht, R., Herrmann, R., and Ugi, I. (1985). *Synthesis*, p. 400.

34. Barton, D. H. R., Bowles, T., Husinec, S., Forbes, J. E., Llobera, A., Porter, A. E. A., *et al.* (1988). *Tetrahedron Lett.*, **29**, 3343.

35. Baldwin, J., Aldinger, R. M., Chandragiamu, J., Edenbarough, M. S., Keeping, J. W., and Ziegler, C. B. (1985). *J. Chem. Soc. Chem. Commun.*, p. 816.

36. Chang, C. W. J. and Scheuer, P. J. (1993). *Top. Can. Chem.*, **167**, 33. Scheuer, P. J. (1992). *Acc. Chem. Res.*, **25**, 433.

37. Miyaoka, H., Shimomura, M., Kimura, H., Yamada, Y., Kim, H.-S., and Wataya, Y. (1998). *Tetrahedron*, **54**, 13467.

38. Lindquist, K. and Sundling, S. (1993). *Xylocain*. Astra, A. B. and Dahlbon, R. (1993). *Det händen på lullen* (Astra 1940–1960). A. B. Astra.

39. Yamada, T., Omoto, Y., Yamanaka, Y., Miyazawa, T., and Kuwata, S. (1998). *Synthesis*, 991.

40. Rossen, K., Pye, P. J., DiMichele, L. M., Volante, R. P., and Reider, P. J. (1998). *Tetrahedron Lett.*, **39**, 6823.

41. Bock, H. and Ugi, I. (1997). *J. Prakt. Chem.*, **339**, 3859.

42. Ugi, I., Offermann, K., Herlinger, H., and Marquarding, D. (1967). *Liebigs Ann. Chem.*, **709**, 1.

43. Bodanszky, M. and Ondetti, M. A. (1966). In *Peptide synthesis* (ed. G. A. Olah). J. Wiley & Sons, New York, p. 127.

44. Ugi, I., Marquarding, D., and Urban, R. (1982). In *Chemistry and biochemistry of amino acids, peptides and proteins* (ed. B. Weinstein), **6**, 245. Marcel Dekker, New York.

45. Ramage, R. and Epton, R. (ed.) (1998). In *Peptides*. Royal Society of Chemistry, Perkin Division, Mayflower Scientific Ltd., Kingswinford 1996, England.

46. Ugi, I. (1982). *Angew. Chem.*, **94**, 826. *Angew. Chem. Int. Ed. Engl.*, **21**, 1221.

47. Ugi, I. and Eckert, H. (1982). *Natural products of chemistry*, **12**, 113. Elsevier, Science Publ., 1000AE, Amsterdam.

48. Ugi, I. and Offermann K. (1963). *Angew. Chem.*, **75**, 917. *Angew. Chem. Int. Ed. Engl.*, **2**, 624.

49. Kunz, H. and Pfrengle, W. (1988). *J. Am. Chem. Soc.*, **110**, 651. (1988). *Tetrahedron*, **44**, 5487.

50. Goebel, M. and Ugi, I. (1995). *Tetrahedron Lett.*, **36**, 6043. Goebel, M., Nothofer, H.-G., Roß, G., and Ugi, I. (1997). *Tetrahedron*, **53**, 3123.

51. Lehnhoff, S., Goebel, M., Karl, R. M., Klösel, K., and Ugi, I. (1995). *Angew. Chem.*, **107**, 1208. (1995). *Angew. Chem. Int. Ed. Engl.*, **34**, 1104. Lehnhoff, S. (1994). *Doctoral Thesis*, Technical University of München.

52. Zychlinski, A. V., in Hippe, Z., and Ugi, I. (ed.) (1998). In *Multicomponent reactions and combinatorial chemistry*; German-Polish Workshop, Rzeszów, 28–30. Sept. 1997; University of Technology, Rzeszów / Technical University, Munich, p. 31.

53. Likhosherstov, L. K., Novikova, O. S., Derivitkava, V. A., and Kochetkov, N. K. (1986). *Carbohydr. Res.*, **146**, C1. Likhosherstov, L. K., Novikova, O. S., Shbaev, V. N., and Kochetkov, N. K. (1996). *Russ. Chem. Bull.*, **45**, 1760.

54. Drabik, J. (1999). *Doctoral Thesis*, Technical University of München, in preparation.

55. Ross, G. (1999). *Doctoral Thesis*, Technical University of München, in preparation.

56. Ugi, I. and Wischhöfer, E. (1962). *Chem. Ber.*, **95**, 136.

57. Asinger, F. (1956). *Angew. Chem.*, **68**, 413. Asinger, F. and Offermann, H. (1967). *Angew. Chem.*, **79**, 953. *Angew. Chem. Int. Ed. Engl.*, **6**, 907.

58. Ugi, I. and Offermann, K. (1964). *Chem. Ber.*, **97**, 2276; accomplished as a 5CR by Lehnhoff, S. and Hanush, C. in 1994.

59. Dömling, A. and Ugi, I. (1993). *Angew. Chem.*, **105**, 634. *Angew. Chem. Int. Ed. Engl.*, **32**, 563. Dömling, A., Herdtweck, E., and Ugi, I. (1998). *Acta Chem. Scand.*, **52**, 107.

60. e.g. (1993). *C&EN*, 32, April 19. Bradley, D. (1993). *New Sci.*, 16, July 3.

61. Ugi, I., Dömling, A., and Hörl, W. (1994). *Endeavour*, **18**, 115.

62. Ugi, I. and Rosendahl, F. K. (1963). *Liebigs Ann. Chem.*, **670**, 80.

63. Armstrong, R. W. (1995). *Combinatorial libraries related to natural products at the ACS National Meeting in Anaheim*, Calif. on April 2, 1995. (1995). *J. Am. Chem. Soc.*, **117**, 7842.

64. Hanusch-Kompa, C. and Ugi, I. (1998). *Tetrahedron Lett.*, **39**, 2725. (b) Hanusch-Kompa, C. (1998). *Doctoral Thesis*, Technical University of München.

65. Rossen, R., Sager, J., and DiMichele, L. M. (1998). *Tetrahedron Lett.*, **39**, 2725.

66. Zychlinski, A. V. and Ugi, I. (1998). *Heterocycles*, **49**, 29.

67. Schöllkopf, U., Porsch, H., and Lau, H. H. (1979). *Liebigs Ann. Chem.*, 95.

68. Dömling, A. and Heck, S. (2000). *Synlett*, **3**, 424.

69. Bossio, K., Marcos, S., Marcaccini, S., and Pepino, R. (1997). *Heterocycles*, **45**, 1589.

70. Bossio, K., Marcaccini, S., and Pepino, R. (1993). *Liebigs Ann. Chem.*, 1229.

71. Dömling, A. and Zu-Chi, K. Unpublished work.

72. Dömling, A., Starnecker, M., and Ugi, I. (1995). *Angew. Chem. Int. Ed. Engl.*, **34**, 2238.

73. Demharter, A., Hörl, W., Herdtweck, E., and Ugi, I. (1996). *Angew. Chem.*, **108**, 185. (1996). *Angew. Chem. Int. Ed. Engl.*, **35**, 173. (a) Demharter, A. (1992). *Doctoral Thesis*, Technical University of München.

Chapter 10

Multi-step solution phase combinatorial synthesis

Dale L. Boger and Joel Goldberg

Department of Chemistry and the Skaggs Institute for Chemical Biology, The Scripps Research Institute, 10550 N. Torrey Pines Road, La Jolla, CA 92037, USA.

1 Introduction

This chapter illustrates the types of libraries that can be synthesized by solution phase combinatorial chemistry. The synthesis of individual compounds, libraries composed of small (10–20 compound) mixtures as well as libraries composed of large mixtures are described. Simple liquid/liquid or liquid/solid extraction procedures are utilized to provide pure products irrespective of the reaction efficiencies. Protocols for deconvolution of the small and large compound mixtures are discussed.

2 Aspects of solution phase combinatorial chemistry

2.1 Solution phase versus solid phase combinatorial synthesis

The use of combinatorial chemistry for the discovery or optimization of novel chemical leads has emerged as a powerful method for the acceleration of the drug discovery process (1–4). Initially explored with peptide or oligonucleotide libraries and related structures, subsequent efforts have been directed at exploiting the diversity and range of useful properties embodied in small molecule libraries (1, 5).

A number of approaches to the generation of chemical libraries have been disclosed, including split and mixed (Chapters 1–3), encoded (Chapter 1 and 4), indexed, or parallel spatially addressed synthesis on pins, beads, chips, and other solid supports (Chapters 5–7), while solution phase synthesis has only recently been embraced as a useful alternative (6) (see Chapters 11–13). In a large measure, this may be due to the natural extension of the methodology from peptide and oligonucleotide synthesis, where solid phase synthesis has emerged as the method of choice for repetitive coupling reactions. The resin-bound product isolation by simple filtration permits the use of large reagent excess to effect high conversions required for each of the repetitive steps.

The disadvantages of solid phase synthesis, however, are well recognized (7). The scale is restricted by the required amount of solid support and its loading capacity. Its use is also restricted to the repertoire of reactions presently extended to the solid phase. It requires functionalized substrates and solid supports, compatible spacer linkers, and orthogonal attachment and detachment chemistries often with the release of spectator functional groups. More problematic, it requires the use of specialized protocols for monitoring the individual steps of a multi-step synthesis, including orthogonal capping strategies for blocking unreacted substrate and does not permit the purification of resin-bound intermediates. This latter feature necessarily produces the released product of a multi-step sequence in an impure state and requires that each reaction on each substrate proceed with an unusually high efficiency.

The optimization of the reactions for assuring the required reaction efficiencies is both challenging and time-consuming. Even a modest criterion for product purity (85% pure), requires that each step of a given two-step sequence proceed in 92% yield or that each step of a three-step reaction sequence proceed in 95% yield. Our experience has been that such generalized reaction efficiencies with a wider range of chemistries and substrates are not routinely obtainable and require both an extensive investment in reaction optimization and purification of the released product.

Several advantages and disadvantages of solution phase combinatorial synthesis are highlighted in *Table 1*. Clearly, solid phase combinatorial chemistry benefits from its ease of product isolation, ability to accommodate a large excess of reagents to drive reactions to completion, and potential for automation. Similarly, solution phase techniques offer several advantages. Various protocols, such

Table 1 Comparison of some advantages (+) and disadvantages (−) of solution and solid phase combinatorial synthesis

	Solid phase		Solution phase
+	Simple removal of excess reagents	+	Chemistry not limited by support or linker
+	Split and mix synthesis	+	Reaction monitoring by traditional techniques
+	Pseudo-dilution effects	+	Purification possible after each step
−	Adapt chemistry to solid phase	+	Large amounts (scale) available
−	Reaction monitoring requires sophisticated equipment	+	Mixture or parallel synthesis
−	No purification possible	+	Convergent or linear synthesis
−	Limited scale	−	Removal of excess reagents limits scope
−	Linear only, cannot be used for convergent synthesis	−	Requires similar reactivity among building blocks when mixture coupling is involved

as those described in this chapter, have been developed which combine the benefits of working in solution with facile isolation of highly pure products making solution phase combinatorial chemistry an important, and in some instances, powerful alternative for library generation (1, 6).

2.2 Synthesis of mixtures versus individual compounds

Solution phase combinatorial chemistry has been applied to both the synthesis of individual compounds and mixtures of compounds. The synthesis of mixtures allows a large number of compounds to be prepared and assayed in relatively few steps (Chapter 3). This is especially important in cases where automated synthesis and high-throughput assays are not available.

Typically, several rounds of synthesis and pharmacological screening are necessary to ascertain the active members of combinatorial mixtures. However, many clever deconvolution strategies (Section 5) have been developed which make the identification of active members of even large libraries achievable in a short period of time (Chapter 3). For lead optimization, where structure–activity relationships are desirable, individual compounds are required. However, the synthesis and screening of mixtures is often a more efficient approach for lead identification (1, 2).

2.3 Development of solution phase combinatorial chemistry

Early reports of the single-step, solution phase synthesis of combinatorial libraries were detailed by at least three groups. Smith and co-workers prepared a library of 1600 amides by reacting 40 carboxylic acid chlorides with 40 nucleophiles (8). The library was screened in a matrix format, allowing immediate deconvolution. A similar sublibrary format was used by Pirrung who prepared a series of carbamate mixtures which were screened for acetylcholinesterase inhibitory activity (9). Prior to these efforts, Rebek reported the single-step construction of libraries presenting amino acid derivatives attached to rigid core templates (10). Because the size of these libraries approached 100 000 members, an iterative selection strategy based on structural groupings of the building blocks was devised. In addition to recent advances in this work, fluorous phase liquid/liquid extractions have been reported (11) (Chapter 11) and substantial progress towards using solution phase multi-component reactions for generating combinatorial mixtures has been disclosed (Chapter 9). Both Ugi and Armstrong have reported Ugi four-component condensations, including the incorporation of a modifiable isocyanide in combination with resin capture strategy to provide useful solution phase library preparations (12). Janda has described a new technology which combines advantages of both solution and solid phase synthesis termed 'liquid phase' combinatorial chemistry (Chapter 12), which entails the use of a polymer support which is soluble under many reaction conditions, but can be precipitated to allow simple product isolation (13).

A number of recent reports have expanded on these approaches including the use of ion exchange resins for solid phase extractions (14) (Section 4.2) and its extension to resin capture for the removal of excess reactants and reagents (15). Several excellent reviews on combinatorial chemistry highlight the advances in solution phase strategies for the preparation of combinatorial libraries (1–3).

3 Cyclic anhydride chemistry with extractive purification

The major limitation to solution phase combinatorial chemistry is the isolation or purification of the library members and intermediates. If the advantages of sample isolation attributed to solid phase synthesis may be embodied in a solution phase synthesis, its non-limiting scale, expanded and non-limiting repertoire of chemical reactions, direct production of soluble intermediates and final products for assay or for purification, and the lack of required linking, attachment/detachment, or capping strategies make solution phase combinatorial synthesis an attractive alternative. One of the most common techniques for solution phase sample isolation and purification is liquid/liquid or liquid/solid extraction (14, 16–23).

Template **1**, a representative of a set of five- and six-membered cyclic anhydride-based templates that have been examined, consists of a densely functionalized core which imposes little structural or conformational bias that might limit its use (14, 17). An interesting example of an alternative template is the generalized dipeptidomimetic template **2**, which contains a rigid bicyclic core with a plane of symmetry that allows it to function as a Gly–X mimetic (16, 17). Its symmetrical structure contains three positions which can be sequentially functionalized enabling the synthesis of libraries with up to three variable groups. This chemistry is applicable to both the parallel synthesis of individual compounds and mixture synthesis. By convergently linking two or more iminodiacetic acid templates, libraries of greater molecular size with a multiplication of diversity are available (Section 4) (20–23).

The anhydride template is activated for the first functionalization by nucleophilic attack which releases a free carboxylic acid group as its second functionalization site. Subsequent deprotection of the secondary amine unmasks the third coupling position. At each step the released functionality aids both the isolation and purification of each of the intermediates and final products from

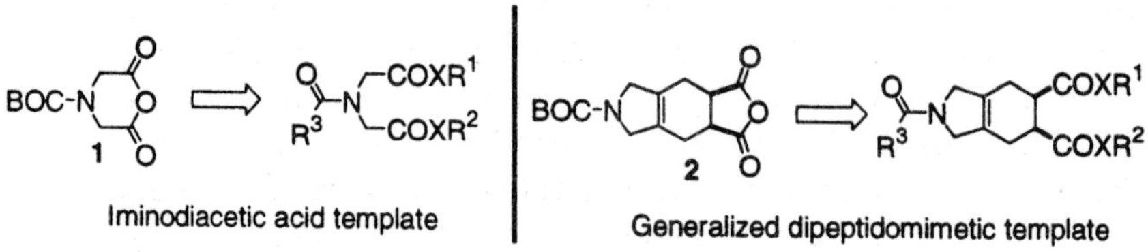

| Iminodiacetic acid template | Generalized dipeptidomimetic template |

Figure 1 Representative cyclic anhydride templates suitable for the solution phase combinatorial chemistry protocols described in this chapter.

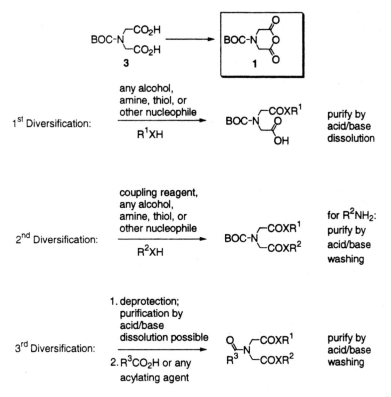

Figure 2 Three-step diversification of the iminodiacetic acid anhydride template. After each step, the products are purified by liquid/liquid or liquid/solid extraction techniques.

starting material, reactants, reagents, and their by-products by simple liquid/ liquid or liquid/solid extraction providing highly pure materials (typically > 95%) independent of reaction efficiencies (*Figure 2*).

Although a number of classes of nucleophiles can be used to functionalize the template, the formation of amide bonds is the most straightforward synthetically, both because the reactions are highly efficient and because excess or unreacted amines can easily be removed with simple acid/base extractions. The extension, however, to other types of linkages can be accomplished using alternative substrates and nucleophiles.

The iminodiacetic acid anhydride template is conveniently protected with the *tert*-butloxycarbonyl (Boc) protecting group. This group is compatible in all of the coupling reactions used in functionalizing the template, and is easily cleaved off by treatment with acid, typically anhydrous HCl, providing the hydrochloride salt which can be directly utilized in the next coupling reaction. The only limitation to the choice of amines used to functionalize the template is that they cannot contain acid-sensitive functionality. N-Boc-iminodiacetic acid anhydride (**1**) is formed by dehydration of N-Boc-iminodiacetic acid (**3**) which is easily prepared from commercially available iminodiacetic acid (*Protocol 1*).

Protocol 1

Synthesis of *N*-Boc-iminodiacetic acid

Reagents

- Iminodiacetic acid (Aldrich)
- Di-*tert*-butyl dicarbonate (Aldrich)

Method

1 Combine iminodiacetic acid (13.3 g, 100 mmol, 1.0 Equiv.), NaOH (8.0 g, 200 mmol, 2.0 Equiv.), water (200 ml), and 1,4-dioxane (200 ml) in a 1 litre flask.

2 Stir until a homogeneous solution is formed (~ 5 min), and then add di-*tert*-butyl dicarbonate (25 ml, 110 mmol, 1.1 Equiv.).[a]

3 Stir for 72 h at room temperature.[b]

4 Pour the mixture into a 1 litre separatory funnel and wash with ethyl ether (2 × 100 ml).

5 Acidify the aqueous layer by adding 10% aqueous HCl (100 ml) and extract with ethylacetate (3 × 150 ml).

6 Combine the organic layers and wash with saturated aqueous NaCl (300 ml).

7 Dry over Na_2SO_4 and remove the solvent by rotary evaporation to afford a viscous oil which upon crystallization by addition of ethyl acetate (30 ml) followed by hexanes (60 ml) affords 17.2 g (74%) of *N*-((*tert*-butyloxy)carbonyl)iminodiacetic acid (*N*-Boc-iminodiacetic acid) (3).

[a] It was found to be easier to carefully melt the di-*tert*-butyl dicarbonate so that it could be added via syringe (m.p. = 23 °C). Alternatively the solid reagent may be added (25 ml = 24 g).

[b] After the addition of the di-*tert*-butyl dicarbonate, the solution becomes cloudy.

The cyclic anhydride template (**1**) may either be prepared *in situ* or isolated and stored for subsequent functionalization. EDCI was found to be an excellent dehydrating agent for the formation of the anhydride, and DMF an appropriate solvent in which nearly all amines or amine hydrochlorides investigated were soluble. Virtually any primary or secondary amine can be used for the first diversification of iminodiacetic acid. Polymer-bound EDCI may also be used to prepare **1** and removed by simple filtration (23).

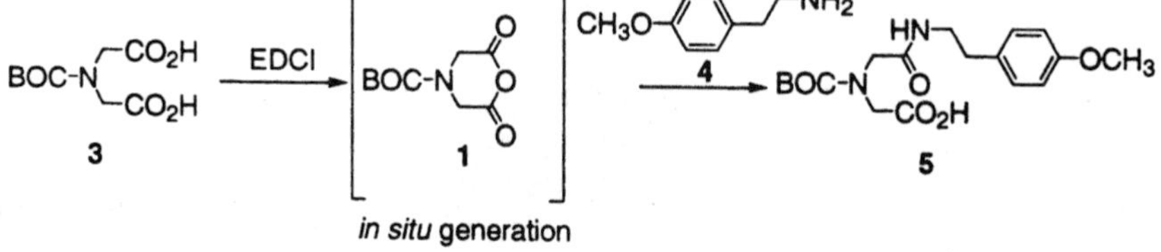

in situ generation

Figure 3 Synthesis and first diversification of the iminodiacetic acid anhydride template.

Protocol 2

First diversification of iminodiacetic acid

Reagents

- 1-(3-Dimethylaminopropyl)-3-ethylcarbodiimide hydrochloride (EDCI) (Aldrich)
- N-Boc-iminodiacetic acid (**3**) (*Protocol 1*)
- 4-Methoxyphenethylamine (**4**) (Aldrich)

Method A

1   Combine N-Boc-iminodiacetic acid (**3**) (4.7 g, 20 mmol, 1.0 Equiv.) and EDCI (3.8 g, 20 mmol, 1.0 Equiv.) in a 100 ml flask.

2   Add dimethylformamide (DMF) (60 ml) and stir for 1 h at room temperature.

3   Slowly add 4-methoxyphenethylamine (**4**) (3.0 ml, 20 mmol, 1.0 Equiv.) via syringe while stirring (slightly exothermic).[a]

4   Stopper the flask, and let stand for 16 h at room temperature.[b]

5   Pour the mixture into a 500 ml separatory funnel.

6   Dilute with ethylacetate (300 ml) and wash sequentially with 10% aqueous HCl (2 × 200 ml) and saturated aqueous NaCl (200 ml).

7   Dry over Na_2SO_4 and remove the solvent by rotary evaporation to afford 7.3 g (99%) of N-Boc-iminodiacetic acid monoamide **5**.[c]

Method B

1   Combine N-Boc-iminodiacetic acid (**3**) (11.7 g, 50 mmol, 1.0 Equiv.) and EDCI (10.6 g, 55 mmol, 1.1 Equiv.) in a 500 ml flask.

2   Add methylene chloride (200 ml), stopper the flask, stir until all solids have dissolved, and allow to stand for 4 h at room temperature.

3   Pour the mixture into a 500 ml separatory funnel.

4   Wash with H_2O (200 ml) and saturated aqueous NaCl (200 ml).

5   Dry over Na_2SO_4 and remove the solvent by rotary evaporation to afford 10.3 g (95%) of N-Boc-iminodiacetic acid anhydride (**1**) as a white crystalline powder.[d]

6   Dissolve the cyclic anhydride (**1**) isolated from step 5 (4.7 g, 20 mmol, 1.0 Equiv.) in DMF (60 ml).

7   Slowly add a slight excess (1.1–1.2 Equiv.)[e] of 4-methoxyphenethylamine (**4**) (3.3 ml, 22 mmol) while stirring (slightly exothermic).[a]

8   Let stand for 16 h at room temperature.[b]

9   Work-up the reaction as described in *Method A* to afford 7.1 g (97%) of N-Boc-iminodiacetic acid monoamide **5**.[c]

[a] If it is more convenient, the amine for this reaction may be added as its hydrochloride salt, followed by one equivalent of diisopropylethylamine.

Protocol 2 continued

[b] These coupling reactions are typically complete in < 1 h, however the extended reaction time ensures complete consumption of starting materials, and when running many reactions in parallel it was found to be more convenient to set up several couplings on one day and conduct the work-ups the following morning. The reactions may be monitored by TLC (5:5:1, ethyl acetate/dichloromethane/methanol eluent, ninhydrin stain) if desired.

[c] Typical yields for this reaction are 90–99%.

[d] This anhydride has also been prepared using less expensive (but highly toxic) triphosgene instead of EDCI (Ni, Z.-J., Maclean, D., Holmes, C. P., Murphy, M. M., Ruhland, B., Jacobs, J. W., *et al.* (1996). *J. Med. Chem.*, **39**, 1601).

[e] Because the anhydride is itself activated, no coupling reagents are present, and an excess of amine can be utilized to guarantee that no unreacted anhydride remains. Any excess amine will be removed during the 10% aqueous HCl washes.

For the second diversification of iminodiacetic acid, EDCI has proven to be less reliable. A wide range of coupling reagents were examined and PyBOP was found to work well in all cases. The reagent and reagent by-products are removed in the aqueous extractions. Again, virtually any primary or secondary amine (or amine hydrochloride) can be coupled to the template in this step. The conversions are usually excellent, and the liquid/liquid extraction protocol provides pure products regardless of reaction efficiencies.

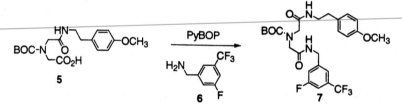

Figure 4 Second diversification of the iminodiacetic acid anhydride template.

Protocol 3
Second diversification of iminodiacetic acid

Reagents

- N-Boc-iminodiacetic acid monoamide **5** (*Protocol 1*)
- 3-Fluoro-5-(trifluoromethyl)benzylamine (**6**) (Aldrich)

- Benzotriazole-1-yl-oxy-*tris*(pyrrolidino)-phosphonium hexafluorophosphate (PyBOP) (Acros)

Method

1 Combine the *N*-Boc-iminodiacetic monoamide **5** (0.64 g, 1.75 mmol, 1.0 Equiv.), 3-fluoro-5-(trifluoromethyl)benzylamine (**6**) (0.28 ml, 1.9 mmol, 1.1 Equiv.), and PyBOP (1.0 g, 1.9 mmol, 1.1 Equiv.) in a 20 ml vial.[a]

Protocol 3 continued

2   Add DMF (10 ml) followed by diisopropylethylamine (0.60 ml, 3.5 mmol, 2.0 Equiv.).

3   Cap the vial, stir until all solids have dissolved, and let stand for 16 h at room temperature.[b]

4   Pour the reaction mixture into a 250 ml separatory funnel.

5   Dilute with ethylacetate (100 ml) and wash sequentially with 10% aqueous HCl (3 × 50 ml), saturated aqueous NaHCO$_3$ (2 × 50 ml), and saturated aqueous NaCl (50 ml).

6   Dry over Na$_2$SO$_4$ and remove the solvent by rotary evaporation to afford 0.85 g (90%) of N-Boc-iminodiacetic acid diamide **7**.[c]

[a] If it is more convenient, the amine for this reaction may be added as the hydrochloride salt, provided that an additional equivalent of diisopropylethylamine is employed in step 2.

[b] These coupling reactions are typically complete in < 1 h, however the extended reaction time ensures complete consumption of starting materials, and when running many reactions in parallel it was found to be more convenient to set up several couplings on one day and conduct the work-ups the following morning. The reactions may be monitored by TLC (5:5:1, ethylacetate/dichloromethane/methanol eluent, ninhydrin stain) if desired.

[c] Typical yields for this reaction are 65–99%.

Although the third diversification of iminodiacetic acid was first described using PyBOP to couple carboxylic acids (14, 17), subsequent work has shown that PyBroP provides superior yields with most substrates. This reagent seems to be particularly efficient for the coupling of secondary amines such as the imino-diacetic acid diamides. The reagent or reagent by-products are removed in the aqueous extraction work-up. Other acylating agents, such as carboxylic acid chlorides and sulfonyl chlorides have also been used for this functionalization (23).

The synthesis of individual compounds is described in *Protocol 4*. However, the same general protocol may be applied for the synthesis of mixtures, by coupling several carboxylic acids to a single iminodiacetic acid diamide. A stock solution of a mixture of ten or more carboxylic acids may be prepared and added to the iminodiacetic acid diamide as described. If mixtures are desired, the only difference in the protocol is to alter the stoichiometry of the reaction so that the iminodiacetic acid diamide and PyBroP are used in a slight excess (1.1–1.5 Equiv.) relative to the total amount of carboxylic acids. This ensures a near-equimolar product mixture regardless of differences in coupling rates for the different carboxylic acid substrates.

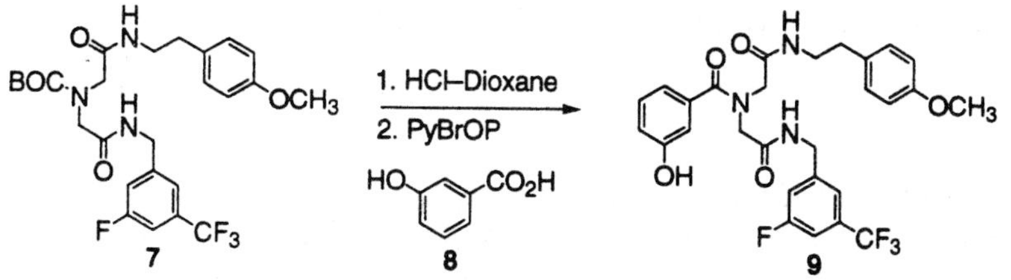

Figure 5 Third diversification of the iminodiacetic acid anhydride template.

Protocol 4

Third diversification of iminodiacetic acid

Reagents

- *N*-Boc-iminodiacetic acid diamide **7** (*Protocol 3*)
- 4 M hydrogen chloride in dioxane (HCl–dioxane) (Aldrich)
- Bromo-*tris*(pyrrolidino)-phosphonium hexafluorophosphate (PyBroP) (Novabiochem)
- 3-Hydroxybenzoic acid (**8**) (Aldrich)

Method

1   Dissolve the *N*-Boc-iminodiacetic acid diamide **7** (54 mg, 0.1 mmol, 1.0 Equiv.) in 4 M HCl–dioxane (2.5 ml) in a 4 ml vial.

2   Stir for 4 h at room temperature and remove the solvent and excess HCl by evaporation under a stream of N_2.

3   Add the carboxylic acid (**8**) (15 mg, 0.11 mmol, 1.1 Equiv.) and PyBroP (51 mg, 0.11 mmol, 1.1 Equiv.).

4   Add DMF (1 ml) followed by diisopropylethylamine (50 μl, 0.3 mmol, 3.0 Equiv.).

5   Cap the vial, and stir for 16 h at room temperature.[a]

6   Pour the reaction mixture into a 125 ml separatory funnel.

7   Dilute with ethylacetate (50 ml) and wash sequentially with 10% aqueous HCl (3 × 50 ml), saturated aqueous $NaHCO_3$ (2 × 50 ml), and saturated aqueous NaCl (50 ml).

8   Dry over Na_2SO_4 and remove the solvent by evaporation to afford 48 mg (85%) of the trifunctionalized iminodiacetic acid template **9**.[b]

[a] These coupling reactions are typically complete in < 1 h, however the extended reaction time ensures complete consumption of starting materials, and when running many reactions in parallel it was found to be more convenient to set up several couplings on one day and conduct the work-ups the following morning. The reactions may be monitored by TLC (5:5:1, ethyl acetate/dichloromethane/methanol eluent, ninhydrin stain) if desired.

[b] The solvent may be evaporated under a stream of N_2 from a gas manifold, by rotary evaporation, or by vacuum centrifuge. Typical yields for this reaction are 16–100%.

4 Higher order libraries

4.1 Dimerization, trimerization, or tetramerization of iminodiacetic acid diamide libraries

Two or more iminodiacetic acid diamides may be linked, producing dimer, trimer, or tetramer libraries. This allows for an increase in diversity, because of the variable elements in the linker itself, and from the multiplication of diversity achieved by combining differently substituted components (hetero-oligomerizations). In addition, the potentially multivalent nature of the generated compounds

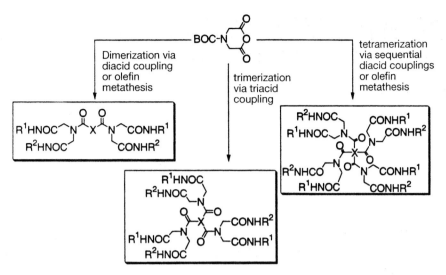

Figure 6 Dimerization, trimerization, and tetramerization of iminodiacetic acid diamides via amide coupling reactions or olefin metathesis.

has proven useful in studying protein–protein interactions including receptor activation by dimerization, trimerization, or oligimerization (20–23).

Here, solution phase combinatorial chemistry provides a unique opportunity when compared to the linear approaches required in solid phase combinatorial chemistry. The convergent linkages are especially suited for solution phase synthesis and would be precluded by typical solid phase techniques since the components are on mutually exclusive solid phases. Thus:

(a) A large number of diverse molecules can be assembled in only a few steps using solution phase techniques.

(b) A further multiplication of diversity can be achieved with each convergent linkage.

(c) Purification at each step is possible by liquid/liquid extractions.

The convergent approach to library generation is particularly appropriate for the discovery of novel antagonists and agonists of biological receptors which are activated by ligand-induced homo- or hetero-dimerization or oligimerization. For example, the convergent approach allows for the identification of binders (receptor antagonists) and subsequently their potential conversion via chemical dimerization into dimerizing (trimerizing) agents (receptor agonists) (20–23).

4.2 Dimerization coupling with dicarboxylic acids

Perhaps the simplest method to dimerize iminodiacetic acid diamides is to couple the free secondary amine with dicarboxylic acids (19, 23). The same liquid/liquid or liquid/solid extraction procedures described previously can be used to provide pure products. Extended reaction times, and an excess of both the coupling reagent and the secondary amine guarantee complete consumption of the limiting dicarboxylic acids and a near–equimolar product library formation if conducted in mixture library synthesis.

The dimerization couplings may be conducted with individual or small mixtures (typically eight to ten) of dicarboxylic acids. If mixtures are to be synthesized and tested, it is advantageous to store and catalogue a portion of the unlinked N-Boc-iminodiacetic acid diamides for future deconvolution. (Section 5.2).

With substrates that are particularly water soluble, conventional extractive work-up with 10% aqueous HCl has been found to occasionally lead to a low recovery of product. In these cases a 1:1 mixture of 20% aqueous HCl and saturated aqueous NaCl, may be substituted which typically improves the yield of isolated product, yet is still effective in removing reagents and reaction by-products.

The acidic liquid/liquid extractions may also be replaced by liquid/solid extractions utilizing ion exchange resin. When in its acidic form, cation exchange resin protonates any basic amine in solution, and the resulting positive ion adheres to the negatively charged resin. Stirring a methylene chloride solution of the crude reaction products in the presence of the resin (e.g. Dowex 50X8–200 ion exchange resin which had previously been washed with methanol and methylene chloride) for several minutes removes all traces of any unreacted iminodiacetic acid diamide (14, 19). This has been found to be true even for large, hydrophobic amines which may sometimes be difficult to remove under the conventional liquid/liquid extraction protocol. Passage through a short column of the same resin (methanol/chloroform eluent) has also been found to be an effective method for purifying a solution of the diamide products contaminated with unreacted secondary amine.

Analogous studies conducted with linking tricarboxylic acids have been used to prepare iminodiacetic acid diamide trimers (22, 23).

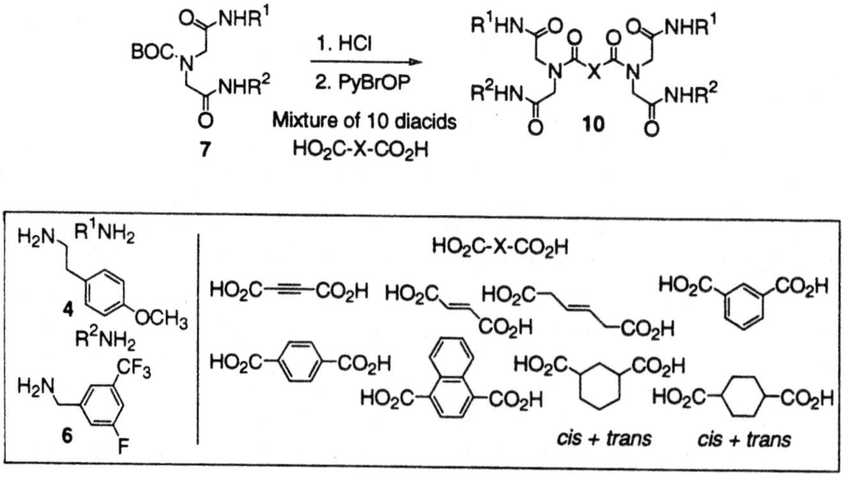

Figure 7 Dimerization of iminodiacetic acid diamides via diacid coupling.

Protocol 5

Dimerization of iminodiacetic acid diamides via dicarboxylic acid coupling

Reagents

- *N*-Boc-iminodiacetic acid diamide **7** (*Protocol 3*)
- PyBroP (Novabiochem)

- Dicarboxylic acid stock solution[a]
- 4 M HCl–dioxane (Aldrich)

Method

1 Dissolve the *N*-Boc-iminodiacetic acid diamide (**7**) (80 mg, 0.15 mmol, 3.0 Equiv.) in 4 M HCl–dioxane (2.5 ml) in a 4 ml vial.

2 Stir for 4 h at room temperature and remove the solvent and excess HCl by evaporation under a stream of N_2.

3 Add 1 ml of diacid stock solution (0.05 mmol, 1.0 Equiv. dicarboxylic acids/0.45 mmol, 9 Equiv. diisopropylethylamine) and PyBroP (70 mg, 0.15 mmol, 3 Equiv.).

4 Cap the vial, shake or stir to effect dissolution, and allow to stand for 16 h at room temperature.[b]

5 Pour the reaction mixture into a 125 ml separatory funnel.

6 Dilute with ethylacetate (50 ml) and wash sequentially with 10% aqueous HCl (3 × 50 ml),[c] saturated aqueous $NaHCO_3$ (2 × 50 ml), and saturated aqueous NaCl (50 ml).

7 Dry over Na_2SO_4 and remove the solvent by evaporation to afford 36 mg (73%) of the mixture of dimerized iminodiacetic acid diamides **10**.[d]

[a] The stock solution is prepared by diluting a mixture of 0.5 mmol of each of ten diacids (*Figure 7*) and diisopropylethylamine (8 ml, 45 mmol) to 100 ml with DMF. Each 1 ml aliquot contains 0.005 mmol of each carboxylic acid (0.05 mmol total) and 0.45 mmol of diisopropylethylamine. All of the diacids are commercially available (Aldrich). Other diacid mixtures have been used resulting in similar yields and purities. If individual compounds are desired, one can simply substitute the diacid mixture stock solution with a solution delivering 0.05 mmol of any individual diacid.

[b] These coupling reactions are typically complete in < 1 h, however the extended reaction time ensures complete consumption of starting materials, and when running many reactions in parallel it was found to be more convenient to set up several couplings on one day and conduct the work-ups the following morning. The reactions may be monitored by TLC (5:5:1, ethylacetate/dichloromethane/methanol eluent, ninhydrin stain) if desired.

[c] In cases where the product has appreciable water solubility it was observed that the successive aqueous washings resulted in poor recovery of product. To offset this, a 1:1 mixture of 20% hydrochloric acid and saturated aqueous NaCl could be used instead of the 10% aqueous HCl. This resulted in improved isolation of product. Alternatively, ion exchange resin (solid/liquid extraction) can be employed to remove any unreacted amine, see text.

[d] The solvent may be evaporated under a stream of N_2 from a gas manifold, by rotary evaporation, or by vacuum-centrifuge. Typical yields for this step are 16–100%.

4.3 Dimerization via olefin metathesis coupling

Olefin metathesis has been established as an excellent reaction in the synthesis of complex organic molecules (24). It is a highly efficient reaction, compatible with most functional groups. Metathesis may also be employed in the field of solution phase combinatorial chemistry, e.g. to join functionalized iminodiacetic acid diamides while randomizing the length of the linking chain (18, 20). A mixture of ω-alkene carboxylic acids is first attached to the iminodiacetic acid diamides. For this step, the conventional liquid/liquid extractive work-up provides the pure products. The subsequent metathesis couplings reactions may be purified by silica gel chromatography although the products are often sufficiently pure to use directly. Although the reaction conditions are optimized to consume all of the metathesis precursors, this simple chromatographic separation provides an additional level of purity control without compromising the integrity of the mixtures. The dimerized products elute in a single band, well after any unlinked starting material, in standard chromatographic separations.

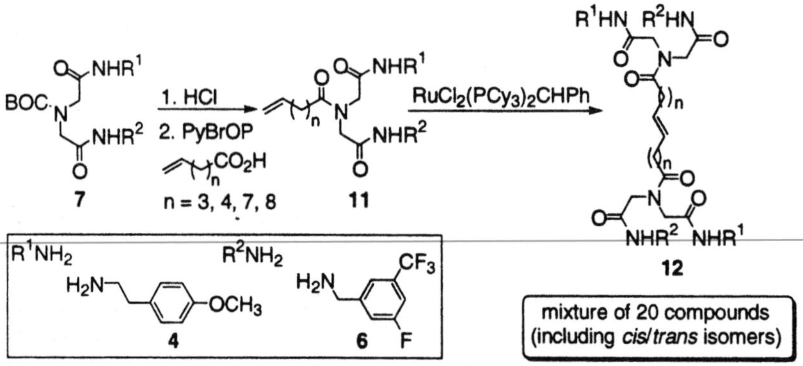

Figure 8 Dimerization of iminodiacetic acid diamides via olefin metathesis coupling.

Protocol 6

Olefin metathesis dimerization coupling

Reagents

- N-Boc-iminodiacetic acid diamide **7** (*Protocol 3*)

- Stock solution of ω-alkene carboxylic acids[a]

- Metathesis catalyst: RuCl$_2$(PCy$_3$)$_2$CHPh[b] (Strem)

- 4 M HCl–dioxane (Aldrich)

- PyBroP (Novabiochem)

A. Preparation of functionalized terminal olefins (metathesis precursors)[c]

1   Dissolve the N-Boc-iminodiacetic acid diamide (**7**) (0.41 g, 0.75 mmol, 1.0 Equiv.) in 4 M HCl–dioxane (2.5 ml) in an 8 ml vial.

2 Stir for 4 h at room temperature and remove the solvent and excess HCl by evaporation under a stream of N_2.

3 Add PyBroP (350 mg, 0.75 mmol, 1.0 Equiv.) and 5.0 ml of ω-alkene carboxylic acid stock solution (0.5 mmol, 0.67 Equiv. ω-alkene carboxylic acids/2.25 mmol, 3.0 Equiv. diisopropylethylamine).[a]

4 Cap the vial, stir until all solids have dissolved, and allow to stand for 16 h.

5 Pour the reaction mixture into a 250 ml separatory funnel. Dilute with ethylacetate (100 ml) and wash sequentially with 10% aqueous HCl (3 × 50 ml), saturated aqueous $NaHCO_3$ (2 × 50 ml), and saturated aqueous NaCl (50 ml).

6 Dry over Na_2SO_4 and remove the solvent by rotary evaporation to afford 0.25 g (87%) of a near-equimolar mixture of 4 ω-alkene-iminodiacetic acid diamides **11**.[d,e]

B. Homodimerization linkage by olefin metathesis

1 Dissolve the mixture of 4 ω-alkene-iminodiacetic acid diamides (47 mg, 0.082 mmol, 1 Equiv.) (part A) and the metathesis catalyst (17 mg, 0.021 mmol, 0.25 Equiv.) in $CHCl_3$ (3 ml) in a 5 ml round-bottom flask equipped with a reflux condenser.

2 Heat to reflux (oil bath) and stir for 16 h under an argon atmosphere (Ar balloon) and then remove the solvent by evaporation.[d]

3 Purify the crude reaction products by flash chromatography: SiO_2, 1.5 × 20 cm, 50–100% ethylacetate/hexanes, and 5% methanol/ethylacetate to afford 31 mg of the ten-component mixture of homodimers **12** (67%).[e,f]

[a] The stock solution was prepared by diluting a mixture of four ω-alkene carboxylic acids (*Figure 8*) (2.5 mmol each) and diisopropylethylamine (8 ml, 45 mmol) to a volume of 100 ml with DMF. Each 5 ml aliquot contains 0.125 mmol of each carboxylic acid (0.5 mmol total) and 2.25 mmol of diisopropylethylamine. These ω-alkene carboxylic acids are commercially available (Aldrich and Lancaster). Other ω-alkene carboxylic acids may be used. However, for best results in metathesis, the olefin should be separated from carbonyl, or other functionality, by at least three methylene groups.

[b] Other metathesis catalysts ($Mo(C_{10}H_{12}) (C_{12}H_{17}N)[OC(CH_3) (CF_3)_2]_2$ and $RuCl_2(PCy_3)_2 = CH - CH = CPh_2$) were found to be less effective, or too sensitive under the reaction conditions.

[c] If individual compounds are desired the individual ω-alkene carboxylic acid (0.67 Equiv.) may simply be used instead of the mixture.

[d] The solvent may be evaporated under a stream of N_2 from a gas manifold, by rotary evaporation, or by vacuum-centrifuge.

[e] Typical yields for this step are 41–99%.

[f] All of the reaction products have a distinct R_f from any starting materials. Although the reaction conditions are optimized to consume nearly all of the metathesis precursors, this simple chromatographic separation provides an additional level of purity control without compromising the integrity of the mixtures. The dimerized products elute in a single band, well after any unlinked starting material. Typical yields for this step are 21–79%.

4.4 Tetramerization of iminodiacetic acid diamides

4.4.1 Tetramerization via dicarboxylic acid couplings

The iminodiacetic acid diamides may be sequentially dimerized producing tetramer libraries (20, 21, 23). The first step is to dimerize the diamides with a dicarboxylic acid which contains a protected reactive functionality capable of a second dimerization reaction. N-Boc-iminodiacetic acid (**3**) was found to be an effective linker. After the first dimerization, the secondary amine may be deprotected, and the molecule is ready for another coupling reaction. The second dimerization may be carried out by either another dicarboxylic acid coupling or olefin metathesis reaction. The liquid/liquid extraction protocols (or liquid/solid extractions, see Section 4.2) are again appropriate for the purification of the diacid coupling reactions.

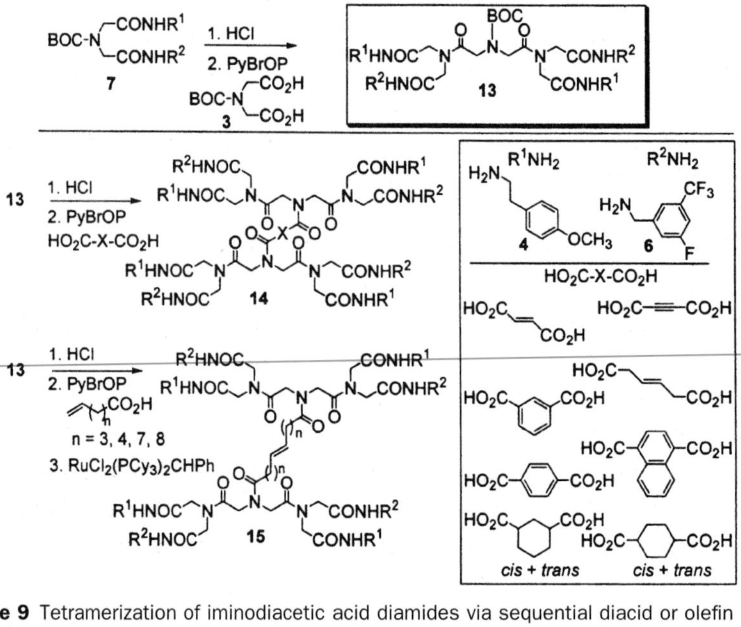

Figure 9 Tetramerization of iminodiacetic acid diamides via sequential diacid or olefin metathesis couplings.

Protocol 7

Tetramerization of iminodiacetic acid diamides by sequential diacid couplings

Reagents

- N-Boc-iminodiacetic acid diamide **7** (*Protocol 3*)

- PyBroP (Novabiochem)

- N-Boc-iminodiacetic acid (**3**) (*Protocol 1*)
- Dicarboxylic acid stock solution[a]
- 4.0 M HCl–dioxane (Aldrich)

A. Preparation of N-Boc-iminodiacetic acid linked dimers

1 Dissolve the N-Boc-iminodiacetic acid diamide **7** (2.7 g, 5.0 mmol, 3.0 Equiv.) in 4 M HCl–dioxane (25 ml) in a 100 ml flask.

2 Stir for 4 h at room temperature and remove the solvent and excess HCl by evaporation under a stream of N_2.

3 Add N-Boc-iminodiacetic acid (**3**) (0.40 g, 1.67 mmol, 1.0 Equiv.) followed by PyBroP (2.32 g, 5 mmol, 3.0 Equiv.), DMF (20 ml), and diisoproplylethylamine (2.6 ml, 15 mmol, 9 Equiv.).

4 Stopper the flask, shake or stir to effect dissolution, and allow to stand for 16 h at room temperature.[b]

5 Pour the reaction mixture into a 250 ml separatory funnel.

6 Dilute with ethyl acetate (200 ml) and wash sequentially with 10% aqueous HCl (2 × 100 ml),[c] saturated aqueous $NaHCO_3$ (2 × 100 ml), and saturated aqueous NaCl (100 ml).

7 Dry over Na_2SO_4 and remove the solvent by rotary evaporation to afford 1.7 g (99%) of dimerized iminodiacetic acid diamide **13**.[d]

B. Dimerization of the iminodiacetic acid diamide dimers (tetramerization)

1 Dissolve the dimerized iminodiacetic acid diamide **13** (108 mg, 0.11 mmol, 3.0 Equiv.) in CH_3Cl (1 ml) in a 4 ml vial and add 4 M HCl–dioxane (1 ml).

2 Stir for 4 h at room temperature and remove the solvent and excess HCl by evaporation under a stream of N_2.

3 Add 1 ml of diacid stock solution (0.033 mmol, 1.0 Equiv. dicarboxylic acids/0.29 mmol, 9 Equiv. diisopropylethylamine) and PyBroP (45 mg, 0.1 mmol, 3 Equiv.).

4 Cap the vial, shake or stir to effect dissolution, and allow to stand for 16 h at room temperature.[b]

5 Pour the reaction mixture into a 125 ml separatory funnel.

6 Dilute with ethylacetate (30 ml) and wash sequentially with 10% aqueous HCl (3 × 50 ml),[c] saturated aqueous $NaHCO_3$ (2 × 50 ml), and saturated aqueous NaCl (50 ml).

7 Dry over Na_2SO_4 and remove the solvent by evaporation to afford 56 mg (82%) of the tetramer mixture **14**.[e]

[a] The stock solution is prepared by diluting a mixture of 0.5 mmol of each of ten diacids (*Figure 9*) and diisopropylethylamine (8 ml, 45 mmol) to 100 ml with DMF. Each 0.64 ml aliquot contains 0.0033 mmol of each carboxylic acid (0.033 mmol total) and 0.29 mmol of diisopropylethylamine. All of the diacids are commercially available (Aldrich). Other diacid mixtures have been used resulting in similar yields and purities. If individual compounds are desired, one can simply substitute the diacid mixture stock solution with a solution delivering 0.05 mmol of any individual diacid.

Protocol 7 continued

[b] These coupling reactions are typically complete in < 1 h, however the extended reaction time ensures complete consumption of starting materials, and when running many reactions in parallel it was found to be more convenient to set up several couplings on one day and conduct the work-ups the following morning. The reactions may be monitored by TLC (5:5:1, ethylacetate/dichloromethane/methanol eluent, ninhydrin stain) if desired.

[c] In cases where the product has appreciable water solubility it was observed that the successive aqueous washings resulted in poor recovery of product. To offset this, a 1:1 mixture of 20% aqueous HCl and saturated aqueous NaCl should be used instead of the 10% aqueous HCl. This resulted in improved isolation of product. Alternatively, acidic cation exchange resin (solid/liquid extraction) can be employed to remove any unreacted amine (Section 4.2).

[d] Typical yields for this step are 57–100%.

[e] The solvent may be evaporated under a stream of N_2 from a gas manifold, by rotary evaporation, or by vacuum-centrifuge, if available. Typical reaction yields are 36–95%.

4.4.2 Tetramerization via olefin metathesis coupling

i. Small mixture libraries

Olefin metathesis may be utilized to covalently link the dimerized iminodiacetic acid diamides (such as **13**), resulting in tetramer libraries. Mixtures of ω-alkene carboxylic acids are first appended and these products are then further dimerized with the metathesis catalyst. The work-ups and purifications are analogous to those described for the dimerization metathesis couplings (Section 4.3).

Protocol 8

Olefin metathesis tetramerizations

Reagents

- Dimerized iminodiacetic acid diamide **13** (*Protocol 7*, part A)

- Stock solution of ω-alkene carboxylic acids[a]

- Metathesis catalyst: $RuCl_2(PCy_3)_2CHPh$[b] (Strem)

- 4 M HCl–dioxane (Aldrich)

- PyBroP (Novabiochem)

A. Preparation of functionalized terminal olefins (metathesis precursors)

1   Dissolve the dimerized N-Boc-iminodiacetic acid diamide, **13** (88 mg, 0.09 mmol, 1.5 Equiv.) in 4 M HCl–dioxane (2.5 ml) in an 8 ml vial.

2   After 4 h at room temperature remove the solvent and excess HCl by evaporation under a stream of N_2.

3   Add PyBroP (41 mg, 0.09 mmol, 1.5 Equiv.) and 0.6 ml of ω-alkene carboxylic acid stock solution (0.06 mmol, 1.0 Equiv. ω-alkene carboxylic acids/0.27 mmol, 4.5 Equiv. diisopropylethylamine).[a]

4 Cap the vial, stir until dissolved, and allow to stand for 16 h.

5 Pour the reaction mixture into a 125 ml separatory funnel. Dilute with ethylacetate (50 ml) and wash sequentially with 10% aqueous HCl (3 × 50 ml),[c] saturated aqueous NaHCO$_3$ (2 × 50 ml), and saturated aqueous NaCl (50 ml).

6 Dry over Na$_2$SO$_4$ and remove the solvent by evaporation to afford 45 mg (87%) of N-alkenyl dimerized iminodiacetic acid diamide (metathesis precursor).[d,e]

B. Dimerization linkage by olefin metathesis

1 Dissolve the mixture of N-alkenyl dimerized iminodiacetic acid diamide dimer (part A, step 6) (39 mg, 0.038 mmol, 1.0 Equiv.) with the metathesis catalyst (6.2 mg, 0.0075 mmol, 0.20 Equiv.) in CHCl$_3$ (3 ml) in a 5 ml round-bottom flask equipped with a reflux condenser.

2 Heat to reflux (oil bath) and stir for 16 h under an argon atmosphere (Ar balloon) and then remove the solvent by evaporation.[d]

3 Purify the crude reaction products by flash chromatography: SiO$_2$, 1.5 × 20 cm, 50–100% ethyl acetate/hexanes, and 5% methanol/ethyl acetate to afford 26 mg (65%) of the tetramer mixture **15**.[f]

[a] The stock solution was prepared by diluting a mixture of four ω-alkene carboxylic acids (Figure 9) (2.5 mmol each) and diisopropylethylamine (8 ml, 45 mmol) to a volume of 100 ml with DMF. Each 0.6 ml aliquot contains 0.015 mmol of each carboxylic acid (0.06 mmol total) and 0.27 mmol of diisopropylethylamine. These ω-alkene carboxylic acids are commercially available (Aldrich and Lancaster). Other ω-alkene carboxylic acids may be used, however for best results the olefin should be separated from the linking carboxylate, or other functionality, by at least three methylene groups. If individual compounds are desired the individual ω-alkene carboxylic acid (0.06 mmol) may simply be used instead of the mixture.

[b] Other metathesis catalysts (Mo(C$_{10}$H$_{12}$) (C$_{12}$H$_{17}$N)[OC(CH$_3$) (CF$_3$)$_2$]$_2$ and RuCl$_2$(PCy$_3$)$_2$=CH–CH= CPh$_2$) were found to not be as effective, or too sensitive under the reaction conditions.

[c] In cases where the product has appreciable water solubility it was observed that the successive aqueous washings resulted in poor recovery of product. To offset this a 1:1 mixture of 20% aqueous HCl and saturated aqueous NaCl should be used instead of the 10% aqueous HCl. This resulted in improved isolation of product. Alternatively, acidic cation exchange resin (solid/liquid extraction) can be employed to remove any unreacted amine (Section 4.2).

[d] The solvent may be evaporated under a stream of N$_2$ from a gas manifold, by rotary evaporation, or by vacuum-centrifuge.

[e] Typical yields are 60–100%.

[f] All the reaction products have a distinct R$_f$ from any starting materials. Although the reaction conditions are optimized to consume nearly all of the metathesis precursors, this simple chromatographic separation provides an additional level of purity control without compromising the integrity of the mixtures. The dimerized products elute in a single band, well after any unlinked starting material. Typical yields for this step are 36–78%.

ii. Large mixture libraries

The olefin metathesis reaction can also be used to generate extremely large libraries containing thousands or even millions of compounds in only a few synthetic steps (20, 21). Notably, the convergent dimerization of mixtures allows a multiplication of diversity, such that when a library of x members is symmetrically dimerized the number of products n for the combination can be determined by the following equation:

$$n = x(x + 1)/2$$

Following the same synthetic route used to generate the previous metathesis products (*Protocol 8*) libraries of large diversity can be achieved by coupling mixtures at each stage (*Figure 10*). The reaction conditions and purification are the same as described (*Protocol 8*), however mixtures of amines are coupled to the template at each stage, and for simplification of deconvolution, the metathesis products (alkenes which are 1:2.2–3.2 mixtures of *cis/trans* isomers) can be later hydrogenated. The screening of such large mixtures is especially useful when combined with the new deconvolution protocol of deletion synthesis (Section 5.3).

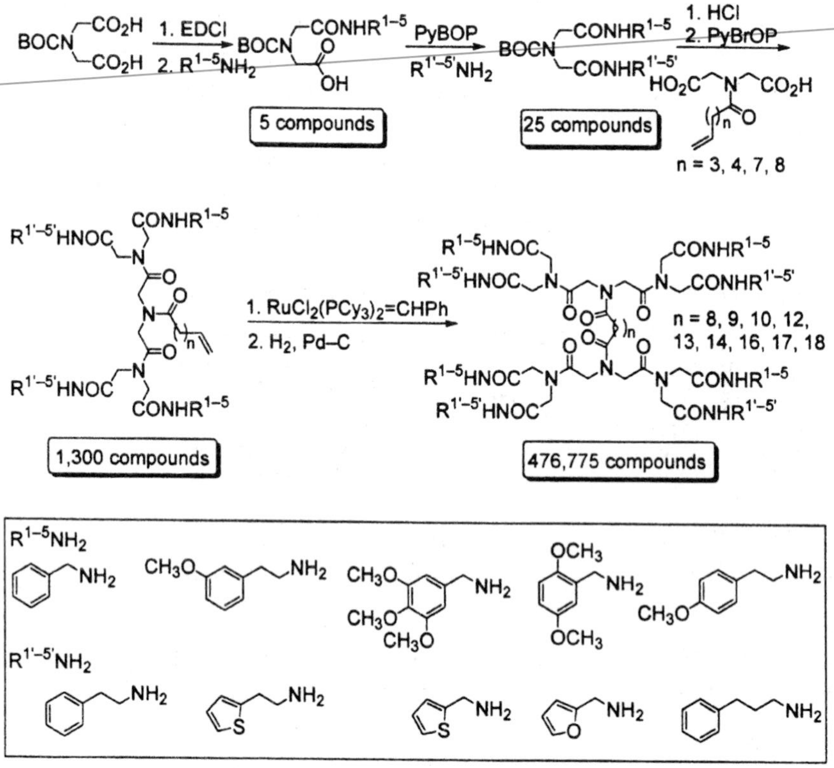

Figure 10 Large mixture libraries are possible by the convergent dimerization of mixtures by olefin metathesis. In only five synthetic steps, a library of 476 775 compounds can be synthesized from only 14 building blocks (ten amines and four ω-alkene carboxylic acids).

5 Deconvolution

5.1 Introduction to deconvolution

Combinatorial chemistry allows diverse mixtures to be synthesized and screened in such a manner that compound identities can be established without complex purification and structure determination steps necessary with natural products screening. Deconvolution is the process of synthesizing and screening a series of sublibraries which are constructed in such a way that the active components of the parent library can be identified (Chapter 3).

Several techniques for the identification of lead compounds from mixtures have been described in recent reviews (1, 2) (Chapter 3). Two techniques that we have found to be the most useful for the deconvolution of the small and large mixtures described in this chapter are iterative (recursive) deconvolution and positional scanning/deletion synthesis.

5.2 Iterative/recursive deconvolution

Iterative deconvolution involves the synthesis and screening of successive rounds of sublibraries. This basic procedure was originally described by Houghten (25) for peptide libraries (Chapter 3), and later by Ecker for nucleotide libraries (known as SURF: synthetic unrandomization of randomized fragments) (26). In 1994 Janda described an iterative procedure for the deconvolution of peptide libraries termed recursive deconvolution (27). At each stage of the linear split synthesis, one-third of the material is stored and catalogued for later use in deconvolution. If activity is detected in the final mixture, the deconvolution sublibraries can be readily prepared from the archived samples. For each position of variability within the library, a separate round of testing is required. From each round of

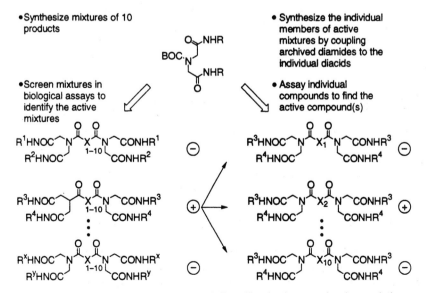

Figure 11 Deconvolution of iminodiacetic acid dimer libraries by recursive deconvolution.

testing, the most active substitution at one position is established, and used in the following testing round. As one approaches the single most active constituent of the mixture, the relative activity increases ensuring the accuracy of the lead identification.

Many of the small mixture iminodiacetic acid libraries described in this chapter have been deconvoluted using this recursive technique. A portion of each of the individual iminodiacetic acid-derived precursors was stored for later deconvolution of the mixture libraries.

5.3 Deletion synthesis deconvolution and scanning deconvolution

One disadvantage of iterative deconvolution procedures is the requirement for several rounds of assays before the identity of the biologically active members is established. Two complementary techniques have been described which allow a series of sublibraries to be concurrently synthesized and tested up front along with the parent mixture library to identify the active components. Positional scanning (*Figure 12*), introduced by Houghten, has been the subject of many reviews (1, 2, 28). A library of compounds containing m positions of variability and n substitutions at each position require $m \times n$ sublibraries (e.g. $4 \times 4 = 16$) to be synthesized. Each position is 'scanned' by synthesizing sublibraries where individual components at one position are coupled to full mixtures at all other positions. These sublibraries are assayed side-by-side and the optimal component at each position is gleaned from the most active sublibrary.

An additional deconvolution technique recently introduced is termed deletion synthesis deconvolution (21, 29). The two identification protocols of positional scanning and deletion synthesis deconvolution are complementary to one another and their combined use provides a powerful method for lead identification within large mixture libraries. As with positional scanning, sublibraries are synthesized alongside the parent library. In each sublibrary, one component is omitted from a single position. A library of compounds containing m positions of variability and n substitutions at each position require $m \times n$ sublibraries (e.g. $4 \times 4 = 16$) to be synthesized (*Figure 12*). The sublibraries with the lowest activity reveal the ideal substitution (which was omitted) at each position.

Our experience has been the two methods are complementary, and that deletion synthesis is more sensitive to establishing the most active lead in a library that contains few hits, albeit at the expense of identifying weaker leads, whereas positional scanning is more effective at identifying the weaker leads especially in libraries with multiple hits, but at the expense of accurately identifying the most potent agent (29). In selected instances, including libraries generated by sequential dimerizations, only deletion deconvolution can survey all library members and is a requisite deconvolution protocol for such cases (21).

Deconvolution of library mixtures up front by positional scanning or deletion synthesis is ideal for depository libraries which are subjected to multiple screening assays. Large mixture libraries, including a 476/775 compound library generated by the convergent olefin metathesis dimerization or bioryl coupling

Positional scanning

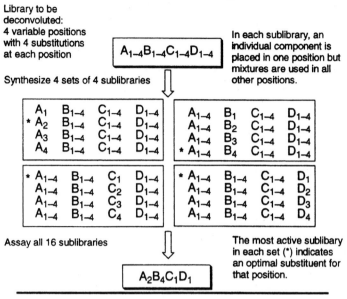

Library to be deconvoluted: 4 variable positions with 4 substitutions at each position

$A_{1-4}B_{1-4}C_{1-4}D_{1-4}$

In each sublibrary, an individual component is placed in one position but mixtures are used in all other positions.

Synthesize 4 sets of 4 sublibraries

Assay all 16 sublibraries

$A_2B_4C_1D_1$

The most active sublibary in each set (*) indicates an optimal substituent for that position.

Deletion synthesis deconvolution

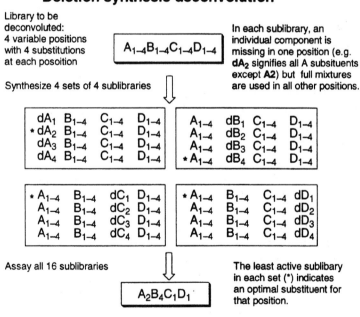

Library to be deconvoluted: 4 variable positions with 4 substitutions at each posoition

$A_{1-4}B_{1-4}C_{1-4}D_{1-4}$

In each sublibrary, an individual component is missing in one position (e.g. **dA$_2$** signifies all A subsituents except **A2**) but full mixtures are used in all other positions.

Synthesize 4 sets of 4 sublibraries

Assay all 16 sublibraries

$A_2B_4C_1D_1$

The least active sublibary in each set (*) indicates an optimal substituent for that position.

Figure 12 General procedure of positional scanning and deletion synthesis deconvolution on a generalized library with four positions of diversity and four substitutions in each position. Both techniques involve the synthesis of 16 sublibraries which are synthesized and screened at the same time as the parent library.

reactions (30, 31) of functionalized iminodiacetic acid substrates (Section 4.4.2) have been deconvoluted using both positional scanning and deletion synthesis techniques (21).

References

1. Balkenhohl, F., von dem Bussche-Hünnefeld, C., Lansky, A., and Zechel, C. (1996). *Angew. Chem. Int. Ed. Engl.*, **35**, 2288.
2. Chaiken, I. M. and Janda, K. D. (ed.) (1989). *Molecular diversity and combinatorial chemistry: Libraries and drug discovery*. ACS, Washington.
3. Czarnik, A. W. and DeWitt, S. H. (ed.) (1997). *A practical guide to combinatorial chemistry.* ACS, Washington.
4. Gordon, E. M., Gallop, M. A., and Patel, D. V. (1996). *Acc. Chem. Res.*, **29**, 144.
5. Ellman, J. A. (1996). *Acc. Chem. Res.*, **29**, 132.
6. Storer, R. (1996). *Drug Disc. Today*, **1**, 248.
7. Crowley, J. I. and Rapoport, H. (1976). *Acc. Chem. Res.*, **9**, 135.
8. Smith, P. W., Lai, J. Y. Q., Whittington, A. R., Cox, B., Houston, J. G., Stylli, C. H., *et al.* (1994). *Bioorg. Med. Chem. Lett.*, **4**, 2821.
9. Pirrung, M. C. and Chen, J. (1995). *J. Am. Chem. Soc.*, **117**, 1240.
10. Carell, T., Wintner, E. A., Sutherland, A. J., Rebek, J., Jr., Dunayevskiy, Y. M., and Vouros, P. (1995). *Chem. Biol.*, **2**, 171.
11. Curran, D. P. (1998). *Chem. Rev.*, **37**, 1174.
12. Armstrong, R. W., Combs, A. P., Tempest, P. A., Brown, S. D., and Keating, T. A. (1996). *Acc. Chem. Res.*, **29**, 123.
13. Han, H., Wolfe, M. M., Brenner, S., and Janda, K. D. (1995). *Proc. Natl. Acad. Sci. USA*, **92**, 6419.
14. Cheng, S., Comer, D. D., Williams, J. P., Myers, P. L., and Boger, D. L. (1996). *J. Am. Chem. Soc.*, **118**, 2567.
15. Kaldor, S. W. and Siegel, M. G. (1997). *Curr. Opin. Chem. Biol.*, **1**, 101.
16. Boger, D. L., Tarby, C. M., Myers, P. L., and Caporale, L. H. (1996). *J. Am. Chem. Soc.*, **118**, 2109.
17. Cheng, S., Tarby, C. M., Comer, D. D., Williams, J. P., Caporale, L. H., Myers, P. L., *et al.* (1996). *Bioorg. Med. Chem.*, **4**, 727.
18. Boger, D. L., Ozer, R. S., and Andersson, C.-M. (1997). *Bioorg. Med. Chem. Lett.*, **7**, 1903.
19. Boger, D. L., Chai, W., Ozer, R. S., and Andersson, C.-M. (1997). *Bioorg. Med. Chem. Lett.*, **7**, 463.
20. Boger, D. L. and Chai, W. (1998). *Tetrahedron*, **54**, 3955.
21. Boger, D. L., Chai, W., and Jin, Q. (1998). *J. Am. Chem. Soc.*, **120**, 7220.
22. Boger, D. L., Ducray, P., Chai, W., Jiang, W., and Goldberg, J. (1998). *Bioorg. Med. Chem. Lett.*, **8**, 2339.
23. Boger, D. L., Goldberg, J., Jiang, W., Ducray, P., Chai, W., Ozer, R. S., *et al.* (1998). *Bioorg. Med. Chem.*, **6**, 1347.
24. Grubbs, R. H., Miller, S. J., and Fu, G. C. (1995). *Acc. Chem. Res.*, **28**, 446.
25. Houghten, R. A., Pinilla, C., Blondelle, S. E., Appel, J. R., Dooley, C. T., and Cuervo, J. H. (1991). *Nature*, **354**, 84.
26. Ecker, D. J., Vickers, T. A., Hanecak, R., Driver, V., and Anderson, K. (1993). *Nucleic Acids Res.*, **21**, 1853.
27. Erb, E., Janda, K. D., and Brenner, S. J. (1994). *Proc. Natl. Acad. Sci. USA*, **91**, 11422.
28. Dooley, C. T. and Houghten, R. A. (1993). *Life Sci.*, **52**, 1509.
29. Boger, D. L., Lee, J. K., Goldberg, J., and Jin, Q. (2000). *J. Org. Chem.*, **65**, 1467.
30. Boger, D. L., Goldberg, J., and Andersson, C.-M. (1999). *J. Org. Chem.*, **64**, 2422.
31. Boger, D. L., Jiang, W., and Goldberg, J. (1999). *J. Org. Chem.*, **64**, 7094.

Chapter 11

Experimental techniques in fluorous synthesis: a user's guide

Dennis P. Curran, Sabine Hadida,[†] Armido Studer,[‡]
Mu He,[§] Sun-Young Kim, Zhiyong Luo,
Mats Larhed,* Anders Hallberg,* and Bruno Linclau

University of Pittsburgh, Department of Chemistry, Chevron Science Center,
Pittsburgh, PA 15260, USA.
[†]CombiChem, Inc., 9050 Camino Santa Fe, San Diego, CA 92121, USA.
[‡]Laboratorium für Organische Chemie, ETH Zentrum - Universitätstrasse 16,
CH-8092 Zürich, Switzerland.
[§]Pimlico Court, Quail Ridge Development, Hockessin, DE 19707, USA.
*Department of Organic Pharmaceutical Chemistry, BMC, Uppsala University,
Box 754, S-75123 Uppsala, Sweden.

1 Introduction

In discovery research, the 'reaction' stage of a synthetic step and the 'separation' stage have traditionally been uncoupled. Reactions are executed and then reaction mixtures are purified by using trial and error (or, perhaps more commonly, experience) to select a suitable purification technique. Experimental techniques for combinatorial chemistry and parallel synthesis need to be simple, fast, and efficient, and these needs have precipitated the development of a number of general new approaches for making small organic molecules (1, 2) that couple reaction and separation at strategy level. The goal of strategy level separation is to plan a reaction that can be purified simply by workup (3, 4). In other words, the desired product of a reaction will partition into (or be a part of) one phase, while all the other products of the reaction will partition into another phase. The commonly used phases for organic separations are the gas phase, the water phase, the organic liquid phase, and the solid phase, and these phases can be separated by evaporation, extraction, filtration (used alone or in combination, as appropriate). A number of imaginative and effective 'workup level' separation techniques have recently been introduced that take advantage of these different phases (3).

Recently, the fluorous phase has been recognized as an additional phase that is generally 'orthogonal' to the other common phases, and accordingly adds a new dimension to strategy level separation techniques. In 1994, Horváth and Rábai (5) introduced the concept of 'fluorous biphasic catalysis' (FBC). This technique places a catalyst in a fluorous liquid phase and a substrate in an organic liquid phase. Reactions can occur at the interface, or the phases can be blended by heating. At the separation stage, the two liquid phases are simply separated. In essence, this is a novel technique for catalyst immobilization, and it is quickly becoming a thriving subdiscipline in catalysis research (6, 7). While FBC is a highly valuable technique in catalysis and process chemistry, its direct application in discovery research can be problematic because fluorous solvents are very non-polar and are extraordinarily poor solvents for organic molecules in general.

Over the last several years, we have begun to describe a series of new fluorous techniques that are designed to be directly applicable to traditional types of problems in discovery-oriented research targeted at small organic molecule synthesis. The aim of this chapter is to bring together the experimental techniques that we have introduced and to try to convey some of the practical details that we have learned over the last few years. The concepts of fluorous synthesis (and strategy level separations in general) have been laid out elsewhere (8), and a reader not familiar with these concepts may want to consult these sources. Compared to standard techniques of small molecule synthesis or solid phase techniques, the development of fluorous synthesis techniques is still in its infancy. Nonetheless, a number of the techniques are ready for prime time, and we are beginning to use them as a means rather than an end in our research program. Fluorous techniques have much to recommend them, and we hope that the information in this article will help to spread these techniques into other laboratories. Patents are issued or pending on a number of the processes and compounds described herein (9).

2 General aspects of fluorous chemistry

2.1 Features of fluorous techniques

In the fluorous approach, one or more reaction components are 'tagged' (or 'labelled') with a highly fluorinated domain. Separation techniques are then applied such that the reaction mixture is (ideally) partitioned into two fractions based on the presence or absence of the fluorous tag (*Figure 1*). Reaction components like catalysts, reagents, and reactants that do not become part of the final product can be tagged 'permanently', and can often be recovered for reuse at the end of a reaction. These types of reaction components have use across the whole field of organic synthesis and offer new options for green chemistry. On the other hand, when substrates are tagged as fluorous, the tag must be designed in a removable fashion so that a 'tag-free' organic product can be produced at some stage. Once again, the tags can often be recovered and reused.

Perhaps the simplest fluorous techniques are based on liquid/liquid extract-

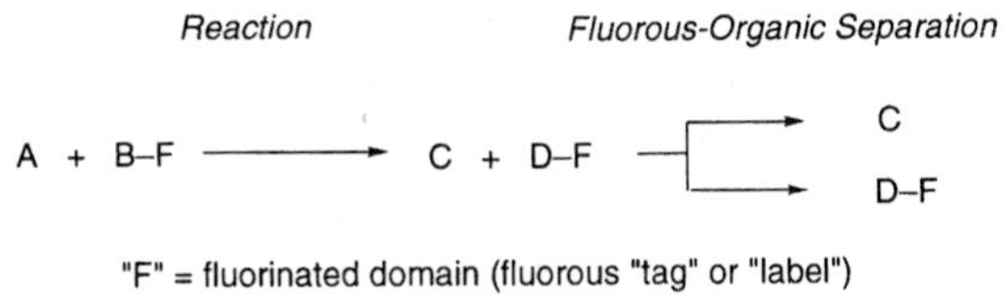

Reaction | Fluorous-Organic Separation

A + B–F ⟶ C + D–F

C

D–F

"F" = fluorinated domain (fluorous "tag" or "label")

Figure 1

ions, which can either be two-phase (fluorous–organic or fluorous–water) or three-phase (fluorous–organic–water). These are very practical for a number of applications, especially where highly fluorous reactants, reagents, and catalysts are employed. However, in a number of applications in our group, the liquid/liquid extractions are now being supplanted by what we call fluorous solid phase extractions (FSPE) (10, 11). These versatile procedures use a fluorinated solid phase (fluorous reverse phase (FRP) silica gel) and they tend to be easier to perform in parallel and to provide better separations when lightly fluorinated reaction components are used.

Conceptually speaking, fluorous techniques and solid phase techniques are related in that each is a 'tagging' strategy. But this similarity aside, the features of the two techniques are different and in many ways complementary. The polymer nature of the beads in solid phase synthesis is a key feature that enables 'one-bead one-compound' split synthesis techniques (12). This feature has no equivalent in fluorous synthesis. However, on the flip side of this detraction is a strong attraction. Fluorous reactions can often be run such that all of the reaction components are soluble (or substantially soluble) in the reaction medium; partitioning of the fluorous material away from the organic material only happens after the reaction. In this respect, fluorous techniques resemble more those of 'soluble polymers' (13), although as noted below there are quite a number of practical differences here as well. However, fluorous tags do affect the solubility of molecules and the selection of a suitable reaction solvent is often crucial for the success of a given transformation.

Fluorous techniques lend themselves especially well to parallel synthesis. The 'tags' in fluorous synthesis are often perfluorocarbon units, and these are among the most stable functional groups in organic chemistry. Thus, stability of the fluorinated domains themselves to the reaction conditions is rarely an issue. For attachment to substrates, fluorinated domains are readily fashioned into new 'protecting groups' which do double duty in both controlling the separation of a molecule and shielding one (or more) of its functional groups from undesired reactions.

While fluorous molecules naturally have a high molecular weight, they are amenable to analysis and purification by standard small molecule techniques. Except for couplings to nearby nuclei, the fluorinated domains are silent in ^{1}H and ^{13}C NMR experiments, and do not mask structural information. The volatility of organofluorine compounds means that standard ionization techniques can be used to obtain mass spectra even for relatively heavy molecules. Reactions can be followed by TLC or other normal means, and fluorous impurities can readily be detected by spectroscopic means (including ^{19}F NMR). The goal of most fluorous techniques is to avoid standard chromatographic purifications; however, when this goal is not reached, fluorous techniques offer the handy alternative that chromatographic separation can be used if needed. This is generally not an option for techniques based on polymers, where separations of different types of bound molecules are never possible until the molecules are cleaved from the polymer.

2.2 Reaction and extraction solvents

Different considerations often apply when choosing reaction and extraction solvents for fluorous synthesis techniques. At the reaction stage, it is frequently desirable to have a mixture which is either homogeneous or has substantial concentrations of all reaction components dissolved. On the other hand, if liquid/liquid extractions are used for separation, then an immiscible pair of solvents must be chosen. A nice overview of some of the properties and features of fluorous solvents can be found in a chapter by Barthel-Rosa and Gladysz (14), and we summarize here only a few qualitative trends that have emerged from our work.

In addition to the fluorous biphasic catalysis approach (5, 7) which uses highly fluorinated solvents or co-solvents, there are several other alternative strategies to approach the problem of choosing a reaction solvent. One of the simplest and most general is to use a 'lightly fluorinated' organic solvent, which is capable of dissolving both organic and fluorous compounds (15). Benzotrifluoride has quickly become one of the favourites. Benzotrifluoride (BTF, trifluoromethyl benzene $C_6H_5CF_3$) is inexpensive and has a number of attractive features (16). It is miscible in common organic solvents, so solvent blends can readily be formulated. In addition, benzotrifluoride can be used to blend otherwise immiscible fluorous and organic solvents. We have also used trifluoroethanol as a solvent, and a number of other candidates (for example, ClF_2CCl_2F) are also available. Ethereal solvents like THF and diethyl ether have good dissolving power for fluorous compounds and can often be used directly without additives.

The structure of the fluorous component naturally affects a molecule's solubility in organic solvents. While quantitative data are still scarce, we have learned that the commonly employed tactic of using reagents or protecting groups with three perfluorohexylethyl ($C_6F_{13}CH_2CH_2$) chains often results in compounds with relatively low solubility in many organic solvents. Addition of a single methylene group, shortening of the perfluorocarbon segment, or reduction of the number of chains can all result in compounds with dramatically improved organic solubility (17).

When liquid/liquid extractions are used, a highly (or fully) fluorinated solvent is needed. Barthel-Rosa and Gladysz provide a nice overview of the available alternatives (14). We have tended to use members of 3M's 'FC' family like FC-72® (a mixture of compounds containing mostly perfluorohexanes) and FC-84® (a mixture containing mostly perfluoroheptanes). All these solvents are relatively expensive and we like to recover and reuse them as much as possible. They are also quite dense (d = 1.7 g/ml) and constitute the bottom phase of all the two- and three-phase extractions that we have done. At room temperature, most organic solvents are immiscible with these liquids, so there is a broad range of options for liquid/liquid extractions at the organic end. We often select polar solvents like methanol or acetonitrile, or aromatic solvents like benzene or toluene, which tend to be fluorophobic and provide good partition coefficients.

Extractive workups benefit from the very poor dissolving power of FC-72 and other fluorous solvents for typical organic molecules. Aside from very small molecules (gases), which have excellent solubility in fluorous solvents, and straight-chain hydrocarbons, which have some solubility, most 'garden variety' organic compounds are essentially insoluble. This means that it is possible to extract the organic liquid phase repeatedly to increase the recovery or purity of one of the products. This repeated extraction continues to remove fluorous-tagged materials from the organic phase without drawing non-tagged products into the fluorous phase. Partition coefficients of about 10 or more are ideal for fluorous liquid/liquid extractions, but in practice one can get by with partition coefficients as low as about 1. For compounds with lower partition coefficients, solid/liquid separation methods are often preferred.

2.3 Fluorous starting materials

While fluorous tags, reagents, protecting groups, etc. could be fashioned from all types of organofluorine compounds, our early work has tended to focus on fluorinated compounds that are readily available and can be quickly fashioned into useful molecules. We have worked mostly with perfluoroalkyl iodides: $C_nF_{2n+1}I$. These reagents are readily available from large fine chemical suppliers although they are less expensive if purchased from some of the smaller companies (suppliers of specific compounds are listed in the protocols below). We also often use the adducts of these iodides to ethylene, $C_nF_{2n+1}CH_2CH_2I$, which are also commercially available. The Grignard and lithium reagents derived from these iodides are well suited for generating a number of classes of fluorous compounds, including fluorous tin and silicon reagents. A number of long chain fluorinated acids are commercially available, and these make useful precursors as well, as do perfluorinated polyethers.

2.4 Fluorous reverse phase silica gel

Separations conducted by solid/liquid methods rather than liquid/liquid methods have a number of attractions. First and foremost, separations can be accomplished even when the fluorous-tagged molecules do not have enough fluorines to give them high partition coefficients in liquid/liquid extractions. In addition, these procedures are effectively filtrations and are easy to execute in parallel or to automate.

We call silica gel with highly fluorinated bonded phases 'fluorous reverse phase (FRP) silica gel'. Fluorinated silica bonded phases have been known for some time, although their applications have been very limited compared to standard bonded phases which have long hydrocarbon chains (like C-18). Analytical HPLC columns can be purchased in the US from Keystone Scientific (Fluofix®, $Si(Me)_2CH_2CH_2CH_2C(CF_3)_2CF_2CF_2CF_3$) and ES Industries (FluoroSep-RP Octyl, $Si(Me)_2CH_2CH_2C_8F_{17}$). These columns are marketed as reverse phase columns, but they are less retentive than standard reverse phase columns. However, they also separate by fluorine content—molecules with multiple fluorines are more strongly retained—and it is relatively easy to develop analytical conditions where all non-tagged molecules come off at or near the solvent front, whereas fluorous tagged molecules have barely begun to move from the head of the column. Such analytical conditions typically translate to unmonitored solid phase extractions in parallel synthesis applications. After loading the reaction mixtures to small tubes containing FRP silica, an elution with a first solvent is conducted to provide an 'organic' (non-tagged) fraction. Then a second, more powerfully eluting solvent, is passed through to provide a fluorous (tagged) fraction. The choice of solvent varies, and we find the analytical columns very useful for methods development and elution solvent and fluorous tag selection.

Unfortunately, as of this writing, there is currently no source of commercial FRP silica gel suitable for solid phase extraction applications. We prepare our own 'home-made' material, and this protocol describes the synthesis of fluorous reverse phase (FRP) silica gel from standard flash chromatography silica (18). The FRP silica is used as the solid phase in the fluorous solid phase extraction to separate fluorous compounds from organic compounds.

Standard 'flash chromatography grade' silica gel (230–400 mesh) is activated by heating under low pressure and then mixed with perfluorohexylethyl dimethylchlorosilane (1 Equiv.), dry pyridine (2 Equiv.), and dry toluene (*Figure 2*). The mixture is heated at 100°C for 24 h. Then the solvent is filtered off and the solid is washed consecutively with toluene, methanol, methanol/water (1:1, v/v), water, methanol, and ether. The resulting silica is dried under low pressure at 120°C for 12 h. Elemental analysis of a typical batch is contained in ref. 19.

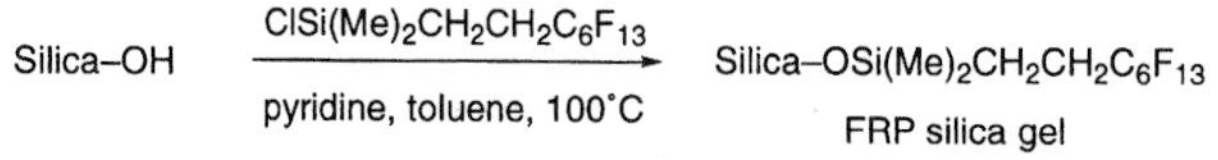

Silica–OH $\xrightarrow[\text{pyridine, toluene, 100°C}]{\text{ClSi(Me)}_2\text{CH}_2\text{CH}_2\text{C}_6\text{F}_{13}}$ Silica–OSi(Me)$_2$CH$_2$CH$_2$C$_6$F$_{13}$

FRP silica gel

Figure 2

Protocol 1
Preparation of FRP silica gel

Reagents

- Tridecafluoro-(1*H*,1*H*,2*H*,2*H*-tetrahydrooctyl)dimethylchlorosilane (perfluorohexylethyl dimethylchlorosilane, Gelest)

- Silica gel (230–400 mesh, commercially available, Merck)

Method

1 Activate 12 g of silica in a vacuum oven (180°C, 0.05 mm Hg) for 12 h.

2 Add the dried silica to a solution of 13.6 ml (45.1 mmol) of tridecafluoro-(1H,1H,2H,2H-tetrahydrooctyl)dimethylchlorosilane and 7.5 ml (92.7 mmol) of dry pyridine in 300 ml of dry toluene. Heat the mixture at 100°C for 24 h and shake occasionally.

3 Filter the mixture and wash the solid sequentially with toluene (200 ml), methanol (200 ml), methanol/water, 1:1 (200 ml), water (200 ml), methanol (200 ml), and ether (300 ml). Dry the silica in a vacuum oven (120°C, 0.5 mm Hg) for 12 h to yield 17 g of FRP silica.

3 Fluorous tin chemistry

Many techniques and reactions in fluorous chemistry have been pioneered through the use of organotin reactants, reagents, and catalysts. This section describes the synthesis, reaction, separation, and recycling of some typical fluorous tin compounds in traditional and parallel settings.

3.1 Synthesis of representative perfluorohexylethyltin reagents

Tin reagents bearing three perfluorohexylethyl chains were the first fluorous tin reagents to be investigated and they are among the most readily available and useful. In this protocol an efficient preparation of *tris*(2-perfluorohexylethyl) tin hydride **3** is reported. The synthesis of **3** is shown in *Figure 3*, and it involves a three-step reaction sequence starting from phenyltin trichloride. Tin hydride **3** behaves like a normal tin hydride reagent in radical reactions but it has significant practical advantages over tributyltin hydride (20–23).

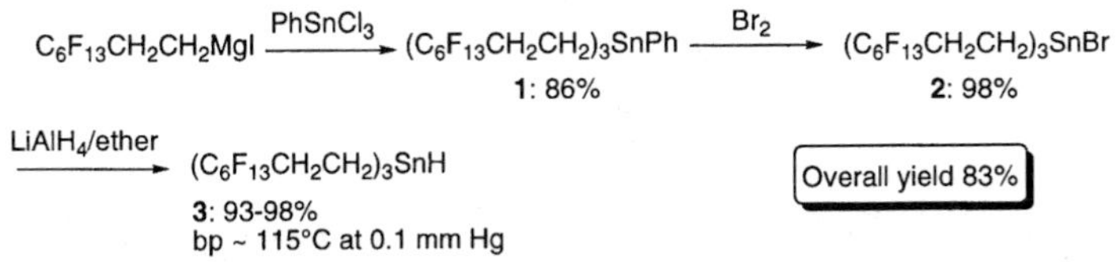

$$C_6F_{13}CH_2CH_2MgI \xrightarrow{\ PhSnCl_3\ } (C_6F_{13}CH_2CH_2)_3SnPh \xrightarrow{\ Br_2\ } (C_6F_{13}CH_2CH_2)_3SnBr$$
1: 86%   2: 98%

$$\xrightarrow{\ LiAlH_4/ether\ } (C_6F_{13}CH_2CH_2)_3SnH$$
3: 93-98%
bp ~ 115°C at 0.1 mm Hg

Overall yield 83%

Figure 3

The Grignard reagent prepared from 2-perfluorohexylethyl iodide (4 Equiv.) and magnesium (5.1 Equiv.) in ether was reacted with a benzene solution of phenyltin trichloride (1 Equiv.). The mixture was refluxed for 4 h and then stirred at 25°C for 16 h. The reaction mixture was poured over a saturated solution of ammonium chloride. The organic phase was washed with 5% solution of $Na_2S_2O_3$ and deionized water, dried over $MgSO_4$, and evaporated to dryness. Purification of the mixture by vacuum distillation followed by column chromatography on neutral alumina with hexane gave product **1** in 86% yield. Bromine (1.05 Equiv.) in ether was added to an ice-cooled solution of **1** in ether. The mixture was warmed to 25 °C over 2 h with stirring. Removal of the ether, bromobenzene, and excess bromine resulted in an orange oil that was purified by vacuum distillation. Bromotin compound **2** was obtained in 98% yield. A 1.0 molar solution of LiAlH$_4$ (1 Equiv.) was added to an ice-cooled ether solution of tin bromide **2** in ether and the reaction mixture was stirred at 0 °C for 3 h. Quenching of the reaction mixture with a 20% solution of sodium potassium tartrate, extraction with ether, and vacuum distillation of the yellowish extract yielded the tin hydride **3** (97%) as a viscous, colourless oil.

Protocol 2

Synthesis of *tris*(2-perfluorohexylethyl)phenyltin (1), *tris*(2-perfluorohexylethyl)tin bromide (2), and *tris*(2-perfluorohexylethyl)tin hydride (3)

Reagents
- Phenyltin trichloride (Aldrich)
- 2-Perfluorohexylethyl iodide (Lancaster)
- Magnesium (powder, 325 mesh, 99.5%, Aldrich)

- Bromine (Aldrich)
- Lithium aluminium hydride (1 M solution in ether, Aldrich)

Method

1 Add 5.17 ml (2.11 mmol) of 2-perfluorohexylethyl iodide to a suspension of magnesium (6.53 g, 269 mmol) in 100 ml of dry ether under N$_2$ and sonicate for 30 min.

2 Add dropwise the rest of the 2-perfluorohexylethyl iodide (46.53 ml, 189.9 mmol) in 50 ml of ether and heat the mixture at reflux for 1.5 h.

3 Dissolve the phenyltin trichloride (15.9 g, 52.7 mmol) in 100 ml of dry benzene and add this dropwise to the reaction mixture.

4 Reflux the reaction for 4 h and then stir at 25 °C for 16 h.

5 Pour the reaction mixture over 75 ml of saturated solution of ammonium chloride. Separate the two layers and wash the organic layer with 50 ml of 5% Na$_2$S$_2$O$_3$ solution, and 50 ml of deionized water. Dry with MgSO$_4$ and evaporate under reduced pressure.

6 Remove the major by-product of the mixture [*bis*(1,4-perfluorohexyl)butane] by vacuum distillation (87–92 °C, 0.2 mm Hg).

7 Purify the residue from the distillation by column chromatography on neutral alumina with hexane as mobile phase. **1** is obtained as a colourless oil (56.1 g, 86%).

8 Dissolve **1** in 105 ml of dry ether. Cool the mixture to 0 °C and add dropwise a solution of bromine (7.62 g, 47.67 mmol) in 15 ml of ether.

9 Warm to 25°C and stir for 2 h.

10 Remove the solvent under reduced pressure and purify by vacuum distillation (150–152°C, 0.5 mm of Hg). **2** is obtained as a colorless oil (55.3 g, 98%).

11 Dissolve **2** in 500 ml of dry ether and cool to 0°C. Add dropwise 44.24 ml of an ethereal solution of LiAlH$_4$ (1 M). Stir for 3 h. Quench by pouring the mixture into 200 ml of 20% solution of sodium potassium tartrate. Separate the layers and extract the aqueous phase three times with 300 ml of ether. Combine the organic extracts and dry over anhydrous MgSO$_4$. Filter and evaporate. Purify by vacuum distillation (145–150°C, 3 mm of Hg) to obtain 50.3 g of **3** as a colorless oil.

3.2 Parallel synthesis with a fluorous tin hydride

In this present protocol, we describe a series of parallel additions of radicals generated from iodoalkanes to activated alkenes using a catalytic amount of fluorous tin hydride **3**. The tin hydride is generated *in situ* by reduction of tin bromide **2** with NaBH$_3$CN. The reaction mixture is purified by liquid/liquid extraction providing the desired addition products in excellent yields and purities.

In the first step, a suspension of the alkyl iodide (1 Equiv.), alkene (5 Equiv.), fluorous tin bromide **2** (0.1 Equiv.), sodium cyanoborohydride (1.3 Equiv.), and AIBN (0.05 Equiv.) in benzotrifluoride (BTF) and *tert*-butanol was heated at reflux in a sealed vial for 12 h (*Figure 4*). The reaction was cooled and FC-72 and dichloromethane (DCM) were added to the mixture. The FC-72 layer was removed (lower layer) and the DCM layer was extracted with FC-72 and water. The organic phase (middle layer) was separated, filtered through neutral alumina, and evaporated to dryness. The yields of the reactions were determined by ^{1}H NMR using CH$_2$Cl$_2$ and hexamethyldisiloxane as internal standards. The purity of the compounds was determined by gas chromatography.

R/E	CN	CO$_2$Me	COMe
C$_{15}$H$_{31}$	72%	92%	67%
c-C$_6$H$_{11}$	75%	65%	75%
Ad	89%	94%	78%

Figure 4

Protocol 3

Parallel experiment with tin hydride 3 under catalytic conditions (9-component library)

Reagents

- 1-Iodopentadecane, 1-iodoadamantane, and cyclohexyliodide (Aldrich)
- Acrylonitrile, methyl acrylate, and methyl vinyl ketone (Aldrich)
- Fluorous tin bromide (**2**) prepared in *Protocol 2*
- Sodium cyanoborohydride (Aldrich)

- AIBN (2,2′-azobisisobutyronitrile, Aldrich)
- FC-72 (3M Company)
- BTF (OxyChem)
- *tert*-Butanol (Aldrich)
- Neutral alumina (Aldrich)

Method

1. In a sealed tube, suspend the alkyl iodide (0.1 mmol), the olefin (0.5 mmol), *tris*-2-(perfluorohexyl)ethyl tin bromide (**2**) (12.0 mg, 0.01 mmol), sodium cyanoborohydride (9.6 mg, 0.13 mmol), and AIBN (catalytic amount) in BTF (0.5 ml) and *tert*-butanol (0.5 ml).

2. Heat at reflux for 12 h.

3. Add FC-72 (2 ml) and DCM (1 ml) and shake for 2 min.

4. Use a Pasteur pipette to remove the FC-72 layer (bottom).

5. Add 1 ml of FC-72 and 1 ml of water to the DCM layer and shake for 2 min.

6. Use a Pasteur pipette to remove the DCM layer (middle layer) and filter through neutral alumina.

7. Evaporate the DCM layer to dryness.

8. Calculate the yield of the reaction by ^{1}H NMR in CDCl$_3$ using DCM and hexamethyldisiloxane as internal standards.

3.3 Synthesis and reaction of a fluorous tin azide

The synthesis of *tris*(2-perfluorohexylethyl)tin azide **5** and the reaction of the azide **5** with benzyl cyanide to give 5-benzyl tetrazole **7** are described (24). Fluorous tin azide **5** is prepared from *tris*(2-perfluorohexylethyl)tin bromide **2** and used directly after workup without further purification (*Figure 5*). Tin azide **5** allows the efficient synthesis of the tetrazole **7** without using potentially explosive sodium azide or tributyltin azide of which by-products are difficult to remove. The desired tetrazole **7** is obtained in a high yield by using excess tin azide **5** and is purified only by extractive fluorous workup.

Fluorous tin bromide **2** was treated with 1.2 Equiv. of sodium azide in a biphasic mixture of ether and water (7:1) while vigorously stirring at room

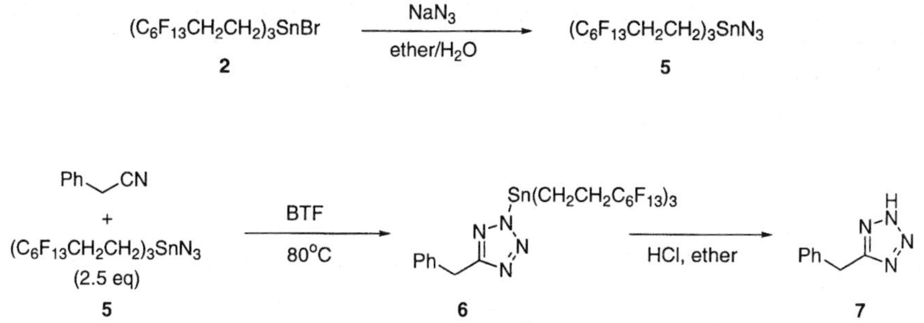

Figure 5

temperature. After stirring overnight, ether (10 ml) and water (10 ml) were added. Phase separation occurred and the ethereal layer was washed three times (10 ml each) with water and dried over anhydrous magnesium sulfate. Evaporation to dryness afforded **5** in 97% yield. The tin azide **5** was a viscous liquid, which partially solidified upon standing.

The synthesis of tetrazole **7** started with the reaction of benzyl cyanide and tin azide **5** (2.5 Equiv.) in BTF. After heating at 80 °C for 12 h in a sealed tube, the BTF was evaporated under reduced pressure, and the residue was partitioned between benzene and FC-72. The benzene layer was extracted twice with FC-72 and the combined FC-72 layers were evaporated to give fluorous tetrazole **6** with excess tin azide **5**. The crude mixture from fluorous phase was dissolved in an ethereal HCl (1 M in ether) and stirred at room temperature. Upon the addition of ethereal HCl, a white solid formed and the mixture turned milky white. After 12 h, the ether was evaporated and the residue was partitioned in FC-72 and acetonitrile. The acetonitrile layer containing the desired tetrazole **7** was separated and washed twice with FC-72 to remove all of the fluorous compounds. Evaporation of acetonitrile layer afforded pure 5-benzyl tetrazole **7** (98%). An excess of tin azide was used to drive the reaction to completion but this did not reduce the purity of the final product because it partitioned into the fluorous phase while the desired tetrazole partitioned into the organic phase. In many cases, the initial liquid/liquid extraction is not needed, and the crude product can be directly treated with HCl.

Protocol 4

Synthesis and reaction of tin azide 5

Equipment and reagents

- *tris*(2-Perfluorohexylethyl)tin bromide (see *Protocol 2*)
- Sodium azide (Aldrich)
- Benzyl cyanide (Aldrich)

- BTF (benzotrifluoride, Aldrich)
- FC-72 (3M Company)
- 1 M HCl in ether (Aldrich)

Protocol 4 continued

A. Synthesis of tin azide 5

1. Add a solution of 315 mg (4.8 mmol) of sodium azide in 1.0 ml of water at 25°C with vigorous stirring to a solution of 5.0 g (4.0 mmol) *tris*(2-perfluorohexylethyl)tin bromide **2** in 7.0 ml of ether and stir overnight.

2. Add 10 ml of water and 10 ml of ether while stirring and transfer the mixture to a separatory funnel. Shake and separate the ether phase. Wash the ether phase twice with 10 ml of water. Dry the ether phase over anhydrous $MgSO_4$ and evaporate ether to afford fluorous tin azide **5**.

B. Reaction of tin azide 5 to yield 5-benzyl tetrazole

1. Dissolve 1.0 g (0.83 mmol) of **5** in 0.6 ml of BTF in a tube. Add 38 μl (0.33 mmol) of benzyl cyanide and purge with argon. Seal the tube and heat at 80°C for 12 h.

2. Transfer the reaction mixture to a flask and remove the solvent. Add 10 ml of benzene and 10 ml of FC-72. Transfer to a separatory funnel and separate the FC-72 phase. Extract the benzene phase twice with 10 ml of FC-72. Combine the fluorous phases and remove the FC-72 under reduced pressure to afford fluorous tetrazole **6** (mixed with **5**).

3. Dissolve crude **6** in 10 ml of 1 M HCl in Et_2O at 25°C, and stir for 12 h.

4. Remove the solvent and partition the residue between 10 ml of FC-72 and 10 ml of acetonitrile. Transfer to a separatory funnel and separate acetonitrile layer. Wash acetonitrile phase twice with 10 ml of FC-72. Remove the solvent to yield 52 mg of tetrazole **7**.

3.4 Rapid fluorous Stille coupling reactions with microwave heating

For more than forty years, people have been using microwave ovens to heat and cook food quickly. Surprisingly, preparative organic chemists have been slow to discover the potential of microwave heating (25). In contrast to conventional heating, microwave heat is generated inside the bulk of the sample (*in situ* heating) and is distributed from inside and out (no wall effects), causing the sample to heat up evenly and rapidly.

The dramatic reduction in reaction times experienced in palladium catalysed cross-couplings using microwave heating (26, 27) encouraged us to explore whether the comparatively slow fluorous Stille cross-coupling reaction (28, 29) might be speeded up with this technology. Fifteen different fluorous Stille couplings were performed on 0.20 mmol scale. In all attempted combinations of organic halide or organic triflate and polyfluorinated aryl tin reactant, micro-wave treatment delivered complete conversion after only 90–120 sec of irradia-tion (see *Figure 6* for examples) (30). Comparable thermal reactions required approximately one day for completion. After cooling and evaporation, the excess

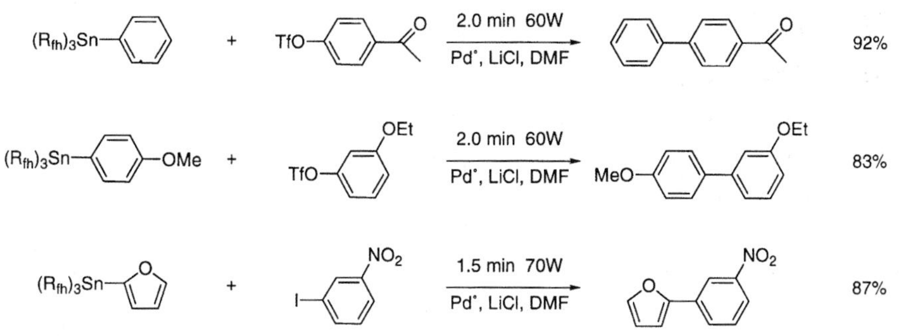

Figure 6

of polyfluorinated tin reactant and polyfluorinated tin halide partition into the fluorous phase in a three-phase extraction with a fluorocarbon solvent, dichloromethane, and water. The fluorous tin compounds are readily recovered and recycled. The evaporated crude organic product from the combined dichloromethane phases was thereafter, if necessary, further purified by flash chromatography.

The fluorous Stille reaction combines disparate aspects of physics and chemistry, such as use of single-mode microwave heating and multi-phase liquid systems with palladium catalysed coupling reactions, to achieve both rapid reactions and efficient purification of the product. Because of the extreme temperature profiles acquired with microwave irradiation (flash heating) (31), it is highly important to find the correct balance between geometry (e.g. volume) of the reaction mixture, irradiation power, and reaction time. Thus, minor alterations of reaction scale, power, or duration of irradiation may result in reduced conversion or yield.

Protocol 5

Rapid fluorous Stille coupling reactions with microwave heating

Equipment and reagents

- MicroWell 10 single-mode microwave cavity (2450 MHz, Labwell AB, Sweden)

- Heavy-walled Pyrex tubes (8 ml, l = 150 mm) sealed with a silicon septum (e.g. 110.623–18, KEBO Lab AB, Sweden)

- Fluorocarbon solvent FC-84 or FC-72 (3M Company)

- *tris*(2-Perfluorohexylethyl)aryltin reactants (see *Protocol 1*) and aryl triflates

- Commercially available chemicals from Aldrich: organo halides, *bis*(triphenylphosphine)palladium(II) chloride, lithium chloride, and DMF

Method

1. Place *tris*(2-perfluorohexylethyl)aryltin (0.24 mmol), an organic halide or an organic triflate (32) (0.20 mmol), *bis*(triphenylphosphine)palladium(II) chloride (0.004 mmol,

Protocol 5 continued

2.8 mg), lithium chloride (0.60 mmol, 25.4 mg), and DMF (1.0 ml) into the heavy-walled Pyrex tube.

2. Close the reaction tube under a nitrogen atmosphere and use a whirlimixer (if available) to mix the contents.

3. Irradiate the mixture for 90–120 sec with an appropriate microwave power.

4. Allow the mixture to cool, add toluene (2 ml), and remove most of the DMF with azeotropic evaporation.

5. Purify the residue by three-phase extraction with FC-84 (10 ml, bottom), dichloromethane (20 ml, middle), and water (10 ml, top). Repeat the washing of the organic layer three more times with FC-84 (3 × 10 ml) and water (3 × 10 ml). Evaporate the dichloromethane phase and, if necessary, purify the crude product by flash chromatography.

    Caution! The microwave heating synthetic method described above should not be attempted unless an appropriate septum or a related device is used for relief of over-pressure.

3.5 Synthesis of 'propylene-spaced' fluorous allyltin reagents

The nature of the spacer can have major effects on the reactivity of fluorous reagents. In some types of applications, ethylene spacers are too short and propylene spacers are preferred. In this protocol, the synthesis of two 'propylene spaced' fluorous allyltin reagents, *tris*(perfluorohexylpropyl)allyltin **8** and *tris*(perfluorobutylpropyl)allyltin **9** is described (*Figure 7*) (33, 34). The synthesis consists of three steps and can be scaled up to make tens of grams of the final products. The use of the reagents is illustrated in a parallel allylation of aldehydes with purification by solid phase extraction over FRP silica gel.

The Grignard reagents, prepared from the known perfluorohexylpropyl iodide and perfluorobutylpropyl iodide, were reacted with phenyltin trichloride to give *tris*(perfluoroalkylpropyl)phenyltin compounds (*Figure 7*). After bromolysis with bromine, the *tris*(perfluoroalkylpropyl)tin bromide was reacted with allylmagnesium bromide to give the desired *tris*(perfluoroalkylpropyl)allyltins **8** and **9**. All the products and intermediates are stable enough to be purified by distillation or column chromatography, if desired.

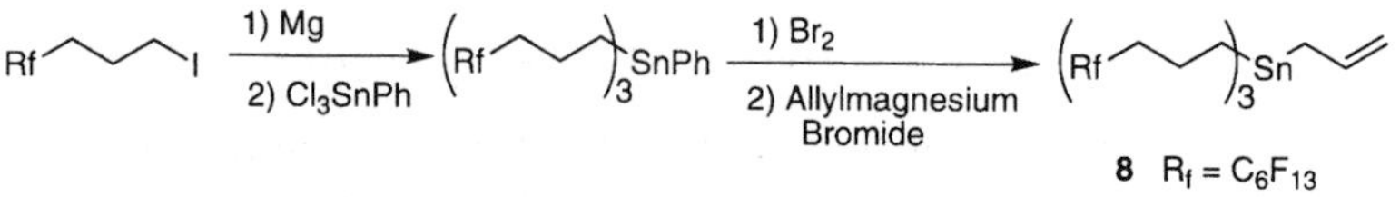

8 $R_f = C_6F_{13}$

9 $R_f = C_4F_9$

Figure 7

Protocol 6

Synthesis of 'propylene-spaced' fluorous allyltin reagents

Reagents

- Perfluorohexylpropyl iodide (see ref. 35)
- Perfluorobutylpropyl iodide (see ref. 35)
- Magnesium powder 100 mesh (Aldrich)
- Phenyltin trichloride (Aldrich)

- Bromine (Aldrich)
- Allylmagnesium bromide, 1.0 M solution in ether (Aldrich)

Method

1. Sonicate 2.68 g perfluorohexylpropyl iodide (35) (5.5 mmol) and 1.50 g of magnesium (50 mesh, 63 mmol) in 100 ml of dry ether for 30 min. Then slowly add 22.75 g of perfluorohexylpropyl iodide (46.6 mmol) and reflux for 2 h.

2. Add 2.2 ml of phenyltin trichloride (13.4 mmol) dropwise and reflux overnight.

3. Pour the mixture into 100 ml 2 M HCl, separate the layers, and dry the ether layer over MgSO$_4$.

4. Remove the solvent by evaporation and heat the residue under high vacuum (0.01 mmHg) for 2 h to remove the homocoupling by-product.

5. Pass the residue through a short pad of silica with hexanes to give pure *tris*(perfluohexylpropyl)phenyltin.

6. Add 1.40 g bromine (8.76 mmol) to an ice-cooled solution of *tris*(perfluohexylpropyl)phenyltin (10.36 g, 8.36 mmol) in dry ether and stir at room temperature for 2.5 h.

7. Evaporate the solvent and partition the residue between DCM and FC-72. Wash the DCM layer with FC-72 three times.

8. Evaporate the combined FC-72 layers to give pure *tris*(perfluorohexylpropyl)tin bromide.

9. Add 8.3 ml of allylmagnesium bromide (1.0 M solution in ether) to a solution of 8.64 g of *tris*(perfluorohexylpropyl)tin bromide (6.74 mmol) in dry ether.

10. Reflux the mixture overnight, cool, and quench with 2 M HCl. Dry the ether phase over MgSO$_4$ and evaporate to dryness.

11. Purify the residue by column chromatography with hexanes to give pure *tris*(perfluorohexylpropyl)allyltin.

3.6 Parallel allylation of aldehydes with fluorous allyltins

In a preliminary experiment, benzaldehyde was allylated with *tris*(perfluorohexylpropyl)allyltin **8** catalysed by platinum *bis*(triphenylphosphine) dichloride at 70 °C (*Figure 8*). The crude mixture, which contains desired product, excess fluorous allyltin **8**, catalyst, and fluorous by-product was purified by simple fluorous solid phase extraction (FSPE). The desired organic product eluted with acetonitrile and fluorous reagents and by-products were retained on the fluorous reverse phase silica (FRPS).

A parallel synthesis was carried out with four aldehydes and two fluorous allyltin reagents in eight sealed tubes in the presence of the platinum catalyst. All of the eight reaction mixtures were purified simultaneously with a SPE-24 column processor using fluorous reverse phase silica gel. The catalyst was not retained on the FRPS, hence the products purified by FSPE were still contaminated. We later found that this problem can be solved by using a platinum catalyst with fluorous phosphine ligand. The yield and conversion of the parallel allylations of aldehydes, calculated from ^{1}H NMR integration by using methyltriphenylsilane as standard are summarized in *Figure 8* (upper half).

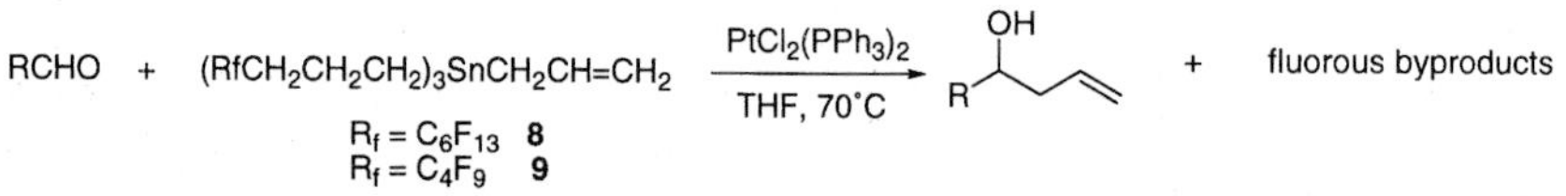

RCHO + (RfCH$_2$CH$_2$CH$_2$)$_3$SnCH$_2$CH=CH$_2$ $\xrightarrow[\text{THF, 70°C}]{\text{PtCl}_2(\text{PPh}_3)_2}$ R-CH(OH)-CH$_2$CH=CH$_2$ + fluorous byproducts

R$_f$ = C$_6$F$_{13}$ **8**
R$_f$ = C$_4$F$_9$ **9**

	PhCHO	p-MeOC$_6$H$_4$CHO	o-NO$_2$C$_6$H$_4$CHO	1-Naphthaldehyde
8	66%(100%)	74% (49%)	90% (100%)	84% (85%)
9	62% (100%)	67% (49%)	81% (100%)	85% (91%)

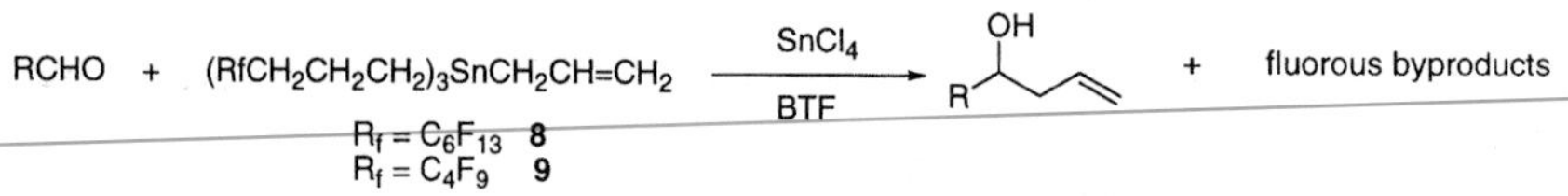

RCHO + (RfCH$_2$CH$_2$CH$_2$)$_3$SnCH$_2$CH=CH$_2$ $\xrightarrow[\text{BTF}]{\text{SnCl}_4}$ R-CH(OH)-CH$_2$CH=CH$_2$ + fluorous byproducts

R$_f$ = C$_6$F$_{13}$ **8**
R$_f$ = C$_4$F$_9$ **9**

	PhCHO	p-MeOC$_6$H$_4$CHO	o-NO$_2$C$_6$H$_4$CHO	1-Naphthaldehyde
8	82%	(a)	81%	81%
9	68%	65%	74%	68%

(a) reaction failed

Figure 8

Lewis acids promote the allylation of aldehydes with allyl stannanes at low temperature. The mild reaction conditions make this reaction synthetically more useful for aldehydes that are not stable at elevated temperature. In a preliminary experiment, the allylation of benzaldehyde with fluorous allyltin promoted by SnCl$_4$, gave pure homoallylic alcohol product in 64% yield after aqueous workup followed by FSPE.

Protocol 7

Parallel allylation with separation by FRP silica solid phase extraction

Equipment and reagents

- SPE-24 solid phase extraction column processor (VWR)
- Benzaldehyde (Aldrich)
- *p*-Methoxybenzaldehyde (Aldrich)
- *o*-Nitrobenzaldehyde (Aldrich)
- 1-Naphthaldehyde (Aldrich)
- $PtCl_2(PPh_3)_2$ (Aldrich)
- Fluorous reverse phase silica (see *Protocol 1*)

A. Platinum catalysed allylation of aldehydes

1 In eight dry tubes, add the aldehydes (0.25 mmol) to fluorous allyltin (0.50 mmol), $PtCl_2(PPh_3)_2$ (0.025 mmol), and 0.5 ml of dry THF. Heat at 70 °C for 24 h, then remove most of the THF by evaporation.

2 Add to eight short glass columns with Teflon frits, 3.5 g fluorous reverse phase silica. Wet the silica with 5 ml diethyl ether and wash with 10 ml of acetonitrile.

3 Charge the reaction mixture onto fluorous silica and elute with 8 ml of acetonitrile.

4 Evaporate the acetonitrite using a Speed vac to yield the alcohol.

B. SnCl₄-mediated allylation of aldehydes

1 Add $SnCl_4$ (1.1 Equiv., 1.0 M in DCM) to a solution of an aldehyde in dry THF at 78 °C. To this slurry, add fluorous allyltin (1.1 Equiv.), stir the mixture at –78 °C for 2 h.

2 Quench the reaction with a few drops of aqueous NaOH solution before evaporating the THF.

3 Purify the eight reaction mixtures as described in part A, steps 3 and 4.

To conduct reactions in parallel, we simplified the workup by quenching the reaction mixture with a few drops of aqueous NaOH solution followed by FSPE. In the parallel allylation experiment, the same aldehydes and fluorous allyltins as shown in *Figure 8* were used. In all eight reactions, the aldehydes were completely consumed and pure products were obtained after FSPE in seven cases. The isolated yields are summarized in *Figure 8* (lower half).

4 Fluorous synthesis

'Fluorous synthesis' is a strategic alternative to traditional small molecule synthesis and solid phase synthesis. In this approach, it is the substrates and/or target products that are rendered fluorous rather than the reagents or reactants. This necessarily involves a temporary tagging strategy because the tagged organic molecule must be released at some point. Work on fluorous synthesis techniques is in its infancy, and there are not many practical fluorous tagging groups yet available. But the principles and concepts have been laid out, and the path to more practically useful groups is often evident. In one of the simplest approaches, a standard protecting group is 'retooled' into a fluorous tag by addition of a suitable fluorinated domain. The group then serves both to protect the tagged molecule and to dictate its separation behaviour in a simple fluorous/organic workup method.

Fluorous synthesis was first demonstrated by making a small library of isoxazolines and isoxazoles (36). This reaction nicely illustrates the basic features of the method on relatively small molecules. The synthesis of larger molecules is illustrated in the context of a fluorous Ugi reaction (37). Simple disaccharides have also been made by fluorous synthesis (38).

4.1 Fluorous isoxazoline synthesis without intermediate purification

In this protocol an efficient synthesis of (3-propyl-4,5-dihydroisoxazol-5-yl) methanol (**12**) starting from allyl alcohol is described (*Figure 9*). The fluorous methodology allows the preparation of highly pure isoxazoline **12** in good yield without using conventional column chromatography. The synthesis comprises a silylation step followed by a dipolar cycloaddition with a subsequent desilylation to liberate the desired isoxazoline.

In the first step, allyl alcohol (4 Equiv.) and NEt_3 (triethylamine, 4 Equiv.) in THF were treated with a THF solution of bromo *tris*(2-perfluorohexylethyl)silane (1 Equiv.). Excess of allyl alcohol was used to drive the reaction to completion. After 3 h at 25 °C, the solvent was removed and the residue was purified by three-phase extraction with FC-72, DCM, and H_2O. The DCM and the H_2O layers containing the unreacted alcohol, the amine, and the triethylammonium salt were discarded. Silyl ether **10** obtained after evaporation of the fluorous extracts was dissolved in BTF and transformed according to Mukaiyama's procedure (39) with nitrobutane (10 Equiv.), phenyl isocyanate (19 Equiv.), and a catalytic amount of NEt_3 to isoxazoline **2**. The excess of nitrile oxide was used to obtain quantitative conversion in the cycloaddition reaction and to deliberately add some impurities in order to test the efficiency of the subsequent fluorous purification strategy. After completion of the reaction (three days, 25 °C), the solvent

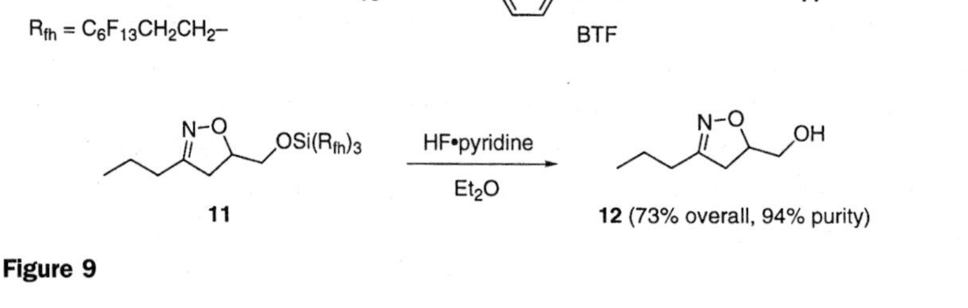

Figure 9

344

was removed and the residue was purified by three-phase extraction using FC-72, H_2O, and benzene. Evaporation of the fluorous phase afforded the silylated isoxazoline **11**. Despite the large amounts of nitrile oxide used, all the 'organic' side-products (for example the furoxan dimer) could easily be removed by the simple three-phase extraction procedure.

Desilylation of **11** was performed with HF•pyridine in Et_2O at 25 °C for 1 h. After removal of the solvent, the residue was dissolved in CH_2Cl_2 and sat. aqueous NH_4Cl was added. To remove the fluorosilane formed in the desilylation, the CH_2Cl_2–aqueous biphase was extracted with FC-72. The isoxazoline **12** was finally obtained from the CH_2Cl_2 phase in high yield and high purity as determined by GC analysis.

Protocol 8

Fluorous isoxazoline synthesis without intermediate purification

Reagents

- Bromo *tris*(2-perfluorohexylethyl)silane (see ref. 1)
- BTF (benzotrifluoride, Aldrich)
- FC-72 (3M Company)
- Phenyl isocyanate (Aldrich)
- HF•pyridine (Aldrich)

Method

1   Add a solution of 437 mg (0.38 mmol) of bromo *tris*(2-perfluorohexylethyl)silane in 3.3 ml THF at 25 °C over 5 min to a solution of 0.10 ml (1.52 mmol) allyl alcohol and 0.22 ml (1.52 mmol) NEt_3 in 3.4 ml THF, and stir at that temperature for 3 h.

2   Remove the solvent under reduced pressure and add FC-72, DCM, and H_2O (16 ml each). Shake and separate the fluorous phase (bottom layer). Wash the organic-aqueous biphase additionally twice with 16 ml FC-72. Combine the fluorous phases and remove the FC-72 under reduced pressure to afford silyl ether **10**.

3   Dissolve **10** in 15 ml BTF. Add 0.38 ml (3.80 mmol) nitrobutane and 0.73 ml (7.20 mmol) phenyl isocyanate followed by two drops of NEt_3. Stir the resulting mixture for three days at 25 °C.

4   Remove the solvent and purify the residue by three-phase extraction using FC-72, H_2O, and benzene (50 ml each), according to step 2 of this protocol, to afford silylated isoxazoline **11**.

5   Dissolve **11** in 12 ml Et_2O at 25 °C, add 0.4 ml HF•pyridine, and stir at 25 °C for 1 h.

6   Evaporate the solvent and dissolve the residue in 60 ml DCM. Add 30 ml sat. aqueous NH_4Cl and wash the organic–aqueous biphase twice with 30 ml FC-72. Separate the layers and extract the aqueous phase twice with 30 ml DCM. Combine the CH_2Cl_2 layers and dry over $MgSO_4$. Filter off the $MgSO_4$ and remove the solvent to yield 39.5 mg of isoxazoline **3**.

4.2 Fluorous Ugi sequence

In the Ugi four-component reaction (40), an acid, an aldehyde, an amine, and an isonitrile are condensed to form the corresponding amino acid amide. In the present protocol, we describe the application of fluorous chemistry to the Ugi reaction. The fluorous acid **13** can be directly used as the 'fluorous labelled' component in the Ugi reaction (*Figure 10*).

The fluorous benzoic acid derivative **13**, propylamine (17 Equiv.), cyclohexa-necarboxaldehyde (17 Equiv.), and cyclohexyl isocyanide (17 Equiv.) in trifluoro-ethanol (TFE) were heated in a sealed tube at 90°C for 48 h to afford the perfluorosilylated amide **14**. TFE serves as a 'hybrid organic/fluorous' solvent with the ability to dissolve both the organic compounds (amine, aldehydes, and isocyanide) and the fluorous compounds (acid **13** and all the products derived therefrom). Large excesses of the organic compounds were used to drive the reaction to completion. Purification was performed after removal of the solvent by a two-phase extraction using FC-72 and benzene. Despite the large excess of organic compounds present in the crude reaction mixture, the organic im-purities could be completely removed by the extraction procedure. Evaporation of the FC-72 phase then afforded the fluorous amino acid amide **14**. The benzene layer (upper layer) containing the excess of aldehyde, isonitrile, and all the non-fluorous products derived therefrom was discarded. For desilylation, amide **14**, dissolved in THF at 25°C, was treated with a THF solution of tetra-butylammonium fluoride (TBAF, 1 M) to provide the benzoylated amino acid amide **15**. After evaporation of the solvent, the crude product was purified by two-phase extraction using benzene and FC-72. The fluorous phase containing the fluorosilane was discarded. The organic phase was additionally washed with 0.1 M HCl, aqueous Na_2CO_3, and brine. The amide **15** was isolated in quanti-tative yield with high purity (> 95% as determined by GC analysis).

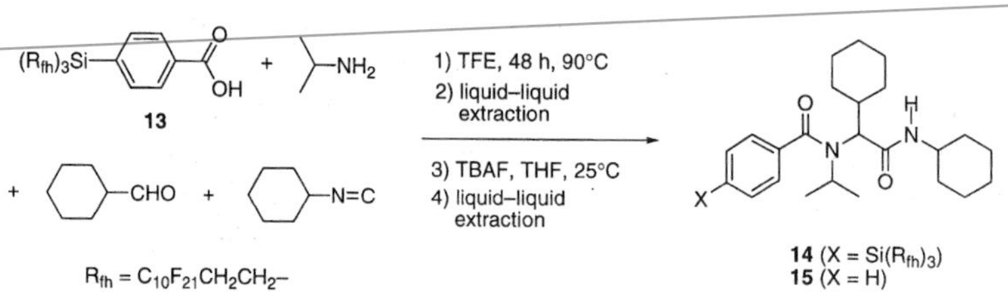

$R_{fh} = C_{10}F_{21}CH_2CH_2-$

Figure 10

Protocol 9

Fluorous Ugi reaction

Reagents

- 4-*tris*(2-Perfluorodecylethyl)silyl benzoic acid **13** (see ref. 40)
- Cyclohexyl isocyanide (Aldrich)
- Cyclohexanecarboxaldehyde (Aldrich)
- FC-72 (3M Company)
- 1 M TBAF solution in THF (Aldrich)

Method

1. Add 26.4 mg (0.015 mmol) of 4-*tris*(2-perfluorodecylethyl)silyl benzoic acid (**13**), 21 μl (0.25 mmol) propylamine, 30 μl (0.25 mmol) cyclohexanecarboxaldehyde, and 31 μl (0.25 mmol) cyclohexyl isocyanide to 0.3 ml TFE and seal the tube.

2. Heat the mixture under argon at 90°C for 48 h.

3. Remove the solvent under reduced pressure and dilute the residue with 15 ml FC-72. Wash the fluorous phase with 15 ml benzene. Extract the benzene layer twice with 15 ml FC-72. Combine the fluorous phases and evaporate to isolate the per-fluorosilylated amide **14**.

4. Dissolve crude amide **14** in 2 ml THF at 25°C and add 0.022 ml (0.022 mmol) of a 1 M TBAF solution in THF. Stir at 25°C for 30 min.

5. Remove the solvent, dilute the residue with 30 ml benzene, and wash twice with 15 ml FC-72. Add 30 ml Et_2O to the benzene layer and wash with 0.1 M HCl, aqueous Na_2CO_3, and brine (15 ml each). Dry the organic phase ($MgSO_4$), filter off the $MgSO_4$, and remove the solvent to yield 5.7 mg of **15** with > 95% purity.

5 Fluorous quenching (scavenging)

The removal of undesired reagents, reactants, side-products, etc. by a chemo-selective reaction is becoming increasingly popular in parallel synthesis. These techniques go by names such as quenching and scavenging (among others), and they are all applications of phase switching techniques (3). Fluorous quenching methods have much to recommend them, and they nicely supplement and complement techniques like resin quenching or ionization-based techniques. Fluorous quenching techniques are illustrated with a quench for excess alkenes (or alkynes) by hydrostannation with a fluorous tin hydride and by quenching of electrophiles with a fluorous amine (41). Both reactions are worked up by the liquid/liquid method. As the second example illustrates, this is especially useful in a library setting since insoluble members of the library are not removed along with the quenched products.

5.1 Tin hydride quench of alkenes by hydrostannation

In this protocol, radical addition of adamantyl iodide to benzyl acrylate and the removal of excess benzyl acrylate by hydrostannation with fluorous tin hydride **3** is described (*Figure 11*). This protocol is useful for the reactions that require excess alkene to form a desired product in good yield. The hydrostannation re-action can be done immediately after the desired reaction without any purifica-tion step and simple fluorous/organic extraction affords the pure addition product **16**.

Adamantyl iodide was reacted with 5 Equiv. of benzyl acrylate in the pres-ence of 10 mol% of the fluorous tin hydride **3**, sodium cyanoborohydride, and

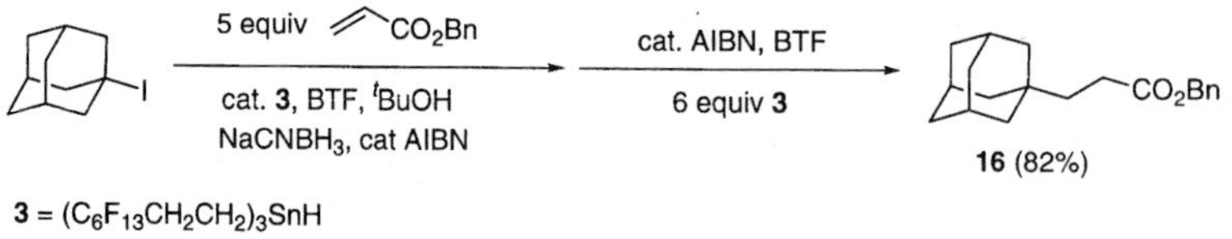

Figure 11

catalytic amount of AIBN in BTF and *tert*-butanol. After heating at reflux for 12 h, 6 Equiv. of **3** and catalytic amount of AIBN in BTF were added to the reaction mixture. The resulting mixture was heated at 90 °C for 24 h. The residue from the evaporation of the solvents was dissolved in chloroform and washed three times with FC-72. The organic layer was filtered through alumina and concentrated to yield the desired product **16** in 82% yield. The fluorous phase gave the hydrostannation product and excess tin compound.

Protocol 10

Tin hydride quench of an excess alkene by hydrostannation

Reagents

- *tris*(2-Perfluorohexylethyl)tin hydride (see *Protocol 2*)
- Adamantyl iodide (Aldrich)
- Benzyl acrylate (synthesized from acroyl chloride and benzyl alcohol)
- Sodium cyanoborohydride (Aldrich)

- AIBN (Aldrich)
- *t*-Butanol (Aldrich)
- BTF (benzotrifluoride, Aldrich)
- FC-72 (3M Company)

Method

1. Add BTF (0.5 ml) and *t*-butanol (0.5 ml) to a mixture of 26.2 mg 1-iodoadamantane (0.10 mmol), 81.1 mg benzyl acrylate (0.50 mmol), 11.6 mg fluorous tin hydride **3** (0.01 mmol), 9.6 mg sodium cyanoborohydride (0.13 mmol), and catalytic amount of AIBN to a reaction tube.

2. Purge with argon and seal the tube. Heat at 90 °C for 12 h.

3. Cool to room temperature and add 696 mg fluorous tin hydride **3** (0.60 mmol) and catalytic amount of AIBN in BTF (0.2 ml).

4. Purge with argon and seal the tube. Heat at 90 °C for 24 h.

5. Transfer the reaction mixture to a flask and evaporate to dryness.

6. Dissolve the residue in chloroform (10 ml) and transfer to a separatory funnel. Wash with FC-72 (3 × 10 ml).

7. Filter the organic layer through alumina and remove the solvent to yield 24.6 mg the product **16** (82%).

5.2 Fluorous amine quenching in robotic parallel synthesis of ureas

This section describes the parallel synthesis of a urea library using a fluorous quenching procedure. The whole process is carried out on a solution phase synthesizer (HP 7686 robot) and is fully automated.

The process consists of mixing THF solutions of three different amines **18** (each 0.5 M, 1 Equiv.) with three different isocyanates **19** (each 0.5 M, 1.5 Equiv.) (*Figure 12*). The preparation of the reagent solutions is done on the robot by diluting pure reagent to obtain a 0.5 M solution. After completion of the reaction, the excess isocyanate reagent is reacted with a fluorous amine quenching reagent **17** bearing six perfluorinated chains. The fluorous amine **17** is dissolved in FC-72. Consequently the quenching mixture is heterogeneous and it must be mixed for at least 30 min to allow complete quenching. The resulting fluorous urea products **21** from the quenching reaction as well as excess **17** are then removed by a series of liquid/liquid extractions in which the organic phase is washed four times with FC-72. The desired organic ureas **20** are not soluble in FC-72 and remain in the organic phase, while any fluorous products **17** and **20** are removed with the fluorous phase. Evaporation of the organic solvent yields nine different ureas **20a-i**, which are dried in a vacuum oven at 50–60 °C prior to analysis. The yields and purities of the products from a typical 9-compound library are contained in *Table 1*.

Table 1 Results of the 3 × 3 library of ureas

Entry	Nr	R^1	R^2	Urea yield (%)	HPLC purity^a (%)	%F^b
1	**20a**	4-Br	4-CF$_3$	87	98.0 (0.5)	0.5
2	**20b**	4-Br	3-CF$_3$	80	98.5 (0.4)	0.3
3	**20c**	4-Br	2-CF$_3$	90	98.4 (0.2)	0.4
4	**20d**	3-Br	4-CF$_3$	79	95.7 (2.4)	0.3
5	**20e**	3-Br	3-CF$_3$	93	96.5 (1.7)	0.3
6	**20f**	3-Br	2-CF$_3$	85	96.8 (0.9)	0.3
7	**20g**	2-Br	4-CF$_3$	73	92.3 (4.4)	0.2
8	**20h**	2-Br	3-CF$_3$	96	94.0 (3.6)	0.2
9	**20i**	2-Br	2-CF$_3$	90	93.6 (3.6)	0.3

^{a} Waters Nova-Pack® C18 3.9 × 150 mm. Flow rate 1.5 ml/min. Detection at 210 nm, solvent, acetonitrile 40–60% in water. The numbers in parenthesis refer to the amount of symmetrical urea.

^{b} Mol%, determined in the ^{19}F NMR spectrum, relative to the product CF$_3$ group.

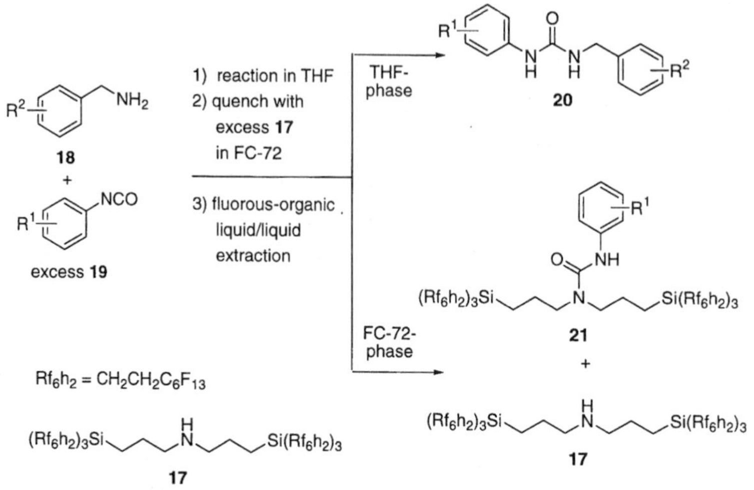

Figure 12

Protocol 11

Fluorous amine quench for 3 × 3 library

Equipment and reagents

- Hewlett Packard 7686 solution phase synthesizer (Hewlett Packard)
- Reaction vials and 'high recovery' vials (Hewlett Packard)
- FC-72 (3M Company)

- *N,N-bis*[3-[*tris*(2-Perfluorohexylethyl)silyl] propyl]amine **17** (see ref. 40)
- *o-*, *m-*, *p-*Trifluoromethyl benzyl amine (Aldrich)
- *o-*, *m-*, *p-*Bromophenyl isocyanate (Aldrich)

A. Preparation of the reagent solutions

1. The following quantities of reagent are diluted with THF to a final volume of 1.5 ml to obtain a 0.5 M solution: 107 μl of *p*-trifluoromethyl benzyl amine; 107 μl of *m*-trifluoromethyl benzyl amine; 105 μl of *o*-trifluoromethyl benzyl amine; 94 μl of *m*-bromophenyl isocyanate; 93 μl of *m*-bromophenyl isocyanate.

2. *p*-Bromophenyl isocyanate (solid): weigh 159 mg, and dissolve in 1.59 ml.

3. Solutions of the fluorous amine are obtained by adding 0.5 ml of FC-72 to each of the nine vials containing 110 mg fluorous amine.

B. Reaction and quenching

1. Dispense 0.35 ml THF in the reaction vial.

2. Transfer 0.1 ml from the amine solution vial (0.05 mmol) to the reaction vial (rinse and mix for 0.5 min).

3. Transfer 0.15 ml from the isocyanate solution vial (0.075 mmol) to the reaction vial (rinse and mix for 0.5 min). Wait for 1 h at room temperature.

4. Transfer the fluorous amine (0.05 mmol) to the reaction vial.

5. Rinse the fluorous amine vial with 0.1 ml FC-72 and transfer to the reaction vial.

6. Mix the reaction vial for 30 min. Wait 30 min.

7. Transfer the bottom fluorous phase from the reaction vial to a fluorous waste vial.

8. Add new FC-72 to the reaction vial (0.6 ml). Mix for 5 min. Wait for 30 min.

9. Transfer 0.60 ml from the reaction vial to a fluorous waste vial. Repeat three times.

10. Evaporate the reaction vial at 50 °C for 15 min.

6 Conclusions

The fluorous techniques described in this chapter are representative of the techniques that we have used in our group during the initial phase of development of fluorous methods (1994–1998). The convenient liquid/liquid separations are limited in that very large numbers of fluorines are sometimes required to render large molecules fluorous. But this problem is being solved by solid/liquid methods, which are very convenient for parallel synthesis as well. Compared to standard small molecule techniques and solid phase techniques, the current scope of fluorous techniques is rather limited. But this limitation is simply due to the youth of the field. As more and more fluorous reagents, reactants, and protecting groups become available, we will begin to learn just how broad the applicability of these techniques is and what niches they can uniquely or advantageously fill in both traditional and parallel synthesis.

Acknowledgements

We thank the National Institutes of Health and the National Science Foundation for funding of this work. We are also most grateful to the following companies for gifts or grants: Hewlet Packard (LCMS and robotic synthesizer), Parke-Davis, CombiChem, Merck, OxyChem (benzotrifluoride), and Elf Atochem (perfluoroalkyl iodides).

References

1. Thompson, L. A. and Ellman, J. A. (1996). *Chem. Rev.*, **96**, 555.
2. Balkenhohl, F., von dem Büssche-Hunnefeld, C., Lansky, A., and Zechel, C. (1996). *Angew. Chem. Int. Ed. Engl.*, **35**, 2289.
3. Curran, D. P. (1998). *Angew. Chem. Int. Ed. Engl.*, **37**, 1175.
4. Curran, D. P. (1996). *Chemtracts–Org. Chem.*, **9**, 75.
5. Horvath, I. T. and Rabai, J. (1994). *Science*, **266**, 72.
6. Cornils, B. (1997). *Angew. Chem. Int. Ed. Engl.*, **36**, 2057.
7. Horvath, I. T. (1998). *Acc. Chem. Res.*, **31**, 641.
8. Studer, A., Hadida, S., Ferritto, R., Kim, S. Y., Jeger, P., Wipf, P., *et al.* (1997). *Science*, **275**, 823.

9. Curran, *et al.* US Patent 5 777 121.

10. Curran, D. P., Hadida, S., and He, M. (1997). *J. Org. Chem.*, **62**, 6714.

11. Kainz, S., Luo, Z. Y., Curran, D. P., and Leitner, W. (1998). *Synthesis*, 1425.

12. Lam, K. S., Lebl, M., and Krchnak, V. (1997). *Chem. Rev.*, **97**, 411.

13. Gravert, D. J. and Janda, K. D. (1997). *Chem. Rev.*, **97**, 489.

14. Barthel-Rosa, L. P. and Gladysz, J. (1999). *Coord. Chem. Rev.*, **192**, 587.

15. Curran, D. P. and Hadida, S. (1996). *J. Am. Chem. Soc.*, **118**, 2531.

16. Ogawa, A. and Curran, D. P. (1997). *J. Org. Chem.*, **62**, 450.

17. Curran, D. P., Hadida, S., Kim, S.-Y., and Luo, Z. (1999). *J. Am. Chem. Soc.*, **121**, 6607.

18. Berendsen, G. E. and Galan, L. D. (1978). *J. Liquid Chromatogr.*, **1**, 403

19. Elemental analysis: carbon 9.73%; fluorine 14.90%; chlorine < 0.5%; hydrogen 0.71%.

20. Horner, J. H., Martinez, F. N., Newcomb, M., Hadida, S., and Curran, D. P. (1997). *Tetrahedron Lett.*, **38**, 2783.

21. Hadida, S., Super, M. S., Beckman, E. J., and Curran, D. P. (1997). *J. Am. Chem. Soc.*, **119**, 7406.

22. Ryu, I., Niguma, T., Minakata, S., Komatsu, M., Hadida, S., and Curran, D. P. (1997). *Tetrahedron Lett.*, **38**, 7883.

23. Ryu, I., Niguma, T., Minakata, S., Komatsu, M., Luo, Z., and Curram, D. P. (1999). *Tetrahedron Lett.*, **40**, 2367.

24. Curran, D. P., Kim, S.-Y., and Hadida, S. (1999). *Tetrahedron*, **55**, 899.

25. Kingston, H. M. and Haswell, S. J. (ed.) (1997). *Microwave-enhanced chemistry.* American Chemical Society: Washington, DC.

26. Larhed, M., Lindeberg, G., and Hallberg, A. (1996). *Tetrahedron Lett.*, **37**, 8219.

27. Larhed, M. and Hallberg, A. (1996). *J. Org. Chem.*, **61**, 9582.

28. Curran, D. P. and Hoshino, M. (1996). *J. Org. Chem.*, **61**, 6480.

29. Hoshino, M., Degenkolb, P., and Curran, D. P. (1997). *J. Org. Chem.*, **62**, 8341.

30. Larhed, M., Hoshino, M., Hadida, S., Curran, D. P., and Hallberg, A. (1997). *J. Org. Chem.*, **62**, 5583.

31. Strauss, C. R. and Trainor, R. W. (1995). *Aust. J. Chem.*, **48**, 1665.

32. Saá, J. M., Dopico, M., Martorell, G., and García-Raso, A. (1990). *J. Org. Chem.*, **55**, 991.

33. Curran, D. P. and Luo, Z. (1998). *Med. Chem. Res.*, **8**, 261.

34. Curran, D. P., Luo, Z., and Degenkolb, P. (1998). *Bioorg. Med. Chem. Lett.*, **8**, 2403.

35. (a) Sugiyama, K., Hirao, A., and Nakahama, S. (1996). *Macromol. Chem. Phys.*, **197**, 3149. (b) Brace, N. O., Marshall, L. W., Pinson, C. J., and van Wingerden, G. J. (1984). *Org. Chem.*, **49**, 2361.

36. Studer, A. and Curran, D. P. (1997). *Tetrahedron*, **53**, 6681.

37. Studer, A., Jeger, P., Wipf, P., and Curran, D. P. (1997). *J. Org. Chem.*, **62**, 2917.

38. Curran, D. P., Ferritto, R., and Hua, Y. (1998). *Tetrahedron Lett.*, **39**, 4937.

39. Mukaiyama, T. and Hoshino, T. (1969). *J. Am. Chem. Soc.*, **82**, 5339.

40. Ugi, I. (1982). *Angew. Chem. Int. Ed. Engl.*, **21**, 810.

41. Linclau, B., Singh, A., and Curran, D. P. (1999). *J. Org. Chem.*, **64**, 2835.

Chapter 12

Combinatorial synthesis on multivalent oligomeric supports

Ronald M. Kim and Jiang Chang

Merck and Co., Inc., PO Box 2000, RY 800–C306, Rahway, NJ 07065, USA.

1 Introduction

This chapter describes a strategy for liquid phase combinatorial chemistry which we term combinatorial synthesis on multivalent oligomeric supports (COSMOS). In this approach, compounds are synthesized on multivalent soluble supports through iterative cycles of solution phase synthesis and size-based product purification. Final products may be cleaved from the support, and isolated from the oligomer by size. A brief comparison of solid phase and liquid phase synthesis will be presented, followed by an introduction to COSMOS methodology. The role of the support in COSMOS, and methods for rapid product purification will be highlighted. Detailed procedures for the preparation of soluble supports, and an illustrative synthesis using the COSMOS approach will be described. Emphasis will be placed on protocols for isolating support-bound and cleaved products. Parallel sample processing and automation of the COSMOS methodology will be discussed.

2 Use of supports in organic synthesis

2.1 Solid phase synthesis (see Chapters 1–7)

The recent renaissance in biology has led to a proliferation of new therapeutic targets and the advent of high-throughput approaches to biological screening. This in turn has placed increased demands on medicinal chemists to supply small molecules with interesting biological activity. To meet this challenge, technological advances have been made in the areas of combinatorial chemistry and high-speed synthesis (1). Perhaps the most significant development in the field has been the application of solid supports toward combinatorial synthesis. Pioneered by Merrifield over three decades ago for the preparation of peptides (2), solid phase synthesis can now be used to build massive libraries of non-peptide small molecules in both mixture and single compound format. The power of the solid phase approach lies largely in the ease by which reaction products are isolated; impurities are simply removed by filtration and washing, which expedites rapid and parallel sample processing, and enables the production of combinatorial libraries via pooling strategies. Large reagent excess may also be employed to drive reactions to completion (Chapters 1–4). In addition, tagging strategies (Chapter 4) may enable the rapid identification of biologically active species from combinatorial libraries comprised of even millions of constituents (3).

Despite the utility of solid phase synthesis, solid supports can introduce non-trivial liabilities to a synthetic scheme (Chapter 6 and 7). As with any supported approach, solid phase synthesis requires a chemical handle by which compounds are attached to the support. In addition to necessitating a suitable linkage site on the target molecule, other issues such as inertness of the handle during the reaction sequence, and stability of the product to the cleavage conditions, come into play. Heterogeneous reaction conditions can also impede translation of classical solution phase reactions to a combinatorial format; poor resin solvation, variable site accessibility, and chemical or mechanical instability of the support itself can all hamper reactivity of bound species. In addition, solid supports can obstruct characterization of bound products, as many routine analytical methods, including NMR, mass spectroscopy, and chromatographic approaches, cannot be brought to bear on heterogeneous samples. While significant strides in the analysis of resin-bound species have been made, application of these approaches typically require specialized techniques or equipment (Chapter 8).

2.2 Liquid phase synthesis

In order to circumvent the drawbacks associated with heterogeneous systems, while maintaining an expedient means of product isolation, soluble supports have been used in chemical synthesis (4). The strategy of soluble-supported synthesis is often referred to as 'liquid phase' synthesis (LPS), to differentiate the approach from classical solution phase methods (Chapter 10). As with resins, soluble supports were first introduced for the production of biopolymers (5); their use in combinatorial chemistry, however, has thus far been quite limited (4, 6).

While LPS also requires the use of a chemical handle to link the target molecule to the support, homogeneous reaction conditions are maintained. This facilitates adaptation of classical solution phase chemistry to the liquid phase regime, and enables the use of heterogeneous reagents and catalysts. In fact, a large role of the support can be to solubilize bound compounds that may otherwise not dissolve in the reaction medium. This 'solubilizing power' can be of critical importance in a combinatorial process, where the solubility of subunits may vary greatly, and where several solvents may be required during a reaction scheme. Another significant advantage gained with soluble supports is the ability to routinely characterize bound species using standard analytical tools such as NMR, IR, and UV. Even mass spectroscopy and chromatography can be utilized, depending on the homogeneity and complexity of the support.

354

The benefits in reactivity and characterizability offered by liquid phase synthesis over solid phase methods can greatly reduce the time investment necessary to develop a combinatorial scheme. For a LPS approach to be practical, however, products must be rapidly and reliably isolable from the reaction mixture. In order to process large numbers of reactions, the means of purification should also be amenable to parallel and automated operation. The second role of the support, then, is to serve as a marker by which bound compounds can be identified and segregated from the rest of the reaction medium. This capacity stands in sharp contrast to the desired 'invisibility' of the support during reaction steps, and reconciling the dichotomous roles of the support during synthesis and purification is the major crux of developing a practical LPS methodology.

3 Overview of combinatorial synthesis on multivalent oligomeric supports (COSMOS)

We recently described a method for liquid phase combinatorial synthesis (7), which we now term combinatorial synthesis on multivalent oligomeric supports (COSMOS). In this approach, multiple copies of the compound are built on a multivalent soluble support through iterative steps of liquid phase synthesis and size-based purification. Similar liquid phase approaches have previously been applied in the synthesis of biopolymers (5b, c, 8). A generic procedure is outlined in *Figure 1*, and is functionally akin to solid phase synthesis. In the initial step, the first subunit (A) is loaded onto an oligomeric support bearing multiple chemical linkers. In step 2, the soluble product, consisting of the support bearing multiple subunit copies, is isolated from low molecular weight

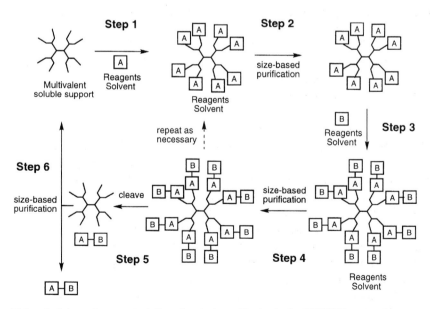

Figure 1 Schematic representation of organic synthesis via the COSMOS approach.

355

impurities by a size-selective process such as size exclusion chromatography or ultrafiltration. The next subunit is then coupled (step 3), and the oligomeric product is again isolated from the reaction mixture based on its greater size (step 4). The process is repeated until the final product is prepared, at which time it can be cleaved from the support (step 5), and separated from the oligomer by size (step 6). Depending on the handle, the support may be regenerated and reused.

Alternatively, products may be screened while still attached to the soluble support. In fact, the resultant multivalent ligand presentation can give rise to advantageous biological activity (9). The avidity achieved by multiple interactions can also lead to greatly enhanced binding and selectivity (10).

COSMOS couples the advantages of solution phase chemistry, including homogeneous reaction conditions and routine product characterization, with those realized in solid phase synthesis, such as the ability to drive reactions using reagent excess, and a means of convenient product isolation. Like solid phase synthesis, COSMOS purification is based on the *intrinsic* size difference between the oligomeric support and low molecular weight impurities, and should therefore be general in scope. The coarse nature of the purification step also enables the development of rapid and efficient procedures which can be used to prepare discrete compound arrays, as well as mixture libraries.

4 Supports for COSMOS

4.1 General considerations

The function of the support in the COSMOS approach is twofold:

(a) It confers solubility to bound intermediates during chemical reaction steps.

(b) It provides sufficient size for attached compounds to be readily separated from reagents and solvents.

A wide variety of compounds can fulfil this role, including star polymers, dendrimers, linear and crosslinked polymers, and block co-polymers. Monovalent supports can also be incorporated into the COSMOS strategy, although since only one compound is synthesized per oligomer, the support loading may be compromised.

A number of factors will influence the performance of a support in a synthetic scheme, which are as yet largely unexplored. Size is of obvious importance, with larger oligomers generally being more readily separated from small molecules. Gains in separation efficiency may be offset by other factors related to the support architecture, however, such as reduced reactivity, compromised characterizability, or decreased mechanical stability. For example, large linear or weakly crosslinked polymers may aggregate in solution, which can adversely affect solubility and reactivity (11). By contrast, highly branched structures of comparable molecular weight, such as dendrimers, may adopt globular conformations with terminal reactive sites extending into solution (12), thereby favouring reactivity.

Because structure, size, homogeneity, and loading of the support can all influence the reactivity of bound intermediates, and the ease by which they can be isolated and characterized, we tend to use structurally simple, highly symmetric oligomers which are only large enough to permit ready isolation by size. In addition to being easier to prepare and characterize, the smaller supports may provide a more extended, 'solution-like' platform for organic synthesis. We also prefer to use discrete oligomers when possible, since the support loading is explicit, uniform reactivity may be obtained, and reaction products are single entities that can be *fully characterized*. Despite our incomplete understanding of what constitutes an 'ideal' soluble support, it should be stressed that compared to resins, the ability to precisely control the structure, loading, and physical properties is greatly enhanced, and efforts by us and others (13) to design optimized supports for given applications are underway. Examples of supports developed for COSMOS are presented in Sections 4.2 and 4.3.

4.2 PAMAM dendrimers

In our original publication (7), we utilized an eight-armed polyamidoamine (PAMAM) dendrimer for the preparation of 2-aryl indole single compounds and a small library by the mix-split approach. Shown in *Figure 2*, PAMAM-HMB **1** embodies several features which are advantageous to a soluble support:

(a) Unlike typical polymers, **1** is a discrete entity; thus, support loading is unambiguous, and reaction products can be fully characterized using traditional methods.

(b) Each handle is identical, affording uniform reactivity and greatly simplifying product characterization, especially by NMR.

(c) The reactive groups are located at the support periphery, maximizing reagent accessibility and minimizing intramolecular reactions.

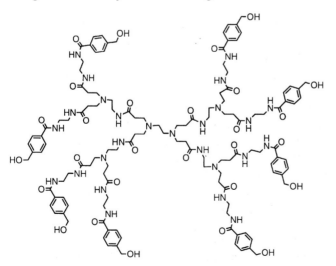

Figure 2 Dendritic soluble support PAMAM-HMB **1**.

357

Despite the successful application of support **1** toward the preparation of single compounds and mixture libraries, PAMAM dendrimers suffer from several shortcomings which may limit their widespread use as soluble supports for combinatorial chemistry. These include inconsistent solubility, an ionizable and potentially reactive amidoamine backbone, and numerous and generally broad ^{1}H NMR signals which can hamper characterization of bound intermediates. We have gone on to investigate alternative polymers as soluble supports for COSMOS; our results with polyethylene glycol derivatives are summarized below.

4.3 Polyethylene glycol-armed supports

Due to their high solubility in a range of solvents, chemical inertness, and ready availability, polyethylene glycol (PEG) oligomers have been the most extensively used supports in LPS. To exploit these qualities, and to explore the effect of support size, structure, and loading toward the COSMOS process, we have synthesized a series of discrete PEG-armed supports ranging in molecular weight from 2–26 kDa. Like PAMAM dendrimers, the PEG oligomers are symmetric, with identical chemical handles located at the end of each arm. Two exemplary structures, derived by grafting hexa(ethylene glycol) onto appropriate core structures, are shown in *Figure 3*. The biphenyl cores provide good spacing of the PEG arms, and also furnish a chromophore to facilitate monitoring of product isolation. While more complete details of the structures and syntheses of the PEG-armed polymers will be presented elsewhere, our studies indicate that the smaller supports (*c.* 2.5–5 kDa) offer good chemical stability, high compound loading, and excellent solubilizing power in a range of solvents. The smaller

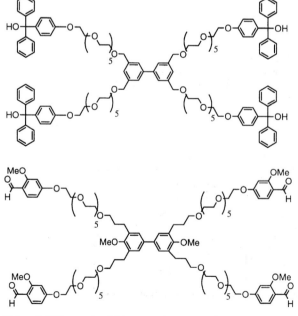

Figure 3 PEG-armed soluble supports.

oligomers are also relatively easy to prepare and characterize, yet are of sufficient size to be readily separated from low molecular weight compounds. In addition, the PEG supports tend to display sharp ^{1}H and ^{13}C NMR signals, and except for the PEG ethyl region near 3.6 ppm, the proton NMR field is largely unobstructed for characterization of bound species.

Recently, the Sunbright PTE polymer series, related four-armed PEG oligomers radiating from a pentaerithrytol core, have become commercially available from Shearwater Polymers. Although consisting of a mixture of oligomers of similar mass, thereby precluding full characterization of reaction products, their high symmetry and simple architecture allows for product analysis by NMR. The core methylene protons also serve as a convenient internal standard to calibrate compound loading. The low polydispersity of the polymers also permit characterization and purification by chromatographic methods. The preparation of a soluble support PEG4-Rink **2**, derived from Sunbright PTE-2000 (avg. MW = 2 kDa) coupled to four Rink (14) handles via amide bonds, is described in *Protocol 1*, and depicted in *Figure 4*. The support was used for the liquid phase synthesis of guanidines (15), as will be described in Section 5.

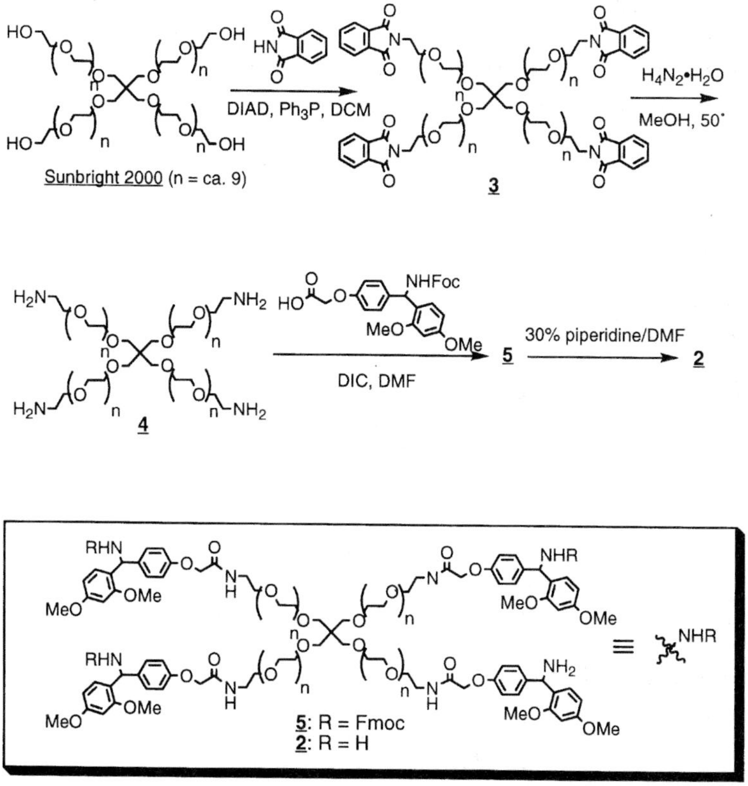

Figure 4 Synthesis of four-armed soluble support PEG4-Rink **2**.

Protocol 1

Synthesis of soluble support PEG4-Rink 2

Equipment and reagents

- Rotary evaporator
- Magnetic stirring plate
- Glassware for chemical synthesis
- Sunbright PTE-2000 (Shearwater Polymers)

- p-[(R,S)-α-[1-(9H-fluoren-9-yl)methoxy-formamido]-2,4-dimethoxybenzyl] phenoxyacetic acid (Fmoc-Rink linker, Novabiochem)
- All other reagents and solvents were obtained from Aldrich

Method

1   **PEG4-phthalimide 3**. In a 250 ml round-bottom flask equipped with a stirbar and a rubber septum, charge 50 ml of anhydrous tetrahydrofuran (THF) with Sunbright PTE-2000 (1.0 mmol, 2.34 g), phthalimide (8.0 mmol, 1.2 g, 2 Equiv. per PEG chain), and diisopropylazodicarboxylate (8.0 mmol, 1.63 g). Using an airtight syringe, add drop-wise a solution of triphenylphosphine (8.0 mmol, 2.1 g) in 10 ml of THF to the stirring reaction mixture at a rate that does not allow the exothermic reaction to become too vigorous. Allow the reaction to proceed at ambient temperature for 2 h. Isolate compound **3** by size exclusion chromatography (SEC), as described in *Protocol 5*.[a]

2   **PEG4-NH$_2$ 4**. In a 100 ml round-bottom flask fitted with a rubber septum, heat a solution of PEG oligomer **3** (1.0 mmol, 2.8 g) in 50 ml of MeOH containing 2 ml of hydrazine hydrate at 50 °C for 3 h. Allow the solution to cool to ambient tempera-ture. A white precipitate (phthalhydrazide) will form during the reaction; remove the precipitate by filtration. Rotary evaporate the filtrate to remove the MeOH. More precipitate may form. Take up the residue in *c*. 20 ml of dichloromethane (DCM), and remove any precipitate by filtration if necessary. Purify the tetraamine by SEC (*Protocol 5*).

3   **PEG4-Rink-NHFmoc 5**. In a 100 ml round-bottom flask, prepare a solution of tetraamine **4** (1 mmol, 2.3 g), Fmoc-Rink linker (10 mmol, 5.4 g), and diisopropyl-ethylamine (10 mmol, 1.3 g, 1.7 ml) in 50 ml of DMF. Add diisopropylcarbodiimide (10 mmol, 1.27 g), and let the reaction stand at ambient temperature for 4 h. Purify the product by SEC (*Protocol 5*).[b,c]

4   **PEG4-Rink 2**. Dissolve Fmoc-protected **5** (0.01 mmol, 35 mg) in a 3 ml solution of 30% piperidine (v/v) in DMF. Let the reaction stand at room temperature for 30 min. Purify support **2** by SEC (*Protocol 5*).

[a] Parallel purification is useful for the isolation of large quantities of product. Compound **3** was purified on a four-column system described in Section 6.6, in four cycles (16 individual runs), loading 150–200 mg of oligomer per run. We have also used larger S-X1 columns to isolate 0.5–1 g quantities of support per run.

[b] Relatively non-polar supports such as **1** and **5** can be purified from incompletely coupled oligomers by chromatography on silica, eluting with DCM containing 5–10% MeOH.

[c] Support to be kept for long periods should be stored as the Fmoc-protected species **5**.

5 Organic synthesis in COSMOS

As discussed in Section 2.2, a major advantage of LPS over solid phase approaches is the maintenance of homogeneous reaction conditions; adaptation of classical reactions is facilitated, traditional synthetic equipment and conditions may be used, and products can be characterized using standard analytical tools. The high loading that can be achieved on multivalent soluble supports also enhances the ability to drive reactions. As a result, often fewer reagent equivalents are required compared to solid phase procedures, although large reagent excess can also be employed in the COSMOS approach. When first attempting to adapt solution phase chemistry to the COSMOS scheme, we typically utilize reaction conditions similar to existing procedures, but incorporate an excess (1.5–3-fold) of reagent. When performing synthesis on the multimilligram scale, we commonly carry out reactions in 16 × 125 cm glass tubes, since the tubes can also be used to collect reaction products isolated by size exclusion chromatography (see Section 6.3–6.6). PEG polymers are not soluble in certain solvents such as diethyl ether, and alternative solvents may be required. However, non-volatile solvents may be used, since the products are separated from the reaction media during purification.

Because soluble supports are free to interact with each other, the same site isolation that can be achieved on solid supports is not realized. Thus, as in solution phase synthesis, precautions must be taken to avoid crosslinking when coupling subunits capable of forming multiple bonds. Multivalent supports carrying linkers that form reactive species such as cations during the cleavage step can also polymerize, due to crosslinking of the handles. Any insoluble polymer thus formed can simply be removed by filtration. Such crosslinking was obtained during cleavage of guanidine products from Rink-derived support **2** (15), as described in *Protocol 2*.

Detachment of final products from the support in COSMOS is analogous to cleavage in other supported schemes, and may be achieved by chemical or photochemical means. Unlike solid phase synthesis, however, heterogeneous catalysts may also be used to affect compound cleavage. Detached products may be isolated from the support based on size, with the product being collected in the small molecule fraction. Selective extraction, either of the support or of cleaved compounds, can also be used to isolate products. During synthesis of 2-aryl indoles on PAMAM support **1**, the cleaved products could be separated from the dendrimer by SEC, or alternatively, by selective extraction of the product in acetonitrile (7).

As an illustrative example of organic synthesis via the COSMOS methodology, the preparation of trisubstituted guanidine **10**, using tetravalent support **2**, is outlined in *Protocol 2* and in *Figure 5*. The compound was synthesized as part of a 48-compound guanidine library (15). Purities of crude cleavage products prepared using the procedure averaged nearly 90%, with overall yields for the five-step reaction averaging *c.* 50%, based on Rink loading of **2**. Of note is the use of $HgCl_2$ to affect guanidine formation; while solid phase guanidine synthesis

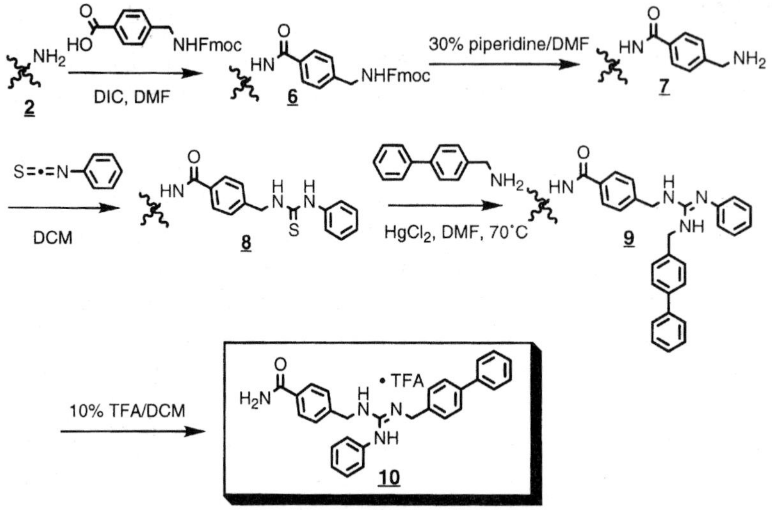

Figure 5 Synthesis of a trisubstituted guanidine using soluble support **2**.

has been described (16), to our knowledge the mercury-mediated transformation has not been reported, presumably due to the inability to remove insoluble mercuric sulfide from resin supports. Only the synthetic procedures will be described in this section; methods for purification of reaction products will be presented in *Protocols 3–5*, in Section 6.

Protocol 2

Synthesis of a trisubstituted guanidine using a soluble support

Equipment and reagents

- Magnetic stirring/heating plate
- PEG4-Rink **2** (see *Protocol 1*)
- Fmoc-(4-aminomethyl)-benzoic acid (Neosystem)

- 4-Phenylbenzylamine (Organix)
- All other reagents and solvents were obtained from Aldrich

Method

1   **Intermediate 6**. Dissolve PEG support **2** (0.01 mmol, 31 mg) and Fmoc-amino-methylbenzoic acid (0.1 mmol, 37 mg, 2.5 Equiv. per handle) in 2 ml of DMF. Add diisopropylcarbodiimide (0.1 mmol, 13 mg). Let the reaction mixture stand at ambient temperature for 6 h. Purify oligomer **6** by SEC, as described in *Protocol 5*.

2   **Intermediate 7**. Dissolve **6** in a 3 ml solution of 30% (v/v) piperidine in DMF. Let the reaction stand at ambient temperature for 30 min. Purify **6** by SEC (*Protocol 5*).

3   **Intermediate 8**. Dissolve **7** in 3 ml of DCM. Add phenylisothiocyanate (0.1 mmol, 14 mg) to the solution, and stir at ambient temperature for 2 h. Purify the product by SEC (*Protocol 5*).

4   **Intermediate 9**. Prepare a solution of **8** and 4-phenylbenzylamine (1 mmol, 107 mg) in 3 ml of DMF. Add $HgCl_2$ (0.1 mmol, 27 mg), and stir the reaction at 70 °C for 16 h. A black precipitate (mercuric sulfide) will form. Allow the reaction to reach room temperature, and add sodium sulfide (0.1 mmol, 78 mg) to the mixture to precipitate the remaining Hg(II). Stir at ambient temperature for 0.5 h. Remove the mercuric sulfide by filtration over celite. Purify the product by SEC (*Protocol 5*).

5   **Guanidine 10**. Dissolve **9** in 2 ml of 10% trifluoroacetic acid in DCM and let the reaction stand for 1 h. The solution will turn deep red, and some insoluble polymer comprised of crosslinked support may form. Remove the solvent by evaporation. Take up the residue in 3 ml of DCM. Filter any polymerized support on celite. Purify the filtrate by SEC (*Protocol 5*), collecting guanidine **10** in the low molecular weight fraction.

6 Size-based purification of homogeneous reaction products in COSMOS

6.1 General considerations

That moderately sized, architecturally simple soluble supports can offer synthetic and analytical advantages over heterogeneous polymers is not surprising. The challenge of LPS, however, lies in the ability to selectively segregate dissolved products from all other components of a reaction mixture using a method that is simple, reliable, and high-yielding. Furthermore, to be competitive with solid phase synthesis, the purification scheme should be congruous with parallel processing and automation.

As in solid phase synthesis, COSMOS purification consists of sequestration of large support-derived oligomers from small molecule reagents, solvents, and detached products; oligomeric species are not separated from each other, although this can often be achieved by other chromatographic means. By contrast, alternative methods of liquid phase combinatorial synthesis exploit the differential solubility between the supports and reagents to drive purification by phase separation (17).

A consequence of size-selective purification is the maintenance of homogeneous conditions throughout the reaction and purification steps. Several methods for the size-based partitioning of solutes exist, including dialysis, ultrafiltration, and size exclusion chromatography. Our examination of ultrafiltration and size exclusion chromatography as methods for product purification are summarized in the following sections.

6.2 Ultrafiltration

Ultrafiltration (UF) was first employed in liquid phase synthesis over two decades ago for the preparation of peptides (5). In simplest terms, UF is a molecular-level filtration (18). A solution is passed through a membrane containing microscopic pores, and molecules that are smaller than the pore diameter pass through the filter, while those that are too large are retained by the membrane. Often a force such as centrifugation is applied to help drive the solution through the filter. Because UF is actually an enrichment technique, several filtrations may be necessary to satisfactorily purify the large molecules. We have evaluated the use of regenerated cellulose, Teflon, and YM (Amicon®) ultrafilters due to their stability toward organic solvents. Our preliminary results indicate that relatively large supports are required for good recovery and clean separation from impurities, and that UF is considerably slower than purification by size exclusion chromatography (see Sections 6.3–6.6), and will not be further elaborated on here.

6.3 Size exclusion chromatography

6.3.1 General considerations

Size exclusion chromatography (SEC), or gel permeation chromatography (GPC), is also a purification technique that separates molecules based on their effective size in solution (19). Common applications include desalting of proteins and analysing the weight distribution of polymers. SEC is typically performed by loading a sample onto a column containing a stationary phase comprised of microporous particles, and eluting the sample with a liquid mobile phase. Molecules which are small enough to enter the pores of the stationary phase permeate the particles, while those too large to enter the pores flow through the interstitial space between the particles. Thus, the largest species are the first to elute, while small molecules are increasingly detained as their partitioning into the stationary phase increases. Like other chromatographic methods, SEC is a true purification technique; however, due to the coarse nature of product isolation (large oligomers from small molecules), COSMOS purifications by SEC are rapid and efficient, resembling solid phase extractions.

By selecting a support that nears or exceeds the size exclusion limit of the stationary phase, the volume in which the oligomer elutes from the column remains relatively constant (beginning near the column void volume), independent of the species coupled to the support. Consequently, both oligomeric and small molecule products can be collected based simply on elution volume. Furthermore, because oligomeric species elute together regardless of attached constituents, the COSMOS approach can be used to prepare combinatorial libraries using pooling strategies. The consistency of the elution profile has important ramifications toward automated and parallel sample processing, as will be discussed in Section 6.6. In addition, because SEC is based on permeation of the stationary phase, and not on affinity, solvent gradients are not required, and columns can be used repeatedly.

6.3.2 Caveats and limitations of SEC

As with most chromatographic procedures, the formation of precipitates can hinder SEC purification. The solubilizing role of the support is therefore important in keeping bound compounds in solution not only during reaction steps, but also during work-up. Precipitation of reagents during SEC, especially when the mobile phase differs from the reaction solvent, can also complicate purification. The mobile phase can therefore be chosen with reagent solubility in mind; since the eluent must be removed after purification, however, it should be reasonably volatile. Conversely, the reaction mixture can be diluted with the mobile phase, and any precipitates removed prior to SEC (see *Protocol 1*, step 2). Because COSMOS purification relies on the separation of large molecules from substantially smaller ones, separation of the support from large solutes such as oligomers, aggregates, colloids, or other high molecular weight compounds, may not be feasible.

Secondary interactions between solutes and the SEC matrix can also be detrimental to product isolation. Both attractive and repulsive interactions can impede size-based purification through mechanisms such as peak tailing, compound retention, or fronting. The choice of mobile phase is important toward minimizing secondary interactions, as exemplified by the use of polystyrene/divinylbenzene resins for SEC in non-polar solvents, and for reverse phase separations under aqueous conditions. We have also obtained good size-based product isolation on a reverse phase gel comprised of C-8 capped porous silica (Whatman® BioPrep, 100 Å pore size), by employing dichloromethane as the mobile phase.

6.4 Size exclusion chromatography resins for COSMOS purification

6.4.1 Sephadex LH-20

In our original publication (7), SEC purifications were performed on Sephadex® LH-20 (Pharmacia Biotech). DMF was utilized as the mobile phase due to the variable solubility of PAMAM derivatives in less polar solvents. While LH-20 was effective for isolating dendritic species, in some cases moderate product losses were observed due to tailing of dendritic peaks into reagent fractions, presumably arising from associative interactions with the support. Tailing of some polar reagents also increased the time and solvent required to flush the columns before they could be reused.

6.4.2 Polystyrene/divinylbenzene SEC gels

An important outgrowth of the high solubility of PEG supports is the ability to use volatile non polar eluents for SEC purification. In addition to being easily removed from reaction products by evaporation, solvents such as THF and DCM are compatible with more hydrophobic polystyrene/divinylbenzene (PS/DVB) SEC gels. These resins have good chemical stability, and in our experience, moderately crosslinked gels do not exhibit significant secondary interactions (20).

The SEC gel which we currently utilize for gravity and low pressure operation is Bio-Beads® S-X1 from Bio-Rad. S-X1 is a 1% DVB crosslinked PS resin with a listed exclusion limit of 14 kDa, although we have found only minor differences in the elution volume of oligomers above *c*. 2.5 kDa. Thus, even moderately sized oligomers are efficiently isolated from small molecules, with product recoveries averaging *c*. 95%. The large particle size (200–400 mesh) also affords low column back pressures (< 5 psi), and permits gravity operation. A disadvantage of SX-1 is its 'softness' due to the low resin crosslinking, which is manifested in compressibility under high flow rates and non-uniform swelling in different solvents, both of which can affect the size exclusion limit. We therefore dedicate each column to a single mobile phase, and employ flow rates not exceeding 5 ml/min. S-X1 also tends to be more fragile than higher cross-linked gels, and we have noted decomposition of the resin with mobile phases containing unstablized THF, perhaps due to oxidative damage from peroxides. In DCM, however, columns can generally be used for more than a dozen purifications.

Our procedure for packing Bio-Beads® S-X1 columns for SEC purification is presented in *Protocol 3*. For convenience, some of the column parts that we use are listed, although many commercial alternatives exist. The columns can be used to purify samples of up to 3 ml, and containing up to 100 mg of oligomeric material, which is appropriate for product synthesis on the multimilligram scale. The mobile phase may be delivered by a pump, as in the system described in *Protocol 4*, or by gravity. A related gel, Bio-Beads® S-X3 from Bio-Rad, is also effective for low pressure SEC purification, and the higher crosslinking (3% DVB) affords less resin swelling and increased durability compared to S-X1. The lower size exclusion limit of S-X3, however, makes for tighter support separation from larger compounds, such as cleaved products.

Protocol 3

Packing a SEC column with Bio-Beads® S-X1

Equipment and reagents

- Bio-Beads® S-X1 (Bio-Rad)
- 2.5 cm (i.d.) glass chromatography column (Omnifit, Cat. No. 6222NS), fixed endpiece (Omnifit, Cat. No. 6240NS), and silicone O-rings (Omnifit, Cat. No. 6267NS)
- Adjustable endpiece (Omnifit, Cat. No. 6265NS)
- Teflon stopcock (Omnifit, Cat. No. 1101)
- Mobile phase: 3% MeOH in DCM

Method

1   In a 250 ml Erlenmeyer flask, swell 5 g of Bio-Beads® S-X1 in 100 ml of the mobile phase.[a] The amount of resin expansion is significant. Follow the manufacturer's instructions for resin swelling.

2   Transfer the slurry into a glass chromatography column fitted with one fixed end-

piece and a stopcock. Rinse the flask with the mobile phase as necessary to transfer all of the resin to the column. Allow solvent to run through the column to help pack the resin bed, always keeping the solvent above the bed level to avoid drying the resin.

3   Keeping solvent above the bed, close the stopcock and gently fit the adjustable endpiece into the column. Adjust the frit until it is in contact with the resin bed. Hook the column to the solvent delivery system (see *Protocol 4*), and pass *c.* 100 ml of solvent through the column to pack the resin bed, and to wash out any resin fines and dissolved impurities. Adjust the level of the adjustable frit as necessary. The column height for a 2.5 cm (i.d.) column should be 10–11 cm.

a The resin will float in DCM, but will pack when solvent is passed through the column bed. To prevent the resin from floating, columns can be packed in 1:1 DCM/THF, and switched over to the mobile phase, as swelling of S-X1 in THF and DCM is similar.

Preparative HPLC columns can also be used for COSMOS purification. SEC gels for HPLC are generally fairly rigid due to higher crosslinking, which provides high durability and consistent swelling in a range of solvents. Thus, different mobile phases can be used with some columns, although manufacturer's guidelines should be observed when switching solvents. We have obtained excellent product isolation using Ultrastyragel® and Syragel® HPLC columns from Waters.

6.5 Methods for COSMOS purification via SEC

As stressed previously, central to the practicality of a LPS methodology is the ease of product isolation. In COSMOS, purification by SEC consists of:

(a) Loading the reaction mixture onto a SEC column.

(b) Eluting the sample with the mobile phase.

(c) Collecting the product in a predetermined volume.

(d) Removing the mobile phase from the product.

The manner by which these steps can be achieved vary greatly, ranging from fully manual to fully automated approaches. In *Protocol 4*, we describe the assembly and operation of a simple and robust system for the sequential purification of reaction products by SEC. Illustrated in *Figure 6*, the system incorporates a solvent pump, a manually-loaded sample loop, an SEC column, a UV detector, and a fraction collector. A manually operated three-way valve diverts column effluent to the fraction collector or to waste. The solvent pump provides delivery of the mobile phase at a constant flow rate, enabling time-based product collection. The UV detector is convenient for monitoring product elution, and obviates the need for column calibration. We have listed the equipment in our assembly, although many commercial alternatives exist; complete

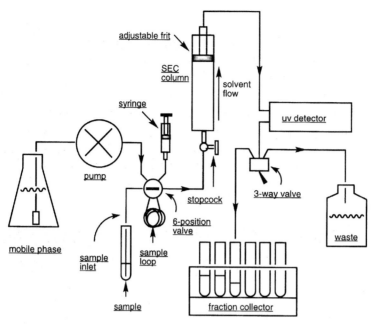

Figure 6 A system for LPS sample purification by SEC.

set-ups can also be purchased. A pared down version consisting of a pump, sample loop, SEC column, and fraction collector can also provide facile product isolation. Similar set-ups are commonly used for normal and reverse phase separations. Simple gravity-driven separations can also be performed on S-X1, using the same guidelines as to column packing, mobile phase, product elution volumes, and flow rates. More complex systems capable of automated and parallel operation can also be assembled, as will be discussed briefly in Section 6.6.

Protocol 4

Assembling a system for sequential product isolation by SEC

Equipment and reagents

- Low/medium pressure solvent pump (Fluid Metering, Inc., model QSY)
- Six-position sample valve (Rheodyne, Cat. No. 5011), and 5 ml sample loop
- SEC column: described in *Protocol 3*
- UV detector (Isco, model V4)
- Three-way valve (Rheodyne, Cat. No. 53010)
- Fraction collector (Isco, model Foxy 200)
- Teflon tubing, 1/16 inch i.d.
- Mobile phase: 3% MeOH in DCM

Method

1   Assemble the components as depicted in *Figure 6*. Do not connect the column until all the upstream lines have been filled with the mobile phase (see step 2).

2   Fill the lines leading to the column, including the sample loop, with mobile phase before hooking up the SEC column. This ensures that air does not enter the column, which could result in drying of the resin bed.

3   Install the column. For columns that have not been used, pass approx. 100 ml of the mobile phase through the column to pack the bed (see *Protocol 3*, step 2).

[a] Note that the column is installed with the inlet entering the column bottom; we have found that flowing the mobile phase against gravity results in slightly tighter elution bands, perhaps due to decreased channelling. Excellent product isolation is also obtained, however, with the opposite solvent flow.

Protocol 5

Product purification by SEC (using the system presented in *Protocol 4*)

Equipment and reagents

- Chromatography system (see *Protocol 4*)
- Mobile phase: 3% MeOH (v/v) in DCM

Method

1   Filter the sample to remove particulates, if necessary. For a 5 g column, the sample should preferably not exceed 3 ml, nor contain more than 100 mg of oligomeric product (see *Protocol 1*, footnote *a*).

2   With the six-position valve in the load position, fill the sample inlet line with mobile phase by pushing a small amount of solvent (0.2–0.3 ml is adequate) from the loop out of the sample inlet. Do this by pressurizing the other inlet line with a syringe. This prevents air from being loaded onto the column, and washes out any residue from the previous sample.

3   Transfer the sample to the sample loop. Load the sample by placing the sample inlet tube directly into the reaction vessel, and drawing the sample onto the loop by pulling with a syringe connected to the other inlet, as depicted in *Figure 6*. Alternatively, the sample may be pushed onto the loop from a syringe.

4   With the flow rate between 4–5 ml/min, switch the six-position valve to inject, thereby diverting the mobile phase through the sample loop, and pushing the sample onto the column.

5   With the fraction collector set to collect 30–45 sec fractions,[a] start the collector. Switch the three-way valve to divert column effluent to the fraction collector. Remember to compensate for the time delay between the detector and the fraction collector, which is dependent on the volume of intervening tubing. The approximate elution volume of oligomeric products is 17–29 ml, and 30–50 ml for small molecules.

Protocol 5 continued

6 Remove the mobile phase from the product fraction(s) by evaporation. For single reactions, this can be done by rotary evaporation, or under a stream of nitrogen. For simultaneous solvent removal from multiple samples, an evaporator such as centrifugal concentrator or a multi-line nitrogen blowdown station may be used.

7 After the product has eluted, divert the eluent to waste. Make sure at least 60 ml (total volume) of mobile phase has passed through the column before performing the next purification, to ensure it is flushed of low molecular weight impurities. When not in use for long periods, close the stopcock to prevent the column from drying.

[a] Products may also be collected by observing the UV trace. Once the column is calibrated, products may be collected based on volume or time.

6.6 Automation and parallelization of COSMOS purification by SEC

The consistency of the SEC elution profile for different reactions greatly simplifies product isolation in the COSMOS scheme, as both oligomeric species and small molecule products can be isolated simply by collecting a predetermined elution volume. As a result, simultaneous purifications can easily be performed, and the process can be automated using existing liquid handling technology.

We have performed simultaneous reaction work-ups by gravity SEC using Bio-Beads® S-X1. Despite different flow rates between columns, the product elution volumes remained constant, and even temporarily stopping solvent flow did not affect product isolation. Due to the robustness and simplicity of SEC purification, significant numbers of compounds can be prepared using completely manual approaches.

We have also developed semi- and fully-automated parallel systems for SEC purification. A four-column unit incorporating manually loaded sample loops and syringe pump mobile phase delivery was used to synthesize a 48-compound combinatorial library (15). Valves were programmed to simultaneously collect products in a predetermined volume. A fully automated eight-column system capable of performing six cycles, thereby processing 48 samples, has also been developed. Both systems will be described in greater detail in the future.

Automated HPLC systems incorporating automated sample loading can easily be adapted for serial SEC purification, and a number of HPLC columns for preparative SEC are commercially available (Section 6.4). In addition, parallel HPLC systems can currently be obtained.

The methods described in this chapter are aimed at compound production on the multimilligram scale. Increasing the column size or performing parallel purifications make gram scale syntheses easily achievable. Due to the advent of miniaturized biological screens, however, milligram or even submilligram product quantities may be adequate for evaluation in multiple assays. Production on this scale presents an opportunity to significantly drive down the size of the purification apparatus, and opens the prospect for massively parallel compound synthesis via the COSMOS methodology.

References

1. For reviews of solid phase synthesis in combinatorial chemistry, see: (a) Thompson, L. A. and Ellman, J. A. (1996). *Chem. Rev.*, **96**, 555. (b) Früchtel, J. S. and Jung, G. (1996). *Angew. Chem. Int. Ed. Engl.*, **35**, 17. (c) Gallop, M. A., Barrett, R. W., Dower, W. J., Fodor, S. P. A., and Gordon, E. M. (1994). *J. Med. Chem.*, **37**, 1233. (d) Terrett, N. K., Gardner, M., Gordon, D. W., Kobylecki, R. J., and Steele, J. (1995). *Tetrahedron*, **51**, 8135. (e) Balkenhohl, F., Bussche-Hunnefeld, C. von dem, Lansky, A., and Zechel, C. (1996). *Angew. Chem. Int. Ed. Engl.*, **35**, 2288.

2. Merrifield, R. B. (1963). *J. Am. Chem. Soc.*, **85**, 2149.

3. (a) Ohlmeyer, M. H. J., Swanson, R. N., Dillard, L. W., Reader, J. C., Asouline, G., Kobayashi, R., *et al.* (1993). *Proc. Natl. Acad. Sci. USA*, **90**, 10922. (b) Tan, D. S., Foley, M. A., Shair, M. D., and Schreiber, S. L. (1998). *J. Am. Chem. Soc.*, **120**, 8565.

4. (a) Gravert, D. J. and Janda, K. D. (1997). *Chem. Rev.*, **97**, 489, and references therein. (b) Mutter, M., Uhmann, R., and Bayer, E. (1975). *Liebigs Ann. Chem.*, 901.

5. (a) Shemyaki, M. M., Ovchinnikov, Y. A., and Kiryushkin, A. A. (1965) *Tetrahedron Lett.*, **27**, 2323. (b) Mutter, M., Hagenmaier, H., and Bayer, E. (1971). *Angew. Chem. Int. Ed. Engl.*, **10**, 811. (c) Mayer, E. and Mutter, M. (1972). *Nature (London)*, **237**, 512.

6. (a) Merritt, A. T. (1998). *Comb. Chem. High Throughput Screening*, **1**, 57. (b) Gravert, D. J. and Janda, K. D. (1997). *Curr. Opin. Biol.*, **1**, 107. (c) Curran, D. P. (1998). *Angew. Chem. Int. Ed. Engl.*, **37**, 1174.

7. Kim, R. M., Manna, M., Hutchins, S. M., Griffin, P. R., Yates, N. A., Bernick, A. M., *et al.* (1996). *Proc. Natl. Acad. Sci. USA*, **93**, 10012.

8. Bayer, E., Gatfield, I., Mutter, H., and Mutter, M. (1978). *Tetrahedron*, **34**, 1829.

9. (a) Tam, J. P. (1988). *Proc. Natl. Acad. Sci. USA*, **85**, 5409. (b) Tam, J. P. and Spetzler, J. C. (1997). In *Methods in Enzymology*, Fields, G. B., ed. (Academic Press, San Diego). (c) Qureshi, S. A., Kim, R. M., Konteatis, Z., Biozzo, D. E., Motamedi, H., and Rodrigues, R. (1999). *Proc. Natl. Acad. Sci. USA*, **96**, 12156. Vol. 289, p. 612.

10. (a) Roy, R., Zanini, D., Meunier, S. J., and Romanowska, A. (1993). *J. Chem. Soc. Chem. Commun.*, 1869. (b) Kiessling, L. L. and Pohl, N. L. (1996). *Chem. Biol.*, **3**, 71.

11. Wooley, K. L., Fréchet, J. M. J., and Hawker, C. J. (1994). *Polymer*, **35**, 4489.

12. (a) Tomalia, D. A., Naylor, A. M., and Goddard III, W. A. (1990). *Angew. Chem. Int. Ed. Engl.*, **29**, 138. (b) Fréchet, J. M. J. (1994). *Science*, **263**, 1710.

13. Gravert, D. J., Datta, A., Wentworth Jr., P., and Janda, K. D. (1998). *J. Am. Chem. Soc.*, **120**, 9481.

14. Rink, H. (1987). *Tetrahedron Lett.*, **28**, 3787.

15. Chang, J., Oyelaran, O., Esser, C. K., Kath, G. S., King, G. W., Uhrig, B. G., *et al.* (1999). *Tetrahedron Lett.*, **40**, 4477.

16. (a) Dodd, D. S. and Wallace, O. B. (1998). *Tetrahedron Lett.*, **39**, 5701. (b) Drewry, D. H., Gerritz, S. W., and Linn, J. A. (1997). *Tetrahedron Lett.*, **38**, 3377. (c) Kearney, P. C., Fernandez, M., and Flygare, J. A. (1998). *Tetrahedron Lett.*, **39**, 2663.

17. (a) Han, H., Wolfe, M. M., Brenner, S., and Janda, K. D. (1995). *Proc. Natl. Acad. Sci. USA*, **92**, 6419. (b) Studer, A., Jeger, P., Wipf, P., and Curran, D. P. (1997). *J. Org. Chem.*, **62**, 2917.

18. Blatt, W. F., Robinson, S. M., and Bixler, H. (1968). *J. Anal. Biochem.*, **26**, 151.

19. Moore, J. C. (1964). *J. Polymer Sci.*, **A-2**, 835.

20. We have observed that extremely highly crosslinked gels, such as Amberchrom® CG 161 fromToso Haas and DVB gels from Jordi, tend to retain both supports and reagents through secondary interactions.

Chapter 13

Automated solution phase synthesis and its application in combinatorial chemistry

Tohru Sugawara and David G. Cork
Takeda Chemical Industries Ltd., Pharmaceutical Research Division,
17–85 Juso Honmachi, 2-chome, Yodogawa-ku, Osaka 532–8686, Japan.

1 Introduction

In this chapter we present examples of syntheses using the automated synthesis apparatus developed in the research laboratories at Takeda Chemical Industries Ltd., Osaka (1). These will show both applications of automated systems, which include units for synthesis, analysis, extraction, purification, and isolation of compounds, as well as automated workstations, which have only synthesis, extraction, and isolation capability.

The automated apparatus are designed for preparing libraries of compounds by combining a series of reactants, in all or selected permutations. The common series of operations used in solution phase synthesis are automated to allow continuous and unattended preparation. Rather than high-throughput the apparatus are focused on medium to large scale synthesis, either for preparing combinatorial optimization libraries, or producing series of reactants for subsequent use in smaller scale library synthesis.

2 Automated synthesis systems and workstations

There are several ways to categorize the various automated apparatus that have been developed for synthesis. One way is to make a division based on the design, with two main categories for:

(a) Those using robotic transfers (Chapters 1, 5, 11).

(b) Those using flow-lines for transfer between fixed reactors (2).

The apparatus we used for the work described here were all developed around a flow-line design (3).

A further division of automated apparatus may be made based on the extent of the operations and functions incorporated. Fully automated systems include

Table 1 Automated synthesis apparatus developed at Takeda Chemical Industries Ltd.

Generation	Years	Name (Type)	Description	Features
1	1985–7	– (System)	The prototype	Two reactors, 0–80 °C, column chromatography, predictive logic control algorithm, fixed control sequence.
2	1988–9	TACOS (System)	Takeda's automated computer operated system	Three reactors, 0–80 °C (×2), 25–180 °C (×1), large reservoirs and measuring for multi/repetitive use, column chromatography, HPLC reaction monitoring, unit-operation control program.
		MAVIS (System)	Multipurpose automated versatile intelligent system	
3	1990–3	VACOS (System)	Versatile automated computer operated system	Three reactors, 0–80 °C (×2), 25–180 °C (×1), reservoirs for single use, thermal sensor for solvent evaporation, conductivity sensor for phase separation, column chromatography, HPLC monitoring, versatile unit operation control.
		MATES (System)	Multipurpose automated tech. equip. for synthesis	
		TARO (System)	Takeda's automated reliable operator	
		EASY (System)	Expert automated synthesizer for you	
4	1994–7	RAMOS (Workstat.)	Reliable automated multireactor org. synthesizer	Three reactors, –15 to 180 °C (×3), synthesis and phase separation units only, quick-fit reaction flasks and reagent reservoirs, precipitate sensor.
		TAFT (Workstat.)	Takeda's automated flow technique	
		ASTRO (Workstat.)	Automated synthesis with totally reliable operation	
		FUTOSHI (Workstat.)	A Japanese nickname	One reaction flask of 2 litres, phase separation.
		ASRA (Workstat.)	Automated supported reagent apparatus	One reactor, –15 to 180 °C, handling of solid powdered reagents, work-up by extraction and filtration from solid. One fixed-bed type react or, 15–80 °C, solid powdered catalysts, multi-batch/cycles possible.
		ASCA (Workstat.)	Automated solid catalyst apparatus	
		SOAP (Workstat.)	Stand-alone or option-type apparatus for purification	Preparative scale column chromatography, with fraction collector.
5	1997–	WOOS (Workstat.)	Windows operated organic synthesizer	One reactor with 1 litre flask, –15 to 180 °C, Windows operating system.
		TACOS-II (Workstat.)	Takeda's automated computer operated system II	Two reactors, –15 to 180 °C, centrifugal two-phase separation, panel PC control.

units for all the stages that may be required, such as synthesis, analysis, extraction, purification, and isolation of compounds. On the other hand automated workstations are more dedicated to a chosen utility, such as synthesis and extraction, chromatographic isolation, intermediate–large scale synthesis, use of solid-supported reagents, or reaction monitoring. *Table 1* lists the five generations of automated apparatus developed at Takeda in the automated synthesis group over the last 14 years.

2.1 The automated synthesis hardware

2.1.1 Synthesis systems

Takeda's automated systems comprise units for performing various tasks:

- carrying out the reactions
- supplying the reagents, reactants, and solvents
- performing two-phase extraction
- separating compounds by chromatography
- monitoring reactions by HPLC
- washing the whole apparatus

A schematic diagram of the unit layout is shown in *Figure 1* and a photograph of one system, EASY, is shown in *Figure 2*.

2.1.2 Synthesis workstations

Automated workstations are dedicated to a chosen utility, such as synthesis and extraction, chromatographic isolation, intermediate–large scale synthesis, use of solid-supported reagents, or reaction analysis. *Figure 3* shows the layout of the units for:

(a) A synthesis/extraction workstation such as RAMOS.

(b) A workstation capable of controlled handling of powdered reagents.

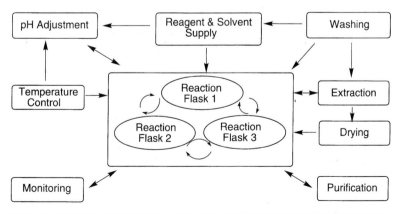

Figure 1 Schematic diagram of the units that make up Takeda's second and third generation automated synthesis systems. (Adapted from ref. 3.)

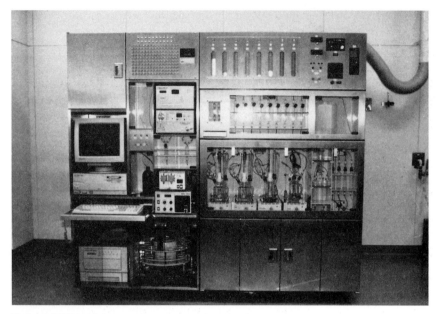

Figure 2 EASY, third generation automated synthesis system.

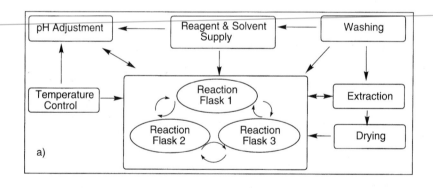

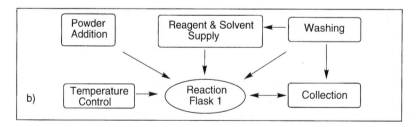

Figure 3 Schematic diagrams of the units in two of Takeda's fourth generation automated synthesis workstations. (a) The upper one corresponds to RAMOS, and (b) the lower one to ASRA.

376

Figure 4 Three fourth generation automated synthesis workstations, ASRA, SOAP, and RAMOS, which can be used separately or linked to one another.

Both workstations and another for automated chromatographic purification, SOAP, are shown in the photograph of *Figure 4*.

2.2 The automated synthesis control software

The computers controlling the first four generations of apparatus used software written in N88-BASIC running under MS-DOS. Whereas the most recent, fifth generation has been developed using Visual Basic operating on a Windows 95 PC. The computers have also varied from the early desktop (first to third generations) design, to compact laptops (fourth generation), and recently to a Windows 95 panel computer (TACOS-II).

Organic chemists often carry out experiments according to synthetic flow charts or reaction procedures. Similarly, the automated synthesis system is operated according to a series of procedures for reaction, reagent addition, product extraction, washing, drying and so on, which are contained in subroutine programs.

For the MS-DOS software we developed five main programs (MAKE, REFORM, SIMU, RUN, and CHART), which are led by menus so as to be fairly self-explanatory. The user can easily 'compose' a reaction program with MAKE by selecting one by one from the reaction subroutine pool which contains about 120 subroutines (see *Table 2*). We found it convenient to separate the reaction work-up subroutines so that the user can create or modify an extraction module by choosing from an extraction subroutine pool of approximately 60 subroutines (see *Table 3*). Rather than have to make every synthesis control program from scratch, the operator can modify an existing one with the editor program called REFORM.

Table 2 Typical pool of subroutines used for the third generation automated systems

Title	Function
START	First subroutine to input all required parameters
RSx-RFy	Measure volume of liquid in RSx in measuring tube, transport to RFy (x = 1–6, y = 1–3)
RRx-RFy	Transport liquid in RRx to RFy (x = 1–3 , y = l; x = 4–6, y = 2; x = 7–9, y = 3)
RFx-RFy	Transport liquid in RFx to RFy (x = 1, y = 2, 3; x = 2, y = 1, 3; x = 3, y = 1, 2)
RFx-PH	Transport liquid in RFx to PH (x = 1–3)
PH-RFx	Transport liquid in PH to RFx (x = 1–3)
RFx-SRy	Transport liquid in RFx to SRy (x = 1–3, y = 2, 3)
RFx-SF	Transport liquid in RFx to SF (x = 1–3)
SF-RFx	Transport liquid in SF to RFx (x = 1–3)
SF-SRx	Transport liquid in SF to SRx (x = 0, 1)
SRx-SF	Transport liquid in SRx to SF (x = 0, 1)
SF-Fx-Fy	Transport half the liquid in SF to RFx and RFy (x, y = combination of two out of 1, 2, and 3)
FRACT-RFx	Transport liquid in fraction tubes to RFx (x = 1–3)
RFx-MIX	Stir vigorously (x = 1–3)
RFx-STR-ON	Start stirring in RFx (x = 1–3)
RFx-STR-OF	Stop stirring in RFx (x = 1–3)
RFx-BUBB	Bubble in RFx (x = 1–3)
SF-BUBB	Bubble in SF
PH-BUBB	Bubble in PH
RFx-REA-n	The nth reaction in RFx (n = 1–3, x = 1–3): monitoring available
RFx-CON-n	The nth concentration in RFX (n = 1–3, x = 1–3)
EXTn	The nth extraction (n = 1–10)
RFx-CL-ON	Start cooling RFx (x = 1, 2)
RFx-CL-OF	Stop cooling RFx (x = 1, 2)
RFx-HT-ON	Start heating RFx (x = 1–3)
RFx-HT-OF	Stop heating RFx (x = 1–3)
A-LC-ON	Switch on monitoring HPLC system
A-LC-OF	Switch off monitoring HPLC system
DE-CO-ON	Switch on preparative HPLC system
DE-CO-OF	Switch off preparative HPLC system
HPLC	Inject from SR3 to HPLC, carry out column chrom.
PH-ADJ	Adjust pH of liquid in PH
MATU	Wait for 5 min
MATU 15	Wait for 15 min
ALARM	Alarm and pause until user's interference
WASH	Clear screen for washing
MR-WASH	Wash MT1, MT2, and RR1-RR9
WS-RFx	Transport washing solvent to RFx via MT1
SF-DR	Drain liquid in SF
SRx-DR	Drain liquid in SRx (x = 0–3)
RFx-DRY	Dry RFx (x = 1–3)
RF-DRY	Dry RF1, RF2, and RF3
R-WASH	Wash specified RS

Table 3 Extraction module subroutine pools for the automated synthesis system

Title	Function
EX-START	Compulsory first subroutine
EXSLCT	Select extracting solvent, measure, transport to RF
EWSSLC	Select washing solvent, measure, transport to RF
ERSn-RFy	Measure volume of liquid in RSn in measuring tube, transport to RFy ($n = 1$–6, $y = 1$–3)
ERRn-RFy	Transport liquid in RRn to RFy ($n = 1$–3, $y = 1$; $n = 4$–6, $y = 2$; $n = 7$–9, $y = 3$)
ERFn-SF	Transport liquid in RFn to SF ($n = 1$–3)
ESF-RFn	Transport liquid in SF to RFn ($n = 1$–3)
ESF-SRn	Transport liquid in SF to SRn ($n = 0$, 1)
ESRn-SF	Transport liquid in SRn to SF ($n = 0$, 1)
ERFn-BUBB	Bubble in RFn ($n = 1$–3)
ESF-BUBB	Bubble in SF
ERFn-MIX	Stir vigorously ($n = 1$–3)
ESEP-SRn	Separate, transport bottom layer to SRn ($n = 0$, 1)
ESF-DT-Fn	Transport through drying tube to RFn ($n = 1$–3)
ESF-DR	Drain liquid in SF
ESRn-DR	Drain liquid in SRn ($n = 0$, 1)
EDT-SLCT	Select drying tube
EDT-RST	Reset drying tube

If the user requires, the software control sequence can be checked, before actual running of the experiment, by using the SIMU program to simulate operation of the hardware and confirm the order and timing of events is appropriate. When the START subroutine is chosen as the first subroutine, the various parameters that will control reaction time, temperature, reagent volumes, etc., can be input, and saved for subsequent use. START also contains a self-diagnosis function. The initial state of the apparatus is checked and if any of the sensors (liquid level, photosensors, microswitches, etc.) indicate a problem it is rectified or the operator is warned. During operation, the user may interrupt the procedure at any point to renew the parameters, either using the keyboard function keys or by manual operation of the units. The program CHART is used to recall the HPLC chromatogram and modify, if necessary, its scale.

The computer software for our automated supported reagent apparatus, ASRA, includes an algorithm for controlling the rate of addition from a powder addition funnel (4).

3 Applications of automated synthesis systems

3.1 Unusual amino acid derivatives

The first generation automated synthesis system had a hardware and software structure that could only be used to perform reactions in one fixed direction of

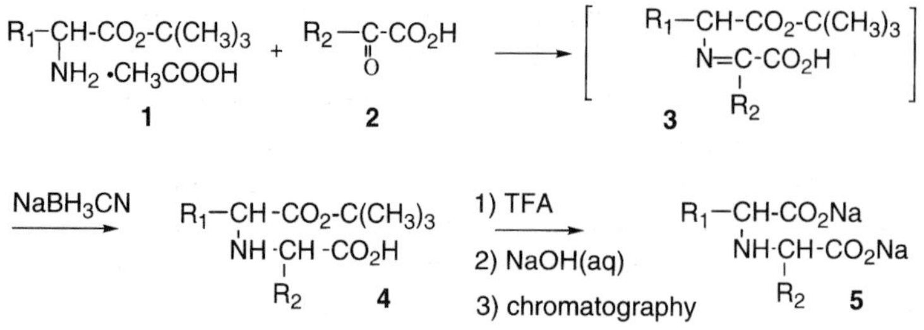

Figure 5 Automated synthetic route used to prepare several hundred substituted *N*-carboxyalkyl amino acid derivatives. (Adapted from ref. 5.)

flow. However, using this prototype system, it was possible to prepare and isolate a wide variety of compounds, including the synthesis of several hundred substituted N-carboxyalkyl amino acid derivatives (5). All processes were automatically performed, from the mixing of the reactants to the isolation of the products as powders or crystals, as well as the washing and drying of the apparatus after each synthetic run.

Figure 5 shows the general reaction of an amino acid *tert*-butyl ester acetic acid salt, **1**, with a 2-keto acid, **2**, which produces an unstable intermediate Schiff base, **3**. This is reduced with sodium cyanoborohydride to give a substituted N-(carboxyalkyl)amino acid *tert*-butyl ester, **4**, and the disodium salt, **5**, upon further treatment with trifluoroacetic acid (TFA) and hydrolysis. The equilibrium and the consecutive reactions were carefully controlled by adding sodium cyanoborohydride at a controlled rate using artificial intelligence software that contained novel kinetic equations and substituent effects. Automation allowed high yields to be obtained through careful control of the reaction conditions and the addition rate of reagents (5d).

3.2 Condensed azole derivatives

A series of N-substituted sulfamoylpropylthioimidazo[1,2-*b*]pyridazines were synthesized, by modifying the sulfonamide moiety, in an effort to find an analogue of 3-(imidazo[1,2-*b*]pyridazin-6-yl)thiopropanesulfonamide with improved anti-asthmatic activity (6).

Protocol 1 describes the procedure for the synthesis of N-substituted sulfonamides using three reaction steps (see *Figure 6*).

(a) Preparation of sulfonamides, **7**, using 3-chloropropanesulfonyl chloride, **6**, and various kinds of amines.

(b) Substitution reaction of **7** by potassium hydrogen sulfide to give 3-mercaptopropanesulfonamide, **8**.

(c) Substitution reaction of **8** with 6-chloroimidazo[1,2-*b*]pyridazine to give the N-substituted sulfonamide, **9**.

A flow chart showing the reactions performed by the automated synthesis apparatus is presented in *Figure 7*. Namely, **6** and an amine are introduced and

380

reacted in one reaction flask, then the reaction mixture is transferred to the extraction drying unit and washed with water. After drying, the organic solution is transferred back into the reaction unit for concentration. Methanol (MeOH) and ethanolic potassium hydrogen sulfide (KSH) solution are added to the residue and the thiol, **8**, is prepared in a second reaction flask. 6-Chloroimidazo[1,2-*b*] pyridazine and sodium methoxide are introduced into the reaction flask and give the desired compound, **9**. After concentration, water and ethyl acetate are added to the residue and the mixture is transferred to the extraction/drying unit. After drying, the organic layer is concentrated, and the residue is transferred to the purification unit. Finally, the desired fraction from column chromatography is concentrated and recrystallized manually.

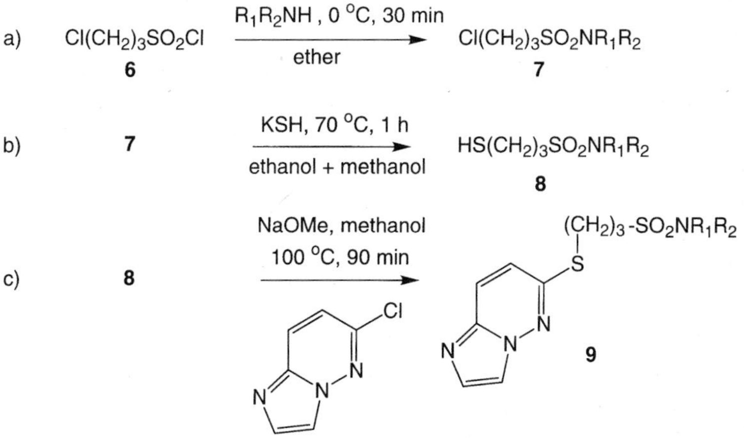

a) $Cl(CH_2)_3SO_2Cl$ $\xrightarrow[\text{ether}]{R_1R_2NH, 0\,°C, 30\ min}$ $Cl(CH_2)_3SO_2NR_1R_2$
           6   **7**

b)   **7** $\xrightarrow[\text{ethanol + methanol}]{KSH, 70\,°C, 1\ h}$ $HS(CH_2)_3SO_2NR_1R_2$
                                                                                 8

c)   **8** $\xrightarrow[\text{100\,°C, 90 min}]{NaOMe, methanol}$

Figure 6 Automated synthetic route used to prepare a series of sulfonamide derivatives. (Adapted from ref. 6.)

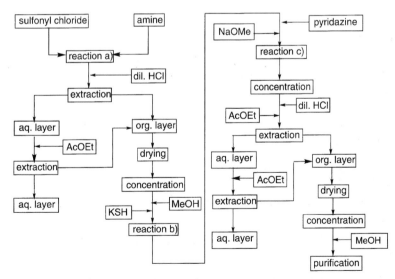

Figure 7 Flow chart showing how the series of sulfonamide derivatives were synthesized using the automated system. (Adapted from ref. 6.)

381

Protocol 1

Use of an automated system for the synthesis of 3-(imidazo[1,2-*b*]pyridazin-6-yl)thiopropanesulfonamide

Equipment and reagents

- Automated synthesis apparatus, TACOS
- Starting material, **6**, prepared according to the known procedure (7)
- 6-Chloroimidazo[1,2-*b*]pyridazine, prepared by the known procedure (8)
- 0.75 M solution of KSH (1.08 g, 15 mmol) dissolved in ethanol (20 ml)
- 0.75 M solution of sodium methoxide (0.81 g, 15 mmol) dissolved in MeOH (20 ml)

- 0.75 M solution of 6-chloroimidazo[1,2-*b*]pyridazine (2.3 g, 15 mmol) dissolved in MeOH (20 ml)
- 1.0 M solution of **6** (2.655 g, 15 mmol) dissolved in ether (15 ml)
- 2.0 M solution of methylamine (0.93 g, 30 mmol) dissolved in ether (15 ml)
- HPLC column of Lichroprop SI-60 (Merck, 25–40 mm, 20 × 500 mm) and eluents of dichloromethane (DCM) followed by DCM/MeOH (40:1)

A. Construction of the control program

1 Select the [START] subroutine to input the reaction variables (time, temperature, etc.).

2 Select the [RF1-ST-ON, RF1-CL-ON] subroutines to stir RF1 and cool the contents.

3 Select the [RR1-RF1] subroutine to add an ether solution of **6** from RR1 into RF1 via the volumetric tubes, and stir the solution at 0 °C for 3 min.

4 Select the [RR3-RF1, RF1-CL-OF] subroutines to add dropwise (10 sec × 100 times) an ether solution of methylamine from RR3 into RF1.

5 Select the [RF1-REA-1, RS1-RF1] subroutines to stir the reaction mixture at 0 °C for 30 min and then add water (10 ml × 3 times) from RS1 to the reaction mixture.

6 Select the [RF1-MIX, RF1-SF] subroutines to transfer the resulting two-phase mixture to the separation funnel, using a diaphragm pump, after stirring for 3 min.

7 Select the [EXT1] subroutine block to:

   (a) Let the two layers stand for 3 min to allow separation.

   (b) Separate the upper organic layer into RF3, through a drying tube (DT) filled with anhydrous sodium sulfate (65 g).

   (c) Introduce ethyl acetate (10 ml × 3 times) from RS2 into RF1 and transfer the obtained organic layer into the separation funnel.

   (d) Bubble up air through the separation funnel to aid extraction and then pass the organic layer through a DT to RF3.

8 Select the [RF1-ST-OF, RF3-ST-ON, RF3-CON-1, RS4-RF3, RF3-BUBB] subroutines to concentrate the organic solution under reduced pressure at 50 °C for 15 min, and then introduce MeOH (10 ml) from RS4 to dissolve the concentrated residue.

9 Select the [RR8-RF3, RF3-REA-1, RS5-RF3] subroutines to introduce an ethanol solution of KSH (5 ml × 4 times, 1.08 g, 15 mmol) from RR8 into RF3. Then, after heating the reaction mixture with stirring at 70 °C for 1 h, introduce a MeOH solution of sodium methoxide (10 ml × 2 times, 0.81 g, 15 mmol), stored in RS5, into RF3.

10 Select the [RR9-RF3, RF3-REA-2, RF3-CON-2] subroutines to introduce a MeOH solution of 6-chloroimidazo[1,2-*b*]pyridazine from RR9, into RF3. Then reflux the reaction mixture at 100 °C for 90 min and concentrate under reduced pressure at 50 °C.

11 Select the [RS1-RF3, RS2-RF3, RF3-MIX] subroutines to add water (30 ml) from RS1, to the residue, followed by introducing ethyl acetate (10 ml × 3 times), stored in RS2, to RF3.

12 Select the [RF3-BUBB, RF3-SF, RF3-ST-OF] subroutines to stir the reaction mixture by bubbling for a short time, and transfer it to the separation funnel.

13 Select the [EXT2] subroutine block to:
   (a) Let the two layers stand for 3 min to allow separation.
   (b) Separate the upper ethyl acetate layer into RF3, through a DT filled with anhydrous sodium sulfate (65 g).
   (c) Introduce ethyl acetate (10 ml × 3 times) from RS2 into RF1 and transfer the obtained organic layer into the separation funnel.
   (d) Bubble up air through the separation funnel to aid extraction and then pass the organic layer through a DT to RF3.

14 Select the [RF3-ST-OF, RF2-ST-ON, RF2-CON-1, RR6-RF2, RF2-BUBB, RF2-MIX] subroutines to collect the anhydrous organic layer in RF2, and concentrate it at 50 °C under reduced pressure. Then introduce DCM (15 ml) from RR6 into RF2, and bubble to stir the mixture to dissolve the residue.

15 Select the [RF2-SR3, RF2-ST-OF, DE-CO-ON, HPLC, DE-CO-OF] subroutines to transfer the resulting DCM solution to SR3, and charge on to the column.

16 Select the [WASH] subroutines to wash all flow-lines, tubes, and flasks with the MeOH and acetone wash solvents.

17 Select the [DRY] subroutines to dry all flow-lines, tubes, and flasks.

B. Operation of the apparatus

1 Fill reservoir RR1 with the ether solution of **6**.

2 Fill reservoir RR3 with the ether solution of methylamine.

3 Fill reservoir RS1 (10 ml, VT1) with water, and fill reservoir RS2 (10 ml, VT1) with ethyl acetate; place MeOH into RS4 (10 ml, VT2).

4 Fill reservoir RR8 with the ethanol solution of KSH.

5 Fill reservoir RS5 with the MeOH solution of sodium methoxide.

6 Fill reservoir RR9 with the MeOH solution of 6-chloroimidazo[1,2-*b*]pyridazine.

7 Fill reservoir RR6 with DCM (15 ml, VT8 and VT4).

8 Fill the two wash solvent tanks with MeOH and acetone.

9 Set the HPLC column and eluents.

10 Start the control program.

11 Collect the final product from RF2 and start the wash and dry programs.

12 Recrystallize the product, **9**, from MeOH/ether to give colourless needles.

3.3 Fragment peptide derivatives

The automated synthesis system is suitable for repetitive syntheses using similar reaction procedures, and was used beneficially to systematically synthesize a library of 25 dipeptides and 125 tripeptides by combining five protected amino acids (9). *Protocol 2* illustrates how the automated system, EASY, was used for the synthesis, purification, and isolation of a dipeptide derivative.

The synthetic route used to prepare a series of dipeptide derivatives is shown in *Figure 8*. The peptide bond formation starts with activating the carboxyl group of Boc-Glu(OcHex)-OH, **10**, in one reaction vessel, while simultaneously removing the protecting Boc group from Boc-Leu-OBn, **11**, in another reaction vessel. The flow chart for the synthesis of peptide derivatives is shown in *Figure 9*.

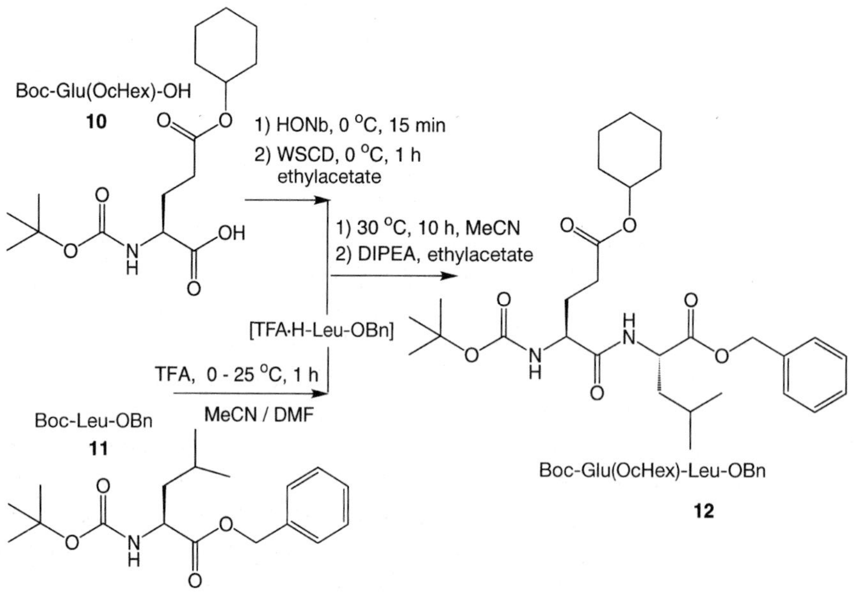

Figure 8 Automated synthetic route used to prepare a series of dipeptide derivatives.

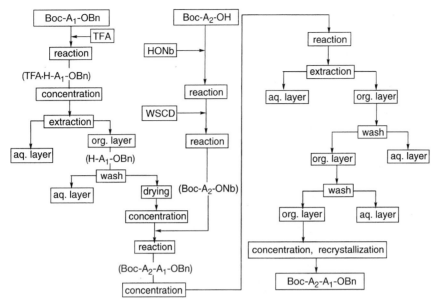

Figure 9 Flow chart showing how dipeptide derivatives were synthesized using the automated system. (Adapted from ref. 9.)

Protocol 2

Use of an automated system for the synthesis of a dipeptide derivative, Boc-Glu(OcHex)-Leu-OBn

Equipment and reagents

- Automated synthesis apparatus, EASY
- 5% (v/v) aqueous sodium bicarbonate solution
- 0.2 M dilute aqueous hydrochloric acid solution
- Solution of *N,N*-diisopropylethylenedia-mine (DIPEA; 0.75 g, 5.8 mmol) dissolved in ethylacetate (10 ml)
- Solution of Boc-Leu-OBn (1.6 g, 5 mmol), **11**, prepared from Boc-Leu-OH according to *Protocol 5*, or the known procedure (9), dissolved in ethylacetate (5 ml)

- Solution of Boc-Glu(OcHex)-OH (1.65 g, 5 mmol), **10**, and *N*-hydroxy-5-norbornene-2,3-dicarboximide (HONb; 0.92 g, 5 mmol) dissolved in a mixture of acetonitrile (MeCN; 15 ml) and dimethylformamide (DMF; 5 ml)
- Column chromatography eluent of ethylacetate/*n*-hexane = 3:1
- Mixture of ethylacetate/*n*-hexane (1:20), for recrystallization
- Water soluble carbodiimide (WSCD)

A. Construction of the control program

1 Select the [START] subroutine to input the reaction variables (time, temperature, etc.).

2 Select the [RF1-ST-ON, RF1-CL-ON] subroutines in order to stir the RF1 contents and cool to 0 °C.

Protocol 2 continued

3 Select the [PAUSE] subroutine to stop the program and ask the operator to add 1.0 g of WSCD hydrochloride salt to the RF1 manually.

4 Select the [RF1-REA-1] subroutine to stir the solution at 0 °C for 1 h and to start the reaction.

5 Select the [RF2-ST-ON, RF2-CL-ON] subroutine to cool RF2 to 0 °C with stirring.

6 Select the [PAUSE] subroutine to stop the program and ask the operator to add TFA (15 ml) to RF2.

7 Select the [RF2-CL-OF, RF2-REA-1] subroutines to stop cooling RF2, stir the solution for 1 h.

8 Select the [RF2-CON-1] subroutine to evaporate TFA in RF2 under reduced pressure at 40 °C.

9 Select the [RF2-ST-ON, RS5-RF2] subroutine to add DCM (50 ml) to RF2 in order to dissolve the residue.

10 Select the [RS2-RF2] subroutine to add aqueous sodium bicarbonate (40 ml) to the reaction mixture.

11 Select the [RF2-MIX, RF2-ST-OF] subroutine to stir vigorously the reaction mixture in the RF2.

12 Select the [EXT1] subroutine to transfer the reaction mixture to the separation funnel and re-extract the separated aqueous layer with DCM (40 ml). Wash the combined organic layer with water (40 ml) and transfer to RF3 through a DT filled with anhydrous sodium sulfate. Wash the separation funnel and flow-lines with DCM (40 ml) and then transfer the washings to RF3.

13 Select the [RF3-ST-ON, RF3-CON-1, RS3-RF3] subroutines to evaporate the solvent in RF3, under reduced pressure at 40 °C, and dissolve the residue in MeCN (10 ml).

14 Select the [RF1-CL-OF, RF1-ST-OF, RF1-RF3] subroutines to transfer the solution in RF1 to RF3 followed by washing with MeCN (10 ml).

15 Select the [WS-RF1, RF1-BUBB, RF1-SF, SF-BUBB, SF-DR, RF1-DRY] subroutines to stir the reaction mixture at room temperature and separate in the separation funnel, SF, and wash and dry RF1.

16 Select the [RF3-REA-1] subroutines to stir the reaction mixture in RF3 for 10 h at 30 °C.

17 Select the [RF3-CON-2, RS4-RF3, RF3-MIX] subroutines to remove excess solvent under reduced pressure at 40 °C, and dissolve the residue in ethyl acetate (50 ml).

18 Select the [RR7-RF3, MATU15] subroutines to add DIPEA, dissolved in ethyl acetate (10 ml), to the solution in RF3 at room temperature, and stir the mixture for 15 min.

19 Select the [RS2-RF3, RF3-ST-OF] subroutine to add aqueous sodium bicarbonate solution (40 ml) to RF3.

20 Select the [EXT2] subroutine to:

   (a) Transfer the mixture to the separation funnel.

   (b) Separate the organic layer and wash it with dilute aqueous hydrochloric acid (40 ml) followed by water (40 ml).

   (c) Transfer the organic layer to RF1 through a DT filled with anhydrous sodium sulfate.

   (d) Wash the separation funnel and flow-lines with ethylacetate (40 ml), and transfer to RF1.

21 Select the [RF1-ST-ON, RF1-CON-1] subroutines to concentrate the solution in RF1 by evaporation under reduced pressure at 40 °C.

22 Select the [RR1-RF1, RF1-MIX, RF1-SR3, RF1-ST-OF] subroutines to add DCM to the residue in RF1, mix, and transfer to the reservoir, SR3, for charging on to the chromatography column.

23 Select the [DE-CO-ON, HPLC] subroutines to switch on the preparative HPLC system and inject the solution from SR3.

24 Select the [DE-CO-OF] subroutine to switch off the preparative HPLC system after the set time for separation.

25 Select the [WASH] subroutines to wash all flow-lines, tubes, and flasks with the two wash solvents.

26 Select the [DRY] subroutines to dry all flow-lines, tubes, and flasks.

B. Operation of the apparatus

1 Fill reservoir RR7 with the solution of DIPEA in ethyl acetate.

2 Place the dilute aqueous hydrochloric acid into RS1 (40 ml, MT1); place the sodium bicarbonate solution into RS2 (40 ml, MT2); place MeCN into RS3 (10 ml, MT3); place water into RS4 (40 ml, MT4); place DCM into RS5 (50 ml, MT5).

3 Fill the wash solvent tanks with MeOH and acetone.

4 Place the ethylacetate solution of Boc-Leu-OBn, **11**, into RF2.

5 Place the starting material, Boc-Glu(OcHex)-OH, **10**, and HONb solution in RF1, and start the control program.

6 Add the 1.0 g of WSCD to RF1 when the program pauses.

7 Add the 15 ml of TFA to RF2 when the program pauses.

8 Collect final product from the fraction collector tubes and start the wash and dry programs.

9 Evaporate the solvent under reduced pressure and recrystallize from ethylacetate/ n-hexane (1:20) to give Boc-Glu(OcHex)-Leu-OBn, **12**, as white crystals (m.p. 58.0–58.5 °C).

3.4 β-Lactam derivatives

Optimization of a selective O-acylation of 3-(1-hydroxyethyl)-4-acetoxyazetidin-2-one, **13**, was carried out using the automated system, VACOS. *Figure 10* shows the three acylated products, **14–16**, which may be obtained when reacting the starting material with an acylating reagent. Acylation at oxygen gives the desired product, **14**, but acylation at nitrogen to give **15**, and diacylation at both oxygen and nitrogen, to give **16**, was also observed. Using the synthesis and reaction monitoring capability of the automated system VACOS we were able to find suitable reaction conditions for giving either **14** or **15** as the major product.

A flow chart showing how the reaction was investigated using the automated system is shown in *Figure 11*, and *Protocol 3* illustrates the use of the system with the reaction conditions that gave high selectivity for O-acylation.

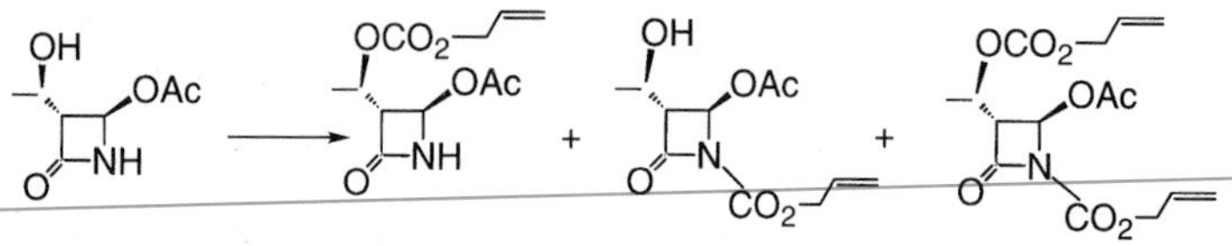

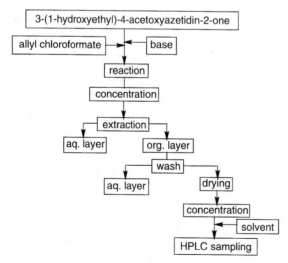

Figure 10 The three products that can be obtained upon acylating carbapenem compounds. With the help of the automated synthesis system the reaction conditions were optimized to yield either 90% O-acylation (allyl chloroformate (4 Equiv.), 2,6-lutidine base (2 Equiv.), + pyridine (0.2 Equiv.), tetrahydrofuran) or 81% N-acylation (allyl chloroformate (2 Equiv.), triethylamine base (2 Equiv.), chloroform).

Figure 11 The flow chart showing how the acylation conditions were automatically investigated and improved.

Protocol 3

Use of an automated system for investigating the selectivity of *O*-acylation of 3-(1-hydroxyethyl)-4-acetoxyazetidin-2-one

Equipment and reagents

- Automated synthesis apparatus, VACOS
- 1.0 M dilute aqueous hydrochloric acid
- Solution of 2,6-lutidine (1.29 g, 12 mmol) and pyridine (94.9 mg, 1.2 mmol) dissolved in 5 ml of THF
- Solution of allyl chloroformate (2.91 g, 24 mmol) dissolved in 5 ml of THF
- 1.0 M aqueous copper sulfate solution

- Analytical HPLC column: reverse phase C-18, 250 × 4.6 mm, 218 nm detection wavelength, water/MeCN (3:1) eluent, 1.0 ml/min flow rate. Retention time for each compound is: 3 min for **13**, 14 min for **14**, 10 min for **15**, and 27 min for **16**.
- Starting material, **13**, (1.04 g, 6 mmol) prepared from 6-aminopenicillanic acid according to a known procedure (10)

A. Construction of the control program

1 Select the [START] subroutine to input the reaction variables (time, temperature, etc.).

2 Select the [RS2-RF3, RF3-ST-ON] subroutines to add 10 ml of THF solvent to the RF3 and start stirring the reaction mixture.

3 Select the [RF3-REA-1, RR8-RF3] subroutines to perform reaction control and add the solution of amine from reservoir RR8 to RF3.

4 Select the [RR9-RF3, RR7-RF3] subroutines to add 5 ml more THF from reservoir RR9 to the reaction flask, via the same delivery line, to wash any residual amine into RF3, and then add a THF solution of allyl chloroformate slowly from RR7 to the reaction mixture, over 3.5 h.

5 Select the [RF3-ST-OF] subroutine to stop stirring of RF3 at the end of the reaction time.

6 Select the [RF1-ST-ON, RF1-CL-ON] subroutines to stir and cool (10 °C) RF1.

7 Select the [RS3-RF1, RS4-RF1] subroutines to add ethyl acetate from RS3 to RF1 and add dilute aqueous hydrochloric acid from RS4 to RF1.

8 Select the [RF3-CON-1, RF1-CL-OF] subroutines to evaporate solvent from RF3 and also stop cooling RF1.

9 Select the [RF1-RF3, RF1-ST-OF, RF3-BUBB] subroutines to transfer the chilled mixture from RF1 to RF3, followed by bubbling nitrogen through the mixture in RF3 in order to completely dissolve all the residue.

10 Select the [EXT1] subroutine block to:

(a) Add dilute aqueous hydrochloric acid from RS4 to RF3.

(b) Separate the organic layer in the separation funnel and return it to RF3.

11 Select the [A-LC-ON] subroutine to start the analytical HPLC pump.

Protocol 3 continued

12 Select the [EXT2] subroutine block to:

(a) Wash the organic layer in RF3 with aqueous copper sulfate.

(b) Wash the organic layer in RF3 with water from RR2.

(c) Pass the organic layer through a DT of sodium sulfate, and collect in RF2.

13 Select the [RF2-ST-ON, RF2-CON-1] subroutine to stir and evaporate solvent from RF2.

14 Select the [RS5-RF2, RF2-BUBB] subroutines to add 10 ml of MeCN from RS5 to RF2 and then bubble nitrogen through the mixture in RF2 in order to completely dissolve all the residue.

15 Select the [RF2-CON-2, RS5-RF2] subroutines to evaporate solvent from RF2, and then again add 10 ml of MeCN from RS5.

16 Select the [RF2-BUBB, RF2-REA-1] subroutines to again bubble nitrogen through the mixture in RF2 to completely dissolve all the residue, and then take a sample from RF2 and inject to the analytical HPLC.

17 Select the [RF2-CL-ON, PAUSE] subroutines to cool RF2 and wait for the user to remove the final product mixture if required.

18 Select the [RF2-CL-OF, RF2-ST-OF] subroutines to stop cooling and stirring RF2.

19 Select the [A-LC-OF] subroutine to stop the HPLC pump.

20 Select the [RF2-ST-OF, PAUSE] subroutine to allow the final product be removed from the RF2.

21 Select the [WASH] subroutines to wash all flow-lines, tubes, and flasks with the three wash solvents.

22 Select the [DRY] subroutines to dry all flow-lines, tubes, and flasks.

B. Operation of the apparatus

1 Fill reservoir RR8 with the mixture of 2,6-lutidine and pyridine in THF; RR2 with 10 ml of water, RR7 with the allyl chloroformate dissolved in THF, and RR9 with 5 ml of THF.

2 Place the aqueous copper sulfate into RS1 (10 ml, MT1); place THF into RS2 (10 ml, MT2); place ethyl acetate into RS3 (20 ml, MT3); place dilute aqueous hydrochloric acid into RS4 (10 ml, MT4); place MeCN into RS5 (10 ml, MT5).

3 Fill the three wash solvent tanks with MeOH, acetone, and THF respectively.

4 Set the analytical HPLC.

5 Place the starting material, **13**, in the RF3 and start the control program.

6 Collect final product from RF2, check the HPLC chart, and start the wash and dry programs.

4 Applications of automated synthesis workstations

4.1 Preparative multigram scale synthesis

The fourth generation of automated synthesis apparatus were made as stand-alone, modular apparatus, each with the minimum functions to be specialist workstations (synthesis, purification, or reaction monitoring). The versatility of their use, separately or in sequence as needed, was considered to make them an effective investment.

A new general method for automated synthesis of peptides was developed, in which use of methanesulfonic acid (MSA) gave improved results compared to the commonly used TFA (see *Protocol 2*). Using this method the following problems were resolved:

- damage to the apparatus caused by strong acid
- the formation of intractable emulsions
- avoiding the use of the restricted solvent DCM

4.1.1 Protection of amino acid derivatives

Figure 12 shows the flow chart for the preparative gram scale synthesis of a series of protected amino acids using the intermediate–large scale synthesis apparatus, FUTOSHI. *Protocol 4* illustrates an automated multigram scale synthesis of a protected amino acid, Boc-D-Ala-OBn, which was subsequently used for preparing a series of oligopeptide derivatives.

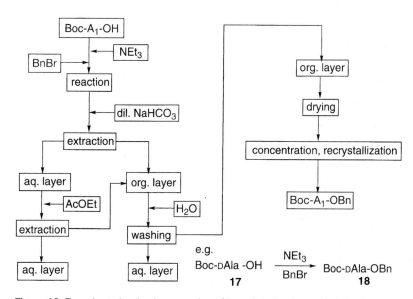

Figure 12 Flow chart showing how a series of benzylated amino acid derivatives were synthesized using the automated workstation, FUTOSHI.

Protocol 4

Use of an automated workstation for the multigram scale synthesis of an intermediate peptide derivative, Boc-D-Ala-OBn

Equipment and reagents

- Automated synthesis apparatus, FUTOSHI
- 2% (v/v) aqueous sodium bicarbonate solution
- Triethylamine (TEA; 56.2 g, 550 mmol) dissolved in DMF (100 ml)

- Benzyl bromide (94.1 g, 550 mmol) dissolved in DMF (100 ml)
- Solution of 94.6 g (500 mmol) of starting material, Boc-D-Ala-OH, **17**, (Peptide Institute, Inc. Japan) dissolved in DMF (200 ml)

A. Construction of the control program

1 Select the [START] subroutine to input the reaction variables (time, temperature, etc.).

2 Select the [RF1-ST-ON, RF1-LF-UP] subroutines to stir the RF1 contents and raise the reaction bath.

3 Select the [RF1-CL-ON] subroutine to cool the bath to 0 °C.

4 Select the [RR1-RF1, RR3-RF1, RF1-CL-OF, RF1-LF-DN] subroutines to add dropwise DMF solutions of TEA and benzyl bromide to the RF1, with stirring at 0 °C, and then lower the reaction bath and let the RF1 warm to room temperature.[a]

5 Select the [RF1-REA-1] subroutine to react for 10 h at 25 °C.

6 Select the [RS4-RF1, RS2-RF1] subroutines to add 450 ml of ethylacetate from RS4 and 810 ml of the sodium bicarbonate solution from RS2.

7 Select the [RF1-MIX] subroutine to stir the mixture vigorously.

8 Select the [RF1-SF, SEP-SR1, SF-SR0, SR1-RF1, RS4-RF1, RF1-MIX, RF1-SF, SEP-SR1] subroutines to transfer the reaction mixture to the separation funnel (SF) and extract with ethylacetate (450 ml × 2 times).

9 Select the [SF-RF1, RS1-RF1, RF1-MIX, RF1-SF, SEP-SR1, SF-RF1, RS1-RF1, RF1-MIX, RF1-SF, SEP-SR1, SF-CF, RS4-RF1, RF1-MIX, RF1-SF, SF-BUBB, SF-CF, RF1-ST-OF] subroutines to wash the obtained organic layer with water (450 ml × 2 times) and transfer to the collection flask (CF). Then wash the used lines with ethyl acetate (60 ml).

10 Select the [WASH] subroutines to wash all flow-lines, tubes, and flasks with the two wash solvents.

11 Select the [DRY] subroutines to dry all flow-lines, tubes, and flasks.

B. Operation of the apparatus

1 Fill reservoir RR1 with the TEA in DMF and RR3 with benzyl bromide dissolved in DMF.

2 Place sodium bicarbonate solution into RS2 (30 ml, MT1); place ethylacetate into RS4 (30 ml, MT2).

3 Fill the wash solvent tanks with MeOH and acetone.

4 Place the starting material, **17**, Boc-D-Ala-OH dissolved in DMF, in the 2 litre reaction flask, RF1.

5 Start the control program.

6 Collect final product from CF and start the wash and dry programs.

7 Dry over anhydrous sodium sulfate, remove the excess solvent under reduced pressure to give Boc-D-Ala-OBn, **18**, as a powder.

a Addition is paused if the temperature in RF1 increases by more than 5 °C.

4.1.2 Intermediate fragment peptide synthesis

Figure 13 shows the gram scale synthesis of several hundred protected tripeptides and *Figure 14* shows the flow chart for preparing these using the MSA method. *Tables 4* and *5* summarize how the TFA and MSA methods differ and how the MSA method was more useful for automated solution phase synthesis, giving excellent yields with increased efficiency. *Protocol 5* illustrates the MSA synthesis of a tetrapeptide derivative.

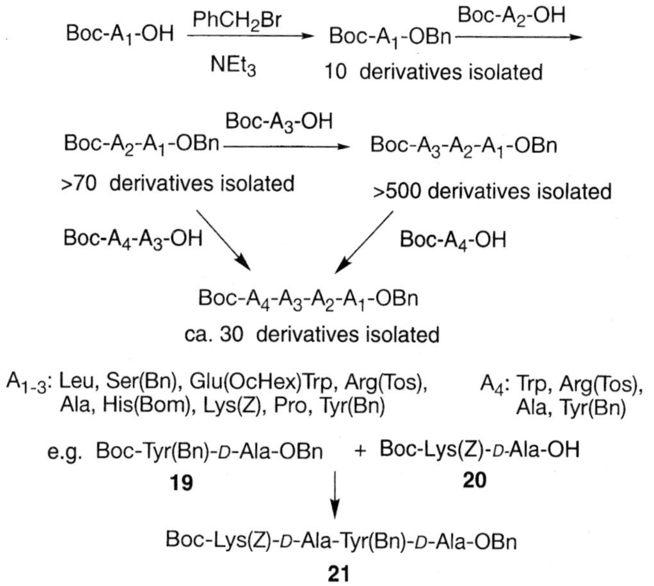

$$Boc-A_1-OH \xrightarrow[NEt_3]{PhCH_2Br} Boc-A_1-OBn \xrightarrow{Boc-A_2-OH}$$

10 derivatives isolated

$$Boc-A_2-A_1-OBn \xrightarrow{Boc-A_3-OH} Boc-A_3-A_2-A_1-OBn$$

>70 derivatives isolated

>500 derivatives isolated

Boc-A$_4$-A$_3$-OH \ / Boc-A$_4$-OH

Boc-A$_4$-A$_3$-A$_2$-A$_1$-OBn

ca. 30 derivatives isolated

A$_{1-3}$: Leu, Ser(Bn), Glu(OcHex)Trp, Arg(Tos), Ala, His(Bom), Lys(Z), Pro, Tyr(Bn)

A$_4$: Trp, Arg(Tos), Ala, Tyr(Bn)

e.g. Boc-Tyr(Bn)-D-Ala-OBn   +   Boc-Lys(Z)-D-Ala-OH

19   **20**

Boc-Lys(Z)-D-Ala-Tyr(Bn)-D-Ala-OBn

21

Figure 13 Peptide derivatives synthesized by the methanesulfonic acid (MSA) method.

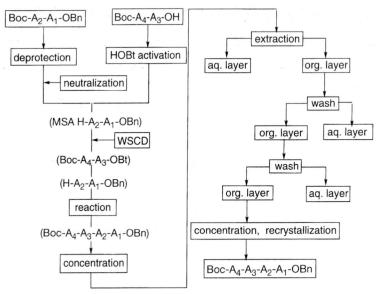

Figure 14 Flow chart showing how the methanesulfonic acid (MSA) method was used to synthesize peptide derivatives on the automated workstation, RAMOS.

Table 4 Comparison of yields for the synthesis of tripeptide derivatives (Boc-A$_3$-A$_2$-A$_1$-OBn)

-A$_3$-A$_2$-A$_1$-[a]	TFA method (%)	MSA method (%)
EEL	75	66
LRW	72	64
KWA	81	77
EWA	85	80
LES	92	71

[a] A = Ala, E = Glu(OcHex), L = Leu, R = Arg(Tos), W = Trp, K = Lys(Z).

Table 5 List of improvements for automated synthesis of tripeptide derivatives using the MSA method

Improvement	Effect
Avoided use of trifluoroacetic acid (volatile, corrosive acid)	Easier handling and reduced damage to apparatus
Avoided use of dichloromethane	Friendlier to the environment
Decreased the number of extraction and drying processes	Increased reliability (from about 90% to 99%)
Generated free amine and activated ester at the same time	High yield (from 70.1% to 77.4%)
Simplify process and set-up	Speedy synthesis

Protocol 5

Use of an automated workstation for the multigram scale synthesis of a tetrapeptide derivative, Boc-Lys(Z)-D-Ala-Tyr(Bn)-D-Ala-OBn

Equipment and reagents

- Automated synthesis apparatus, RAMOS
- Boc-Tyr(Bn)-D-Ala-OBn, **19**, (6.39 g, 12 mmol), prepared in a similar manner to *Protocol 2*
- Boc-Lys(Z)-D-Ala-OH, **20**, (5.42 g, 12 mmol), prepared in a similar manner to *Protocol 2*, followed by alkaline hydrolysis using FUTOSHI
- Solution of 1-hydroxy-1H-benzotriazole monohydrate (HOBt; 2.02 g, 13.2 mmol) dissolved in 10 ml of DMF

- 0.9 M solution of *N,N*-diisopropylethyl-amine (DIEA; 2.33 g, 18 mmol) dissolved in MeCN (20 ml)
- 2 M solution of MSA (2.88 g, 30 mmol) dissolved in MeCN (15 ml)
- Solution of WSCD (2.05 g, 13.2 mmol) dissolved in MeCN (5 ml)
- 0.2 M dilute aqueous hydrochloric acid

A. Construction of the control program

1 Select the [START] subroutine to input the reaction variables (time, temperature, etc.).

2 Select the [RF1-ST-ON, RF2-ST-ON] subroutines to stir RF1 and RF2 contents.

3 Select the [RR1-RF1, RF1-LF-UP, RF1-REA-1] subroutines to add the MSA in MeCN from RR1 at room temperature, and stir the mixture at 40 °C for 1 h.

4 Select the [RF1-CL-ON, RR2-RF1, MATU] subroutines to cool RF1 and add DIEA dissolved in MeCN from RR2, and then stir the reaction mixture for 5 min.

5 Select the [RF2-RF1, RF2-ST-OF] subroutines to transfer the solution from RF2 to RF1 at 0 °C with stirring.

6 Select the [RR3-RF1, RF1-CL-OF, RF1-REA-2, RF1-CON-1, RF1-LF-DN] subroutines to add WSCD in MeCN from RR3 to RF1, and stir the reaction mixture in RF1 at 25 °C for 10 h, and then remove the excess solvent under reduced pressure at 40 °C for 1 h.

7 Select the [RS5-RF1, RS4-RF1, RS2-RF1] subroutines to dissolve the obtained residue in THF from RS5, followed by ethyl acetate from RS4 (30 ml), and then treat with sodium bicarbonate solution (40 ml) from RS2.

8 Select the [RF1-MIX, RF1-ST-OF, EXT1, PAUSE] subroutines to stir the mixture vigorously and transfer to the SF, followed by extracting again with ethyl acetate (40 ml), and washing with both dilute aqueous hydrochloric acid (40 ml) and water (40 ml), followed by transferring the organic layer to RF3.

9 Select the [WASH] subroutines to wash all flow-lines, tubes, and flasks with the MeOH and acetone wash solvents.

10 Select the [DRY] subroutines to dry all flow-lines, tubes, and flasks.

Protocol 5 continued

B. Operation of the apparatus

1 Fill reservoir RR1 with the MSA dissolved in MeCN.

2 Fill reservoir RR2 with the DIEA dissolved in MeCN.

3 Place the aqueous sodium bicarbonate solution into RS2 (40 ml, MT2); place ethylacetate into RS4 (30 ml, MT4); place THF into RS5 (20 ml, MT5).

4 Fill the two wash solvent tanks with MeOH and acetone, respectively.

5 Place the starting material, Boc-Tyr(Bn)-D-Ala-OBn, **19**, into RF1.

6 Place the starting material, Boc-Lys(Z)-D-Ala-OH, **20**, and HOBt dissolved in DMF into RF2, and start the control program.

7 Collect the final product from RF3 and start the wash and dry programs.

8 Evaporate the solvent to give crystals, wash with diisopropyl ether, and dry over anhydrous sodium sulfate under reduced pressure to obtain Boc-Lys(Z)-D-Ala-Tyr(Bn)-D-Ala-OBn, **21**.

4.2 Automated synthesis with supported reagents

The number of solid reagents, particularly supported reagents (11) (see Chapter 7), that are available for carrying out novel, selective, or mild reactions, continues to grow in the literature. Automated synthesis has generally avoided these reagents because of handling problems. However, we designed and constructed an automated apparatus (ASRA) compatible with the common series of operations encountered in supported reagent reactions (4). ASRA can be used as an independent single batch reactor or be combined with our other workstations to expand their capability and the scope of syntheses being performed in the automated laboratory.

Furthermore, the yields for many of the reactions that use supported reagents in stoichiometric amounts are increased, since the reagents can be used in near optimum conditions by slowly adding them to the reaction mixture to minimize the effect of competing reagent decomposition, inactivation, or side-reactions. *Protocol 6* illustrates how 9-bromophenanthrene gave 84% yield of 9-cyanophenanthrene on slow addition of charcoal-supported copper cyanide (12) using ASRA, compared to 65% by a single batch addition under the same conditions (see *Figure 15*).

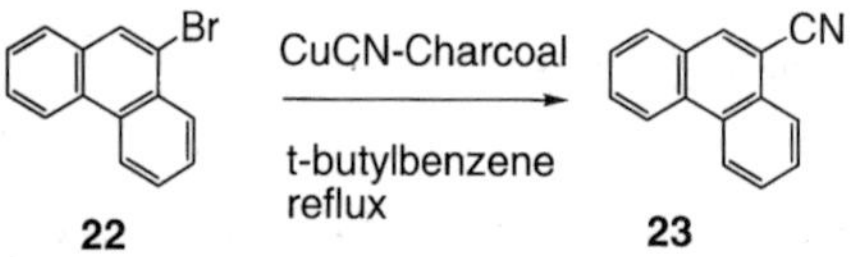

Figure 15 Conversion of 9-bromophenanthrene to 9-cyanophenanthrene using charcoal-supported copper cyanide. Use of the activated solid reagent in a non-polar solvent allows for easy work-up by simple filtration.

Protocol 6

Preparation of 9-cyanophenanthrene using slow addition of a powdered supported reagent

Equipment and reagents

- Automated synthesis apparatus, ASRA (4)
- Starting material, **22**, purchased from Aldrich Chemicals Ltd.
- Charcoal-supported copper cyanide (10 g,

3.3 mmol), prepared by dissolving the salt in hot N-methylpyrrolidone, suspending the charcoal, and gradually cooling the mixture over several hours (12)

A. Construction of the control program

1 Select the [START] subroutine to input the reaction variables (reaction time, temperature, addition time (6 h), bulk density (0.25 g/ml), amount of powdered reagent (10 g), etc.).

2 Select the [RF1-ST-ON, RF1-LF-UP, RF1-HT-ON] subroutines to stir the RF1 contents, cool bath, the condenser, and the neck of the RF1, raise the reaction bath and heat to 155 °C.

3 Select the [RF1-REA-1] to control the slow powder addition and continue reaction for 15 h.

4 Select the [CF-ST-ON, CF-HT-ON] subroutines to start the stirrer in the CF and heat the bath.

5 Select the [CF-PG-ON, CF-RF1, CF-PG-OF] subroutines to bubble nitrogen gas from CF into RF1 via the Teflon ball filter to facilitate subsequent filtration.

6 Select the [RS1-RF1, F1-EX-CF] subroutines to add ether to RF1 from RS1, and extract the product from the solid reagent in F1 into CF.

7 Select the [CF-HT-OF, CF-ST-OF] subroutines to stop heating CF.

8 Select the [CF-PG-ON] subroutine again to bubble nitrogen gas through the Teflon ball filter, to prevent the solid reagent compacting around it.

9 Select the [CF-ST-ON, RR1-CF, CF-F3, CF-ST-OF] subroutines to remove the spent powdered reagent to the waste bottle, F3, as a suspension in MeOH.

10 Select the [WASH] subroutines to wash all flow-lines, tubes, and flasks with MeOH.

11 Select the [DRY] subroutines to dry all flow-lines, tubes, and flasks.

B. Operation of the apparatus

1 Place the charcoal–copper cyanide into the hopper reservoir.

2 Fill the wash solvent tank with MeOH.

3 Place the 9-bromophenanthrene starting material, **22**, into the reaction flask, RF1 (1.29 g, 5 mmol) and add 50 ml of t-butylbenzene.

4 Start the control program.

Protocol 6 continued

5 Collect the final product from CF and start the wash and dry programs.

6 Remove the excess solvent under reduced pressure to give 9-cyanophenanthrene, **23**, as a white solid of greater than 95% purity.

4.3 Catalytic hydrogen transfer

Powdered catalysts have a diverse range of applications in organic synthesis, and their importance has grown as a result of ever stricter environmental legislation and the search for cleaner technologies (13). Since they are often used in a repetitive sequence of operations suitable for automation, we developed an automated apparatus, ASCA (see *Table 1*), capable of effectively handling solid catalysts.

The workstation has one column-shaped reactor to hold the solid catalyst and allow it to be shaken, heated, and cooled, and reservoirs for reactants and solvents. Other features include; HPLC monitoring of samples from the reaction flask for automated reaction-trace, work-up of products by extraction and filtration, and washing to allow repeated use. ASCA can be used as an independent single batch reactor, continuous multi-batch reactor (same reaction), or continuous multi-batch library reactor (different reactions). ASCA may be combined with our other workstations to expand their capability and the scope of syntheses being performed in the automated laboratory.

Figure 16 shows the multigram scale, batch-type, synthesis of a deprotected peptide derivative, Boc-Glu(OcHex)-D-Ala-OH, using catalytic hydrogen transfer.

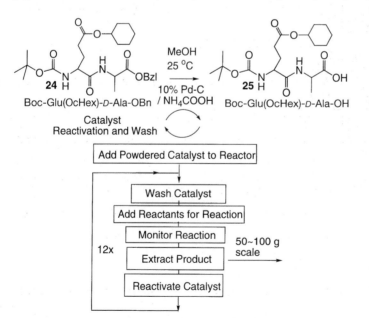

Figure 16 Batch-type synthesis of a deprotected peptide derivative using catalytic hydrogen transfer on the ASCA automated apparatus.

Protocol 7 illustrates how ASCA was used for the synthesis, thus avoiding the potential hazards of using hydrogen gas in an automated, unattended, apparatus.

Protocol 7

Batch preparation of Boc-Glu(OcHex)-D-Ala-OH using catalytic hydrogen transfer

Equipment and reagents

- Automated synthesis apparatus, ASCA
- Starting material, **24**, prepared in a similar manner to *Protocol 2*
- Solution of Boc-Glu(OcHex)-D-Ala-OBn (37 g, 75 mmol) dissolved in 130 ml of MeOH
- Solution of ammonium formate (22.5 g, 360 mmol) dissolved in 130 ml of MeOH
- 10% (v/v) formic acid in MeOH solution
- 10% palladium–carbon (Degussa)

A. Construction of the control program

1 Select the [START] subroutine to input the reaction variables.

2 Select the [COOL-ON, MF-SET] subroutines to cool the RF1 flask and set the nitrogen gas flow rate.

3 Select the [R1-F1] subroutine to add 2 × 10 ml of the protected dipeptide solution.

4 Select the [R4-F1] subroutine to add 2 × 10 ml of the ammonium formate solution.

5 Select the [F1-HT-ON, SWING-ON, REA-1] subroutines to set the reaction temperature, shake the reaction flask, and control the reaction, including HPLC monitoring.

6 Select the [SWING-OF, F2-ST-ON, F2-HT-ON] subroutines to stop shaking, start the stirrer in the CF, and heat the bath.

7 Select the [F1-EX-F2] subroutine to extract and filter the product from the catalyst, and transfer to the CF.

8 Select the [F2-ST-OF] subroutine to stop stirring in the CF.

9 Select the [CAT2WASH] subroutine to add the methanolic formic acid from RS2 reservoir.

10 Select the [RE-CYCLE] subroutine to count the number of synthesis cycles until the pre-set number is reached, and also allow the user to stop or continue.

11 Select the [WASH] subroutines to wash all flow-lines, tubes, and flasks with MeOH.

12 Select the [DRY] subroutines to dry all flow-lines, tubes, and flasks.

B. Operation of the apparatus

1 Suspend the palladium–carbon catalyst in MeOH and transfer it to the reaction flask.

2 Fill reservoir RR1 with the MeOH solution of Boc-Glu(OcHex)-D-Ala-OBn, **24**.

Protocol 7 continued

3 Fill reservoir RR4 with the MeOH solution of ammonium formate.

4 Place the 10% (v/v) formic acid in MeOH solution into RS2 (30 ml, MT1).

5 Fill both the wash and extraction solvent tanks with MeOH.

6 Drain the MeOH from RF1 by operating the apparatus manually and then start the control program.

7 Collect each batch of product from the CF at the interval between cycles.

8 Start the wash and dry programs when all batches have been prepared.

9 Remove the solvent under reduced pressure to give Boc-Glu(OcHex)-D-Ala-OH, **25**, as a white solid.

References

1. Sugawara, T. and Cork, D. G. (1996). *Lab. Robotics Autom.*, **8**, 221.

2. Sugawara, T. and Cork, D. G. (1997). *Yuki Gosei Kagaku Kyokaishi*, **55**, 466.

3. Sugawara, T., Kato, S., and Okamoto, S. (1994). *J. Auto. Chem.*, **16**, 33.

4. Cork, D. G., Kato, S., and Sugawara, T. (1995). *Lab. Robotics Autom.*, **7**, 301.

5. (a) Hayashi, N. and Sugawara, T. (1988). *Chem Lett.*, 1613. (b) Hayashi, N., Sugawara, T., Shintani, M., and Kato, S. (1989). *J. Auto. Chem.*, **11**, 212. (c) Hayashi, N., Sugawara, T., and Kato, S. (1991). *J. Auto. Chem.*, **13**, 187. (d) Hayashi, N. and Sugawara, T. (1989). *Tetrahedron Computer Methodology*, **1**, 237.

6. Kuwahara, M., Kato, S., Sugawara, T., and Miyake, A. (1995). *Chem. Pharm. Bull.*, **43**, 1511.

7. White, E. H. and Lim, H. M. (1987). *J. Org. Chem.*, **52**, 2162.

8. Stanovnic, B. and Tisler, M. (1967). *Tetrahedron*, **23**, 387.

9. Sugawara, T., Kobayashi, K., Okamoto, S., Kitada, C., and Fujino, M. (1995). *Chem. Pharm. Bull.*, **43**, 1272.

10. (a) Fujimoto, K., Iwano, Y., and Hirai, K. (1986). *Bull. Chem. Soc. Jpn.*, **59**, 1363. (b) Shah, N.V. and Cama, L. D. (1987). *Heterocycles*, **25**, 221.

11. (a) Clark, J. H., Kybett, A. P., and Macquarrie, D. J. (1992). *Supported reagents: preparation, analysis and applications*. VCH Publishers, New York. (b) Smith, K. (ed.) (1992). *Solid supports and catalysts in organic synthesis*. Ellis Horwood, Chichester.

12. Clark, J. H., Duke, C. V. A., Miller, J. M., and Brown, S. J. (1986). *J. Chem. Soc. Chem. Commun.*, 877.

13. Clark, J. H. (1994). *Catalysis of organic reactions by supported inorganic reagents*. VCH Publishers, New York.

Chapter 14

Combinatorial discovery and optimization of electrocatalysts

Erik Reddington, Jong-Sung Yu, Benny C. Chan, Anthony Sapienza, Guoying Chen, Thomas E. Mallouk, Bogdan Gurau,[†] Rameshkrishnan Viswanathan,[†] Renxuan Liu,[†] Eugene S. Smotkin,[†] and S. Sarangapani[‡]

The Pennsylvania State University, Department of Chemistry, University Park, PA 16802, USA.

[†]Department of Chemical and Environmental Engineering, Illinois Institute of Technology, Chicago, IL 60616, USA.

[‡]ICET, Inc., 916 Pleasant Street 12, Norwood, MA 02062, USA.

1 Introduction

1.1 Combinatorial materials discovery

Combinatorial methods have been used for over a decade to discover interesting new organic and biological molecules (1). Although no products discovered by a combinatorial technique have reached the market place, some are now in clinical trials (2). While it may seem that combinatorial chemistry is relatively new, its origins are in fact nearly three decades old. Hanak used a combinatorial method, which he called 'the multiple-sample approach', to discover new super-conducting materials (3). The 'high Tc' superconducting alloys he discovered had a critical temperature of less than 10 K, and were quickly forgotten as they found no practical use. Also forgotten was Hanak's novel, but efficient approach to materials discovery. He identified these materials in a period of five months without the benefit of desktop computers or elaborate array fabrication devices. Partially inspired by organic and biochemists, many inorganic materials chemists have implemented combinatorial schemes of their own within the past five years. During this time, several interesting new materials have been discovered including oxide superconductors (4), materials exhibiting colossal magnetoresistance (5), high dielectric materials (6), phosphors (7), sensors (8), heterogeneous catalysts (9), and homogeneous catalysts (10).

Combinatorial materials discovery is best suited to problems in which limited regions of elemental composition or processing space have been explored for a specific property, and in which mixtures of several components might be expected to improve this property. The likelihood of finding superior new materials can be substantially increased if a theory is available to guide the Edisonian combinatorial searches. This approach allows one to focus on the most promising regions of phase space, when the entire space is too vast even for state-of-the-art combinatorial fabrication and screening. With these considerations in mind, electrocatalysts for the oxidation of methanol and the reduction of oxygen seemed a good area for potential discovery.

1.2 Electrocatalysts for the direct methanol fuel cell (DMFC)

Direct methanol fuel cells are not yet commercially viable devices, in large part because of poor kinetics at the anode and cathode (11). Fuel cells convert chemical energy directly into electrical energy, avoiding the thermal losses associated with internal combustion engines. Like all fuel cells, DMFCs are galvanic cells that derive power from the thermodynamically favourable combination of two half-cell reactions, in this case *Reaction 1* and *Reaction 2*. The net reaction is equivalent to the combustion of methanol (*Reaction 3*). Over the past thirty-five years, much has been learned about the electrochemical oxidation of methanol on the surfaces of noble metals, particularly platinum (12–14). Since the discovery that alloying platinum with ruthenium or other metals improves the performance of the catalyst significantly (15), there has been a steady research and development effort aimed at understanding and optimizing this surface reaction.

$$CH_3OH + H_2O \rightarrow CO_2 + 6e^- + 6H^+ \qquad\qquad 1$$

$$O_2 + 4H^+ + 4\,e^- \rightarrow 2H_2O \qquad\qquad 2$$

$$2CH_3OH + 3O_2 \rightarrow 2CO_2 + 4H_2O \qquad\qquad 3$$

Although many groups have investigated binary anode electrocatalysts, few have looked into ternary or higher compositions. As the number of elemental components increases, the number of possible compositions to test grows in a geometric progression. At a resolution of 11% (ten different compositions along any binary line in a phase diagram), there are 55 different materials to synthesize, wash, characterize, fashion into electrodes, and test per ternary phase diagram, which includes all binaries and single elements. This number becomes 220 for a quaternary diagram at the same resolution, and 715 for all combinations of five elements. Smotkin and co-workers have developed a model for predicting the activity of ternary alloys, using heuristic rules derived from phase equilibria and metal oxide bond strengths (16). They were able to identify ternary catalysts containing Pt, Ru, and Os that performed better than $Pt_{50}Ru_{50}$ prepared by the same methods. However, serial testing required that they obtain only a sparse activity map of the Pt/Ru/Os composition space. Using combinatorial synthesis and testing, these compositions can be explored much more rapidly, and the area of highest activity can be focused on. In our combinatorial studies, we have screened for activity beyond the ternary phase space originally investigated, and have expanded the search to logical choices of four- and five-component catalysts (17).

Another problem with DMFCs is that the polymer electrolyte membrane is permeable, and allows methanol to diffuse from the anode to the cathode compartment. A substantial fraction of the fuel is therefore oxidized at the cathode, greatly reducing the fuel efficiency of the cell. This problem is currently mitigated by operating DMFCs at low methanol concentration (c. 0.5 M); however, this limits the anode current density and therefore the power that can be obtained from the cell. A better solution to the crossover problem would be to develop a cathode electrocatalyst that efficiently reduces oxygen, but does not oxidize methanol.

Several kinds of macrocyclic complexes have been investigated for use in methanol-tolerant cathodes, but their catalytic activity is generally too low for practical fuel cells (18). Alonso-Vante and co-workers have developed an alternative strategy, using Ru- and Se-containing catalysts of as-yet uncertain composition (19, 20). The addition of small amounts of Mo and other elements improves the activity of these catalysts significantly, while retaining methanol tolerance (21). This class of catalysts seems tailor-made for combinatorials since there is presently very little known about how they work, and mixtures of different elements affect the performance in fuel cells substantially. In the studies described here, we have screened three- and four-component ruthenium alloys as methanol-tolerant cathode materials.

Anode and cathode electrocatalysts for the DMFC will be used throughout the text to illustrate our approach to the combinatorial discovery of electrode materials. Since the aim of this book is to provide instructive, 'how-to'-protocols for combinatorial chemistry, many considerations and learning experiments are also described. There are many points to consider before launching a combinatorial materials program. Most considerations deal with the synthetic method and screening parameters, and how to design and automate the synthesis of the arrays.

2 Optimization of the catalyst processing conditions

Combinatorial screening is most effective when the screening and processing conditions closely match the conditions of end use. Deviations in these parameters can give skewed results, and may not necessarily yield better catalysts or other types of electrode materials. The approach that has worked for us is to follow the 'golden rule' of combinatorial materials chemistry: first learn how to make good materials (in this case catalysts) that work under real operating conditions, and then use that method to prepare arrays for combinatorial analysis.

Chemists who work with catalysts are constantly cursing, because their results can seem not to be very reproducible. One can make a perfectly good catalyst one day, and seemingly repeat everything in an identical manner the next day with substantially worse results. These experiences reflect the fact that a catalyst's activity is closely related to its surface chemistry, and the surface can become poisoned or deactivated at any one of the many different processing steps. There are several surface parameters (oxidation state, surface area, surface charge, etc.) that are important, and these parameters are very sensitive to the processing and testing conditions of the catalyst.

2.1 Reduction of metal salts

In our attempts to find improved DMFC anode materials, we sought to make noble metal alloys or mixed metal catalysts. Arc-melting and sputtering are common methods for making smooth alloy electrodes, which can be used in various kinds of electrochemical and spectroscopic experiments. Unfortunately, smooth alloy films are significantly different from the catalysts used in fuel cells. Fuel cell catalysts are typically rough metal blacks that have a high specific surface area. These blacks are usually prepared by reduction of mixed metal salts, in one of several procedures: reduction of the dry salts under flowing hydrogen, Adams' method (22), which is oxidation in a nitrate flux followed by hydrogen or electrochemical reduction, or aqueous solution phase reduction by formaldehyde (23) or borohydride (24). Although Adams' method yields the highest surface area materials, and therefore makes good catalysts, the oxidation step causes some elements (particularly Ru and Os) to sublime as oxides. Volatilization of osmium tetroxide not only changes the composition of the catalyst in an unpredictable way, but also poses a health risk. Dry reduction of metal salts under hydrogen produces low surface area, grey metallic materials. Of the remaining reduction methods, sodium borohydride is the most vigorous reducing agent, and makes relatively high surface area catalysts (25–35 m^2/g). Testing of bulk catalysts made with different reducing agents supports the idea that the borohydride method makes reasonably good catalysts for a given composition.

2.2 Optimization of reduction parameters

While it compares favourably with other 'in-house' catalysts of the same elemental composition, borohydride-reduced $Pt_{50}Ru_{50}$ is less active as a catalyst for methanol oxidation than a commercially available high surface area $Pt_{50}Ru_{50}$ catalyst (platinum ruthenium black; Alfa Æsar). The 'in-house' synthetic method was therefore optimized before any combinatorial experiments were done. Many experimental parameters were varied including the rate of addition of the reducing agent, solution pH, concentration of the reagents before reduction, and sources of reagents. The pH of the solution was found to be the single most important parameter in making highly active anode catalysts by the borohydride method. There is a significant improvement in performance as the pH of the precursor solution is increased from 2 to 8 immediately before the reduction step.

It is interesting to note that borohydride reduction of noble metals does not necessarily reduce the metals to a single-phase zero-valent metal alloy. In agreement with earlier findings by Rolison, *et al.* (25), XPS (X-ray photoelectron spectroscopy) and XRD (X-ray powder diffraction) data show that borohydride reduction gives a multi-phase catalyst containing oxides of the oxophilic elements (Ru and/or Os) (26). We note that the oxidation states of metals in these as-prepared catalysts may change once they are operating in the DMFC. These experiments underscore the difference between catalyst blacks and smooth alloys, and provide another note of caution about drawing conclusions from combinatorial screening of the latter. The example of making and testing a high surface area $Pt_{50}Ru_{50}$ anode electrocatalyst is given in *Protocol 1*.

Protocol 1

Synthesizing and testing individual anode catalysts

Equipment and reagents

- Potentiostat (EG&G PARC Model 363 Potentiostat/galvanostat, driven by an EG&G PARC Model 175 universal programmer, EG&G Princeton Applied Research)[a]
- Dynamic hydrogen reference electrode[b] (DHE) (platinum gauze, 100 mesh, Aldrich)
- Platinum counter electrode (platinum gauze, 100 mesh, from Aldrich)
- Analytical rotator (Pine Instruments)

- Custom-made fastener (Pine)
- Nafion perfluorinated ion exchange powder, 5% solution (Aldrich)
- $H_2PtCl_6 \bullet 6H_2O$ (38.88% Pt), $RuCl_3 \bullet xH_2O$ (41.01% Ru), and $NaBH_4$ (Alfa)
- Glassy carbon rod (7 mm in diameter) with one end polished smooth, and fashioned to fit into the fastener (Alfa)
- Magnetic stirring plate

Method

1. Dissolve 0.254 g of $H_2PtCl_6 \bullet 6H_2O$ and 0.125 g of $RuCl_3 \bullet xH_2O$ in *c.* 500 ml of water in a 1 litre Erlenmeyer flask, to make a solution with a total metal concentration of 2 mM. Place the flask on a magnetic stir plate, and stir at medium speed.

2. Adjust the pH of the solution to pH ~ 8 dropwise with a 1 M NaOH solution, or buffer the solution with 1 M $NaHCO_3$.

3. Dissolve 0.41 g $NaBH_4$ (40-fold excess) in 4 ml of water, and quickly pour the entire volume into the metal salt solution.

4. Allow the bubbling black reaction mixture to stir for 1–2 h.

5. Turn off the stir plate, and allow the solid to collect in bottom of the flask overnight.

6. Pour the slurry through a 0.2 μm nylon filter paper in a filter flask.

7. Rinse the catalyst thoroughly with deionized water. This step is repeated several times to eliminate adsorbed chloride ions, which poison Pt alloy catalysts.

8. Remove the filter paper from the filter flask, and sonicate the filter paper with collected solid in a beaker in 15 ml of water.

9 Remove the filter, and pour the catalyst slurry through a new 0.2 μm nylon filter paper in the filter flask again. Save all the filtrates for analysis.

10 Rinse, sonicate, remove the filter paper, and dry the solid in oven at 80 °C.

11 Weigh the solid and place 5–10 mg aside for analysis.

12 Place the remaining solid in small vial, and add 100 μl of water to wet catalyst.[c]

13 Add 10% of the weight of the catalyst of the dissolved Nafion solution. For 100 mg of catalyst, add 265 ml of 5% Nafion solution.

14 Place a magnetic stirring 'flea' bar in vial, and seal the vial with a cap and Parafilm. Allow the catalyst slurry to stir for at least one day before testing. This will make an 'ink' out of the catalyst.

15 Using a paint brush, paint a small amount of the catalyst 'ink' onto the smooth surface of the carbon electrode. After making a continuous film, heat to 120 °C for 1 h to cure the Nafion.

16 Secure the carbon electrode into the fastener, and secure the fastener into the rotator. Immerse the electrode in an argon-purged 0.5 M H_2SO_4 solution. The carbon rod is the working electrode in a one-compartment three-electrode cell. Immerse the counter and reference electrodes in solution. Measure the current as a function of potential from cathodic to anodic in 50 mV steps from −150 mV to +600 mV versus DHE.

17 Repeat step 16 using a 2 M methanol, 0.5 M H_2SO_4 solution.

18 Disassemble electrode. Rinse carbon rod, and allow it to dry. Weigh the rod. Wipe off the catalyst material at the end of the rod, and weigh it again.[d]

[a] The current-potential data were collected with EChem v.1.3 software using the MacLab v.3.4 interface.

[b] See ref. 27.

[c] Note: if the catalyst is not wetted before adding the Nafion solution, a flame may result, and the catalyst will become deactivated.

[d] This gives the weight of the catalyst, and allows for mass normalization of the current.

3 Preparation of electrode arrays

3.1 Selecting a substrate

The substrates for combinatorial screening of electrocatalysts must be electrically conductive, and must allow materials to adhere to the surface. Initial experiments in which borohydride-reduced catalysts were prepared on smooth platinum-coated surfaces were unsuccessful because the catalyst was non-adherent. The same experiment was performed with fibrous Toray carbon paper (ElectroChem, Inc.), and the materials adhered well. Carbon paper is conductive, and does not catalyse the electrochemical oxidation of methanol in the potential range used for screening. It is also slightly flexible, which is important for its use in a commercial printer (see Section 3.3). Also, the same carbon paper is used as an electrode backing in real DMFCs, and therefore it allows the combinatorial screening conditions to more closely match the ultimate application.

3.2 Mapping the arrays

Screening experiments (see Section 4) have shown that relatively broad regions of composition space are active as catalysts. Therefore, an array of discrete electrodes that gradually change in composition is a useful way to map the parameter space. The simplest way to vary the composition is to base the design of the arrays on phase diagrams. Ternary phase diagrams are triangular, with the mole fraction of each element decreasing with distance from the vertex representing that element. Ternary catalyst arrays can therefore be mapped as a 'billiard ball' array of discrete compositions, in which the total number of moles of metal is the same at every spot. A quaternary phase diagram, on the other hand, is a tetrahedron—a three-dimensional object that cannot be mapped in a continuous fashion onto a planar array of electrodes. In order to preserve most of the connectivity of the three-dimensional diagram, we unfolded the three concentric shells of the 220-spot tetrahedron, and removed redundant binary lines, as shown in *Plate 3*. Here, the edges of the outermost shell represent binary lines, the faces represent ternary arrays, and the vertices represent the pure elements. The two inner tetrahedral shells represent quaternary compositions. Currently, we are screening 715-member discrete pentanary (four-dimensional) composition diagrams, unfolded as shown at the bottom of *Plate 3*. In our early experiments, electrode arrays were made by hand pipetting, and only small ternary arrays were made. At higher levels of screening, it is important to use a less labour-intensive way of synthesizing arrays, which is described in the next section.

3.3 Automating the synthesis of arrays

A convenient solution to the problem of fabricating large, complex arrays is to use an ink-jet printer to deliver varying amounts of metal salt solutions to the appropriate spots on the carbon substrate (28, 29). At the time we began our experiments, there were no commercial products available for printing arrays, so we sought to adapt an existing one, i.e. to fill the ink reservoirs of an office ink-jet printer with metal salt 'inks'. There were several points to consider when selecting a commercial printer, including adaptability, reproducibility, robustness, and geometry. With some alterations, we found that a Macintosh Color StyleWriter 2500 driven by MacDraw Pro 1.0 software can print arrays of metal salt solutions with adequate resolution, registry of spots, and control of composition.

3.3.1 Adaptability

An important feature of the Color StyleWriter 2500 is that it can be tampered with and still operate normally. The ink-jet portion of the printer is composed

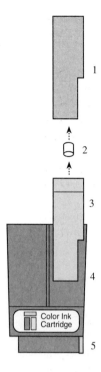

Figure 1. Schematic drawing of the Apple Color StyleWriter 2500 ink cartridge. 1. Large sponge. 2. Small cylindrical sponge. 3. Ink container. 4. Print cartridge with inkwells. 5. Print head.

of two main pieces: the print cartridge and the ink containers, as shown in *Figure 1*. The print cartridge is a plastic shell containing resistive heating devices which force ink out of microscopic holes, in order to print the desired image. The ink containers are sealed plastic pieces which are easily removed from the print cartridge. This easy separation of pieces is especially convenient when adapting the printer to delivery of different metal salt solutions.

3.3.2 Reproducibility

Although the Color StyleWriter 2500 makes use of four ink reservoirs, the colour ink containers cannot easily be used in the combinatorial printing procedure, because the software adjusts the ink levels in order to save colour ink. That is, a command to deliver an equal mixture of three colours (magenta, yellow, and cyan) is interpreted by the driver software as a command to deliver grey (i.e. a black-and-white speckle pattern). However, if only the black reservoir is used, the printed amounts are very predictable and reproducible. One can adjust the amount of 'ink' delivered by changing the greyscale for each electrode drawn onto the array, and the number of times that the substrate is passed through the printer. In order to print out several different inks, a map for each component must be drawn with the appropriate greyscale pattern. For example, to prepare all binary, ternary, and quaternary mixtures of four inks, four different 'masks' are prepared as MacDraw files. These files contain the same pattern in exact registry, except that the greyscale is the appropriate one for a particular element. Each mask represents a different metal salt solution, and each dot on the array represents an individual electrode.

In most cases, 1 mg or more material is desired per spot, and several passes through the printer for each metal salt are necessary. The vertical and horizontal registry of the Color StyleWriter 2500 is within 0.1 mm each time. When making spots that are 1–2 mm in diameter, this is acceptable. The array shown in the middle of *Plate 3* has passed through the printer 40 times, and registry errors are invisible to the naked eye (which is the detection device used in the screening experiments, see below).

We also determined that if the viscosity of the metal salt solution does not match the real ink solution, then the solution will cause the printer to malfunction—usually by leaking out of the print head. This is easily fixed by adding glycerol to the salt solution to increase its viscosity. The real ink solution has a viscosity of 1.79 cSt. A typical metal salt solution contains 12% glycerol, and is 0.8 M in metal salt.

3.3.3 Robustness

The Color StyleWriter 2500 was designed to have pH neutral, non-ionic aqueous solutions pass through it, not concentrated solutions containing corrosive acids and halide ions. When using corrosive solutions (such as chloroplatinic acid), the small stainless steel mesh screens inside the print cartridge must be protected from corrosion. The screen is essential in maintaining flow from the reservoir to the print head. An inert metal, such as gold, can be electroplated onto the screens following *Protocol 2*.

Protocol 2

Electroplating gold onto the stainless steel screen

Equipment and reagents

- Potentiostat
- DHE
- Platinum counter electrode
- Copper wire
- Orotemp 24 gold plating solution (Technic, Inc.)
- Wood's nickel strike solution: 240 g/litre $NiCl_3$ and 125 ml/litre 37% HCl[a]

Method

1   Remove ink containers from print cartridge, and rinse inside of print cartridge thoroughly with deionized water, removing all commercial ink.

2   Contact the stainless steel screen by tightly pressing a piece of copper wire to it. Make the screen the working electrode of a three-electrode electrochemical cell. Use the black ink chamber as the cell to contain the liquid.

3   Activate the surface of the stainless steel with Wood's nickel strike. Hold the working electrode at a potential of –0.9 V versus DHE for 2 min.

4 Move contact position of copper wire to activate 'Achilles' heel' from previous step, and repeat step 3.

5 Thoroughly wash the chamber to remove the strike solution, and fill with Orotemp gold plating solution.

6 Hold the working electrode at -1 V versus DHE (current $= 1$ mA) for 30 min.

7 Move contact position to plate gold onto 'Achilles' heel' from previous step, and repeat step 6.

8 Remove gold plating solution, and thoroughly rinse cartridge with deionized water.

[a] See ref. 30.

3.3.4 Geometry

The Color StyleWriter 2500 was also chosen for making electrode arrays since the paper is fed through at about a 60° angle. Most other ink-jet printers require the paper to feed at a 180° angle. Toray carbon paper is brittle and breaks under severe strain. When taped to a piece of ordinary paper and run through the Color StyleWriter 2500, it does not break.

3.4 **General rules**

Using a commercial ink-jet printer to synthesize arrays is an economical solution for research groups doing combinatorial chemistry, provided that a few simple rules are followed:

(a) The solution to be deposited must be soluble and unreactive in an aqueous glycerol solution. Insoluble particles can clog the microscopic nozzles.

(b) The substrate must be somewhat flexible, and able to turn through a 60° angle without breaking. Toray carbon paper works well for electrocatalyst arrays. Fibrous alumina or flexible polymer sheets should work for other types of arrays.

(c) It is convenient, though not essential, that the active material be synthesized from the printed array in a single step, either by heating or by addition of a reagent. In the case of electrode arrays, the metal salts are reduced by sodium borohydride (anode electrocatalysts) or by hydrogen gas (cathode electrocatalysts). Printed arrays of metal salts are not particularly amenable to the kind of complex, multi-step processing procedures that are routinely performed with polymer bead libraries. However, it is possible to isolate the printed spots in individual 'wells' for solution phase processing, as described below for borohydride reduction of anode catalysts.

Protocol 3

Printing an array of electrocatalysts

Equipment and reagents

- Macintosh computer
- Macintosh Color StyleWriter 2500
- Viscometer with small charge volume (Cannon Instruments, Inc.)
- Metal chloride inks: 0.8 M metal salt with c. 12% glycerol

- 1 M $NaHCO_3$ solution
- 10% wt. $NaBH_4$ solution
- Toray carbon paper
- Waterproof silcone sealant
- Home-built 'reduction chamber' shown in Figure 2[a]

Method

1   Protect the metal screen in ink cartridge as described in *Protocol 2*.

2   Using a hacksaw, remove the top portion of the black ink container.

3   There are two sponges inside: a large one shaped like the ink container, and a small cylindrical one (see *Figure 1*). Remove the sponges from the container and rinse them free of all commercial ink with distilled water.

4   Adjust the viscosity of the metal salt ink to 1.79 cSt. Measure the viscosity using the small charge volume viscometer. Add glycerol or water as necessary.

5   Immerse the small, cylindrical sponge in the metal salt 'ink'. Return the small sponge to the ink container.

6   Immerse the large sponge in the metal salt ink. Return the sponge to the container.

7   Dropper the ink solution onto the large sponge until the sponges are saturated. The sponges are saturated when a drop comes out of the bottom immediately upon placing a drop into the container. If little solution is available, the large sponge can be cut smaller, and thus saturate with a smaller volume of ink.

8   Fasten the top of the ink container on with tape.

9   Dropper one to two drops of the ink solution onto the gold-plated metal grid. Apply suction to the microscopic holes to fill the chamber of the cartridge.[b] Repeat until a cloth touched to the surface of the nozzles causes the solutions to bleed out.

10   Return the ink container to the cartridge, and place the cartridge in the printer.

11   Open the printing file ('mask') for the appropriate ink. Load the printer with Toray carbon paper taped to a piece of ordinary paper. Two or more identical arrays can be printed in the same operation by using multiple carbon strips. Be sure to save room on the carbon paper for making electrical contact.

12   Print.

13   Remove the print cartridge from the printer, remove the ink container from the ink cartridge, and remove the sponges from the ink container. Store the sponges in the metal ink solution.

Protocol 3 continued

14 Rinse the cartridge and the ink container. Load a clean, new ink container with different metal salt ink and repeat steps 2–11 until all metal salts are printed. Use a different container and sponges for each metal ink to prevent cross-contamination. If all operations are not performed in a single day, store the arrays in a desiccated container as hygroscopic materials will cause seeping of the electrode spots.

15 Carefully remove the piece(s) of Toray carbon paper from the piece of ordinary paper, and fasten it into the home-built reducing chamber. Fasten tightly to avoid leaking.

16 Add 2 µl of 2 M $NaHCO_3$ into each well in the chamber, followed by 1 µl of 10% $NaBH_4$. If the solution spills out of well, spot with a lint-free wipe to prevent spill-over into other wells.

17 Allow the array to dry overnight. Remove the array from the chamber using a razor blade, as necessary.

18 Cut a 20 cm square piece of aluminium, and fold one side of it around the portion of the Toray carbon paper that does not contain any electrode spots. Fold any slack, and squeeze tightly to make good electrical contact.

19 Coat the aluminium with silcone sealant to insulate it from the solution to be used in *Protocol 4*.

[a] Reduction chamber consists of two pieces of Plexiglas (25 cm × 8 cm) with holes drilled in the pattern of the electrodes. Two pieces of gasket rubber are sandwiched in between, and contain the equivalent holes. The array is fastened in with screws to create solution phase 'wells' for redissolving and reducing the metal salts.

[b] Use a small pipette bulb. Do *not* mouth pipette osmium solution.

Figure 2 Home-built 'reduction chamber'. Plexiglas top and bottom pieces and rubber gaskets encapsulate the electrode array. Holes drilled in the top Plexiglas piece and top gasket define each individual well.

4 Optimizing the screening and testing conditions

4.1 Optimization of screening conditions of anode electrocatalysts

DMFCs normally operate at high temperature (60–90 °C) in a flow cell at low methanol concentration (< 0.5 M), make use of a solid polymer electrolyte, and have a small amount of soluble Nafion mixed in with the catalyst. The ideal situation would be to address each electrode on an array individually, and measure the individual currents as a function of potential, under conditions that exactly mimic the fuel cell. Unfortunately, the wiring and data handling become unwieldy for arrays of even moderate size. Instead of measuring current directly, we devised an indirect method in which current is converted to a fluorescence signal. This allows parallel screening of arbitrarily large arrays, and allows one to ignore the uninteresting majority of inactive compositions. According to half-reaction (*Reaction 1*), the oxidation of one mole of methanol generates six moles of protons. This implies that there will be a substantial drop in pH at the surface of any electrode spot that oxidizes methanol. An acid–base indicator in solution that fluoresces in its acidic form points to the active areas of composition space. Using *Protocol 4*, we screened all quaternary, ternary, and binary combinations of Pt, Ru, Rh, Os, and Ir for activity as anode electrocatalysts.

Protocol 4

Screening anode electrocatalyst arrays

Equipment and reagents

- Potentiostat
- Large evaporating dish
- Large area (4 cm^2 or more) platinum gauze counter electrode
- DHE reference electrode

- Hand-held UV lamp
- Indicator solution: 6 M methanol, 0.5 M NaClO$_4$, 30 mM NiClO$_4$, 100 μM 3-pyridin-2-yl-(4,5,6)triazolo-(1,5-a)pyridine (PTP)^{a,b} solution (pH adjusted to 3)

Method

1   Immerse the array in the indicator solution, such that the array is parallel to the bottom of the evaporating dish. The solution should be less than 0.5 cm deep over the electrode array.^{c}

2   Make the array the working electrode of a one-compartment three-electrode cell.

3   Hold the array at the thermodynamic potential for the reaction, 0 V versus DHE in the case of methanol electro-oxidation.

4   Position a hand-held UV lamp (366 nm) above the surface of the electrode array and watch for any fluorescence.

5   Step to increasingly anodic potentials in 25–50 mV steps.

6   Repeat steps 4 and 5, noting any fluorescence. At low over-potentials, fluorescence appears as cloudy regions of solution which can be washed away with a pipette. At higher over-potentials, brighter, more obvious fluorescence persists.

7   In order to get the most accurate values of composition, prepare 'zoom' arrays in the same fashion as in *Protocol 3*, expanding the regions of interest. In the case of a zoom array, the 'inks' consist of mixtures of metal salts that define the boundaries

413

Protocol 4 continued

(vertices) of the composition region. Repeat *Protocol 4* with the zoom array to refine the lead compositions for individual testing.

[a] See ref. 31.

[b] PTP is soluble in methanol, so dissolve it in the methanol before adding water.

[c] A small vial filled with indicator solution at one corner of the array may assist in keeping it from floating off the bottom of the dish.

Combinatorial screening should be combined with serial testing of leads in the region of interest. These serial experiments help to eliminate 'duds'—compositions that appear active in the screening experiments, but which are not active as bulk catalysts, presumably because of differences between the screening conditions and the operating conditions of the real fuel cell.

4.2 DMFC testing experiments

DMFC experiments were performed with bulk samples of the appropriate anode catalyst, prepared by borohydride reduction, and Pt black (Alfa) as the cathode catalyst. Membrane electrode assemblies (MEAs) were made by decal transfer of *c.* 4 mg/cm^2 catalyst ink (prepared as described in *Protocol 1*) onto Nafion 117 (DuPont Polymers) membranes (32). A methanol/water solution was delivered to the 5 cm^2 anode at 12.5 ml/min (0 psig back pressure), while dry air was delivered to the cathode at 0 psig, 400 ml/min. The cell temperature was maintained at 60 °C, and the MEAs were conditioned by holding the cell at a voltage of 200 mV for three days before acquiring I-V data. Fuel cell performance curves were measured using a Scribner Associates Series 890 fuel cell load (Scribner Associates, Inc.). Anode polarization curves were measured using a reversible hydrogen electrode (RHE) in the fuel cell, and were iR-compensated. The RHE consisted of a small spot of unsupported Pt on the Nafion membrane, *c.* 0.5 cm from edge of the anode catalyst. Electrical contact was made to the RHE with a Pt electrode affixed to the end of a Teflon plug, which was press-fitted into a hole drilled through a graphite block (ElectroChem, Inc.). Zero-grade hydrogen flowed into the reference port at a rate of 1 ml/min. An o-ring prevented leakage of hydrogen from the RHE to the anode.

4.3 Testing and screening of methanol-tolerant cathode materials

Cathode catalysts screened for oxygen reduction activity included binary, ternary, and quaternary mixtures of ruthenium, selenium, molybdenum, and tin. These catalysts were screened in much the same way as described in *Protocol 4*, with some minor differences. Electro-reduction of oxygen *increases* the pH at the surface of the electrode, and therefore a base-fluorescent indicator was needed. Phloxine B (Aldrich) or eosin Y (Aldrich) both provide visibly detectable fluor-

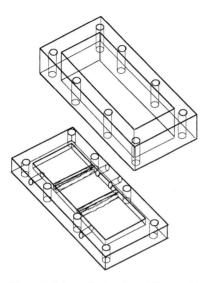

Figure 3 Schematic drawing of the combinatorial gas diffusion cell. The electrode array is sandwiched between the upper and lower halves. Gases are introduced from the bottom, and the upper compartment is filled with an electrolyte/fluorescent dye mixture. The entire assembly can be thermostatted in a constant temperature bath.

escence in these experiments. In screening for anode catalysts, the analyte (methanol) could be dissolved in the indicator solution at high concentration (6 M) in order to increase the current density and thereby increase the fluorescence. However, even in cold solutions, oxygen has a solubility of only about 30 mM. In order to achieve a higher concentration of gas at the electrode surface, we constructed a gas diffusion electrode assembly, which was adapted to both the screening of cathode catalyst arrays and the testing of bulk catalysts.

The screening cell consists of two pieces of Plexiglas (0.5 cm and 1 cm thick, 8 cm wide, and 25 cm long), as shown in *Figure 3*. In the thinner, bottom piece of Plexiglas, a gas reservoir was created in the shape of the array using a router. Two gas inlet and outlet holes were drilled into the sides to allow oxygen to pass through the cell. In the top piece of Plexiglas, the centre was hollowed out into a rectangular shape the size of an electrode array. Both pieces were fitted with appropriately sized gasket rubber and screw holes. The array is placed between the two pieces of Plexiglas and electrically contacted with a thin piece of gold foil. For oxygen reduction, the electrode array is printed on a piece of Toray carbon paper that had been made hydrophobic (ElectroChem, Inc.). The top of the cell serves as a container for the screening solution (0.5 M $NaClO_4$, 100 μM indicator, pH adjusted with $HClO_4$). Oxygen is humidified, and then passed through the cell at a flow rate of 140 ml/min. The screening experiments are performed in the same way as for the anode catalysts, except that the potential is first held at the thermodynamic reduction potential, and then made more *cathodic* in 10–50 mV increments. Any fluorescence is noted, and lead materials are tested individually.

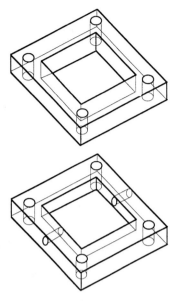

Figure 4 Gas diffusion test cell. Like the combinatorial cell, the electrode is sandwiched between the two Plexiglas plates, and oxygen gas is fed through the back side. The centre hole in the top plate contains the electrolyte, and the reference and counter electrodes.

The gas diffusion cell used to test bulk samples of cathode catalysts is similar to the screening cell, except it has a smaller cell volume, as shown in *Figure 4*. Individual samples (100 mg) are prepared by reduction under flowing hydrogen at 200–250 °C. An ink of the solid catalyst is prepared as described in *Protocol 1*, and *c*. 5 mg is painted as a 1 cm square onto a piece of hydrophobized carbon paper. The electrode is placed into the cell, and is contacted with a piece of gold foil. Oxygen is humidified, and passed through the cell at a flow rate of 140 ml/min. The electrolyte, which is contained in the face of the cell, is an aqueous 0.5 M H_2SO_4 solution. Using a potentiostat, the current is then measured as a function of potential.

5 Results and conclusions

5.1 Results

Many of our results have been published elsewhere, and therefore only the highlights underscoring the successes of the approach will be presented here.

5.1.1 Anode results

When a composition is found that efficiently oxidizes methanol in fuel cells, several individual compositions within the region of interest are individually tested for activity. Using this approach, we identified several new Pt/Ru/Os/Ir catalysts that are more active than $Pt_{50}Ru_{50}$ made by the same method, as verified by anode polarization curves (*Figure 5*). As an interesting side note, the kinetics of

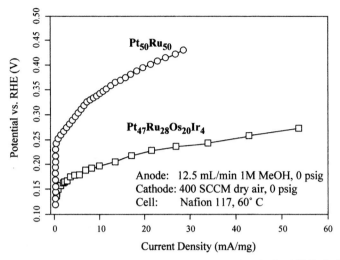

Figure 5 DMFC anode polarization curves, comparing the best Pt-Ru-Os-Ir quaternary catalyst to $Pt_{50}Ru_{50}$. Both catalysts were prepared by borohydride reduction.

methanol oxidation with the new catalysts appear to be first order in methanol (26), compared to zero-order with $Pt_{50}Ru_{50}$. If DMFCs can ultimately be run at higher methanol concentrations (e.g. using low crossover membranes or methanol-tolerant cathode catalysts), the benefits of using Pt/Ru/Os/Ir catalysts instead of $Pt_{50}Ru_{50}$ are expected to increase.

In addition to the best quaternary Pt/Ru/Os/Ir catalysts, the optical screening method identified several other compositions that fluoresced at lower over-potential than $Pt_{50}Ru_{50}$. These included Pt/Rh/Os and Pt/Ru/Rh ternaries, and materials in the Pt/Ru/Os/Rh quaternary diagram. However, the Rh-containing catalysts rapidly lost activity, as determined by their poor performance in RDE experiments. The deactivation, which may arise from catalyst poisoning or corrosion, is less obvious in the screening experiments, because screening time is short (10–15 min). By comparison, RDE voltammetry takes minutes to hours, and testing in actual DMFCs is conducted over periods of days to weeks. Because the screening for anode catalysts is performed at much higher methanol concentration (6 M) than one typically uses in an RDE or DMFC experiment (< 0.5 M), it serendipitously selects catalysts that have a non zero-order rate dependence in methanol concentration. It is therefore possible that the Rh-containing catalysts are not 'duds', but like the Ir-containing catalysts, perform better at very high concentrations of methanol.

5.1.2 Cathode results

The combinatorial screening of all quaternary, ternary, and binary combinations of Ru, Se, Mo, W, and Sn turned up several regions of high activity. Represent-ative data from a quaternary region of this composition space is shown in *Figure 6*. After the active compositions were further refined in zoom arrays, individual

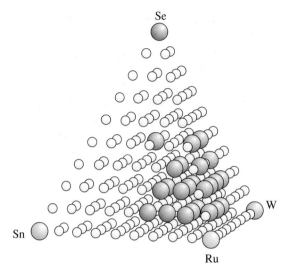

Figure 6 Re-folded activity map of a quaternary array of carbon-supported oxygen electro-reduction catalysts. Active spots (large grey spheres) are found in two distinct regions within this quaternary. For pentanary arrays, the activity map is visualized as computer-generated tetrahedral sections of the four-dimensional space.

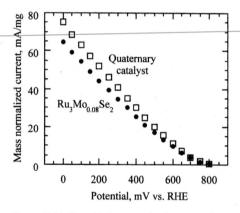

Figure 7 Methanol-tolerant cathode materials tested in a gas diffusion cell. Squares represent the quaternary catalyst, and circles represent a reference Alonso-Vante type ternary catalyst.

samples were prepared and tested as described in Section 4.2. One particular region has been identified as containing the best catalysts for oxygen reduction, as shown by the current–voltage curves in *Figure 7*. The physical characterization of these improved cathode catalysts is still in progress.

5.2 Conclusions

Combinatorial electrochemistry can be used to find the optimum electrode material, synthesized by a particular method. If the conditions of screening closely match the conditions of end use, the likelihood of finding useful new materials is higher. In the Pt, Ru, Os, Ir, Rh composition space, when the catalysts are prepared by borohydride reduction and screened at high methanol concentration, the best catalysts found to date are a Pt/Ru/Os/Ir quaternaries. Detailed physical characterization of these materials shows that they lie in a region of phase space near the solubility limit of Os, Ru, and Ir in face-centred cubic Pt (26). In the Ru, Se, Sn, Mo composition space, when the catalysts are prepared by hydrogen reduction, several improved cathode catalysts have also been found. We are continuing to search for better methanol electro-oxidation catalysts, as well as for methanol-tolerant cathode electrocatalysts. Many other applications of the screening method can be imagined including electrocatalysts for other types of fuel cells, battery materials, and elecrocatalysts for synthetic organic reactions.

Acknowledgements

This work was supported by the US Army Research Office and DARPA under grants DAAH04-94-G-0055, DAAH04-95-1-0330, and DAAH04-95-1-0570.

References

1. Lebl, M. (1999). *J. Comb. Chem.*, **1**, 3.
2. Borman, S. (1998). *Chem. Eng. News*, **76**, 47.
3. Hanak, J. J. (1970). *J. Mater. Sci.*, **5**, 964.
4. Xiang, X.-D., Sun, X., Briceno, G., Lou, Y., Wang, K.-A., Chang, H., *et al.* (1995). Science, **268**, 1738.
5. Briceno, G., Chang, H., Sun, X., Schultz, P. G., and Xiang, X.-D. (1995). *Science*, **270**, 273.
6. van Dover, R. B., Schneemeyer, L. F., and Fleming, R. M. (1998). *Nature*, **392**, 162.
7. Danielson, E., Devenney, M., Giaquinta, D. M., Golden, J. H., Haushalter, R. C., McFarland, E. W., *et al.* (1998). *Science*, **279**, 837.
8. Dickinson, T. A., Walt, D. R., White, J., and Kauer, J. S. (1997). *Anal. Chem.*, **69**, 3413.
9. Moates, F. C., Somani, M., Annamalai, J., Richardson, J. T., Luss, D., and Willson, R. C. (1996). *Ind. Eng. Chem. Res.*, **35**, 4801.
10. Taylor, S. J. and Morken, J. P. (1998). *Science*, **280**, 267.
11. Fuller, T. F. (1997). *Electrochem. Soc. Interface*, 26.
12. Parsons, R. and VanderNoot, T. (1988). *J. Electroanal. Chem.*, **257**, 9.
13. Beden, B., Leger, J.-.M., and Lamy, C. (1992). In *Modern aspects of electrochemistry* (ed. J. O'M. Bockris, B. E. Conway, and R. E. White), Vol. 22, p. 97. Plenum Press, New York.
14. Sun, S.-G. (1998). In *Electrocatalysis* (ed. J. Lipkowski and P. N. Ross), p. 243. John Wiley and Sons, New York.
15. (a) Bockris, J. O'M. and Wroblowa, H. (1964). *J. Electroanal. Chem.*, **7**, 428. (b) Cathro, K. J. (1967). *Electrochem. Tech.*, **5**, 441. (c) Andrew, M. R., Drury, J. S., McNicol, B. D., Pinnington, C., and Short, R. T. (1976). *J. Appl. Electrochem.*, **6**, 99.
16. Ley, K. L., Liu, R., Pu, C., Fan, Q., Leyarovska, N., Segre, C., *et al.* (1997). *J. Electrochem. Soc.*, **144**, 1543.
17. Reddington, E., Sapienza, A., Gurau, B., Viswanathan, R., Sarangapani, S., Smotkin, E. S., *et al.* (1998). *Science*, **280**, 1735.

18. Adzic, R. (1998). In *Electrocatalysis* (ed. J. Lipkowski and P. N. Ross), p. 197. John Wiley and Sons, New York.
19. Alonso-Vante, N. and Tributch, H. (1986). *Nature*, **323**, 431.
20. Alonso-Vante, N., Giersig, M., and Tributsch, H. (1991). *J. Electrochem. Soc.*, **138**, 639.
21. Romero, T., Rivera, R., Solorza, O., and Perez, R. (1998). *Int. J. Hydrogen Energy*, **23**, 1031.
22. Adams, R. and Shriner, R. L. (1923). *J. Am. Chem. Soc.*, **45**, 2171.
23. Bond, G. C. (1962). *Catalysis by metals*. Academic Press, New York.
24. McKee, D. W. (1969). *J. Catal.*, **14**, 355.
25. Hagans, P. L., Swider, K. E., and Rolison, D. R. (1997). *Proc. Electrochem. Soc. 97*, **13**, 86.
26. Gurau, B., Viswanathan, R. K., Lafrenz, T. J., Liu, R., Ley, K. L., Smotkin, E. S., *et al.* (1998). *J. Phys. Chem. B*, **102**, 9997.
27. Giner, J. (1964). *J. Electrochem. Soc.*, **111**, 376.
28. Lemmo, A. V., Fisher, J. T., Geysen, H. M., and Rose, D. J. (1997). *Anal. Chem.*, **69**, 543.
29. Sun, X.-D., Wang, K.-A., Yoo, Y., Wallace-Freedman, W. G., Gao, C., Xiang, X.-D., *et al.* (1997). *Adv. Mater.*, **9**, 1046.
30. Dini, J. W. and Helms, J. R. (1970). *Plating*, 906.
31. Mori, H., Sakamoto, K., Mashito, S., Matsuoka, Y., Matsubayashi, M., and Sakai, K. (1993). *Chem. Pharm. Bull.*, **41**, 1944.
32. Wilson, M. S. and Gottesfeld, S. (1992). *J. Appl. Electrochem.*, **22**, 1.

Chapter 15

Combinatorial library synthesis using polymer-supported catalysts

Shū Kobayashi

Graduate School of Pharmaceutical Sciences, The University of Tokyo, Hongo, Bunkyo-ku, Tokyo 113–0033, Japan.

1 Introduction

Combinatorial synthesis (Chapters 1–7), a synthetic strategy which leads to large numbers of structurally distinct molecules, in association with high-throughput screening techniques is making a significant impact not only on the field of drug discovery (1, 2) but also on the development materials sciences and catalysis (Chapters 14, 16).

Owing to the availability of a myriad of high-yielding and highly selective reactions, complex and structurally diverse natural and synthetic compounds can be prepared by traditional organic synthesis strategies. However, these reactions do not necessarily provide useful methods for the synthesis of large chemical libraries where each member must be obtained in high yield and selectivity. Thus, we are facing the challenge of developing new reliable synthetic methodologies to prepare these libraries, just as new methodologies for natural product synthesis were required forty years ago (3).

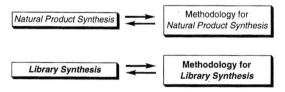

Scheme 1 Methodology for library synthesis. Just like natural product synthesis, each chemical library requires the development of a new synthetic methodology.

Combinatorial synthesis, where the substrates to be synthesized are often tethered to a polymeric support (2) has several drawbacks. First, the reactions of polymer-supported reactants are often slower. Secondly, differences in reactivity between the support-bound reactants lead to incomplete or impure libraries.

Table 1 Tetrahydroquinolines synthesis[a]

| 75% (90/10) | 80% (86/14) | 83% (62/38) |

| 65% (nd) | 84%[b] | 92% (95/5) |

| 85% (99/1) | quant. (100/0) | 71% (100/0) |

| 70% (100/0) | 96% (100/0) | 91% (100/0) |

| 78% (100/0) | 80% (100/0) | 99% (100/0) |

[a](a) Diastereomer ratio in parenthesis. Relative stereochemical assignment was made by ^{1}H NMR analysis. (b) Butyl ethynyl sulfide was used as a dienophile.

Thirdly, the amount of polymer-supported reactants is limited by the loading of the support which is generally low ($\leq$ 0.8 mmol/g). Thus, large scale syntheses are difficult. To overcome these problems we combined multi-component reactions (see Chapter 9) and polymer-supported catalysis for the rapid synthesis of a large number of structurally distinct compounds in hundreds of milligrams quantities. The following sections illustrate this methodology.

2 Tetrahydroquinoline library (4)

Polymer-supported catalysis (8) offers several advantages such as simplification of product work-up, separation, isolation, catalyst recycling, and may be useful for combinatorial as well as parallel library construction. On the other hand, one of the drawbacks of this approach is the generally low reactivity of polymer-supported catalysts, which may be ascribed to their heterogeneous nature. This limitation was alleviated using a new polymer-supported scandium-based catalyst, which is partially soluble in certain solvents and subsequently recovered by precipitation in a suitable solvent after completion of the reaction. The synthetic route for such catalyst, PA-Sc-TAD (polyallylscandium trifylamide ditriflate), is shown in *Scheme 2*. Treatment of polyacrylonitrile with $BH_3 \bullet SMe_2$ (borane•dimethylsulfide) in diglyme at 150 °C results in amine (1) which upon reaction with Tf_2O (trifluoromethanesulfonic acid anhydride) in the presence of Et_3N (triethylamine) in 1,2-DCE (1,2-dichloroethane) at 60 °C affords sulfonamide 2 (9). Scandium is introduced using a potassium salt of 2 and $Sc(OTf)_3$ (scandium triflate). The gummy PA-Sc-TAD (3) thus prepared can be dispersed and partially solubilized in a mixture of CH_2Cl_2 (dichloromethane) and CH_3CN (acetonitrile), then precipitated upon addition of hexane and recovered for further uses.

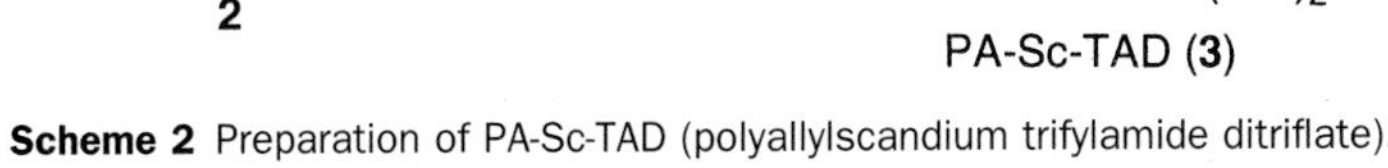

Scheme 2 Preparation of PA-Sc-TAD (polyallylscandium trifylamide ditriflate).

Recently, we developed lanthanide triflate catalysed quinoline synthesis based on three-component coupling reactions (5, 6). Through the combination of many aldehydes, amines, and olefins a large quinoline library could be generated (7). This method was especially useful for the construction the tetrahydroquinoline library shown in *Table 1*. The procedure is very simple. Typically, an aldehyde, an aromatic amine, and an alkene (or alkyne) in CH_2Cl_2–CH_3CN (2:1) and in the presence of catalytic amounts of PA-Sc-TAD (3) are reacted at 60 °C for 12 h.

After precipitation and recovery of the catalyst by filtration, the filtrates are concentrated *in vacuo* to give essentially pure tetrahydroquinoline derivatives. If necessary the crude adduct is purified by chromatography to yield the desired quinoline derivative in high yield and purity. Dehydrating agents (e.g. molecular sieve 4A or magnesium sulfate) are unnecessary. Furthermore, PA-Sc-TAD is water tolerant (10) since hydrated substrates can be used without prior drying as shown in *Scheme 3*.

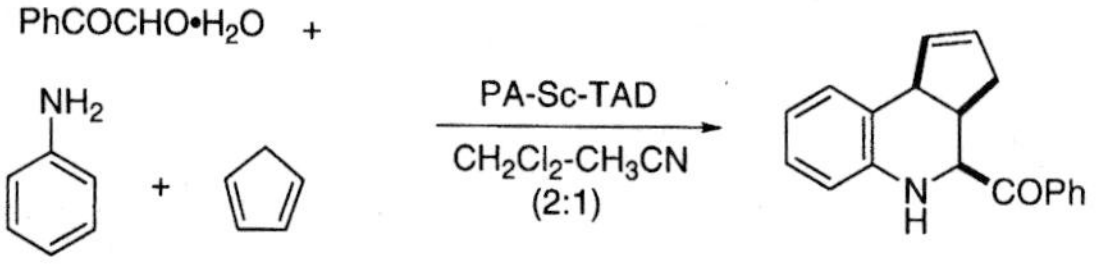

Scheme 3 Tetrahydroquinoline synthesis using phenylglyoxal. The catalyst can be precipitated, recovered, and recycled without apparent loss of activity (first use, 90%; second use, 91%; third use, 93%).

Protocol 1

Preparation of polyallylscandium trifylamide ditriflate

Equipment and reagents

- Two-necked, round-bottom flask (30 ml) fitted with a three-way stopcock and connected to a vacuum/argon source
- Magnetic stirrer or a shaker
- Vacuum/inert gas source (argon source may be a balloon filled with argon)
- Polyacrylonitrile ([–$CH_2CH(CN)$–] = 53.06), 265.3 mg, 5.0 mequiv.
- $BH_3 \bullet SMe_2$ (FW 75.97), 1139.6 mg, 15.0 mmol
- Distilled Et_3N (FW 101.19), 1517.9 mg, 15.0 mmol

- Distilled Tf_2O (FW 282.13), 4232.0 mg, 15.0 mmol
- KH (potassium hydride, FW 40.11), 200.6 mg, 5.0 mmol
- $Sc(OTf)_3$ (FW 492.16), 2460.8 mg, 5.0 mmol
- Dry, distilled dyglime
- Dry, distilled CH_2Cl_2
- Dry, distilled THF (tetrahydrofuran)

Method

1 Flame dry the reaction vessel with a stirring bar under dry argon.

2 Add polyacryronitrile and 10 ml of dyglime to the flask with stirring. Slowly add $BH_3 \bullet SMe_2$ at 0 °C.

3 After stirring the mixture for 5 min at room temperature, heat to 40 °C and then 150 °C.

4 After stirring for 36 h, cool to 0 °C.

5 Add 6 M HCl (27 Equiv.) and stir for 2 h under reflux.

6 Cool to 0 °C, add 7 M NaOH (54 Equiv.), and stir the mixture for 1 h.

7 Filter, wash with water, dioxane, and ether, and dry the solid (**1**) at room temperature for 10 h.

8 Under a stream of argon dissolve **1** in 5 ml of dichloromethane in a flame-dried reaction vessel equipped with a stirring bar.

9 Add Et_3N and Tf_2O at $-20\,°C$, and stir at $60\,°C$ for 10 h.

10 Filter the solid and wash with H_2O, dioxane, and ether, and dry the solid (**2**) at room temperature for 10 h.

11 Under a stream of dry argon add KH to a flame-dried reaction vessel equipped with a stirring bar, then add **2** in 3 ml of THF, and stir at room temperature for 0.5 h.

12 Add $Sc(OTf)_3$ and stir at room temperature for 48 h.

13 Filter the solid (**3**) and wash with water, dioxane, and ether, then dry at room temperature for 10 h.

Protocol 2

Tetrahydroquinoline synthesis

Equipment and reagents

- One-necked, round-bottled flask (50 ml) fitted with a magnetic stirring bar
- Magnetic stirrer or a shaker
- Distilled benzaldehyde (FW 106.12), 42.4 mg, 0.40 mmol
- PA-Sc-TAD (2% Sc, w/w), 56.0 mg
- Distilled aniline (FW 93.13), 37.3 mg, 0.40 mmol
- Distilled 2-methylindene (FW 130.19), 57.3 mg, 0.44 mmol

Method

1 Combine PA-Sc-TAD, benzaldehyde, aniline, and 2-methylindene in CH_2Cl_2:CH_3CN (2:1, 3.0 ml) and stir at $40\,°C$ for 16 h.

2 Add hexane (20 ml) and filter out the polymer-supported catalyst.

3 Concentrate the filtrate from step 2 under reduced pressure, then purify the desired material by silica gel preparative TLC (thin-layer chromatography) using an eluent mixture of hexane:ethylacetate (4:1).

A characteristic feature of polymer-supported reagents compared to conventional combinatorial synthetic technologies is that more than hundred milligram scale syntheses with a large array of diverse reactants can be achieved in high yields and selectivities. The number of commercially available aromatic-, aliphatic-, and heterocyclic-aldehydes, glyoxals, and glyoxylates that could be employed in this system is over 200, and more than 200 aromatic amines and

50 alkenes (and alkynes) (11) are also commercially available. Therefore, a tetra-hydroquinoline library of more than a million compounds could potentially be obtained using an automated version of this methodology. In addition, the tetrahydroquinoline library obtained could be easily oxidized to dihydroquino-line or quinoline derivatives, which could double or triple the size of the library.

3 β-Amino ketone and ester library (12)

The Lewis acid catalysed reaction of imines with silyl enolates is one of the most efficient methods for the preparation of β-amino esters (13). Although the re-action generally requires stoichiometric amount of a Lewis acid catalyst such as TiCl$_4$ (titanium tetrachloride) (14), a small amount of either TMSOTf (trimethyl-silyltriflate) (15), a diphosphonium salt (16), FeI$_2$ (iron iodide) (17), a trityl salt (17), montmorillonite (18), or B(C$_6$F$_5$)$_3$ (19) (tripentafluorophenyl borane) promote the reaction. On the other hand, many imines are hygroscopic, unstable at high temperature, and difficult to purify by distillation or chromatography. From a synthetic point of view it is desirable that imines, generated from aldehydes and amines react *in situ* with silyl enolates and provide β-amino esters in a one-pot reaction, but most Lewis acid catalysts decompose or deactivate under such conditions.

We found that three-component reactions between aldehydes, amines, and silyl enolates proceed smoothly in the presence of a catalytic amount of lantha-nide triflate (Ln(OTf)$_3$) (20). PS-Sc-TAD was then used in these reactions. When benzaldehyde, aniline, and the silyl enol ether of propiophenone (4) (molar ratio 1:1:1.1) were combined in the presence of PA-Sc-TAD, it was found that the reaction proceeded a little slower at room temperature compared to that using Ln(OTf)$_3$ as a catalyst, but the reaction proceeded smoothly and cleanly to afford the corresponding β-amino ketone without any detectable side-product. Hetero-cyclic and aliphatic aldehydes, and glyoxals worked as well with various amines and **4** to give β-amino ketone derivatives in high yields as shown in *Table 2* (entries 1–12). After the reaction is completed, the catalyst is filtered out and the filtrate concentrated *in vacuo* to afford in most cases pure β-amino ketone.

We then examined this reaction using ketene silyl acetal of methyl isobuty-late (**5**) as a silylated nucleophile. It was expected that a β-amino ester could be produced by the reaction of cyclohexanecarboxaldehyde, *p*-chloroaniline, and ketene silyl acetal **2** under standard conditions (*Protocol 3*). However, only a trace amount of the product was obtained after 19 h at room temperature. It was assumed that water was produced during imine formation from the reaction of the aldehyde with the amine, thereby promoting the hydrolysis of the ketene silyl acetal. In the presence of MgSO$_4$, as a dehydrating agent, the yield dram-atically improved to afford the desired adduct in 74% yield. Under these reaction conditions, several β-amino ester derivatives were obtained in high yields (*Table 2*, entries 13–18).

Table 2 β-Amino ketone and ester synthesis[a]

Entry	Aldehyde	Amine	Silyl nucleophile	Product	Yield/%
1	PhCHO	PhNH₂	**4**		91 (1:1)
2	PhCHO	p-MeO-PhNH₂	**4**		87 (1:2)
3	PhCHO		**4**		92 (2:2)
4		PhNH₂	**4**		84 (1:2)
5		p-MeO-PhNH₂	**4**		85 (1:2)
6			**4**		91 (4:0)
7		p-Cl-PhNH₂	**4**		91 (1:2)
8		PhNH₂	**4**		87 (1:1)
9	PhCOCHO	p-Cl-PhNH₂	**4**		95 (2:6)
10	PhCOCHO	p-MeO-PhNH₂	**4**		91 (1:9)
11	PhCOCHO		**4**		84 (1:6)

427

Table 2 *Continued*

Entry	Aldehyde	Amine	Silyl nucleophile	Product	Yield/%
12	cyclohexyl–CHO	PhNH$_2$	**4**	O NHPh Ph-C(=O)-CH(CH$_3$)-CH(NHPh)-cyclohexyl	77 (2:1)
13	PhCHO	p-Cl-PhNH$_2$	MeO-C(=CH$_2$)-OSiMe$_3$ **5**	O NHPh-p-Cl MeO-C(=O)-C(CH$_3$)$_2$-CH(NHPh-p-Cl)-Ph	88
14	furyl–CHO	PhNH$_2$	**5**	O NHPh MeO-C(=O)-C(CH$_3$)$_2$-CH(NHPh)-furyl	89
15	furyl–CHO	p-Cl-PhNH$_2$	**5**	O NHPh-p-Cl MeO-C(=O)-C(CH$_3$)$_2$-CH(NHPh-p-Cl)-furyl	87
16	furyl–CHO	p-MeO-PhNH$_2$	**5**	O NHPh-p-MeO MeO-C(=O)-C(CH$_3$)$_2$-CH(NHPh-p-MeO)-furyl	85
17	Ph–CH=CH–CHO	PhNH$_2$	**5**	O NHPh MeO-C(=O)-C(CH$_3$)$_2$-CH(NHPh)-CH=CH-Ph	73
18	cyclohexyl–CHO	p-Cl-PhNH$_2$	**5**	O NHPh-p-Cl MeO-C(=O)-C(CH$_3$)$_2$-CH(NHPh-p-Cl)-cyclohexyl	74

[a] All the reactions were carried out at room temperature. Magnesium sulfate was added whenever compound **5** was used. The yields given correspond to those of the pure isolated compounds. The diastereomeric ratio indicated in parenthesis under the yield was determined by ^{1}H and/or ^{13}C NMR. The relative stereochemical assignment was not made.

Protocol 3

β-Amino ketone and ester synthesis

Equipment and reagents

- One-necked, round-bottom flask (50 ml) fitted with a magnetic stirring bar
- Magnetic stirrer or a shaker
- PA-Sc-TAD (2% Sc, w/w), 56.0 mg
- Distilled benzaldehyde (FW 106.12), 42.4 mg, 0.40 mmol

- Distilled 1-phenyl-1-trimethylsiloxy-propene (FW 206.36), 90.8 mg, 0.44 mmol
- Distilled aniline (FW 93.13), 37.3 mg, 0.40 mmol

Method

1　Combine PA-Sc-TAD, benzaldehyde, aniline, and 1-phenyl-1-trimethylsiloxypropene in CH_2Cl_2:CH_3CN (2:1, 3.0 ml) and stir at room temperature for 19 h. Magnesium sulfate (125 mg) is added whenever ketene silyl acetals are used.

2　Add hexane (20 ml) and filter out the polymer-supported catalyst.

3　Concentrate the filtrate from step 2 under reduced pressure, then purify the desired material by silica gel preparative TLC using an eluent mixture of hexane: ethylacetate (4:1).

4　α-Amino nitrile library (12)

α-Amino nitriles are useful intermediates for the synthesis of amino acids (21) and nitrogen-containing heterocycles such as thiadiazoles and imidazole derivatives (22). The synthetic routes to α-amino nitriles using cyanotrimethylsilane (TMSCN) are divided into two categories: the reaction of O-silylated cyano-hydrines (prepared from aldehydes and TMSCN) with amines (23), and the reactions of imines with TMSCN (24). While these methods have both merits and demerits, we opted for the latter one mainly due to its potential in asymmetric synthesis and because the presence of a Lewis acid is key to a higher reaction efficiency.

While $Ln(OTf)_3$ was found to be an excellent catalyst in the above α-amino acid synthesis (25), the three-component reactions of aldehydes, amines, and TMSCN proceeded smoothly in the presence of PA-Sc-TAD to afford various α-amino nitrile derivatives (*Table 3*). TMSCN was used as a cyano anion source because it provides promising and safer alternative to the highly toxic HCN (21).

Protocol 4

α-Amino nitrile synthesis

Equipment and reagents

- One-necked, round-bottom flask (50 ml) fitted with a magnetic stirring bar
- Magnetic stirrer or a shaker
- Distilled benzaldehyde (FW 106.12), 42.4 mg, 0.40 mmol
- PA-Sc-TAD (2% Sc, w/w), 56.0 mg
- Distilled aniline (FW 93.13), 37.3 mg, 0.40 mmol
- Distilled TMSCN (FW 99.21), 43.7 mg, 0.44 mmol

Protocol 4 continued

Method

1 Combine PA-Sc-TAD, benzaldehyde, aniline, TMSCN, and magnesium sulfate (125 mg) in CH_2Cl_2:CH_3CN (2:1, 3.0 ml), and stir at room temperature for 19 h.

2 Add hexane (20 ml) and filter out the polymer-supported catalyst.

3 Concentrate the filtrate from step 2 under reduced pressure, then purify the desired material by silica gel preparative TLC using an eluent mixture of hexane: ethylacetate (4:1).

Table 3 α-Aminonitrile synthesis

Entry	Aldehyde	Amine	Silyl nucleophile	Product	Yield/%
1	PhCHO	PhNH₂	Me₃SiCN **6**	NHPh; NC—Ph	86
2	PhCHO	p-Cl-PhNH₂	**6**	NHPh-p-Cl; NC—Ph	94
3	furyl-CHO	PhNH₂	**6**	NHPh; NC—furyl	83
4	cyclohexyl-CHO	PhNH₂	**6**	NHPh; NC—cyclohexyl	83
5	cyclohexyl-CHO	p-Cl-PhNH₂	**6**	NHPh-p-Cl; NC—cyclohexyl	99
6	cyclohexyl-CHO	p-MeO-PhNH₂	**6**	NHPh-p-MeO; NC—cyclohexyl	96

5 Conclusion

Three-component reactions between aldehydes, amines, and silylated nucleo-philes have been successfully carried out by using a polymer scandium catalyst to afford β-amino ketones, β-amino esters, and α-amino nitriles in high yields. The reactions are very clean and the procedure very easy: a simple mixing of the catalyst (PA-Sc-TAD) and almost equimolar amounts of an aldehyde, an amine, and a silylated nucleophile, followed by precipitation and recovery of the catalyst. The filtrates are concentrated to give almost pure products in most cases and the PA-Sc-TAD catalyst can be recycled without any apparent loss of activity. Finally these reactions provide a useful route to a large number of structurally distinct nitrogen-containing compounds in high quality and quantity.

Acknowledgements

This work was partially supported by CREST, Japan Science and Technology Corporation (JST), and a Grant-in-Aid for Scientific Research from the Ministry of Education, Science, Sports, and Culture, Japan. The author expresses his deep gratitude to his co-workers whose names appear in the references.

References

1. Borman, S. (1996). *Chem. Eng. News*, **12**, 29.
2. (a) Jung, G. (ed.) (1996). *Combinatorial peptide and nonpeptide libraries*. VCH, Weinheim. (b) Balkenhohl, F., von dem Bussche-Hünnefeld, C., Lansky, A., and Zechel, C. (1996). *Angew. Chem. Int. Ed. Engl.*, **35**, 2288. (c) Thompson, L. A. and Ellman, J. A. (1996). *Chem. Rev.*, **96**, 555. (d) Früchtel, J. S. and Jung, G. (1996). *Angew. Chem. Int. Ed. Engl.*, **35**, 17. (e) Terrett, N. K., Gardner, M., Gordon, D. W., Kobylecki, R. J., and Steele, J. (1995). *Tetrahedron*, **51**, 8135. (f) Lowe, G. (1995). *Chem. Soc. Rev.*, **24**, 309. (g) Gallop, M. A., Barrett, R. W., Dower, W. J., Fodor, S. P. A., and Gordon, E. M. (1994). *J. Med. Chem.*, **37**, 1233. (h) Gordon, E. M., Barrett, R. W., Dower, W. J., Fodor, S. P. A., and Gallop, M. A. (1994). *J. Med. Chem.*, **37**, 1385.
3. Kobayashi, S. (1999). *Chem. Soc. Rev.*, **28**, 1.
4. Kobayashi, S. and Nagayama, S. (1996). *J. Am. Chem. Soc.*, **118**, 8977.
5. (a) Kobayashi, S., Araki, M., Ishitani, H., Nagayama, S., and Hachiya, I. (1995). *Synlett*, 233. (b) Kobayashi, S., Ishitani, H., and Nagayama, S. (1995). *Chem. Lett.*, 423. (c) Kobayashi, S., Ishitani, H., and Nagayama, S. (1995). *Synthesis*, 1095. See also, (d) Makioka, Y., Shindo, T., Taniguchi, Y., Takaki, K., and Fujiwara, Y. (1995). *Synthesis*, 801. (e) Kobayashi, S., Komiyama, S., and Ishitani, H. (1998). *Biotech. Bioeng.*, **1**, 23.
6. As for quinoline synthesis from *N*-aryl imines, (a) Lucchini, V., Prato, M., Scorrano, G., and Tecilla, P. (1988). *J. Org. Chem.*, **53**, 2251. (b) Grieco, P. A. and Bahsas, A. (1988). *Tetrahedron Lett.*, **29**, 5855. (c) Boger, D. L. (1983). *Tetrahedron*, **39**, 2869. (d) Kametani, T. and Kasai, H. (1989). *Studies in Natural Product Chem.*, **3**, 385. (e) Kametani, T., Takeda, H., Suzuki, Y., Kasai, H., and Honda, T. (1986). *Heterocycles*, **24**, 3385. (f) Cheng, Y. S., Ho, E., Mariano, P. S., and Ammon, H. L. (1985). *J. Org. Chem.*, **50**, 5678. (g) Worth, D. F., Perricine, S. C., and Elslager, E. F. (1970). *J. Heterocyclic Chem.*, **7**, 1353. (h) Joh, T. and Hagihara, N. (1967). *Tetrahedron Lett.*, 4199. (i) Povarov, L. S. (1967). *Russ. Chem. Rev.*, **36**, 656. (j) Povarov, L. S., Grigos, V. I., and Mikhailov, B. M. (1963). *Izd. Akad. Nauk SSSR, Ser. Khim.*, 2039. (k) Nomura, Y., Kimura, M., Takeuchi, Y., and Tomoda, S. (1978). *Chem. Lett.*, 267. (l) Narasaka, K. and Shibata, T. (1993). *Heterocycles*, **35**, 1039.
7. As for utility of multiple-component reactions for combinatorial synthesis, (a) Ugi, I., Dömling, A., and Hörl, W. (1994). *Endeavour*, **18**, 115. (b) Armstrong, R. W., Combs, A. P., Tempest, P. A., Brown, S. D., and Keating, T. A. (1996). *Acc. Chem. Res.*, **29**, 123. (c) Tempest, P. A., Brown, S. D., and Armstrong, R. W. (1996). *Angew. Chem. Int. Ed. Engl.*, **35**, 640. (d) Wipf, P. and Cunningham, A. (1995). *Tetrahedron Lett.*, **36**, 7819. See also, ref. 5 (b), (c).
8. Reviews: (a) Bailey, D. C. and Langer, S. H. (1981). *Chem. Rev.*, **81**, 109. (b) Akelah, A. and Sherrington, D. C. (1981). *ibid.*, **81**, 557. (c) Frechet, J. M. J. (1981). *Tetrahedron*, **37**,

663. Quite recently, we have developed a scandium catalyst immobilized onto Nafion (Nafion-Sc). (d) Kobayashi, S. and Nagayama, S. (1996). *J. Org. Chem.*, **61**, 2256.

9. (a) Takahashi, H., Kawakita, T., Ohno, M., Yoshioka, M., and Kobayashi, S. (1992). *Tetrahedron*, **48**, 5691. (b) Corey, E. J., Imwinkelried, R., Pikul, S., and Xiang, Y. B. (1989). *J. Am. Chem. Soc.*, **111**, 5493.

10. We have found that rare earth triflates are water-tolerant Lewis acids and efficient catalysts in several synthetic reactions in aqueous media. (a) Kobayashi, S. (1994). *Synlett*, 689. (b) Kobayashi, S. (1991). *Chem. Lett.*, 2187. (c) Kobayashi, S., Hachiya, I., Araki, M., and Ishitani, H. (1993). *Tetrahedron Lett.*, **34**, 3755. (d) Kobayashi, S. and Hachiya, I. (1994). *J. Org. Chem.*, **59**, 3590. (e) Kobayashi, S. and Ishitani, H. (1995). *J. Chem. Soc. Chem. Commun.*, 1379. (f) Kobayashi, S., Nagayama, S., and Busujima, T. (1996). *Tetrahedron Lett.*, **37**, 9221. (g) Kobayashi, S., Wakabayashi, T., Nagayama, S., and Oyamada, H. (1997). *Tetrahedron Lett.*, **38**, 4559. (h) Kobayashi, S., Wakabayashi, T., and Oyamada, H. (1997). *Chem. Lett.*, 831. (i) Kobayashi, S., Busujima, T., and Nagayama, S. (1998). *J. Chem. Soc. Chem. Commun.*, 19. (j) Kobayashi, S., Busujima, T., and Nagayama, S. (1998). *J. Chem. Soc. Chem. Commun.*, 981. (k) Kobayashi, S., Nagayama, S., and Busujima, T. (1998). *J. Am. Chem. Soc.*, **120**, 8287. (l) Kobayashi, S. (1998). In *Organic reactions in water* (ed. P. Grieco), pp. 262–305. Chapman & Hall. (m) Kobayashi, S. (1998). In *Aqueous phase organometallic catalysis* (ed. B. Cornils and W. A. Herrmann), pp. 519–28. VCH. (n) Kobayashi, S. (1999). *Eur. J. Org. Chem.*, 15.

11. Electron deficient dienophiles will not react under the conditions.

12. Kobayashi, S., Nagayama, S., and Busujima, T. (1996). *Tetrahedron Lett.*, **37**, 9221.

13. Kleinnman, E. F. (1991). In *Comprehensive organic synthesis* (ed. B. M. Trost), Vol. 2, Chapter 4.1. Pergamon Press, New York.

14. Ojima I., Inaba, S., and Yoshida, K. (1977). *Tetrahedron Lett.*, 3643.

15. Guanti, G., Narisano, E., and Banfi, L. (1987). *Tetrahedron Lett.*, **28**, 4331.

16. Mukaiyama, T., Kashiwagi, K., and Matsui, S. (1989). *Chem. Lett.*, 1397.

17. Mukaiyama, T., Akamatsu, H., and Han, J. S. (1990). *Chem Lett.*, 889.

18. Onaka, M., Ohno, R., Yanagiya, N., and Izumi, Y. (1993). *Synlett*, 141.

19. Ishihara, K., Funahashi, M., Hanaki, N., Miyata, M., and Yamamoto, H. (1994). *Synlett*, 963.

20. Kobayashi, S., Araki, M., and Yasuda, M. (1995). *Tetrahedron Lett.*, **36**, 5773.

21. Shafran, Y. M., Bakulev, V. A., and Mokrushin, V. S. (1989). *Russ. Chem. Rev.*, **58**, 148.

22. (a) Weinstock, L. M., Davis, P., Handelsman, B., and Tull, R. (1967). *J. Org. Chem.*, **32**, 2823. (b) Matier, W. L., Owens, D. A., Comer, W. T., Deitchman, D., Ferguson, H. C., Seidehamel, R. J., *et al.* (1973). *J. Med. Chem.*, **16**, 901.

23. Mai, K. and Patil, G. (1984). *Tetrahedron Lett.*, **25**, 4583.

24. Ojima, I., Inaba, S., and Nakatsugawa, K. (1975). *Chem. Lett.*, 331.

25. Kobayashi, S., Ishitani, H., and Ueno, M. (1997). *Synlett*, 115.

Chapter 16
Combinatorial approaches to chiral catalyst discovery

Marc L. Snapper and Amir H. Hoveyda
Merkert Chemistry Center, Boston College, Chestnut Hill, MA 02467-3860, USA.

1 Introduction

In most combinatorial approaches to drug discovery, lead compounds are generally identified from large, structurally diverse libraries (1). Preliminary findings are then often optimized through the subsequent design and examination of more limited libraries that focus and expand on the initial results. In a similar manner, this layered approach to drug discovery and optimization can be adapted to catalyst development (*Scheme 1*, see also Chapters 14 and 15). In the first phase, a wide range of catalyst candidates can be screened to select specific complexes that effect a reaction of interest. Once a catalyst has been identified that demonstrates the desired *reactivity*, efforts can then turn toward optimizing the conditions and system to yield the desired *selectivity*. This two-tiered development strategy can offer distinct advantages in the discovery and identification of catalysts for asymmetric reactions. If the catalyst discovery and optimization protocols are sufficiently rapid and reliable, there is no prerequisite for finding general solutions to catalytic reactions; each reaction can enjoy a catalyst that is designed specifically for the substrate and transformation of interest.

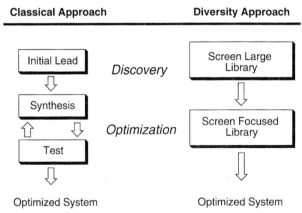

Scheme 1 Classical and diversity approaches to reaction/drug development.

1.1 Background

Given its early state of development, the use of combinatorial methods to discover new catalysts has been limited (2). In addition to work that has been carried out in our laboratories, some notable exceptions are summarized below. In 1995, Ellman examined a combinatorial approach toward an asymmetric diethylzinc addition to aldehydes (3). In this study, ligands attached to a Merrifield resin were synthesized from a 4-hydroxyproline precursor. The enantioselectivity for the reactions catalysed by the resin-bound ligands proved high, but slightly lower than those obtained with the free ligands in solution (e.g. 89% versus 94% *ee*). The following year, Gilbertson and co-workers prepared a 63-member library of chiral bisphosphines, built within a helical scaffold that could be screened for catalytic and enantioselective olefin hydrogenation. Folding of the polypeptide backbone was expected to bring together two donor phosphine units to present a chiral environment for the transition metal (4). Based on precedence, Rh(I) was selected and examined with the bisphosphine library in an enantioselective hydrogenation of an α-amino acid. Likewise, Burgess and Sulikowski set up a grid of chiral ligands and metal salts to optimize an asymmetric carbene insertion reaction (5). A third dimension of diversity was introduced through variation of solvent. In total, five chiral ligands coupled with six metal salts in four solvents were screened for their ability to direct the asymmetric C–H insertion reaction. The most effective catalyst, a Cu(I)•(bis) oxazoline ligand complex, was found to give a 3.9:1 diastereomeric ratio. In this study, unexpected catalysis by Ag(I) was also observed, underlining the additional benefit of serendipity in these high-throughput approaches. More recently, Jacobsen and co-workers have published on several catalytic systems that were discovered and optimized through diversity-based strategies (6). For example, in addition to an asymmetric Strecker reaction, they have also reported their preliminary findings on new enantioselective epoxidation catalysts.

1.2 Lewis acid catalysis

Our initial efforts have focused on Lewis acid catalysed processes (7). We envisioned that these diversity-based catalyst discovery and optimization protocols should be applicable to a wide range of chemical transformations. The development of two new asymmetric catalytic processes using a two-tiered diversity-based search for catalysts is described below. Specifically, an enantioselective opening of *meso*-epoxides with TMSCN (*Equation 1*) and an asymmetric Strecker reaction (*Equation 2*), both catalysed by a ligand-modified Ti(IV) complex are detailed.

(eq 2)

*catalyst discovery and optimization
through diversity-based techniques*

To identify effective new catalysts, the first series of protocols are designed to screen a wide range of metal–ligand candidates. During the development of this chemistry, some advancements were made in our ability to carry out reactions in a high-throughput fashion. The earlier screenings are less reaction intensive requiring only a minimal investment in equipment, but involve some questionable assumptions about co-operativity in ligand structure and the influence of catalyst mixtures. Subsequent screening methods involve fewer untested assumptions but call for more specialized instrumentation. The catalyst discovery approaches described in this chapter are divided, therefore, into the easier, less reaction intensive protocols that anyone can do with typical laboratory equipment, and the higher-throughput methods that require a greater investment in technology.

The subsequent series of catalyst optimization protocols build on the findings from the initial catalyst discovery phase. These methods focus on improving the selectivity of a particular catalytic process. While the catalyst discovery phase is amenable to a wide range of reactions and catalytic systems, the optimization protocols are more limited in scope. That is, ligand systems that are readily synthesized in a parallel fashion from enantiomerically pure building blocks are preferred. Furthermore, given the typical sequential and time intensive nature of evaluating enantioselectivity, the number of transformations that can be readily evaluated in a library of catalysts is usually more limited.

2 Catalyst discovery (optimization of reactivity)

The starting point in any reaction development strategy is not absolute and, at times, perhaps difficult to define. Much depends on the goals of the research and prior effort in the area. Is there any precedence for the reaction of interest? Is the desired transformation a known reaction that just needs some improvements in selectivity and turnover? Are there reaction conditions or reagents that can not be included in the study? Depending on the objectives, the starting point will need to be tailored to the program. In the protocols described below, the goal was to identify novel metal–ligand complexes that effect the asymmetric addition of cyanide to various electrophiles. Efforts in other laboratories on related transformations or racemic versions of the same reaction suggested that Lewis acids should provide the desired reactivity. In this regard, electrophilic metal salts modified by chiral ligands were selected in our initial screening protocols.

One of the first steps in developing a new asymmetric transformation is to determine which reaction parameters will remain invariant, which are to be studied through traditional methods, and which are amenable to optimization through high-throughput strategies. In the cyanide addition reactions, diversity-based methods were explored to optimize the metal salt, chiral ligand, and reaction additives for a range of substrates. Before attempting these high-throughput optimizations, some preliminary studies were carried out in regard to the cyanide source, solvent, reaction temperature, and time. During these initial studies, efforts were also required to optimize some of the necessary analytical techniques that were to be used to examine the reaction libraries.

2.1 Indexed grid

The evaluation of reaction efficiency for a large number of parameters and library components can quickly become an unmanageable undertaking. Even with the appropriate resources, the serial evaluation of enantioselectivity for a large metal–ligand library (through analytical methods such as nuclear magnetic resonance (NMR), gas liquid chromatography (GLC), high performance liquid chromatography (HPLC)), can be a significant bottleneck in an investigation. Accordingly, reaction development technologies that substantially reduce the number of evaluations can significantly accelerate catalyst discovery and optimization. For example, if procedures are designed where mixtures, rather than individual components are examined, the catalyst search protocol might be notably expedited. The following protocols describe a screening process where mixtures of metals and ligands are employed in an 'Indexed Grid' format to identify specific metal–peptide combinations that catalyse the addition of TMSCN to cyclohexene oxide (*Equation 1*).

Peptidyl-based ligands represent an attractive system for modifying metal environments. Di- and tripeptides are easily prepared on solid support from readily available, optically pure amino acid building blocks (8). The peptides have a modular structure where asymmetry can be systematically varied. Furthermore, modifications to the peptide system allow for the introduction of metal binding sites that can be electronically tuned for optimal selectivity. An example of a Schiff base modified peptide ligand system used extensively in these studies is illustrated in *Figure 1*.

Initial observations indicated that complications may arise when mixtures of metals or ligands are tested for enantioselectivity; we have found that subtle, non-stereochemical changes in ligand structure can lead to unpredictable variations in enantioselectivity (10). Accordingly, our efforts to establish whether

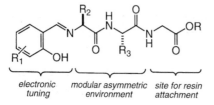

Figure 1 Peptidyl Schiff base ligand.

mixtures of metal–ligand complexes can be used for catalyst discovery have focused on searching for peptide–metal combinations that provide effective ligand accelerated catalysis (LAC) (11). Once an active metal–ligand combination has been identified, efforts will then turn toward optimizing the ligand structure for enantioselectivity.

Protocol 1

Solid phase synthesis of peptidyl Schiff base ligands

Equipment and reagents

- Polypropylene Bio-Spin Chromatography Columns (Bio-Rad Laboratories, Inc.)
- Thermolyne Roto-mix
- Wang-Gly-Fmoc resin, 0.59 mmol/g loading, and Fmoc-protected amino acids (Advanced ChemTech)
- Piperidine (Advanced ChemTech)

- N,N-dimethylformamide (DMF, Aldrich)
- 1,3-Diisopropylcarbodiimide (DIC, Aldrich)
- Aldehydes (Aldrich)
- Catecholborane (Aldrich)

Method

1. Solid phase synthesis is carried out on Wang-Gly-Fmoc resin and the reactions are performed in polypropylene Bio-Spin Chromatography Columns. Place the Wang-Gly-Fmoc resin (100 mg, 0.06 mmol) in a polypropylene reaction vessel and wash with DMF (3 × 1.0 ml).

2. Swell the resin by agitating in DMF (1.0 ml) on a Thermolyne Roto-mix for 1 h, wash with additional DMF (3 × 1.0 ml).

3. Deprotect the resin by agitating for 1.5 h in 20% piperidine/DMF (1.0 ml), followed by washing with DMF (10 × 1.0 ml) making sure to rinse the caps as well.

4. Activate the Fmoc-protected amino acid (0.24 mmol) as the symmetrical anhydride by treating with excess DIC (74 µl, 0.36 mmol) in DMF (0.5 ml, 10 min). Add the resulting solution to the Wang-Gly-NH$_2$ resin, mix for 1 h, and then wash with DMF (10 × 1.0 ml).

5. To check if the reaction is complete, subject a sample of the resin to a ninhydrin test (9). Samples testing positive are resubjected to coupling conditions (step 4).

6. Deprotect the Wang-Gly-AA1-Fmoc resin by agitation for 1.5 h in 20% piperidine/ DMF (1.0 ml) and then wash the resin with DMF (10 × 1.0 ml).

7. The second amino acid coupling and deprotection is performed as described in steps 4–6.

8. Following coupling and deprotection, the Wang-Gly-AA1-AA2-NH$_2$ resin is treated with 4.0 Equiv. (0.24 mmol) of aldehyde (0.5 ml DMF) for 2 h and washed with DMF (10 × 1.0 ml).

9. (Optional) To reduce the imine, the appropriate ligands are agitated with catecholborane (0.60 mmol, 10 Equiv.) in toluene (0.7 ml) for 15 min and then washed with DMF (10 × 1.0 ml) and MeOH (10 × 1.0 ml).

The catalyst discovery plan, therefore, is to use mixtures of the ligands with individual metals (metal library) and mixtures of the metal salts with individual ligands (ligand library) such that all possible metal–ligand complexes are prepared at least twice in the mixtures, once in the ligand library and once in the metal library. The metal and ligand libraries can then serve as indices to a grid where each hypothetical cross-peak represents a unique metal–ligand combination. It is important to note that when mixtures are examined for rate acceleration, every active system may not be discernible (12); however, this plan should allow for the detection of those complexes whose catalytic activities stand far above others in both the metal and ligand libraries (13). Once a reactive metal–ligand complex is identified from the mixtures, the enantioselectivity of the catalytic system may then be optimized.

Our preliminary results indicate that when the appropriate measures described below are taken into consideration, the 'Indexed Grid' format can be used to identify specific metal–ligand combinations that possess beneficial catalytic activity. To ensure the predictive value of the libraries, the following guidelines are suggested:

(a) Aggregation of ligands about inactive metal centres, which may mask ligand accelerated catalysis by more active metals that bind less effectively can be minimized by immobilizing the ligands on a solid support.

(b) The effects of different counter ions in metal mixtures can be minimized by using a common counter ion amongst the metal salts examined. In these studies, the counter ion was maintained as an isopropoxide.

(c) Since a high background reaction can mask significant LAC of metals with little background reactivity, libraries are of greater predictive value if metal salts are grouped according to their background reactivity.

(d) Examining the background reactions for co-operativity amongst the metal salts (higher than expected backgrounds for metal mixtures than the individual metal salts) may reveal a co-operativity in the metal mixture that could conceal the desired reaction acceleration of a ligand modified metal (14).

Failure to observe these concerns can lead to unreliable indications about the relative catalytic activity of specific metal–ligand combinations. Nevertheless, the advantage of using mixtures is that a range of ligand–metal complexes can be examined rapidly without the separate synthesis and analysis of every metal–ligand system.

Based on these considerations, the following metal salts and immobilized peptide ligands were evaluated in indexed grid format for relative LAC in the opening of cyclohexene oxide with TMSCN (*Equation 3*) (15).

Metal salts

1. Al(i-PrO)$_3$ 2. Ba(i-PrO)$_2$ 3. Ti(i-PrO)$_4$ 4. Hf(i-PrO)4 5. Zr(i-PrO)$_4$

Peptide ligands

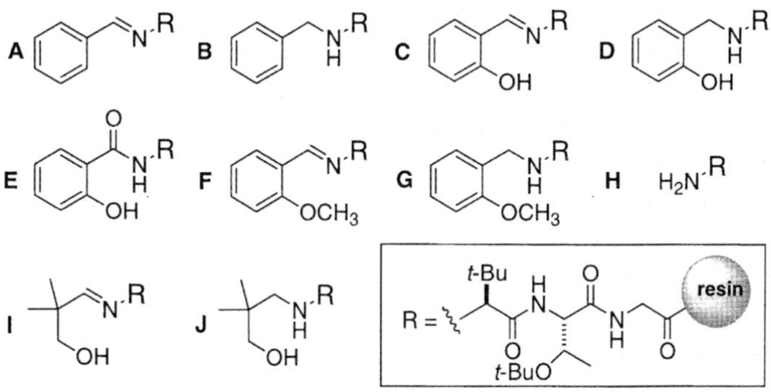

Figure 2 Metal salts and immobilized peptide ligands evaluated in the opening of cyclohexane oxide wth TMSCN (*Equation 3*).

Protocol 2

Indexed grid

1 The ligands and metal salts are organized into three groups of experiments to initiate the screening process (see *Protocol 3*).

2 In the first experiment (metal library), reaction vessels, each containing a single metal salt and a mixture of peptidic ligands (equal molar equivalent of all ligands combined with respect to metal) are assayed for yields of the nitrile product (**2**).

3 In the second experiment (ligand library), individual peptides are combined with a mixture of the metal salts (equal molar equivalent of each salt with respect to ligand) and are similarly analysed for yields of the nitrile.

4 A third set of transformations is performed to measure the individual and combined metal salt background reactions against which the above two experiments can be quantified for ligand acceleration.

5 All experiments are carried out for an identical length of time (4 h at 4°C), such that the most active systems proceed to partial conversion (= 50%) (16).

Charts 1 and *2* summarize the results from screening metal and ligand mixtures that incorporate the indexed grid library guidelines noted above. As *Chart 1* indicates, mixture of ligands (**A–J**) have a notably positive influence on the reaction catalysed by Ti(i-PrO)$_4$. Moreover, as illustrated in *Chart 2*, ligands **C** and **D** demonstrate greater ligand accelerated catalysis with a mixture of metals compared to any of the other individual ligands (19).

439

Protocol 3

Solid phase ligands in the epoxide ring opening reaction (17)

Equipment and reagents

- Thermolyne Roto-mix
- Gas liquid chromatograph (GLC) equipped with a flame ionization detector (FID)
- Test-tubes with rubber septa
- Toluene (Aldrich)
- Titanium isopropoxide (Ti(Oi-Pr)$_4$, Aldrich)
- Trimethylsilyl cyanide (TMSCN, Aldrich)
- Dodecane (Aldrich)
- Diethyl ether (Et$_2$O, Aldrich)

Method

1   Synthesize the ligands on the Wang resin (18) as described in *Protocol 1*.

2   Wash the ligand–Wang conjugates repeatedly (10 × 1.0 ml) with toluene and evaporate to dryness twice from toluene to remove any remaining solvent and water.

3   Weigh and place the resin-supported ligand (0.02 mmol) in a dried test-tube with a rubber septa under argon.

4   Add Ti(Oi-Pr)$_4$ (0.1 M in toluene, 200 μl, 0.02 mmol) and allow to stand for 10 min.

5   Add cyclohexene oxide (1.0 M in toluene, 200 μl, 0.2 mmol) and TMSCN (1.0 M in toluene, 200 μl, 0.2 mmol). Dodecane is also added to the reactions to serve as an internal standard.

6   Gently agitate the reactions with a Thermolyne Roto-mix for 4 h (4 °C).

7   Quench the reactions by adding wet Et$_2$O (1.0 ml) and passing the resulting solution through a silica gel plug with an additional 1.0 ml of Et$_2$O as a wash.

8   GLC yields were corrected for FID response factors relative to dodecane.

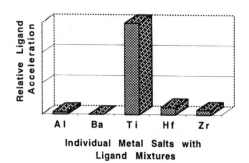

Chart 1 Relative reaction acceleration for TMSCN addition to **1** by metal isopropoxides with mixtures of ligands (reaction acceleration axis data clipped for clarity). Conditions: 10 mol% metal–ligand, 4 °C, 4 h, toluene. When product was not observed, the lowest detectable yield (0.02%) was used (reaction acceleration = reaction yield/yield with metal only).

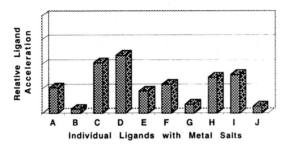

Chart 2 Relative reaction acceleration for TMSCN addition to **1** by ligands with mixtures of metal isopropoxides.

In principle, if the results obtained for the metal and ligand libraries of *Charts 1* and *2* represent the sum of the LAC contributions from the individual complexes, these libraries could function as grid indices where the cross-products represent the unique metal–ligand combination (20). As such, the two indices should be complementary; offering a facile identification of particular metal–ligand complexes that possess high catalytic activity (21). If the data in *Charts 1* and *2* are of predictive value, the combination of ligand **C** or **D** with Ti(*i*-PrO)$_4$ should represent the individual metal–ligand complexes displaying the greatest ligand accelerated catalysis. As described in the high-throughput screening of individual metal–ligand complexes, this is indeed the case.

An advantage of this indexed grid strategy is that examining only 20 reactions (five from the metal library, five background, and ten from the ligand library) provides information on the effective LAC of all 50 individual metal–ligand complexes. That is, insight is made into which are the most active complexes by testing less than 40% of all the possible metal–ligand combinations. This is particularly important when the assay of the individual reactions is rate limiting. The value of this method become even more apparent as the library grows. For example, a library of 20 metals and 20 ligands (400 unique complexes) could be surveyed by only running less than 15% of all the possible individual reactions.

Alternatively, high-throughput screening of all the individual components of a metal–ligand grid can avoid some of the problems and limitations of working with mixtures. For example, you are no longer limited to immobilized ligands. This significantly widens the range of possibilities. On the other hand, a greater investment in equipment is required to proceed efficiently with the complete screening of larger libraries. To be able to examine a wide range of reactions at a rate of 200–400 per day, it is recommended that automated liquid handler in a double sized glove box is used. In addition, depending on the reaction of interest, a centrifugal evaporator with a microtitre plate rotor and several HPLCs with auto-samplers that can accommodate microtitre plates will probably be necessary.

2.2 High-throughput screening

While we have run complete metal–ligand libraries with a variety of ligand types, the data selected was shown to support the findings described in the indexed grid

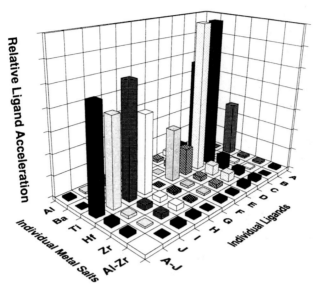

Chart 3 Evaluation of a metal isopropoxide and immobilized ligand library for LAC in formation of nitrile **II** (ligand acceleration axis data clipped for clarity).

search. *Chart 3* is the grid summarizing the LAC for the individual complexes represented by these metal and ligand libraries (the outer row and column or indices represent the metal and ligand libraries shown in *Charts 1* and *2*). Screening of the individual cross-peaks represented by the metal and ligand indices was carried out to confirm that the titanium salt, together with ligands **C** or **D**, constitute the most reactive complexes. These data support that examining metal and ligand mixtures for LAC can be an efficient means for the identification of specific metal–ligand combinations that may serve as effective catalysts.

These efforts have shown that screening combinations of peptidic ligands and metal salts can be used to identify new catalytic systems. This indexed grid strategy, while limited (22), does offer a novel and potentially efficient means of discovering new catalytically active metal–ligand combinations. Alternatively, with a greater investment in equipment, complete searches of all metal–ligand combinations are possible. In either case, while these methods allow for the discovery of reactive metal–ligand combinations, a second screening protocol is required to optimize the metal–ligand system for the desired enantioselectivity.

3 Catalyst optimization (optimization of selectivity)

With 20 amino acids and 20 different aldehydes as possible subunits for the Schiff base peptidyl ligand system (*Figure 1*), a library of 8000 (20^3) unique chiral structures is possible. Given a limited ability to assay enantioselectivity, strategies to reduce these numbers are helpful. In this regard, several abbreviated search methods can be employed to effectively sample the molecular diversity

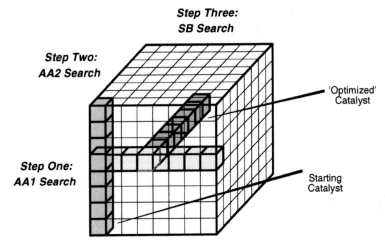

Figure 3 Representational search strategy adopted for catalyst screening allows identification of effective ligands without examining all possibilities.

of a large library. The strategy used to optimize the ligand for the opening of *meso*-epoxides, as well as the Strecker reaction, is illustrated in *Figure 3*.

3.1 Positional scanning of ligand structure (see Chapter 3)

In the first optimization sequence, 20 or so reactions are run with a different amino acid in one position of the ligand (the other ligand positions are held constant with arbitrary subunits). The amino acid that provides the highest selectivity is identified and used in that position for subsequent searches. In the second series of reactions, the first amino acid is fixed and the second amino acid is then varied. The optimal results are determined and the second position is set to the best case amino acid. In the final series of reactions, the aldehyde is varied with the best case amino acids. This last sequence is evaluated to establish the highest selectivity in the desired reaction. Ignored in this approach is the co-operativity that is possible in the untested ligand structures; nonetheless, these three series of 20 reactions (60 reactions total) lead to a ligand that affects the opening of cyclohexene oxide in 89% *ee* (and the Strecker addition in > 95% *ee*).

Scheme 2 illustrates the results for the opening of cyclohexene oxide. *Protocol 5* discusses the experimental aspects of the ligand screening. Note that epimerization of the ligand upon removal from the resin (as detailed in *Protocol 4*) limited the number of amino acids in the first search. Subsequent studies indicated that the use of immobilized ligands in the search could circumvent this problem without compromising the results. The ligand optimization results detailed in *Scheme 2* are summarized in *Scheme 3*. In this ligand optimization strategy, over 2000 possible ligand structures were sampled by running just 39 reactions. It is not clear whether this method directs us to the *best* results for this ligand class; nevertheless a ligand structure that effects the desired opening of *meso*-epoxides in nearly a 95:5 enantiomeric ratio has been identified.

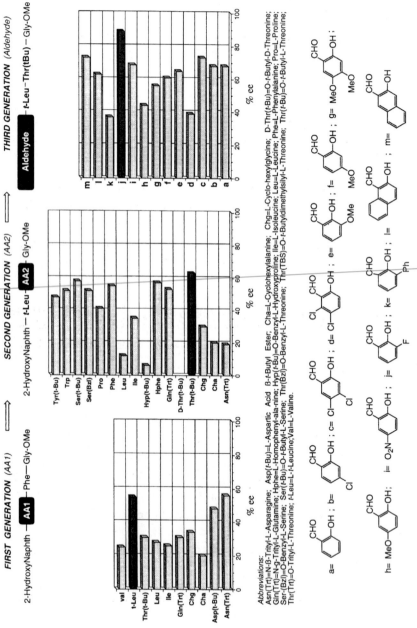

Scheme 2 Ligand optimization results by positional scanning for cyclohexene oxide.

444

Protocol 4

Cleavage of peptidyl Schiff base ligands prepared on resin

Equipment and reagents

- RC10.10 Jouan centrifugal evaporator
- Ligand library
- Triethylamine (Et$_3$N, Aldrich)

- Dichloromethane (DCM, Aldrich)
- Tetrahydrofuran (THF, Aldrich)
- Ethylacetate (EtOAc, Aldrich)

Method

1 Cleave the Schiff base ligands from the Wang resin by treating with Et$_3$N/DMF/MeOH (1:1:9, 1.0 ml) for 60 h.

2 Filter the solutions and wash the resins with THF (3 × 1.0 ml).

3 Remove the solvent under vacuum using a RC10.10 Jouan centrifugal evaporator.

4 Dissolve the resulting solids in DCM (1.0 ml) and load onto a pipette packed with a cotton plug and silica gel. Elute the ligands with EtOAc (~ 10 ml).

5 Dissolve the isolated compounds in toluene (5 ml) and concentrate (3 ×) to give white to bright yellow products (overall yield: 80–100%).

Protocol 5

Testing of the solution phase ligands in the epoxide ring opening reaction

Equipment and reagents

- GLC equipped with an FID and a chiral column (BETADEX-120, from Chiral Technologies)
- 5 ml test-tubes with septa and stir bars
- Argon source
- Cyclohexane oxide (Aldrich)
- Dodacane
- Ligand library

- Chloroform (CHCl$_3$, Aldrich)
- Trimethylsilylcyanide (TMSCN, Aldrich)
- Et$_2$O (diethyl ether, Aldrich)
- Toluene (Aldrich)
- Hexanes
- Silica gel (Aldrich)

Method

1 Dissolve the ligands (0.01 mmol) in CHCl$_3$ (1.0 ml) and transfer the resulting solutions to a 5 ml test-tube.

2 Add toluene (2.0 ml) and then concentrate the solution to a solid. Equip the reaction tubes with stir bars and septa. The reaction vessel are then evacuated and flushed with Ar (3 ×).

> **Protocol 5** continued
>
> **3** Add the metal salts in toluene (208 µl, 0.05 M, 0.01 mmol) and stir for 15 min at 22 °C.
>
> **4** Add cyclohexene oxide (11 µl, 0.10 mmol), followed by TMSCN (14 µl, 0.10 mmol). Dodecane is also added to the reactions to serve as an internal standard.
>
> **5** Stir the reaction mixtures for 4 h (4 °C).
>
> **6** Quench the reactions by the addition of wet Et$_2$O (1.0 ml) and then pass the resulting solutions through a plug of silica gel (0.5 cm) with Et$_2$O (1.0 ml) wash.
>
> **7** (Reactivity) GLC yields are corrected for FID response factors relative to dodecane.
>
> **8** (Enantioselectivity) Quench the reactions by adding 1.0 ml of ether/hexanes (1:1) and passing through a silica gel plug with an additional 1.0 ml of ether/hexanes (1:1) as a wash. The reaction selectivity is determined with chiral GLC analysis (BETADEX-120 chiral column).

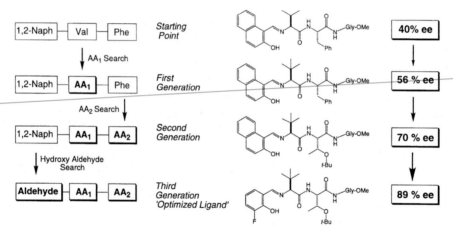

Scheme 3 Summary for optimization of peptidyl Schiff base ligand. Ligand optimization results for cyclohexene oxide opening shown on right.

3.2 Positional scanning (see Chapter 3) for optimal ligand structure in Strecker reaction

In a similar fashion, peptidyl ligands were screened by a positional scanning in an asymmetric Strecker reaction (*Equation 4*). *Protocol 6* provides the experimental details and *Scheme 4* illustrates the results from the screening protocol. In this search, a ligand pool of 15 000 unique structures was surveyed by examining just 74 reactions. As mentioned earlier, greater diversity was possible by screening the first amino acid position with ligands that were still bound to the solid support. Also of interest is the ability of the second amino acid to influence the product enantioselectivity. Note, different amino acids of the same relative stereochemistry at the α-position provided products with opposing absolute configurations. These results indicate an important mechanistic role for this second amino acid. Again, it is not certain whether the best catalytic solution was obtained, nevertheless, the strategy provided ligands affecting the asymmetric Strecker reaction in high yield and in 87–95% *ee* for a range of imine substrates.

(eq 4)

Protocol 6

Positional scanning of Strecker reaction on solid support

Equipment and reagents

- HPLC equipped with a chiral column (Chiralpak AD, from Chiral Technologies)
- Solvent rotary evaporator
- Shaker plate
- Vacuum line
- 96-well plates and adhesive sealers with aluminium liners
- 96-well filter plate
- Silica gel (Aldrich)
- Diphenylmethyl amine (Aldrich)

- Aldehydes (Aldrich)
- Resin-bound ligand library
- Ti(O*i*-Pr)$_4$ (Aldrich)
- Magnesium sulfate (MgSO$_4$)
- Cyclohexane oxide (Aldrich)
- Trimethylsilylcyanide (TMSCN, Aldrich)
- Et$_2$O (diethyl ether, Aldrich)
- Toluene (Aldrich)
- Benzene (Aldrich)

A. Imine substrate synthesis

1   Stir the aldehyde (50 mmol), diphenylmethyl amine (50 mmol), and MgSO$_4$ (~ 1 g) in benzene (100 ml) for 8 h at 22 °C.

2   Filter and concentrate the solution under vacuum.

3   Purify the products by recrystallization.

B. Screening

1   Weigh the resin-bound ligands (0.007 mmol) into individual wells of a 96-well plate.

2   Add toluene (100 μl) to each well followed by Ti(O*i*-Pr)$_4$ (0.007 mmol, 100 μl of a 0.07 M solution in toluene).

3   Agitate the plate on a shaker plate for 10 min at 22 °C.

4   Add the imine (0.07 mmol, 100 μl of a 0.7 M solution in toluene), followed by TMSCN (0.14 mmol, 100 μl of a 1.4 M solution in toluene).

5   Seal the plate with an adhesive aluminium liner and agitate the reaction mixtures for 15 h at 22 °C.

6   Quench the reactions by addition of wet ether (0.5 ml).

7   Filter the reactions through a 96-well filter plate containing silica gel. Wash the silica gel with 0.5 ml of ether.

8   Determine reaction selectivity by chiral HPLC (Chiralpak AD).

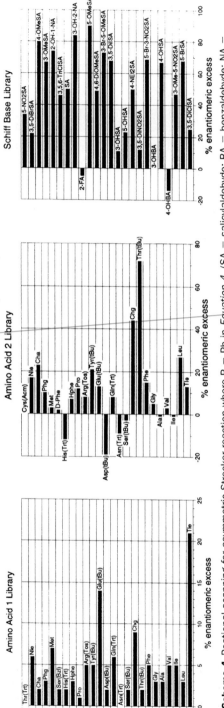

Scheme 4 Postional scanning for asymmetric Strecker reaction where R = Ph in *Equation 4*. (SA = salicylaldehyde; BA = benzaldehyde; NA = naphthaldehyde; FA = furylaldehyde.)

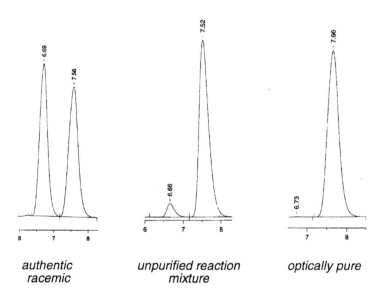

A representative example of the asymmetric Strecker reaction is illustrated in *Equation 5*. In this case, 3,5-dichloro-salicyl-*t*-Leu-Thr(*t*-Bu)-Gly(OMe) (53 mg, 0.1 mmol), Ti(O*i*-Pr)₄ (0.1 mmol, 1.0 ml of a 0.1 M solution in toluene), *o*-chlorobenzaldehyde diphenylmethyl imine (306 mg, 1.0 mmol), TMSCN (267 ml, 2.0 mmol) in toluene (2.5 ml) are used in the procedure outlined in *Protocol 7*. Isopropanol (115 ml, 1.5 mmol) in toluene (2 ml) was added over 3 h with additional stirring for 21 h. Upon addition of hexanes (4.5 ml) the product precipitated and was collected by filtration to afford (285 mg, 85% yield, > 99% *ee*) of a white solid. (The crude reaction mixture shows 96% conversion and 93% *ee* by HPLC.) The optical purity of the cyanide addition product is established by the HPLC analysis; related chromatograms are illustrated in *Figure 4*.

Figure 4 Examples of a chiral HPLC analysis of Strecker addition products.

An important application for the asymmetric Strecker reaction is the preparation of enantiomerically pure, unnatural amino acids. As illustrated in *Equation 6*, hydrolysis of the Strecker addition product with aqueous acid provides the desired amino acid derivative in good yield and high enantiomeric purity.

449

Protocol 7

Preparative Ti catalysed addition of TMSCN to imines

Equipment and reagents

- HPLC equipped with a chiral column (Chiralpak AD, from Chiral Technologies)
- Glove box under nitrogen atmosphere
- 10 ml round-bottom flask
- Argon source
- Syringe pump
- Shaker plate
- Vacuum line
- Teflon tape

- Silica gel (Aldrich)
- Resin-bound ligand library
- Ti(Oi-Pr)$_4$ (Aldrich)
- Imine
- Isopropanol (Aldrich)
- Trimethylsilylcyanide (TMSCN, Aldrich)
- Et$_2$O (diethyl ether, Aldrich)
- Toluene (Aldrich)

Method

1 Set up the reactions inside of a glove box under a nitrogen atmosphere. Weigh ligand (0.05 mmol) into a flame-dried round-bottomed flask and add toluene (2.0 ml).

2 Charge the flask with Ti(Oi-Pr)$_4$ (0.05 mmol, 0.5 ml of a 0.1 M solution in toluene) and stir the yellow solution for 10 min at 22 °C.

3 Add the imine (0.5 mmol). Cap the reaction vessel with a septa and seal with Teflon tape. Remove the reaction from the glove box and place in an area maintained at 4 °C.

4 Place a balloon of argon over the flask and add TMSCN (134 μl, 1.0 mmol) to the stirred solution.

5 Add isopropanol (58 μl, 1.0 mmol) in toluene (2 ml) via a syringe pump over 3–20 h, with additional stirring for another 4–20 h.

6 Quench the solution with wet ether (5 ml) and pass through a plug of silica gel.

7 Purify the products by crystallization or silica gel chromatography.

8 Analyse the products for conversion and enantioselectivity by chiral HPLC (Chiralpak AD).

Protocol 8

Conversion of amino nitriles to BOC-protected amino acids (*Equation 6*)

Equipment and reagents

- Solvent rotary evaporator
- 25 ml round-bottom flask
- Shaker plate

- Vacuum line
- *tertio*-Butyloxycarbonyl (*t*-Boc)$_2$O (Aldrich)
- Dioxane (Aldrich)

- EtOAc (Aldrich)
- 1 M sodium hydroxide (NaOH, Aldrich)
- 6 M hydrochloric acid (HCl, Aldrich)
- Et_2O (diethyl ether, Aldrich)

- Sodium hydrogen sulfate ($NaHSO_4$, Aldrich)
- $MgSO_4$ (Aldrich)

Method

1 Heat a solution of the amino nitrile (1.4 mmol) in 6 M HCl (10 ml) to reflux for 1–2 h (23).

2 Cool the solution to ambient temperature and wash the aqueous layer with diethyl ether (3 × 10 ml).

3 Concentrate the aqueous solution under vacuum to a white solid.

4 Dissolve the solid in water (10 ml). Neutralize the aqueous layer with 1 M NaOH and add an additional 1.5 ml of 1 M NaOH.

5 Wash the solution with Et_2O (3 × 10 ml) to remove benzhydrylamine.

6 Add to the cooled (0 °C) basic aqueous layer, dioxane (10 ml) and (t-Boc)$_2$O (327 mg, 1.5 mmol).

7 Warm the solution to ambient temperature and stir for 1 h.

8 Remove the dioxane under vacuum and wash the remaining aqueous layer with EtOAc (15 ml).

9 Acidify the aqueous layer to pH = 2 with 1 M $NaHSO_4$ and extract with EtOAc (2 × 15 ml).

10 Dry the organic phase with $MgSO_4$ and concentrate under vacuum.

11 Purify the crude white solid by recrystallization from EtOAc/Et_2O.

4 Discussion

4.1 Overview of current status

Whereas traditional methods to invent, build, and optimize catalysts for new reactions can take years, the catalyst discovery and optimization protocols outlined in this chapter can significantly accelerate the process. Instead of depending on the stepwise design, construction and testing of a single ligand, the diversity strategy allows for the parallel examination of many potential catalysts and relies on the high-throughput screening to *select* the optimal catalytic solution.

One of the most difficult aspects of traditional catalyst development is the need for a general catalyst; one that will function with high reactivity and selectivity for a wide range of substrates. This need is, in part, due to the difficulty in designing and constructing the catalyst. In reality, the typical user of an asymmetric transformation is only interested in one or, at most, a few substrates. The generality of a reaction will only suggest whether the existing catalyst could

Table 1 Optimized ligands for catalytic enantioselective addition of TMSCN to *meso*-epoxides[a]

Entry	Substrate	Product	ee[%]	Yield[%]	Optimized Ligand
1			83	72	
2			84	68	
3			78	69	

[a] Conditions: 20 mol% Ti(O*i*Pr)$_4$, 20 mol% ligand, 4 °C, toluene, 6–20 h.

function effectively on the desired, but untested substrates. For diversity-based catalyst development methods, generality is not a requirement; a catalyst is rapidly tailored and selected for a specific transformation. The catalyst, much like an enzyme, is optimized in selectivity and reactivity for a particular substrate. As illustrated in *Table 1*, whereas the opening of cyclopentene oxide with TMSCN is catalysed most effectively by one ligand (entry 1), the opening of cycloheptene oxide and *cis*-octene oxide are catalysed by different ligands (entries 2 and 3).

In as much as traditional catalyst optimization strategies depend on our imperfect understanding of the reaction mechanism, diversity approaches are not straddled by this requirement. By probing the reaction of interest with fewer mechanistic biases, unanticipated opportunities are more frequently noted. For example, co-operatively between several metals in the opening of *meso*-epoxides was observed that perhaps would not have been found if we had not explored mixtures of metals. That is not to say that rational design does not play a significant role in library development, it only suggests that there are more opportunities to benefit from serendipity. Because of the rapid acquisition of results obtained from structurally diverse catalyst libraries, in fact, a deeper mechanistic understanding of a process is likely to develop. It is even conceivable that those interested in exploring the mechanistic fine points of a transformation could benefit from the diversity protocols outlined in this chapter.

4.2 Future opportunities

Screening large libraries of possible catalysts presents significant and unique obstacles in analysis. If a 30 second GC run is required to evaluate every component of a one million-membered library, approximately one year would be required to perform a serial analysis of the collection. A potential solution to this problem was recently reported by Morken and Taylor (24), who described a parallel assay of an encoded polymer bead-bound catalyst library for a solution phase catalytic transformation through infrared thermal imaging. In this screening method, beads were monitored for the largest enthalpy changes, reflecting the catalytic structures that effected the highest reaction turnover. Alternatively, a parallel colorimetric assay recently reported by Crabtree and co-workers also offers opportunities for rapid analysis in high-throughput catalyst screenings (25).

The general usefulness of this catalyst discovery strategy will become evident with the introduction of numerous new and useful transformations. The early findings are encouraging, but numerous challenges remain and the scope of the strategy still remains to be defined. As with drug development, these combinatorial approaches will not replace the classical catalyst discovery techniques, but should provide the practitioner with an additional tool to address their chemical needs.

Acknowledgements

We are grateful for the extensive intellectual and experimental contributions made by current and former undergraduate, graduate, and postdoctoral co-workers who have collaborated with us on this chemistry (Bridget M. Cole, Ken D. Shimizu, Clinton A. Krueger, Kevin W. Kuntz, Carolyn D. Dzierba, Wolfgang G. Wirschun, Joseph P. A. Harrity, Ramon Abola, and John D. Gleason). Numerous helpful and insightful suggestions from Dr David Casebier (ArQule) are also acknowledged.

This research was supported by the NIH (GM-57212 to A. H. H. and M. L. S.; postoloctoral fellowships for B. M. G. and K. D. S.), ArQule, and the NSF (CHE-9632278 to A. H. H.). Additional funds were provided by Albemarle and DuPont (to A. H. H.). M. L. S. is a Sloan Research Fellow, a DuPont Young Investigator, an Eli Lilly Grantee, and a Glaxo-Wellcome Chemistry Scholar. A. H. H. and M. L. S. are Camille Dreyfus Teacher-Scholars.

References

1. For recent reviews and discussions on combinatorial chemistry, see: (a) Balkenhohl, F., von dem Bussche-Hünnedfeld, C., Lansky, A., and Zechel, C. (1996). *Angew. Chem.*, **108**, 2436; (1996). *Angew. Chem. Int. Ed. Engl.*, **35**, 2288. (b) Borman, S. (1998). *Chem. Eng. News*, **76**(14), 47.

2. (a) Borman, S. (1996). *Chem. Eng. News*, Nov. 4, 37. (b) Service, R. F. (1997). *Science*, **277**, 474. For other recent examples, see: (c) Menger, F. M., Eliseev, A. V., and Migulin, V. A. (1995). *J. Org. Chem.*, **60**, 6666. (d) Gao, X. and Kagan, H. B. (1998). *Chirality*, **10**, 120. (e) Reetz, M. T., Zonta, A., Schimossek, K., Liebeton, K., and Jaeger, K.-E. (1997). *Angew. Chem. Int. Ed. Engl.*, **36**, 2830.

3. Liu, G. and Ellman, J. A. (1995). *J. Org. Chem.*, **60**, 7712.

4. Gilbertson, S. R. and Wang, X. (1996). *Tetrahedron Lett.*, **37**, 6475.

5. Burgess, K., Lim, H.-L., Porte, A. M., and Sulikowski, G. A. (1996). *Angew. Chem.*, **108**, 192; (1996). *Angew. Chem. Int. Ed. Engl.*, **35**, 220.

6. (a) Francis, M. B., Finney, N. S., and Jacobsen, E. N. (1996). *J. Am. Chem. Soc.*, **118**, 8983. (b) Sigman, M. S. and Jacobsen, E. N. (1998). *J. Am. Chem. Soc.*, **120**, 4901.

7. (a) Cole, B. M., Shimizu, K. D., Krueger, C. A., Harrity, J. P. A., Snapper, M. L., and Hoveyda, A. H. (1996). *Angew. Chem.*, **108**, 1776; (1996). *Angew. Chem. Int. Ed. Engl.*, **35**, 1668. (b) Shimizu, K. D., Cole, B. M., Krueger, C. A., Kuntz, K. W., Snapper, M. L., and Hoveyda, A. H. (1997). *Angew. Chem.*, **109**, 1781; (1997). *Angew. Chem. Int. Ed. Engl.*, **36**, 1704.

8. Yet, with a few exceptions, dipeptide systems have not been used in this context. See: (a) Nitta, H., Yu, D., Kudo, M., Mori, A., and Inoue, S. (1992). *J. Am. Chem. Soc.*, **114**, 7969, and references cited therein. (b) Hayashi, M., Miyamoto, Y., Inoue, T., and Oguni, N. (1993). *J. Org. Chem.*, **58**, 1515.

9. Kaiser, E., Colescott, R. L., Bossinger, C. D., and Cook, P. I. (1970). *Anal. Biochem.*, **34**, 595.

10. For example, in the reaction shown in *Equation 1*, a peptide ligand containing a protected L-Thr provides an enantiomeric ratio of 12:1, whereas the same ligand with a protected L-Arg reverses the enantiomeric ratio to 1:4. Therefore, a mixture that contains active catalysts providing products with high but opposite sense of enantiocontrol would be indistinguishable from one that affords products with little selectivity.

11. (a) Han, H. and Janda, K. D. (1996). *J. Am. Chem. Soc.*, **118**, 7632. (b) Berrisford, D. J., Bolm, C., and Sharpless, K. B. (1995). *Angew. Chem. Int. Ed. Engl.*, **34**, 1059, and references cited therein.

12. Burgess, K., Lim, H.-J., Porte, A. M., and Sulikowski, G. A. (1996). *Angew. Chem. Int. Ed. Engl.*, **35**, 220.

13. Konings, D. A. M., Wyatt, J. R., Ecker, D. J., and Freier, S. M. (1996). *J. Med. Chem.*, **39**, 2710.

14. Interestingly, Sr(i-PrO)$_2$ in combination with Ti(i-PrO)$_4$ was found to provide a higher level of background reactivity than would be predicted by the background of the individual metal salts. Additional studies are underway to explore and exploit this co-operativity effect. Since this co-operativity effect may not have been predicted, it provides further support for the benefits of examining mixtures.

15. Variations in peptide ligand structure were restricted to electronic and structural perturbations to what is presumed to be the key 'metal-binding' region of the ligand. (See: Nitta, H., Yu, D., Kuo, M., Mori, A., and Inoue, S. (1992). *J. Am. Chem. Soc.*, **114**, 7969.) Metal salts were initially selected due to their availability and their solubility in toluene.

16. Because the reactions were carried out to low conversion, the present results do not identify catalyst systems that effect reaction slowly but with relatively high turnover frequency.

17. Enantiomeric ratios were determined by chiral HPLC. Analytical liquid chromatography (HPLC) was performed on a Shimadzu chromatograph with a Chiralpak AD (4.6 × 250 mm) and a Crownpak CR(+)(4.0 × 150 mm) chiral column by Chiral Technologies. All reactions were conducted in oven (135 °C) and flame-dried glassware under an inert atmosphere of dry argon or nitrogen. Trimethylsilylcyanide, Ti(Oi-Pr)$_4$ (99.99%), DMF (99.9%), diphenylmethyl amine, *tert*-leucine, and all commercially available aldehydes were purchased from Aldrich and used without further purification. All other amino acids, functionalized Fmoc-Gly-Wang resin, DIC, and piperidine were purchased from Advanced Chemtech and used without further purification. Toluene was distilled from sodium/benzophenone ketal. 1,1,1-Trichloroethane and isopropanol were distilled from CaSO$_4$. n-Butanol was distilled from Mg activated with I$_2$.

18. We find that Fmoc-Gly-PAC resin (purchased from PerSeptive Biosystems) may also be used to obtain similar results.

19. The small but reproducible ligand acceleration observed for the ligand library is a consequence of the range of background reactions included in the library. In this case, the LAC displayed by the Ti, a metal with little background, is significantly

attenuated by the nearly 100-fold larger background reactions for the Al, Hf, and Zr salts.

20. (a) Pirrung, M. C. and Chen, J. (1995). *J. Am. Chem. Soc.*, **117**, 1240. (b) Déprez, B., Williard, X., Bourel, L., Coste, H., Hyafil, F., and Tartar, A. (1995). *J. Am. Chem. Soc.*, **117**, 5405. (c) Beck, S. C. and Chapman, K. T. (1997). *Bioorg. Med. Chem. Lett.*, **7**, 837.

21. In a classical approach a total of 196 reactions are required to fully examine 13 ligands and 14 metals (13 ligands $\times$ 14 metals = 182 combinations + 14 controls reactions = 196 reactions). In an ideal indexed library, the same 13 ligands and 14 metals could be evaluated in only 41 reactions (13 ligands + 14 metals + 14 controls). This represents up to a 79% savings in the total number of transformations necessary to identify potentially effective catalytic systems. The efficiency factor grows as the size of the grid increases; for example, a 20 metal by 20 ligand grid could lead to reaction efficiency coefficient of 86%. The ability to deconvolute the library by assessing ligand acceleration in a mixture of reactions is responsible for the effectiveness of this strategy.

22. Among potential limitations to be investigated are the inclusion of other reaction variables, such as turnover frequency information into the screening protocol, as well as defining the as-yet unclear relationship between reactivity and selectivity in these catalytic processes.

23. Extended heating in HCl leads to partial racemization. After heating optically pure *o*-chlorophenyl Gly for 18 h, the derived amino acid is recovered in 86% *ee*. This phenomenon is substrate dependent: *t*-Leu shows < 2% racemization after heating for 48 h in refluxing 6 M HCl.

24. Taylor, S. J. and Morken, J. P. (1998). *Science*, **280**, 267. Also, see: Moates, F. C., Somani, M., Annamalai, J., Richardson, J. T., Luss, D., and Willson, R. C. (1996). *Ind. Engl. Chem. Res.*, **35**, 4801.

25. Cooper, A. C., McAlexander, L. H., Lee, D.-H., Torres, M. T., and Crabtree, R. H. (1998). *J. Am. Chem. Soc.*, **120**, 9971.

List of suppliers

Acros/Fischer Scientific
Acros (Fisher Scientific), 600 Business Center Drive, Pittsburg, PA 15205-9913, USA.
Acros (Fisher Scientific), Bishop Meadow Road, Loughborough, Leicestershire LE11 5RG, UK.

Advanced ChemTech, Inc., 5609 Fern Valley Road, Louisville, KY 40228, USA.

Aldrich
Aldrich Chemical Company, Inc., 1001 West Saint Paul Avenue, Milwaukee, WI 53233, USA.
Aldrich Chemical Company, The Old Brickyard, New Road, Gillingham, Dorset SP8 4JL, UK.

Alfa Æsar, Ward Hill, MA, USA.

Anderman and Co. Ltd., 145 London Road, Kingston-upon-Thames, Surrey KT2 6NH, UK. Tel: 0181 541 0035 Fax: 0181 541 0623

Bachem
Bachem Bioscience, Inc., 3700 Horizon Drive, Renaissance At Gulph Mills, King of Prussia, PA 19406, USA.
Bachem California Inc., 3132 Kashiwa Street, Torrance, CA 90505, USA.
Bachem AG, Hauptstrasse 144, CH-4416 Bubendorf, Switzerland.

Beckman Coulter Inc.
Beckman Coulter Inc., 4300 N Harbor Boulevard, PO Box 3100, Fullerton, CA 92834-3100, USA.
Tel: 001 714 871 4848 Fax: 001 714 773 8283
Web site: www.beckman.com

Beckman Coulter (UK) Ltd., Oakley Court, Kingsmead Business Park, London Road, High Wycombe, Buckinghamshire HP11 1JU, UK.
Tel: 01494 441181 Fax: 01494 447558
Web site: www.beckman.com

Becton Dickinson and Co.
Becton Dickinson and Co., 21 Between Towns Road, Cowley, Oxford OX4 3LY, UK.
Tel: 01865 748844 Fax: 01865 781627
Web site: www.bd.com
Becton Dickinson and Co., 1 Becton Drive, Franklin Lakes, NJ 07417-1883, USA.
Tel: 001 201 847 6800 Web site: www.bd.com

Bio 101 Inc.
Bio 101 Inc., c/o Anachem Ltd., Anachem House, 20 Charles Street, Luton, Bedfordshire LU2 0EB, UK.
Tel: 01582 456666 Fax: 01582 391768 Web site: www.anachem.co.uk
Bio 101 Inc., PO Box 2284, La Jolla, CA 92038-2284, USA.
Tel: 001 760 598 7299 Fax: 001 760 598 0116
Web site: www.bio101.com

Bio-Rad Laboratories Ltd.
Bio-Rad Laboratories Ltd., Bio-Rad House, Maylands Avenue, Hemel Hempstead, Hertfordshire HP2 7TD, UK.
Tel: 0181 328 2000 Fax: 0181 328 2550
Web site: www.bio-rad.com

Bio-Rad Laboratories Ltd., Division Headquarters, 1000 Alfred Noble Drive, Hercules, CA 94547, USA.
Tel: 001 510 724 7000 Fax: 001 510 741 5817
Web site: www.bio-rad.com

Calbiochem
Calbiochem-Novabiochem Corp., 10394 Pacific Center Court, San Diego, CA 92121, USA.
Calbiochem-Novabiochem UK Ltd., Boulevard Industrial Park, Padge Road, Beeston, Nottingham NG9 2JR, UK.
Novabiochem, Weidenmattweg 4, CH-4448 Läufelfingen, Switzerland.

Cannon Instruments, Inc., State College, PA, USA.

Costar, Corning Incorporated, Science Products Division, 45 Nagog Park, Acton, MA 01720, USA.

CP Instrument Co. Ltd., PO Box 22, Bishop Stortford, Hertfordshire CM23 3DX, UK.
Tel: 01279 757711 Fax: 01279 755785
Web site: www.cpinstrument.co.uk

Dupont
Dupont (UK) Ltd., Industrial Products Division, Wedgwood Way, Stevenage, Hertfordshire SG1 4QN, UK.
Tel: 01438 734000 Fax: 01438 734382
Web site: www.dupont.com
Dupont Co. (Biotechnology Systems Division), PO Box 80024, Wilmington, DE 19880-002, USA.
Tel: 001 302 774 1000 Fax: 001 302 774 7321
Web site: www.dupont.com

Dupont Polymers, South Duart, NC, USA.

Eastman Chemical Co., 100 North Eastman Road, PO Box 511, Kingsport, TN 37662-5075, USA. Tel: 001 423 229 2000
Web site: www.eastman.com

EG & G Princeton Applied Research, Princeton, NJ, USA.

ElectroChem, Inc., Woburn, MA, USA.

Fisher Scientific
Fisher Scientific UK Ltd., Bishop Meadow Road, Loughborough, Leicestershire LE11 5RG, UK.
Tel: 01509 231166 Fax: 01509 231893
Web site: www.fisher.co.uk
Fisher Scientific, Fisher Research, 2761 Walnut Avenue, Tustin, CA 92780, USA.
Tel: 001 714 669 4600 Fax: 001 714 669 1613
Web site: www.fishersci.com

Fluid Metering, Inc., 29 Orchard Street, Oyster Bay, NY 11771, USA.

Fluka
Fluka, PO Box 2060, Milwaukee, WI 53201, USA.
Tel: 001 414 273 5013 Fax: 001 414 2734979
Web site: www.sigma-aldrich.com
Fluka Chemical Co. Ltd., PO Box 260, CH-9471 Buchs, Switzerland.
Tel: 0041 81 745 2828 Fax: 0041 81 756 5449
Web site: www.sigma-aldrich.com
Fluka Chemical Corp., 1001 West St. Paul Avenue, Milwaukee, WI 53233, USA.

Hybaid
Hybaid Ltd., Action Court, Ashford Road, Ashford, Middlesex TW15 1XB, UK.
Tel: 01784 425000 Fax: 01784 248085 Web site: www.hybaid.com
Hybaid US, 8 East Forge Parkway, Franklin, MA 02038, USA.
Tel: 001 508 541 6918 Fax: 001 508 541 3041
Web site: www.hybaid.com

HyClone Laboratories, 1725 South HyClone Road, Logan, UT 84321, USA.
Tel: 001 435 753 4584 Fax: 001 435 753 4589
Web site: www.hyclone.com

Invitrogen
Invitrogen BV, PO Box 2312, 9704 CH Groningen, The Netherlands.
Tel: 00800 5345 5345 Fax: 00800 7890 7890
Web site: www.invitrogen.com

Invitrogen Corp., 1600 Faraday Avenue, Carlsbad, CA 92008, USA.
Tel: 001 760 603 7200 Fax: 001 760 603 7201
Web site: www.invitrogen.com

Irori, 11149 North Torrey Pines Road, La Jolla, CA 92037, USA.

Isco, Inc., Box 82531, 4700 Superior Avenue, Lincoln, NE 68504, USA.

Lancaster Synthesis
Lancaster Synthesis, Inc., PO Box 1000, Windham, NH 03087-9977, USA.
Lancaster Synthesis, Eastgate, White Lund, Morecambe, Lancashire LA3 3DY, UK.

Life Technologies
Life Technologies Ltd., PO Box 35, Free Fountain Drive, Incsinnan Business Park, Paisley PA4 9RF, UK.
Tel: 0800 269210 Fax: 0800 838380
Web site: www.lifetech.com
Life Technologies Inc., 9800 Medical Center Drive, Rockville, MD 20850, USA.
Tel: 001 301 610 8000 Web site: www.lifetech.com

Merck Sharp & Dohme
Merck Sharp & Dohme Research Laboratories, Neuroscience Research Centre, Terlings Park, Harlow, Essex CM20 2QR, UK.
Web site: www.msd-nrc.co.uk
MSD Sharp and Dohme GmbH, Lindenplatz 1, D-85540 Haar, Germany.
Web site: www.msd-deutschland.com

Millipore
Millipore (UK) Ltd., The Boulevard, Blackmoor Lane, Watford, Hertfordshire WD1 8YW, UK.

Tel: 01923 816375 Fax: 01923 818297 Web site: www.millipore.com/local/UK.htm
Millipore Corp., 80 Ashby Road, Bedford, MA 01730, USA.
Tel: 001 800 645 5476 Fax: 001 800 645 5439
Web site: www.millipore.com

Neosystem Laboratoire, 7 rue de Boulogne, 67100 Strasbourg, France.

New England Biolabs, 32 Tozer Road, Beverley, MA 01915-5510, USA.
Tel: 001 978 927 5054

Nikon
Nikon Corp., Fuji Building, 2-3, 3-chome, Marunouchi, Chiyoda-ku, Tokyo 100, Japan.
Tel: 00813 3214 5311 Fax: 00813 3201 5856
Web site: www.nikon.co.jp/main/index_e.htm
Nikon Inc., 1300 Walt Whitman Road, Melville, NY 11747-3064, USA.
Tel: 001 516 547 4200 Fax: 001 516 547 0299
Web site: www.nikonusa.com

Nycomed
Nycomed Amersham plc, Amersham Place, Little Chalfont, Buckinghamshire HP7 9NA, UK.
Tel: 01494 544000 Fax: 01494 542266 Web site: www.amersham.co.uk
Nycomed Amersham, 101 Carnegie Center, Princeton, NJ 08540, USA.
Tel: 001 609 514 6000 Web site: www.amersham.co.uk

Ominfit Ltd., 2 College Park, Coldhams Lane, Cambridge DB1 3HD, UK.

Organix, Inc., 65 Cummings Park, Woburn, MA 01801, USA.

Peninsula
Peninsula Laboratories Inc., 601 Taylor Way, San Carlos, CA 94070, USA.
Peninsula Laboratories Europe Ltd., Box 62, 17K Westside Industrial Estate, Jackson Street, St. Helens, Merseyside WA9 3AJ, UK.

Peptide Institute, Inc. Japan, 4-1-2 Ina Minoh-Shi, Osaka 562, Japan.

Peptides International, Inc., 11621 Electron Drive, Louisville, KY 40299, USA.

Perkin Elmer Ltd., Post Office Lane, Beaconsfield, Buckinghamshire HP9 1QA, UK.
Tel: 01494 676161 Web site: www.perkin-elmer.com

Pharmacia
Pharmacia Biotech (Biochrom) Ltd., Unit 22, Cambridge Science Park, Milton Road, Cambridge CB4 0FJ, UK.
Tel: 01223 423723 Fax: 01223 420164
Web site: www.biochrom.co.uk
Pharmacia and Upjohn Ltd., Davy Avenue, Knowlhill, Milton Keynes, Buckinghamshire MK5 8PH, UK.
Tel: 01908 661101 Fax: 01908 690091 Web site: www.eu.pnu.com

PharMingen,10975 Torreyana Road, San Diego, CA 92121, USA.

Pine Instruments, Grove City, PA, USA.

Promega
Promega UK Ltd., Delta House, Chilworth Research Centre, Southampton SO16 7NS, UK.
Tel: 0800 378994 Fax: 0800 181037
Web site: www.promega.com
Promega Corp., 2800 Woods Hollow Road, Madison, WI 53711-5399, USA.
Tel: 001 608 274 4330 Fax: 001 608 277 2516
Web site: www.promega.com

Qiagen
Qiagen UK Ltd., Boundary Court, Gatwick Road, Crawley, West Sussex RH10 2AX, UK.
Tel: 01293 422911 Fax: 01293 422922
Web site: www.qiagen.com
Qiagen Inc., 28159 Avenue Stanford, Valencia, CA 91355, USA.
Tel: 001 800 426 8157 Fax: 001 800 718 2056
Web site: www.qiagen.com

Rapp Polymere GmbH, Ernst-Simon-Str. 9, D-72072 Tübingen, Germany.

Rheodyne, 6815 Redwood Drive, PO Box 996, Cotati, CA 94931, USA.

Roche Diagnostics
Roche Diagnostics Ltd., Bell Lane, Lewes, East Sussex BN7 1LG, UK.
Tel: 01273 484644 Fax: 01273 480266 Web site: www.roche.com
Roche Diagnostics Corp., 9115 Hague Road, PO Box 50457, Indianapolis, IN 46256, USA.
Tel: 001 317 845 2358 Fax: 001 317 576 2126
Web site: www.roche.com
Roche Diagnostics GmbH, Sandhoferstrasse 116, 68305 Mannheim, Germany.
Tel: 0049 621 759 4747 Fax: 0049 621 759 4002 Web site: www.roche.com

Schleicher and Schuell Inc., Keene, NH 03431A, USA.
Tel: 001 603 357 2398

Scientific Marketing Associates, 189/191 High Street, Barnet, Hertfordshire EN5 5SU, UK.

Scribner Associates, Inc., South Pines, NC, USA.

Shandon Scientific Ltd., 93-96 Chadwick Road, Astmoor, Runcorn, Cheshire WA7 1PR, UK.
Tel: 01928 566611 Web site: www.shandon.com

Shearwater Polymers, Inc., 2305 Spring Branch Road, Huntsville, AL 35801, USA.

Sigma-Aldrich
Sigma-Aldrich Co. Ltd., The Old Brickyard, New Road, Gillingham, Dorset XP8 4XT, UK.
Tel: 01747 822211 Fax: 01747 823779 Web site: www.sigma-aldrich.com
Sigma-Aldrich Co. Ltd., Fancy Road, Poole, Dorset BH12 4QH, UK.
Tel: 01202 722114 Fax: 01202 715460
Web site: www.sigma-aldrich.com

Sigma Chemical Co., PO Box 14508, St Louis, MO 63178, USA.
Tel: 001 314 771 5765 Fax: 001 314 771 5757
Web site: www.sigma-aldrich.com

Stratagene
Stratagene Europe, Gebouw California, Hogehilweg 15, 1101 CB Amsterdam Zuidoost, The Netherlands.
Tel: 00800 9100 9100
Web site: www.stratagene.com
Stratagene Inc., 11011 North Torrey Pines Road, La Jolla, CA 92037, USA.
Tel: 001 858 535 5400
Web site: www.stratagene.com

Strem Chemicals
Strem Chemicals, Inc., 7 Mulliken Way, Dexter Industrial Park, Newburyport, MA 01950, USA.
Strem Chemicals, 48 High Street, Orwell, Royston SG8 5QN, UK.

Technic, Inc., Cranston, RI, USA.

United States Biochemical, PO Box 22400, Cleveland, OH 44122, USA.
Tel: 001 216 464 9277

Index